orticultural Society

at-Zone Map

70°W

90°W

85°W

80°W

75°W

45°N

40°N

35°N

30°N

80°W

25°N

Average Number
of Days per Year
Above 86°F (30°C)

	Zone
< 1	1
1 to 7	2
> 7 to 14	3
> 14 to 30	4
> 30 to 45	5
> 45 to 60	6
> 60 to 90	7
> 90 to 120	8
> 120 to 150	9
> 150 to 180	10
> 180 to 210	11
> 210	12

Scale for the mainland U.S.

0   250   500   Miles

0   250   500   Kilometers

N
E
S

PRACTICAL HORTICULTURE

FIFTH EDITION

PRACTICAL HORTICULTURE

FIFTH EDITION

Laura Williams Rice
Robert P. Rice, Jr.

Prentice
Hall

Upper Saddle River, New Jersey 07458

Library of Congress Cataloging-in-Publication Data
Rice, Laura Williams.
    Practical horticulture / Laura Williams Rice, Robert P. Rice, Jr.—5th ed.
      p. cm.
    Includes bibliographical references (p.).
    ISBN 0-13-094634-6 (alk. paper)
    1. Horticulture. 2. Gardening. 3. Indoor gardening. I. Rice, Robert P. II. Title.

SB318.R48 2003
635—dc21
                                                      2002019823

Editor-in-Chief: Stephen Helba
Executive Editor: Debbie Yarnell
Editorial Assistant: Sam Goffinet
Production Editor: Emily Bush, Carlisle Publishers Services
Project Coordinator: Mary Carnis
Director of Manufacturing and Production: Bruce Johnson
Managing Editor: Mary Carnis
Marketing Manager: Jimmy Stephens
Manufacturing Buyer: Cathleen Petersen
Design Director: Cheryl Asherman
Senior Design Coordinator: Miguel Ortiz
Cover Photo: Tim Saville
Cover Designer: Carey Davies
Composition: Carlisle Communications, Ltd.
Printing and Binding: Courier Westford

Pearson Education LTD.
Pearson Education Australia PTY, Limited
Pearson Education Singapore, Pte. Ltd.
Pearson Education North Asia Ltd.
Pearson Education Canada, Ltd.
Pearson Educación de Mexico, S.A. de C.V.
Pearson Education—Japan
Pearson Education Malaysia, Pte. Ltd.

10 9 8 7 6 5 4 3 2
ISBN 0-13-094634-6

BRIEF CONTENTS

CONTENTS

PREFACE TO THE FIFTH EDITION

The fifth edition of *Practical Horticulture* marks over 20 years that the book has been in print, as well as a major turning point in broadening the text. It was originally conceived as a university text for nonmajors courses in horticulture with the idea that the students would aquire a sound (if limited) foundation in aspects of horticultural science such as physiology, nomenclature, anatomy, growth and development, and so on (Part 1). In addition to this, they would learn of the applications of the principles of horticulture to home horticulture both outdoors (Part 2) and indoors (Part 3)—applications that might eventually be useful in their lives after university.

But the clarity and readability of the text, combined with the great number of instructional photos and illustrations, have brought it to the attention of horticulture instructors who teach this course in a myriad of ways and at a variety of levels.

For this reason, in the fifth edition I have chosen to not only update but to *broaden* the text, keeping the basic core horticultural science in Part 1, but *enhancing* it with more information about the profession of horticulture and with descriptions of the techniques in commercial production horticulture. You will find the following improvements in this substantially revised fifth edition.

- A new section with the current (but ever-changing) definition of what a "plant" is in the light of modern investigation into the genetic relationships among earth's organisms

- A list of the national and international professional organizations for the specializations within horticulture, and a list of trade and professional journals

- An introduction to ethics in horticulture, and an example of one professional organization's ethical guidelines

- Sections on commercial production technology for container-grown landscape plants, bedding plants, field vegetables, high-density tree fruit production, commercial seed production, and greenhouse pot plants.

In addition, there is new material in the text on *safety* in regard to the mixing and application of pesticides, pesticide residues on fruits and vegetables, and power landscape maintenance equipment. A sample pesticide label diagram (in the form required by the Environmental Protection Agency) and a summary of national legislation over the past 30 years in regard to pesticides in agriculture have been included also.

Allied information on *nonchemical pest and disease control* and the increasing role of beneficial insects (with photos of some of them in action), a table of biological control agents such as nematodes and predator wasps, and a section on fungicidal inoculants for prevention of root diseases round out this enhancement.

The *soils chapter* has been enlarged to include sections on the role of soil organisms in plant growth, discussions of *field capacity* and *matric potential* in irrigation, current and traditional soil conservation techniques, and a traditional soil texture diagram.

Appreciation is expressed to Erric Ross, Mt. Hood Community College; Craig A. Tolley, County College of Morris; Dawn Gatherum, Weber State University; Ellen B. Peffley, Texas Technical University who reviewed the text.

The authors hope that the updating and broadening of the text will be appreciated by those who currently use it, and will make it more applicable to a wider range of courses taught by professors and teachers of horticultural science in all of North America.

Laura Williams Rice
Sept 21, 2001

PRACTICAL HORTICULTURE

FIFTH EDITION

PART ONE

FUNDAMENTALS
OF HORTICULTURE

INTRODUCTION

The word *horticulture* originates from Latin. The Latin word *hortus,* meaning garden, is the root of the first part; *culturea,* which composes the second part, means cultivation. Together they give a fairly accurate translation: garden cultivation.

Today horticulture is only one branch of the broad field of agricultural plant sciences. But it is among the most important, dealing with the growing of fruits, vegetables, and both indoor and outdoor ornamental plants. Two other branches of agricultural plant sciences, forestry and crop science, encompass the remaining plants in cultivation. Trees grown for other than ornamental use are usually covered under forestry. Grains, pasture grasses, and fiber crops such as cotton are discussed under agronomy or crop science.

Plant pathology (the study of plant diseases), entomology (the study of insects), and plant breeding are allied to all branches of agriculture. Their contribution to the successful cultivation of plants is immeasurably important.

Contrasted to these applied plant sciences is botany, the pure science of plants. Botanists frequently deal with noncultivated plants, researching plant classification and internal plant processes, among other topics. However, they also sometimes work with food or ornamental plants as do horticulturists.

THE DIVISIONS OF HORTICULTURE

Because plants are grown for a variety of uses, the art and science of horticulture has many divisions. The most familiar are listed here, along with samples of types of employment opportunities found in each area.

Fruit Production

Fruit production, called *pomology,* covers the growing of tree fruits such as apples and citrus. It also deals with the "small fruits," including blueberries, grapes, and strawberries.

In the pomology industry, orchard supervisors establish and maintain fruit orchards throughout the country. Other pomologists produce young fruit trees for sale to commercial orchards. Vineyard supervisors specialize in the cultivation of grapes (*viticulture*), and their crops can be destined for either fresh eating or winemaking. In universities, research pomologists dedicate their efforts to improving fruit quality or yield.

Vegetable Production

Vegetable production is called *olericulture.* Olericulture covers the cultivation of all of the vegetables and also such crops as melons and rhubarb that are customarily grown in vegetable gardens.

Because most vegetables are cultivated outdoors, most vegetable growers work in field vegetable production. They are responsible for the crop from planting through harvest, and their tasks may include weed control, insect and disease control, and irrigation. However, in some areas of the country vegetables are grown in greenhouses. Vegetable growers in this specialization must understand both greenhouse operation and vegetable cultivation to perform their jobs.

Landscape Horticulture

The outdoor ornamental-plant facet of horticulture is called *environmental horticulture* or *landscape horticulture.* The landscape horticulture industry employs thousands of people in hundreds of job specialties. Workers in nursery production propagate and grow millions of ornamental trees, shrubs, and groundcovers annually and ship them throughout the country. Horticulturally trained designers then develop landscape plans for residences and commercial buildings that are implemented by landscape installers or crews.

Later, landscape maintenance personnel maintain the plants in healthy condition by watering, pruning, fertilizing, and controlling pests.

In garden centers, nursery salespeople furnish plants and gardening supplies to the public. They in turn are supplied with plants by salespeople representing the wholesale plant growers.

The field of landscape horticulture also includes nonindustry occupations. Botanic gardens and arboretums employ landscape horticulturists for teaching and maintaining the gardens. City park and street tree programs may require people having training in landscape horticulture.

Arboriculture

Arboriculture literally means "tree culture" and includes both large-scale forestry for commercial purposes (generally under a forestry department at a university) and "urban forestry," a subcategory of landscape horticulture.

Arborists are responsible for tree selection, removal of large trees, and maintenance, including pruning, and disease and insect control. Most cities have at least one "city arborist."

Urban forestry focuses on the use of trees in inhabited environments such as cities and suburbs.

Recent research has shown the beneficial psychological effects on residents of trees planted in inner-city environments: decreased violence, increased feelings of security, greater socialization, more creative play among children, and more interaction between children and adults. (See internet reference at the end of this chapter.) In addition, trees planted in urban situations decrease air pollution, notably of particulates. They also reduce summer cooling costs for air conditioning.

Turf

Turf is the branch of horticulture dealing with turfgrass culture. People who work in this field can be involved in breeding new turfgrass varieties; controlling insects and diseases; or maintaining turf for golf courses, parks, and lawns.

Careers in turfgrass management have expanded greatly in the past 10 years, coinciding with the expansion of professional sports and personal sports that use turf as a playing surface and the increasing number of homeowners opting for professional lawn care services in lieu of maintaining their lawns themselves. City and suburban parks are also sites with large grassy areas that require maintenance, as do school grounds, cemeteries, and school athletic fields. Sod production is an allied supply industry for the turfgrass management business.

Many colleges and universities now offer two-year certificate programs in turfgrass management, and employment opportunities after graduation are excellent. Some courses can be done by independent study. Students interested in working out of doors and who enjoy sports such as golf, soccer, football, and the like would find employment in this sector of the horticulture industry very rewarding.

The courses that students take in this program include basic plant sciences; the identification, production, and maintenance of trees and shrubs; drainage techniques; irrigation methods and principles; insects and diseases that attack turf and ornamentals; weed control; fertilization; irrigation; and pesticide usage and handling. There are also courses in record keeping, computers, and personnel management to prepare students for managerial positions. On-the-job training with potential employers is often a part of the education process.

Floriculture

The production and sale of field-grown flowers for cutting or greenhouse-grown flowers and plants is a branch of horticulture known as *floriculture*.

Floriculturists can be involved in either the wholesale or the retail parts of the industry. Many wholesale growers are in charge of greenhouses where they raise flowers and indoor plants throughout the winter and bedding plants (seedings of flowers and vegetables) in the spring. But in warm areas like Florida, California, and Texas, outdoor production of cut flowers and foliage plants is common.

Much of the wholesale growers' output reaches the public through retail florists, where floral designers create arrangements, bouquets, and corsages.

Indoor plant shops and open-air flower kiosks are dynamic small businesses in the retail portion of floriculture. They are one way for a floriculturist to start his or her own business. Some indoor plant shops expand into plant rental businesses and employ people to care for leased plants in offices and hotels.

THE PROFESSION OF HORTICULTURE

A professional horticulturist has many options in employment within the divisions in horticulture, and examples of horticultural careers are found throughout the text in green boxes featuring this symbol:

Professional Ethics

Regardless of whether a person works in retail, wholesale, production, maintenance, advising, design, or any other area, an ethical framework that governs the individual's daily conduct of his or her professional life is an essential foundation. Ethical issues address the way things "should be" rather than the way things are. Ethics involve discussions of moral obligations and an individual's feelings of "right" and "wrong", but they do not necessarily depend on religious beliefs as a foundation. One can consider ethics as derived from several philosophical approaches:

Utilitarian Approach

A problem must be analyzed from several different perspectives, and the solution must favor the greatest good.

The Rights Approach

The rights approach is based on the belief that humans have the right to choose paths that affect their destinies, and that they are justified in expecting that their rights (such as the right to privacy, to truth, to freedom from injury, and the like) should be respected.

Common Good Approach

The common good approach states that ethical actions are those that benefit all members of the community.

The Natural Approach

The natural approach is a belief that nature is inherently valuable in and of itself, regardless of its relation to human beings, although the *perception* of this value is obviously human. As another way of saying this, all objects in nature are valuable simply by their existence, completely independent of the utility or benefits they may confer upon humans. Therefore, the welfare of nature is given priority in ethical decisions, even to the detriment of humanity, which is regarded as only one part of nature.

When a person becomes a professional horticulturist, ethics (regardless of their derivation) and the application of ethical standards of behavior come into play. Some of the many examples of ethical standards and conduct that would be incorporated into daily life of a horticulturist are listed below:

- Respecting the rights of his or her customers and others with whom business is transacted.
- Refraining from publicly condemning, criticizing, or devaluing another horticulturist's work or reputation.
- Ensuring that every customer is notified in writing of what the charges will be for all work carried out, and making certain that the customer accepts these charges before they become liable for payment.
- Maintaining professional competence by keeping abreast of new information and developments.
- Recognizing one's responsibility to the community and the environment in protecting each from exposure to actual or potential hazards.

As an example of a full set of professional standards for landscape and nursery professionals, the code adopted by the Landscape Alberta Nursery Trades Association (LANTA) and used with their permission in Figure I-1.

When production of food (fruit and vegetables) is shown the subject of the ethics discussion, additional questions evolve. For example, philosophical questions can deal with the balance in crop production as it relates to:

Economic efficiency (yield per amount of land, taking into account costs of inputs such as fertilizer, pesticides, and the like.)

versus

Social responsibility (For example, could the crop be knowingly nutritionally inferior because of certain production practices?)

and

Environmental compatibility (Will the production techniques that were used to maximize yield in the short run result in long-term soil degradation or other environmental degradation?)

Professional Publications and Organizations

Part of remaining a competent professional involves keeping current with changes in the field in which one works. A professional horticulturist does this by membership in professional organizations, by attending courses to update his or her skills, and by reading professional publications whose aim is the dissemination of the most up-to-date information in that specific field.

Table I-1 is a list of many of the professional organizations to which horticulturists belong. There are organizations based on specialization (For example floral production, turf, nursery, and the like). In addition, some are "trade" organizations for those involved in the retail end of horticulture, and some are "academic" organizations that deal with university research in horticulture before it

Figure I-1   Ethical Standards of LANTA, (Used with permission of Landscape Alberta, Edmonton, Alberta, Canada)

The Code of Ethics:

To serve our clients with integrity, knowledge, and creative ability.

To consistently maintain the confidence and trust in the profession.

To encourage education and research within the horticultural field.

To perform work or supply goods according to accepted industry standards, and the Canadian Nursery Landscape Association Specifications and/or specifications written in the contract.

To attract to the profession those who possess a high degree of honesty, courtesy, integrity, and competence.

To continually strive toward the betterment of our environment through our dedication to the profession of ornamental horticulture.

To protect, at all times, the integrity of the profession and the interests of the client and the general public.

The Code of Conduct:

1. Members shall act fairly, honestly, and in a manner they would be prepared to defend publicly in all aspects of the practice of ornamental horticulture.
2. Members shall endeavor to serve clients with integrity, knowledge, and creative ability. They shall provide only the highest level of professional services to clients and others in the industry.
3. Members have a responsibility to provide constructive professional criticism when necessary to ensure the continued quality of goods and services.
4. Members shall not make a false or malicious statement that may injure the professional reputation of another member.
5. Members should work to maintain confidence and trust in the profession of ornamental horticulture.
6. Members' advertising shall be neither false or in any way misleading.
7. Members have an obligation to render goods and services in an efficient and cost-effective manner in order to protect the client's interest while maintaining acceptable standards.
8. Members have an ongoing responsibility to improve their own professional knowledge and skills and keep abreast of new developments in their industry.
9. Members shall encourage education and research within the horticultural field.
10. Members shall endeavor to attract to the profession individuals with a high degree of honesty, courtesy, integrity, and competence.
11. Members have a responsibility to pay their suppliers.
12. Members will perform work or supply goods according to accepted industry standards and the Canadian Nursery Landscape Association Specifications and/or specifications written in the contract.
13. Members shall comply with all federal, provincial, and municipal laws governing the operation of a business.
14. Members shall continually strive toward the betterment of the environment through their dedication to the profession of horticulture.

is in the public sector. These organizations are frequently the sponsors of "short courses" to continually educate their members.

Table I-2 is a list of professional journals or trade magazines that address the interests of particular specializations in horticulture (such as city trees) or allied fields such as entomology (the study of insects). Many of these journals are published by the professional organizations listed in Table I-1.

RECREATIONAL HORTICULTURE

Two aspects of horticulture emphasize the emotional and recreational value of working with plants: horticulture therapy and home horticulture.

Horticulture Therapy

The value of horticulture as therapy for those with physical, intellectual, or emotional disabilities has been recognized for many years. The blind, deaf, wheelchair confined, or otherwise physically handicapped can all participate in some horticultural activity because of the many senses that are involved when growing a plant. A blooming flower can be appreciated by sight by many people, but through touch by others, and through smell by nearly all.

When they are taught skills such as potting seedlings, transplanting, making cuttings, watering, and moving pots, those with mental health problems or limited intellectual development that precludes them from other employment opportunities can work and be productive members of society. The elderly and those with limited mobility can find pleasure and the gratification

Table I-1 Professional Organizations for Horticulturists

General and Esoteric Organizations

American Association of Botanical Gardens & Arboreta
American Horticultural Society
American Seed Trade Association
American Society of Horticultural Science
Commercial Horticultural Association (CHA)
Composting Council
Florida State Horticultural Society
Growers, Inc.
Horticultural Research Institute
Hydroponic Society of America
International Herb Growers and Marketers
International Society for Horticultural Science (ISHS)
Iowa State Horticulture Society
Irrigation Association
Massachusetts Horticultural Society
Minnesota State Horticultural Society
Pennsylvania Horticultural Society
Professional Plant Growers Association
Soil & Water Conservation Society

Floriculture

Floral Marketing Association (FMA)
Florists' Transworld Delivery (FTD)
Roses, Inc.
Society of American Florists
Wholesale Florists Association

Fruit Growing

American Vineyard Foundation
American Vintners Association
British Columbia Blueberry Council

California Fruit

The Apple Hill Growers Association
California Apple Commission
California Association of Winegrape Growers
California Avocado Commission
California Avocado Society
California Canning Peach Association
California Cherry Advisory Board
California Date Administrative Committee
California Fig Advisory Board
California Fresh Apricot Council
California Kiwifruit Commission
California Olive Industry
California Olive Oil Council
California Pear Advisory Board
California Pomegranate Council
California Prune Board
California Rare Fruit Growers, Inc.
California Strawberry Commission
California Table Grape Commission
California Tree Fruit Agreement
Citrus Research Board

Dried Fruit Association of California
The Prune Grower's Page
Raisin Administrative Committee
Sunkist Growers
Sun-Maid Growers of California
Sunsweet Dryers
Valley Fig Growers

California Nuts

Almond Board of California
California Macadamia Society
California Pistachio Commission
California Rare Fruit Growers, Inc.
California Strawberry Commission
California Table Grape Commission

Cape Cod Cranberry Growers' Association
Concord Grape Association
Diamond Walnut Growers
Grower, Processor and Marketing Associations
Idaho Farm Bureau
Maine Cranberry Growers Association
New York Wine and Grape
North American Blueberry Council
North American Bramble Growers Association
North American Strawberry Growers Association
O.I.V. - Office International de la Vigne et du Vin
Ontario Berry Growers Association
The Oregon Blueberry Commission
Oregon Blueberry Growers' Association
The Oregon Strawberry Commission
The Oregon Wine Advisory Board
Oregon Winegrowers' Association
Puget Sound Winegrowers Association
Raspberry & Blackberry Commission
Strawberry Growers Association
U.S. Resources
Walnut Marketing Board
Washington Association of Wine Grape Growers
Washington Red Raspberry Commission
Washington State Grape Society
Washington Wine Commission
Wild Blueberry Association of North America
Wine Institute
Wisconsin State Cranberry Growers Association
Women for WineSense

Landscaping/Nursery/Arboriculture Associations

American Landscape Maintenance Association
American Nursery and Landscape Association
American Society of Consulting Arborists
American Society of Golf Course Architects
American Society of Irrigation Consultants
American Society of Landscape Architects

(continued)

Table I-1   Professional Organizations for Horticulturists *(continued)*

Associated Landscape Contractors of America	New York State Turfgrass Association
Associated Professional Landscape Designers	North Texas Golf Course Superintendents Association
Big Island Association of Nurserymen	Nursery Sod Growers Association
Florida Nurserymen and Growers Association	Ontario Golf Superintendents' Association
International Society of Arboriculture	Ontario Parks Association
Landscape Contractors Association	Ontario Turfgrass Research Foundation
Landscape Maintenance Association	Professional Lawn Care Association of America
Landscape Ontario	Professional Lawn Care Association of Ontario
Louisiana Association of Nurserymen	Royal Canadian Golf Association
Massachusetts Nursery and Landscape Association	South Texas Golf Course Superintendents Association
National Arborist Association	Southern California Turfgrass Foundation
National Landscape Association	Sports Turf Managers Association
Ohio Nursery and Landscape Association	Texas Turfgrass Association
Ontario Horticultural Society	Turfgrass Council of North Carolina
Professional Grounds Management Society	Turfgrass Producers International
Southern Nurserymen's Association	USGA Green Section

Turfgrass Associations

Alabama Turfgrass Association
American Society of Golf Course Architects
Association des Surintendents de Golf du Québec
  (Quebec Golf Superintendents Association)
Atlantic Golf Superintendent Association
Atlantic Turfgrass Research Foundation
Canadian Golf Superintendents Association
Canadian Irrigation Association (Eastern Region)
Colorado Sports Turf Managers Association
Florida Turfgrass Association
Georgia Golf Course Superintendents Association
Georgia Turfgrass Association
Golf Course Association
Golf Course Builders Association of America
Golf Course Superintendents Association of America
  (GCSAA)
Illinois Turfgrass Foundation
The Indiana Golf Course Superintendents Association
International Turfgrass Society
Lawn & Garden Dealers Association
The Michigan Turfgrass Foundation
The Mid-Atlantic Association of Golf Course
  Superintendents
Midwest Regional Turf Foundation
Mississippi Turfgrass Association
New Jersey Turfgrass Association

Vegetable Growing Organizations

Access to Asian Vegetables
  http://www.nre.vic.gov.au/trade/asiaveg/
California Artichoke Advisory Board
  http://www.artichokes.org
Florida Fruit & Vegetable Association
  http://www.ffva.com
Florida State Horticultural Society
  http://valencia.lal.ufl.edu/jkbu/fshs/index.html
Leafy Greens Council
  http://www.leafy-greens.org/
Michigan Asparagus Advisory Board
  http://www.asparagus.org/
Minnesota Fruit and Vegetable Growers Association
  http://www.mfvga.org/
National Onion Association
  http://www.onions-usa.org/
National Potato Council
  http://www.npcspud.com
National Potato Promotion Board
  http://www.potatohelp.com/
Ontario Produce Marketing Association
  http://www.opmaassn.com
Vegetable Improvement Center
  http://vic.tamu.edu
Western Growers Association
  http://www.wga.com/

of accomplishment in working with plants. The incarcerated can produce their own food or grow plants for sale that will let them save money for the time of their eventual release.

Home Horticulture

As a recreational activity, horticulture is the nation's most popular pastime: The satisfaction of working with soil and plants to produce home-grown vegetables, lush houseplants, or a beautiful landscape is the hobby of millions of people. Balcony gardens made up of dozens of small pots flourish in the cities, dispelling the notion that gardening is a suburban or farm activity.

There are only two prerequisites one must have to assure success in gardening: a sustained interest in growing plants and information on how to grow them. Contrary to popular belief, green thumbs are not born. They are made by experience and observation.

Table I-2 Professional and Trade Journals in Horticulture

Acta Horticulturae
Agriculture, Ecosystems and Environment
Agronomy Journal
American Fruit Grower
American Gardener
American Journal of Alternative Agriculture
American Journal of Enology and Viticulture
American Nurseryman
Arboricultural Journal
Arborist News
Australian Journal of Agricultural Research
Australian Journal of Experimental Agriculture
BioCycle
Biological Agriculture and Horticulture
Biology and Fertility of Soils
Canadian Journal of Forest Research
Canadian Journal of Plant Pathology
Canadian Journal of Plant Science
Canadian Journal of Soil Science
City Trees
Communications in Soil Science and Plant Analysis
Crop Protection
Crop Science
Diversity
Economic Botany
Environmental Entomology
European Journal of Forest Pathology
European Journal of Soil Science
Experimental Agriculture
Fruit Varieties Journal
Garden Design
Grape Grower
Greenhouse Grower
Greenhouse Management and Production
Grounds Maintenance
Grower Talks
Growing Edge
Hortensia
Horticultural Reviews
Horticulture
Horticulture Abstracts
Horticulturist
HortScience
HortTechnology
Irrigation Science
Journal of Agricultural and Food Chemistry
Journal of Agricultural Entomology
Journal of Applied Horticulture
Journal of Arboriculture
Journal of Citriculture
Journal of Economic Entomology

Journal of Environmental Quality
Journal of Food Science
Journal of Horticultural Science and Biotechnology
Journal of Invertebrate Pathology
Journal of Production Agriculture
Journal of Small Fruit and Viticulture
Journal of Soil and Water Conservation
Journal of the American Society for Horticultural Science
Journal of Tree Fruit Production
Landscape Architecture
Landscape Design
Landscape Journal
New Zealand Journal of Crop and Horticultural Science
Nursery Management and Production
Nutrient Cycling in Agroecosystems
Organic Gardening
Ornamental Horticulture
Pacific Horticulture
Pecan South Including Pecan Quarterly
Pesticide Biochemistry and Physiology
Pesticide Science
Physiological and Molecular Plant Pathology
Phytoparasitica
Phytopathologische Zeitschrift (Journal of Phytopathology)
Phytopathology
Phytoprotection
Plant and Soil
Plant Breeding
Plant Breeding Reviews
Plant Disease/Plant Pathology
Plant Physiology
Plant Physiology and Biochemistry
Plant Science
Postharvest Biology and Technology
Proceedings of the Western Society of Weed Science
Proceedings (Southern Weed Science Society)
Proceedings of the Annual Convention—Southeastern Pecan
    Growers Association
Scientia Horticulturae
Seed Science Research
Soil Science
Soil Science and Plant Nutrition
Soil Science Society of America Journal
Sportsturf
Tree Physiology
Trees
Tropical Agriculture
Vineyard and Winery Management
Weed Research
Weed Science
Weed Technology

A local library or bookstore is one source of horticultural information. Because of the popularity of home horticulture, many books are currently available. They include information from basic gardening to such specialized topics as orchid and bonsai growing. Local nurseries are a second source of gardening information. Their employees are usually very willing and qualified to answer gardening questions.

Perhaps the most overlooked source of gardening help is the Cooperative Extension Service. This government agency is a federal, state, and country cooperative whose purpose is to offer free agricultural and home economics information. Almost every country in the United States has a Cooperative Extension Office, listed under county government in the phone book. The office title will vary among states but will usually be County Extension Service, Cooperative Extension Service, Agriculture Extension Service, or Farm Advisor.

Some of the many services that may be provided by the Extension Service are free gardening publications, phone advice regarding gardening questions, and diagnosis of disease and insect problems. If the county office is large enough, it may provide additional services such as soil testing, public lectures, and recorded dial-a-garden-tip messages.

The County Extension Service satisfies the information requests of most gardeners. However, each state also has at least one Agricultural Experiment Station. The Experiment Station conducts research on agricultural topics, sometimes including those of interest to home gardeners. The results are then published and made available to the public free of charge or for a small fee. Lists of these publications are issued regularly and can be ordered by writing or using the World Wide Web to reach the Agriculture Extension Service of the state (see Table I-3).

Table I-3   State Extension Services (http://ceinfo.unh.edu/state_sites.html)

Alabama
Cooperative Extension Service
Auburn University
Auburn, AL 36830
http://www.ag.auburn.edu/landscape/
www.aces.edu/dept/com_veg/esimonne/

Alaska
Cooperative Extension Service
University of Alaska
Fairbanks, AK 99775-5200
http://zorba.uafadm.alaska.edu/coop-ext/index.html

Arizona
Cooperative Extension Service
Forbes Building
University of Arizona
Tucson, AZ 85721
http://www.ag.arizona.edu/extension/needinfo/plants.html

Arkansas
Cooperative Extension Service
University of Arkansas
P.O. Box 391
Fayetteville, AR 72703
http://www.uaex.edu/

California
Cooperative Extension Service
University of California
Davis, CA 95616
http://envhort.ucdavis.edu/index.htm

Colorado
Cooperative Extension Service
Colorado State University
Fort Collins, CO 80523
http://www.colostate.edu

Connecticut
Cooperative Extension Service
University of Connecticut
1376 Storrs Road U-36 Room 205
Storrs, CT 06269-4036
www.canr.uconn.edu/ces/index.htm

Delaware
Delaware Cooperative Extension Service
University of Delaware
Townsend Hall
Newark, DE 19717-1303
http://ag.udel.edu/extension/home.htm

Florida
Cooperative Extension Service
University of Florida
Institute of Food and Agricultural Sciences
McCarty Hall
Gainesville, FL 32611
http://www.ifas.ufl.edu/www/extension/ces/htm

Georgia
Cooperative Extension Service
University of Georgia
College of Agriculture
Athens, GA 30602
http://www.ces.uga.edu/Agriculture/horticulture/horthome.htm

Hawaii
Cooperative Extension Service
University of Hawaii at Manoa
3050 Maile Way
Honolulu, HI 96822
http://agrss.sherman.hawaii.edu/bookshelf/fb.shtml

Idaho
Cooperative Extension Service
University of Idaho

Table I-3 State Extension Services *(continued)*

Moscow, ID 83843
http://info.ag.uidaho.edu/index.html

Illinois
Cooperative Extension Service
University of Illinois
College of Agriculture
122 Mumford Hall
Urbana, IL 61801
http://www.extension.uluc.edu/home/homelawn.html

Indiana
Cooperative Extension Service
Purdue University
Agriculture Administration Building
West Lafayette, IN 47907
http://persephone.agcom.purdue.edu/AgCom/Pubs/
hort.htm

Iowa
Cooperative Extension Service
Iowa State University
110 Curtis Hall
Ames, IA 50011
http://www.exnetiastate.edu/Pages/pubs/ga.htm

Kansas
Cooperative Extension Service
Kansas State University
Umberger Hall
Manhattan, KS 66506
http://www.oznet.ksu.edu/horticulture/extensn/horticul.htm

Kentucky
Cooperative Extension Service
University of Kentucky
S-1027 Agriculture Science Building North
Lexington, KY 40545-0091.
www.ca.uky.edu/search/

Louisiana
Cooperative Extension Service
Louisiana State University
Knapp Hall
Baton Rouge, LA 70803-1900
http://www.agctr.lsu.edu/nar/garden.htm

Maine
Cooperative Extension Service
University of Maine
Orono, ME 04469
http://www.umext.maine.edu/

Maryland
Cooperative Extension Service
University of Maryland
Symons Hall
College Park, MD 20742
http://www.agnr.umd.edu./oes/

Massachusetts
Cooperative Extension Service
University of Massachusetts
Stockbridge Hall
Amhurst, MA 01003
http://www.umass.edu/umext/programs/agro

Michigan
Cooperative Extension Service
Michigan State University
106 Agriculture Hall
East Lansing, MI 48824-1039
http://www.msue.msu.edu/msue/imp

Minnesota
Minnesota Extension Service
University of Minnesota
St. Paul, MN 55108
http://www.extension.umn.edu/garden/

Mississippi
Cooperative Extension Service
Mississippi State University
P.O. Box 5446
Mississippi State, MS 39762
http://msucares.com

Missouri
Cooperative Extension Service
University of Missouri-Columbia
2-70 Agriculture Building
Columbia MO 65211
http://www.muext.missouri.edu/xplor/agguides/hort/
index.htm

Montana
Cooperative Extension Service
Montana State University
Bozeman, MT 59717
http://extn.msu.montana.edu/About_Us/plant_soils/pla

Nebraska
Cooperative Extension Service
University of Nebraska
Institute of Agriculture and Natural Resources
Lincoln, NE 68583
http://www.ianr.unl.edu/pubs/horticulture

Nevada
Cooperative Extension
University of Nevada Reno
Reno, NV 89557-0004
http://www.extension.unr.edu/horticulture/archives.html

New Hampshire
University of New Hampshire Cooperative Extension
University of New Hampshire
Durham, NH 03824
http://ceinfo.unh.edu/

New Jersey
Rutgers Cooperative Extension
Cook College
P.O. Box 231
New Brunswick, NJ 08903
http://www.rce.rutgers.edu/

New Mexico
Cooperative Extension Service
New Mexico State University
Box 3AE
Las Cruces, NM 88003
http://www.cahe.nmsu.edu/ces/

(continued)

Table I-3 State Extension Services *(continued)*

New York
Cornell Cooperative Extension System
Cornell University
Ithaca, NY 14853
http://www.cce.cornell.edu/

North Carolina
Agriculture Experiment Station
North Carolina State University
Box 7602
Raleigh, NC 27695-7602
http://www.cals.ncsu.edu/hort.sci/ext_home.html

North Dakota
North Dakota State University
Extension Service
Fargo, ND 58105
http://www.ag.ndsu.nodak.edu/subhort.htm_soils.html

Ohio
Cooperative Extension Service
2120 Fyff Road
The Ohio State University
Columbus, OH 43210-1099
http://www.webgarden.html

Oklahoma
Cooperative Extension Service
Oklahoma State University
139 Agriculture Hall
Stillwater, OK 74078
http://www.dasnr.okstate.edu/oces/

Oregon
Extension Service
Oregon State University
Ballard Extension Hall, 101
Corvallis, OR 97331-3606
http://eesc.orst.edu/agcomwebfile/garden/
http://orst.edu/dept/hort/index.html

Pennsylvania
Cooperative Extension Service
Pennsylvania State University
323 Agriculture Administration Building
University Park PA 16802
http://search.cas.psu.edu/cas/

Rhode Island
Cooperative Extension Service
University of Rhode Island
Woodward Hall
Kingston, RI 02881
http://www.edc.uri.edu/

South Carolina
Cooperative Extension Service
Clemson University
107 Barre Hall
Clemson, SC 29634
http://virtual.clemson.edu/groups/hort/HomeHort.htm

South Dakota
Cooperative Extension Service
South Dakota State University
Box 2207
Brookings, SD 57007
http://www.abs.sdstate.edu/hort/exthort.htm

Tennessee
Agriculture Extension Service
University of Tennessee
P.O. Box 1071
Knoxville, TN 37901
http://www.utextension.utk.edu/

Texas
Agricultural Extension Service
Texas A & M University
College Station, TX 77843
http:/agextension.tamu.edu/

Utah
Cooperative Extension Service
Utah State University
Logan, UT 84322-4900
http://ag.extension.usu.edu/coop/index.htm

Vermont
Cooperative Extension Service
University of Vermont
Burlington, VT 05405
http://ctr.uvm.edu/ext/ag/

Virginia
Cooperative Extension Service
Virginia Polytechnic Institute and State University
Blacksburg, VA 24061
http://www.ext.vt.edu/resources/

Washington
Cooperative Extension
Washington State University
Pullman, WA 99164
http://cru.cahe.wsu.edu/
http://ext.wsu.edu/

West Virginia
Cooperative Extension Service
West Virginia University
Morgantown, WV 26506
http://www.wvu.edu/vagexten/

Wisconsin
Cooperative Extension Service
University of Wisconsin-Madison
Madison, WI 53706
http://www.uwex.edu/

Wyoming
Cooperative Extension Service
College of Agriculture
The University of Wyoming
P.O. Box 3354
Laramie, WY 82071
http://uwyo.edu/ag/ces/ceshome.htm

The USDA has many national and regional positions in research, inspection, statistics, economics, and other fields. Internationally, agricultural attaché positions can be filled by horticulturists with a broad knowledge of agriculture and, preferably, foreign language skills. These people assemble information and report on farming techniques, consumption patterns, prices, and other topics.

The United States Department of Agriculture issues many publications for gardeners that are either free or inexpensive. A list may be obtained by writing to the Superintendent of Documents, U.S. Government Printing Office, Washington, DC 20402.

Those whose interest has narrowed to a specific group of plants should note that there are over 60 plant societies ranging from the African Violet Society of America to the Terrarium Association. These societies usually issue newsletters dealing with their area of specialization, and members frequently cooperate in exchanging plants. See the appendix at the end of the book for a list of these plant societies.

 A cooperative extension specialist specializing in horticultural plants answers questions on production, diagnoses problems, provides publications on plant growing, refers growers to sources of information, gives talks to growers, and generally offers his or her services to private growers and the public.

SELECTED REFERENCES FOR ADDITIONAL INFORMATION

American Horticultural Society Staff. *North American Horticulture*. 2nd ed. New York: Macmillan, 1992.

American Horticultural Therapy Association, *AHTA Newsletter,* Gaithersburg, MD: The American Horticultural Therapy Association.

Anonymous. *Choose Horticulture*. Chicago: American Nurseryman Publishing Co., 1991.

Flagler, J., and Raymond P. Poincelot. *People-Plant Relationships*. Binghamton, NY: Haworth Press, 1994.

Flagler, Joel and Raymond P. Poincelot, editors. *People-Plant Relationships: Setting Research Priorities*. Binghamton, NY: Food Products Press, 1993.

Flemer, William, III. *Rewarding Careers in the Nursery Industry*. Chicago: American Nurseryman Publishing Co., 1990.

Golberg, Jan. *Opportunities in Horticulture Careers*.net Library, Inc. http://www.netLibrary.com.

Janick, J. *Horticultural Science*. 4th ed. San Francisco: W. H. Freeman, 1986.

Refl, Diane, ed. *The Role of Horticulture in Human Well-Being and Social Development*. Portland, OR: Timber Press, 1992.

Simson, Sharon P., and Martha C. Straus. *Horticulture as Therapy: Principles and Practice*. Binghamton, NY: Hawthorn Press, 1997.

Simson, Sharon P. and Martha C. Straus. *Horticulture as Therapy: Principles and Practice*. Haworth Press, Incorporated, 1997.

Wells, Suzanne. *Horticultural Therapy and the Older Adult Population*. Binghamton, NY: Food Products Press, 1997.

World Wide Web Addresses

http://www.floriculture.com. The Horticulture Web. 1995.

http://www.ahta.org American Horticultural Therapy Association website.2001.

http://www.ashs.org The American Society for Horticultural Science website.

http://www.chta.ca Canadian Horticultural Therapy Association.2001.

http://plants.usda.gov:80/plants/ This is the Plants National Database for the USDA.

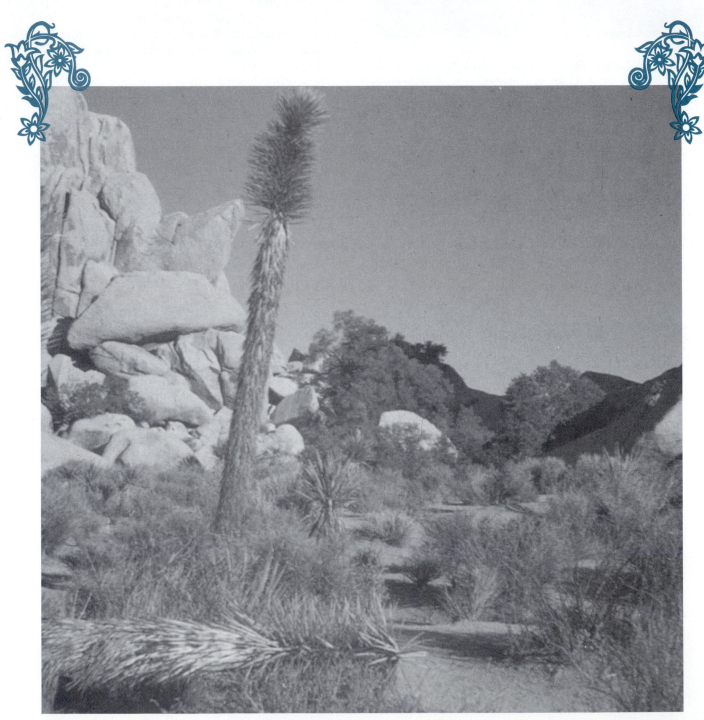

(Photo by the author)

CHAPTER 1

CLIMATE AND PLANT GROWTH

All plants outdoors (and even those indoors to some extent) are affected by the climate associated with the geographic area in which they grow. Humans can alter growing conditions in a limited way (such as by irrigating to compensate for scanty rainfall), but for the most part plant growth is constrained by natural climatic restrictions. Therefore a knowledge of how basic climate factors affect plant growth is of prime concern for the gardener.

CLIMATE AND SEASONS

With the exception of Hawaii and Puerto Rico, all parts of the United States and Canada are considered to be in the "temperate zone" and have to some degree the normal four seasons: spring, summer, fall, and winter. Some areas are considered to be "tropical" or "subtropical" inasmuch as they do not experience frosts and freezes. These areas are not technically in the tropics, however, in that they are not located between the latitudes of the tropics of Cancer and Capricorn, which are 23 degrees north and south of the equator, respectively. Those true tropics experience negligible temperature variation throughout the year, unlike the U.S. mainland tropical zones in California and Florida, which may have variations of 20 to 30°F (-7 to -1°C) in daytime highs and nighttime lows throughout the year.

In addition to being classified according to the normal four seasons, seasons can be classified as "wet" and "dry" in some parts of the country. California in particular has a "Mediterranean climate" with a wet winter season, an abbreviated spring starting in February during which the rain slowly tapers off, and an extended summer and fall until October during which little or no rain falls for months.

The same is true of the desert areas in the Southwest. Winter rains give way to a short, spectacular flowering spring followed by a prolonged, hot, and completely dry summer.

The growth patterns of plants vary widely from season to season depending on the plant type. While some are spring bloomers, others don't flower until fall. Some, such as asparagus, begin to grow as soon as the snow has left the ground. Others, such as lilies, don't produce new growth until all possible chance of frost is past.

Spring

For the majority of plants growing in temperate zones, spring is the period during which the majority of their yearly growth takes place. This yearly "flush," as the growth spurt is called, includes leafing out of trees, emerging and blooming of bulbs, blooming and fruit setting in fruit trees, new shoot growth in shrubs and trees, and the like. Some plants, such as grapes, make all the new growth they will produce for the year during this one- or two-month period. The leaves and shoots produced during the spring mature during the summer, but no new leaves are produced after the initial burst of growth. Others, such as lawn grasses, continue to produce new growth as long as the weather is favorable.

Summer

Summer is a period of maturing in plants. During this period some plants (such as tomatoes and corn) have their flowering period, and others, which flowered in the spring (such as fruit trees), enlarge and ripen their fruits. The new tender leaves produced in spring toughen to a leathery texture. Summer is, for many plants, a time in which energy is accumulated to enable the plant to live through the long winter during which it will have no leaves and be completely dependent on the food stored in its roots and branches during the previous summer.

Fall

Fall is a transitional season for plants that live from year to year and the end of life for those that live only one season. Those that will be surviving the winter are now engaging the mechanisms that will assure their winter survival: dropping leaves, stopping new growth, and entering a near static state during which they will await the return of favorable growing conditions.

Plants that live only one year must now have produced seeds capable of sprouting during the next growing season, and must have dispersed those seeds throughout the area by their own unique evolutionary means. For example, fluffy milkweed seeds will float to new ground; berry seeds will reach new locations in the droppings of birds; and snaggly cockleburs will ride on the fur of animals to new growing areas.

Winter

For most plants, winter is a time of waiting. When temperatures are below freezing, water is solid and therefore cannot travel readily throughout the roots and stems. This (combined with the lack of leaves on most plants in winter) drastically slows down the biochemical processes within the plant and puts most plants in a state of suspended animation until the return of favorable growing conditions.

An exception to this are the evergreen plants such as pines, spruces, firs, junipers, and the like. These plants keep their foliage during winter and continue to function at a higher rate than that of plants that have lost their leaves. Bearing this in mind, it is advisable to water evergreens during the winter during periods of thaw when the soil is able to take in moisture. The plants will take the water through their roots as long as the ground remains unfrozen. To some extent this can help evergreens avoid unsightly "winter burn" (needle-tip death)

due to excess moisture loss as a result of cold, dry winter winds moving over the foliage.

ELEMENTS OF CLIMATE

The elements combining to make up a climate include primarily (1) temperature, (2) precipitation, (3) humidity, (4) light, and (5) wind. Each factor has a wide range of variation and can have a dramatic effect on plant growth.

Temperature

Temperature, and particularly minimum winter temperature, largely determines the geographic range over which a plant will grow. The lowest temperature that a plant can withstand is called its *cold hardiness* or *cold tolerance* and is usually expressed in degrees Fahrenheit or centigrade.

For many plants a temperature of about 28°F (-2°C) is the minimum they can survive because at this temperature the liquid contents of the plant cells freeze and the plant dies. These plants are designated as *frost-tender* and include many vegetables and flowers.

Plants that are able to survive temperatures lower than freezing are called *frost-hardy*. They vary widely in their tolerance to subfreezing temperatures. Some will live to 10°F (-12°C), others to -5°F (-21°C) or even -30°F (-35°C). In some cases the woody portions or root system may be frost-hardy but not the flowers and leaves. For example, a late spring frost may kill the blossoms and young leaves of a fruit tree, but the branches will leaf out again and continue growing. Some garden flowers die to the ground in fall but grow back from the roots the following spring.

Although frost-hardy and frost-tender are the main classifications of cold tolerance, some houseplants and many tropical plants suffer from *chilling injury* at temperatures less than 50°F (10°C). This is why bananas turn black in the refrigerator.

Minimum winter temperature limits the areas where many plants can live, but for some plants lack of cold prevents survival. Fruit trees such as apple and cherry and flowers such as peony and tulip require a period of cold temperatures to grow normally. They cannot be readily grown in tropical areas because of the comparatively warm winters. (The requirement of plants for cold is discussed in Chapter 3 under "Causes of Flower Induction.")

For temperate-zone fruits such as apples and cherries, an adequate number of cold days per winter, called the plant's *chilling requirement*, is essential for normal growth and fruit production. Each type of fruit and each variety within that type has its own specific requirement. Apples, for example, vary from 250 to 1,700 hours of chilling; peaches vary from 50 to 400. This shows why peaches are grown in the South and apples in the North.

Looking at the question of temperature from the opposite perspective, each crop also needs a certain number of adequately warm days in order to mature. This is called the *growing degree-day* requirement. A baseline temperature below which very little growth occurs is established for each crop. This baseline temperature is subtracted from the average temperature for a given day. For example, if a crop has a baseline of 50°F (10°C) and the average temperature on July 10 was 82°F (28°C), then 32 growing degree-days would be accumulated on July 10. If the temperature were 50°F (10°C) or less, no growing degree-days would be accumulated. These daily measurements can then be added and compared to an established standard for that crop to determine when the crop will mature. Information on the growing degree-day requirement of a particular plant can also be compared with norms for a given locale to determine whether the crop could be expected to grow successfully in that location.

Frost

An understanding of frost types and the weather conditions that favor each are important because of the damage done to frost-tender plants.

A frost can be one of two types. The first type is *radiation frost* that occurs when the air is cool and calm and skies are clear. With this frost, warmth from the sun accumulated by the soil and plants during the day is lost at night as heat radiating upward. If the day was warm and the night only slightly below freezing, the reservoir of heat in the soil and plants may last through the night. If not, frost damage will result.

When plants are covered by a barrier that blocks the flow of heat to the sky, radiation frost can often be prevented. In nature, clouds form this barrier, and radiation frosts seldom occur on cloudy nights. On clear nights when a radiation frost is likely, protection of the plants by layers of newspaper, cloth, smoke, or plastic film can serve the same function.

Sprinkler irrigation to leave a protective film of water on plants is also used. Sprinklers are turned on as soon as the temperature drops to freezing and left on until the temperature rises above freezing in the morning. The warmth contained in the water is given off as the water freezes and transferred by conduction to the plants, preventing frost injury.

The second type of frost is caused by a cold air mass moving into an area. The air flows over the earth, car-

rying off heat from the plants, and the plants are frosted. This type of frost occurs irrespective of cloud cover.

Frost often leaves ice crystals on plants and the ground in the early morning. These crystal-depositing frosts are called *hoar frosts.* They occur when moisture from humid air condenses onto the plants because of the drop in temperature.

If the moisture content of the air is low, the air temperature can reach freezing without moisture condensation, but plants will still be injured. These frosts are called *black frosts,* because the first sign of their occurrence is the blackening of injured plants.

The appearance of frost in some areas but not in others close by is usually due to elevation differences. Cold air is heavier than warm air and will flow downward and settle in the lowest area. Accordingly, a low-lying frost pocket may experience a frost two or three weeks earlier than the surrounding area.

Freezing of the Soil

The northern parts of the United States and almost all of Canada experience freezing of the ground during the winter. The period of time during which the ground is frozen, the depth to which it is frozen, and the temperature in the plant root area vary by climate zone and of course from year to year due to variations in winter severity. In unseasonably cold years many plants may "freeze out" as a result of the temperature around the root zone being too cold for the plant to survive.

The freezing of the soil causes the water contained in it to expand, and this may cause a condition known as "frost heaving," in which the soil pushes up in chunks from around the roots of plants leaving deep cracks over the root area. This condition is harmful to plants because it exposes the roots to drying winter winds. For this reason, gardeners frequently lay a thick layer of shredded wood, bark chips, or similar material over the root areas of plants that they know to beborderline cold-hardy in the area. This layer insulates the ground and lessens both frost heaving and water loss from the soil during the winter.

Precipitation

Precipitation can take many forms including rain, snow, hail, and sleet. The precipitation that falls as rain is of greatest value to outdoor plants.

Rain

Lack of rainfall all year or during the period when other weather conditions (mainly temperature) favor growth is a limiting factor to plant growth, particularly in the western United States. In these areas irrigation is essential to grow plants other than those naturally adapted to a drought period.

Excessive rain or rain occurring out of season can be as detrimental to plants as lack of rainfall. Excessive rain can kill plants adapted to dry areas, particularly if the water cannot drain away from the roots. Unseasonable rain near fruit harvest can make strawberries watery-tasting and can split unpicked apples and grapes. Frequent rains also spread plant diseases by splashing microorganisms from one leaf to the next and trigger the growth of dormant disease organisms present on plants.

Snow

Snow cover plays an important role in plant survival in areas with cold winters. When there is persistent snow cover all winter, the snow protects low-growing plants from wind that could otherwise cause low-temperature damage due to the "chill factor" or low-temperature and drying injury due to "sublimation" of the frozen water from the plants. The snow surrounds the plants with a moist insulating wind barrier.

The *chill factor* is the decrease in temperature that occurs when air such as wind passes over a wet object. The air causes evaporation of the water, a chemical process that requires energy input in the form of heat. The energy is provided by the small amount of heat present in the plant. When temperatures are already low, this additional heat loss incurred because of evaporation may be the deciding factor in whether the plant is able to survive.

In *sublimation,* frozen water turns directly into its gaseous state without ever melting. This is the process used in the food preservation technique known as "freeze drying." This process also requires energy in the form of heat, and that heat is taken from the plant. Damage is caused by the water lost from the internal parts of the plant, as well as from the surface.

Sleet, Hail, and Freezing Rain

Sleet, hail, and freezing rain can be very injurious to plant health. Sleet and hail injure plants by tearing leaves and bruising or knocking off developing fruits. These forms of precipitation can fall with such force that tender plants will be completely beaten to the ground. Freezing rain frequently causes branch breakage on trees. As the rain hits cold plants, it quickly freezes. Layers of ice increase the weight of branches, causing them to bend and eventually to break. Heavy snowfall can also cause breakage due to its weight.

Dew

Dew is most likely to occur when the air is warm and humid. After sunset, the air temperature drops, and the atmosphere is unable to contain the amount of water it held during the daytime. The water then precipitates out over the area in the small droplets called *dew.* As the sun shines on the plant leaves early in the morning, it warms them and causes the dew to evaporate.

Although its effect on the growth of most plants is not great, it can be a factor in the spread of diseases in turfgrass. It is also an important water source for plants that absorb water readily through their leaves.

Humidity

Humidity may also determine how well a plant will grow. *Humidity* is defined as the amount of water air contains *relative* to the amount it is capable of holding at that specific temperature. The higher the temperature, the greater the amount of water the air can hold. (See Chapter 17 for an in-depth discussion of humidity.) Often, higher humidity improves plant growth by reducing the rate at which plants lose water from their leaves. But low humidity will not seriously hinder the growth of most outdoor plants, provided there is adequate water around the roots.

Plants growing indoors in cold-winter climates are commonly damaged by insufficient humidity. Heated indoor air has an average relative humidity of only 10 to 25 percent. Because many plants used indoors are native to the humid tropics, the drop from the accustomed 50 to 90 percent humidity to 25 percent can cause problems (see Chapter 17).

Fog and Mist

Two other forms of humidity, fog and mist, can greatly influence plant growth. In coastal California climates, the moisture from fog and mist is absorbed by plants and supplements the meager rainfall (Figure 1-1). At the same time this vapor retards loss of water from leaves.

Light

Light duration, intensity, and quality affect growth of outdoor and indoor plants, and may control flowering and growth rate. Light duration is dependent on distance from the equator and season. Directly at the equator, day and night are equal in length all year. North or south of the equator (the distance being marked by latitude lines), days become longer in the

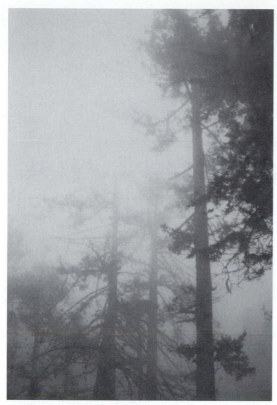

Figure 1-1 Fog in a redwood forest of coastal California. The moisture-laden air is the only source of water during the dry summer months. *(Photo by Rick Smith)*

summer and shorter in winter. Also, the further from the equator, the greater the difference between the longest summer day and the shortest winter day. In the northernmost areas of the United States, 16-hour days in the summer are normal.

The duration of daily light a plant receives is extremely important because it determines for what period of time the plant can manufacture the carbohydrate necessary for growth. This process, called *photosynthesis,* requires light. The more daylight hours, the more carbohydrate will be produced. The duration of the night also affects the start of some plant processes such as flowering (Chapter 3).

Light intensity or brightness also affects photosynthesis because, in general, the brighter the light intensity, the more photosynthesis occurs. Moderate to bright intensities are most beneficial to the majority of plants. Latitude will alter intensity. At the equator the sun passes directly overhead all year. But in the northern United States it is nearly overhead in summer but low in the southern sky in winter. Thus the winter intensity is lower because of the distance through the atmosphere the rays must travel. (See Chapter 16 for further discussion.)

The lesser amount of light in winter does not seriously affect outdoor plants. Some enter a resting phase at the onset of cooler temperatures. For others the light intensity is sufficient to continue photosynthesis and growth continues. However, indoor plants can be seriously retarded by the decrease in light because many are growing at subsistence light levels already. The dual handicaps of short daylight duration and low light intensity combine to make the climatic factor of light critical for indoor plants.

Full Sun, Partial Shade, and Full Shade Designations

When buying plants for the garden, it is essential for the gardener to know the light preferences of the plants being purchased. Similarly, if the gardener is seeking a plant for a particular location around the home, he or she must know the approximate amount of light that the area receives in order to choose a plant that will grow healthily. If not already familiar with how much light a particular location receives, the gardener should look at the area to be planted at two-hour intervals throughout one full day and note how long the area receives sun.

Areas around the home can be classified into three categories: (1) full shade, (2) partial shade/partial sun, and (3) full sun. *Full shade* areas are those in which the sun rarely shines directly, perhaps for only an hour per day at early morning or late afternoon when the sun's rays are the weakest. These areas would include the northern sides of houses, narrow areas between houses, under large trees, and any other areas that are shaded for almost all of the day by trees, buildings, or other structures.

Partial shade (also known as *partial sun*) areas receive direct sunlight for about a third to half of the day. Such areas include the east- and west-facing walls of houses (provided that the eaves are not so wide that they cast shade all day) and sites under small or finely foliaged trees that admit a dappled "filtered sun."

Full sun would be found on the south sides of houses or wherever there are no trees, shrubs, or structures that block the light. Full sun can also be on the west side of houses in the western United States, because the afternoon sun in these areas can be very strong even though it is only shining on the area for a few hours.

Most plants purchased in a nursery are sold with identification and care tags attached that specify the amount of sun the plant should be grown in. Planting a plant that requires full sun in a partial-shade area will generally reduce its blooming and cause it to produce weak new shoots, leading to an unattractive plant without the natural character it should have. Conversely, planting a shade-requiring plant in full sun will result in a stunted plant with bleached yellow leaves.

Wind

Wind is the fifth climatic element governing plant growth. Ocean wind limits the number of plants that can be grown near the shoreline, both by the salt spray, which browns plant leaves, and by the intensity, which whips and breaks leaves and branches. In dry areas, hot desert winds increase the rate of moisture loss from the soil and plant leaves, intensifying existing drought conditions.

Even in cold northern areas wind is an inhibiting factor to plant health. Subzero winds damage evergreens by removing moisture from the foliage, essentially freeze-drying it. In frozen ground the plant cannot replace the water, and injury similar to drought results.

FACTORS MODIFYING CLIMATE

Natural factors modifying climate include distance from the equator, elevation, terrain, and the nearby presence of large bodies of water.

Elevation

Changes in elevation can give areas only a small distance apart completely different climates for plant growth. The higher the elevation of an area, the colder the average year-round temperature, with every rise of 300 feet (91.4 meters) causing an average temperature decrease of 1°F (0.5°C).

Terrain (Topography)

Changes in terrain also alter climate, particularly rainfall. Most rain storms move from west to east in North America. In a narrow range of hills, the west-facing slopes often receive considerably more rainfall than do the east-facing slopes. The heaviness of the water in rain clouds prevents them from passing over the mountains until most of the precipitation has been released.

Sloping terrains can also vary several degrees in temperature from the top portion of the slope to the lower part or valley. Cool air, which is heavier than warm air, will flow down a slope and collect at the bottom, a phenomenon called *air drainage* (Figure 1-2). Consequently, plants growing higher on the slope are less likely to be damaged by cold than those at the bottom.

Bodies of Water

Large bodies of water such as the Great Lakes and oceans exert strong effects on the climate of nearby land. The

Figure 1-2 Typical cold-air drainage on sloping land.

enormous volume of water absorbs heat in the summer, making adjacent land cooler, and gives it off in the winter, raising the temperature. Thus they moderate temperatures all year, making the summer cooler and the winter warmer than in inland areas.

On the North American continent, the influence of large bodies of water on agriculture is considerable. Warmer winter temperatures found along the Great Lakes make these regions large commercial growing areas for fruits such as grapes, cherries, and peaches, which do not grow as well farther from the shores. Similarly, most of the Pacific coast is warm enough to grow certain vegetables all winter, but inland these crops would freeze.

Climate Modification by Humans

People can modify climate either intentionally or unintentionally. Among the most widespread unintentional causes of climate change are air pollution and consequent smog.

Air pollution can be composed of both gases (carbon monoxide, for example) and small airborne pieces of matter called *particulates*. Combined together they become smog, a persistent brown or yellow layer that hangs over some metropolitan areas.

One way smog affects plant growth is by temperature modification. A smog cloud acts like an insulation layer over an area and prevents heat from escaping to the atmosphere. Thus areas with smog cover have warmer summers than nearby areas without smog, and plants may be damaged by excess heat.

Smog also decreases the intensity of sunlight that reaches the ground, giving plants less light from which to manufacture carbohydrate. Particulates settle onto leaves and further decrease the light intensity reaching the plants.

The main way smog affects plant growth is through foliage injury by toxic gases. The gases enter the leaves, causing leaves to turn partially brown and sometimes to die. Plants vary in their susceptibility to smog injury, ranging from smog-tolerant to highly smog-intolerant. Plants classified as intolerant (many pines, for example) cannot be grown in smoggy areas.

MICROCLIMATES

Microclimates are small areas that have slightly different climate characteristics than the surrounding land. They may be less windy, shadier, moister, warmer, or in any other way different from the typical climate. These differences affect plants growing in the microclimate, sometimes helping and sometimes hindering growth.

Microclimates may be either artificially made or naturally occurring. They can be formed by the natural terrain and vegetation or unintentionally created by the erection of structures such as buildings, fences, and roadways. Although they often go unnoticed, they can be used by the observant gardener to provide the specialized growing conditions favored by different plants (Figure 1-3). In some cases they make it possible to grow plants not normally considered cold-hardy in that area.

Outdoor Microclimates

One example of a natural microclimate is a frost pocket. In addition to being colder, this microclimate might also be moister due to water runoff and would be suitable for moisture-loving plants not easily damaged by cold.

Another natural microclimate can be found under trees. The area is shadier and cooler in summer and would be good for shade-loving plants. It would also provide ideal conditions for growing indoor foliage plants outdoors in summer.

The area under the eaves of a house is a manmade microclimate found around most homes; plants there live under different environmental conditions than plants living away from the house. First, they are shielded from rainfall by an eave, so the soil will be drier. Moreover, the wall may afford protection from cold or drying winds, and it radiates warmth from the heated dwelling. During cool periods this additional warmth speeds up or prolongs growth, but if the wall faces south it may be excessively hot in summer and plant damage can occur (Figure 1-4). This is also frequently the case with parking lot trees, which become overheated due to heat radiating from the asphalt.

At night when temperatures drop, objects that absorbed heat during the day begin losing it to the cooler night air. Heat radiating from the ground and buildings can protect plants from frost and extend the growing season.

Figure 1-3 A microclimate for growing ferns. The north side of this wall provides shelter from excess sun and wind. *(Photo courtesy of Barry Glick, Sunshine Farm and Gardens, Renick, WV. www.sunfarm.com)*

Figure 1-4 Scorch on maple due to excess heat. *(Photo courtesy of Dr. R.E. Partyka, Columbus, OH)*

Wind can be intensified or lessened in a microclimate. A plant by a wall will generally receive wind protection unless it is directly in the path of prevailing winds. However, when two walls are parallel, a wind tunnel can be formed. Wind conditions will be worse there than in a completely unprotected area. Such wind tunnels are frequently formed between closely spaced buildings.

Indoor Microclimates

Microclimates are usually thought of as being outdoors. However, indoor microclimates with varying temperature, light, and humidity can be found throughout a house or apartment. Because heat rises, upstairs rooms may be several degrees warmer than downstairs ones. A basement may be up to 20°F (11°C) cooler than the remainder of the home on a hot day, and the north side of the house may stay considerably cooler than the south side. Even within a single room, cool microclimates will be found adjacent to windows in cold winter areas because of cold entering through the glass.

Microclimates with varying humidity levels are also common. In the kitchen and bath, where water is used, evaporation raises the humidity. A basement is also usually humid.

Light microclimates are easy to detect, being governed by the size and location of windows. Obviously the further from a window a plant is, the less light it will receive. However, each foot or meter of distance can be thought of as a different microclimate. In addition, south-facing windows are maximum-light microclimates, east and west moderate, and north-facing low-light microclimates because of the angle of the sun in the northern hemisphere.

PLANT-GROWING ZONES OF NORTH AMERICA

The question of where in the country a specific plant can be grown is important to every gardener. There is little to worry about when plants are bought at local nurseries because nurserymen seldom sell plants that do not grow in their area. But when a plant is ordered from a mail-order nursery, the consumer must check to make sure the plant he has selected will grow in his area.

Because minimum winter temperature is often what determines where a plant can be grown, knowing the plant's minimum survival temperature is essential. This and other valuable information is usually given in the written descriptions of the plant. It can also be found in a gardening book from a library.

United States Department of Agriculture Plant Hardiness Zones

After determining the minimal survival temperature of a plant, it is then necessary to determine the minimum winter temperature of one's area. To simplify this procedure, the United States Department of Agriculture (USDA) has prepared maps showing zones of minimum winter temperatures throughout the United States and Canada (see inside cover). Eleven zones are outlined, ranging from completely frost-free areas in Florida, Texas, and California to a −50°F (−46°C) or colder zone in Canada. The map takes into account cold mountain climates in warmer zones, as shown in Colorado. But for the most part it is made up of bands that correlate minimum winter temperature with distance from the equator.

The USDA zone map is not the only plant hardiness map that exists, but it is the most widely used. Some states whose climate is influenced by many factors have designed more detailed maps. California is one such state, and the University of California at Davis lists 21 separate climate zones in the state due to the influences of the Pacific Ocean and mountain ranges.

American Horticultural Society Heat-Zone Map

The other map on the inside cover is the opposite of the USDA Plant-Hardiness Zone Map in that it addresses the issue of plant hardiness in excessively hot temperatures instead of excessively cold temperatures. Because of the warming of the planet, gardeners' knowledge of their heat zones may become much more important in the future, especially during drought periods.

Heat damage is less drastically apparent than damage from cold, and the death of the plant due to heat is generally more gradual, occurring over months or years. In addition to scorch illustrated in Figure 1-4, symptoms such as failure of flower buds to open, bleaching of leaves to a paler color, and failure to grow at a normal rate may all occur. Leaves may droop even though the plant is receiving regular watering. In its weakened state, the plant may become more attractive to insects, which will hasten its death.

The American Horticulture Society Plant Heat-Zone Map divides the United States into 12 zones by the average number of days per year that the areas experience temperatures over 86°F (30°C). The range is from 0 to 210 "heat days" per year.

The rating of a plant in regard to heat tolerance will soon be listed next to its USDA cold-hardiness number on care tags, in gardening reference books, and in catalogs. A plant will have two ranges, for example:

Lily 3-8, 8-1.

The first range (3-8) will designate the USDA low-temperature hardiness by zone. The second range (8-1) will designate the heat tolerance. So, if you live in USDA zone 7 and AHS zone 7, you will know that the lily will survive both the cold winter temperatures and the hot summer temperatures of your area. When reading the new designations, you will find that plants you previously thought lived only one growing season (such as eggplant) actually can live for years in warmer climates.

As of this writing, not all plants have been coded by AHS zone. Even when they are so coded, some plants may survive out of, or fail to thrive within, the zone area specified for them due to unseasonably warm or cool years. In addition, stresses such as light deprivation, microclimates, certain soil characteristics, and nutritional status of the plant will alter independently and together the ability of a plant to grow within a specific heat zone. In all cases, the classification assumes that the plant is receiving adequate watering.

Zone Maps Based on Overall Climate Characteristics

Several zone maps divide the country into general climate types. Such maps are helpful because they include such climatic factors as precipitation and light intensity as well as temperature. They give a more general view of the basic climate in large sections of the country. Such information is particularly valuable today when people move so frequently.

One such general climate classification developed by the late Dr. J.B. Edmond, Professor Emeritus, Mississippi State University, divides the continental United States into five regions (Figure 1-5).

1. *Northeast and Great Lakes Region.* This area is characterized by comparatively cool summers, long cold winters, and moderate rainfall. Light intensity is moderate in the summer and dim in the winter owing to a persistent cloud cover. Summer daytime

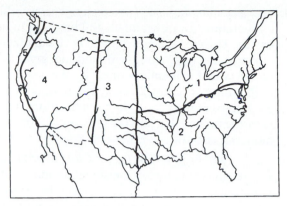

Figure 1-5 Climates of the continental United States: (1) cool-humid of the Northeast and Great Lakes, (2) warm-humid of the Southeast and Gulf Coast, (3) subhumid of the Great Plains, (4) arid of the intermountain region, (5) summer dry of the Pacific Coast. *(Map from* Fundamentals of Horticulture *by J. B. Edmond. Philadelphia: Blakiston, 1951. Reprinted by permission of McGraw-Hill)*

temperatures may reach 90–95°F (32–35°C), but they are for the most part cool enough to grow deciduous fruits (from trees that drop their leaves) such as apples, pears, and cherries. At the same time, temperatures in the summer are warm enough to grow most warm-season vegetables (such as tomatoes and beans). But summers are not long enough to grow many varieties of watermelons and sweet potatoes that require a long maturation. The rainfall throughout this region is usually sufficient, and no irrigation is needed.

2. *Southeast and Gulf Coast Region.* This region has long warm summers, short mild winters, bright light intensity, and moderate to heavy rainfall. The temperature is quite variable: in the northern half yearly temperatures are cool enough for deciduous fruits; in the southern part they are warm enough for semi-tropical fruits like citrus and avocado. Winter temperatures in southern Florida are mild enough to grow vegetables all year, and much of the winter produce for the East is from there.

3. *Great Plains Region.* Wide fluctuation in temperature, scarce rainfall, and bright light intensity are typical of this region. Summer temperatures in the northern part of this large region may equal those in the southern section, but winter temperatures differ greatly. In the north they can drop as low as −30°F (34°C), whereas near the Gulf they are quite mild. The short growing season in summer limits the commercial production of crops in the north, but parts of Oklahoma and Texas produce vegetable crops throughout the year with the aid of irrigation.

4. *Intermountain Region.* This region has variable temperatures, scarce rainfall, and bright light intensity. The area is characterized by a wide range of elevations that, along with latitude, gives this region variable temperatures. In the southern part, summer temperatures are hot, and the growing season is long. In the remainder, temperatures are cooler, and the growing season is of moderate length.

5. *Pacific Coast Region.* The Pacific Coast divides into two sections lengthwise: the shoreline and the intermountain valleys directly over the coastal mountains. Summer temperatures are cool along the coast due to the Pacific Ocean but hot in the valleys east of the coastal mountains. Winter temperatures are mild throughout the region but particularly so along the ocean, where frost is infrequent. Winter is the rainy season in this area, with seasonal rainfall ranging from very plentiful in the northern half to very scarce toward the Mexican border. The amount of rainfall also influences light intensities, making them moderate in the north but bright through the south. This region is very productive agriculturally, producing fruits and vegetables all year.

SELECTED REFERENCES FOR ADDITIONAL INFORMATION

Cathey, Marc. *Successful Summer Gardening: Heat-Zone Gardening.* Alexandria, VA: Time-Life, 1999.

Chawan, D.D., ed. *Environment and Adaptive Biology of Plants.* New York: 1994.

Hay, R K. and Alastair H. Fitter. *Environmental Physiology of Plants.* San Diego, CA.: Academic Press, Incorporated, 1988.

Roberts, Edith A., Elsa Rehmann, and Darrel G. Morrison. *American Plants for American Gardens: Plant Ecology—The Study of Plants in Relation to Their Environment.* Athens: University of Georgia, 1996.

World Wide Web Addresses

http://www.swswis.com/lifecycles/what.html. Phenology, The Study of Nature's Cycles of Life.

http://www.urbanext.uiuc.edu/ecosystems/teacherguide1.html University of Illinois Extension. "Ecosystems and Climate" website. (Teacher's Guide)

(Photo courtesy of Agricultural Research Service, USDA)

CHAPTER 2

BOTANICAL NOMENCLATURE, ANATOMY, AND PHYSIOLOGY

WHAT IS A PLANT?

There was a time when it was easy to tell what was a plant and what was an animal, and there was the plant kingdom and the animal kingdom. The invention of the microscope, however, revealed organisms that were neither animals nor plants, but possessed the qualities of both. Eventually the simple two-kingdom (plant and animal) model of life was replaced by a four- and then a five-kingdom one:

Animalia

Plantae

Fungi

Protista (mainly bacteria and blue-green algae)

Monera (one-celled organisms that share characteristics of both plants and animals).

In the four- and five-kingdom models "plants" only referred to higher plants, not algae—in other words, land plants. But even this model proved unworkable, and at present there is no universally accepted definition of what is a plant. Instead, there are three "domains" based on genetic analysis and many kingdoms. But for purposes of this text, the original definition of a plant as a land plant will be mainly used.

PLANT NOMENCLATURE AND CLASSIFICATION

Botanical nomenclature is the orderly classification and naming of plants. It is a subject that unnecessarily intimidates people unfamiliar with it because they believe one must know Latin in order to use the system. But the botanical naming system is not overly complex, and it does not require any background in Latin. In fact, a surprising number of common names are the same as botanical names, such as iris, fuchsia, and citrus.

There are several reasons for using botanical names in place of common names. The first reason is the universality of botanical names. Latin names are the standard worldwide system for communicating the identity of plants. The Latin name *Pyrus communis* would be recognized as the common pear by people in Asia, Africa, the Americas, or Europe, provided they were familiar with botanical nomenclature.

Precision is the second reason. Common names such as "daisy" or "ivy" can refer to any number of similar plants. If only common names were used to identify these plants, confusion would result.

Knowing the botanical name of a plant can also give clues to its growing requirements. For example, the common names "bunny ears", "beaver tail", and "prickly pear" do not indicate that these plants have anything in common, but their botanical names are similar and so is their culture.

The Origin and Construction of Botanical Names

The branch of botany that deals with the naming of plants is called *taxonomy*, and people doing the work are *taxonomists*. The naming system used dates back 250 years to the Swedish botanist Carolus Linnaeus, the "father of botany." He named and published the first references to many plants, using a naming method called the "binomial system." This system specifies that a plant name must have at least two parts.

Taxonomists are generally botanists, but some specialists work only with horticultural plants. A taxonomist generally has a Ph.D. in horticulture and spends his or her time examining plants (including through microscopes) and comparing them to related plants to determine the correct botanical name. The name the taxonomist has determined to be the correct one then serves as the standard name that all horticulturists will use.

For example, in the botanical name *Tagetes patula* (French marigold), *Tagetes* is called the *genus* (*genera*, plural). The second word, *patula*, is called the *specific epithet*. When combined, these two words form the plant *species*.

For ease of understanding, the genus can be thought of as the last name or surname of a plant. It indicates that the plant is related to other plants in that same genus. The specific epithet can be thought of as the first name of the plant, the name that distinguishes it from its relatives in the same genus.

In the two plants *Tagetes patula* and *Tagetes erecta*, for example, *Tagetes patula* is a small marigold with thumbnail-sized blossoms. *Tagetes erecta*, on the other hand, is a tall marigold with fist-sized blossoms. It is apparent that both plants are marigolds. But because they grow so unlike each other, they are given different specific epithets and hence become separate species.

For many wild plants, a two-part name is sufficient to distinguish between similar, related plants. However, new plants that are mutations or bred from existing species are constantly being developed. To distinguish them from their parent plants, a third part called the *variety* or *cultivar* is added to the name. Common marigold cultivars, for example, would be *Tagetes patula* 'Lemondrop' and *Tagetes patula* 'Petite Harmony.'

A plant *variety* is a naturally occurring mutation or offspring that is significantly different from the parent and that becomes established in nature. For example, a species that normally only has white flowers might spontaneously mutate and a new variety with pink flowers would appear.

A *cultivar* is a manmade and/or man-maintained variety, hence, the name, which is short for "cultivated variety." As an example, an orange tree might develop one branch that produces seedless fruit. But in nature this new kind of orange would not be able to reproduce because it has no seeds, and so the new type of orange would therefore not become established and so would not be called a variety. But through the actions of man, who is able to reproduce the branch without seed, the cultivar would become established, and in fact this was the actual origin of seedless navel oranges.

The botanical names given to plants are not assigned haphazardly. A set of rules called the *International Code of Botanic Nomenclature* is adhered to strictly when new species are named. Although the genera and specific epithets must be in Latin-like form, the words are not always derived from that language. The names of prominent botanists, statesmen, and even explorers were often Latinized and incorporated in plant names. Frequently though, a Latin name conveys information about the plant it represents. The specific epithets *rubra, alba, atropurpurea,* and *variegata* are used with many plant names to connote colors. Table 2-1 defines some common specific epithets.

Table 2-1 English Translations of Common Specific Epithets and Cultivars

Latin epithet	English translation
alba	white
atropurpurea	dark purple
aureum	golden
carnosa	fleshy
compacta	compact
esculentus	edible
fastigiata	erect
floribunda	free-flowering
glauca	with white or gray coating
grandiflora	with large or showy flowers
horizontalis	horizontal
nanus	dwarf
nidus	nest
occidentalis	from the Western Hemisphere
officinalis	medicinal
pendula	hanging
rotundifolia	round-leaved
rubrum	red
sativus	cultivated
semperflorens	ever (semper)-flowering (florens)
stellata	starlike
sylvestris	of the forest
tuberosum	bearing tubers
variegata	variegated
vulgaris	common

Writing and Pronouncing Botanical Names

The writing of botanical names follows a prescribed pattern. The genus is always capitalized, followed by the specific epithet beginning with a lowercase letter, and both are italicized or underlined. Any variety name follows and may be set off from the species by "v." or "var." The abbreviation "cv." is used to designate a cultivar. Single quotes around the variety or cultivar name can substitute in either case. For example, the variety of ash tree known as Modesto ash may be written as:

Fraxinus velutina 'Modesto'

F. v. v. Modesto

F. v. var. Modesto

Note that in a listing, the genus or specific epithet is written out once and abbreviated thereafter.

Another botanical abbreviation is the use of the genus name followed by the word "species," "sp.," or "spp." "Species" or "sp." indicates an unknown species in that genus. "Spp." is plural and is used to refer collectively to all species of that genus. The latter abbreviation technique is useful when discussing a broad topic such as the diseases of oaks (*Quercus* spp.).

Unlike the standardized method of assigning and writing botanical names, pronouncing the names is somewhat arbitrary. Botanists themselves disagree on certain pronunciations. Common usage has become the basis for pronunciation.

Botanical Classification of Plants

Seed- and Spore-bearing Plants

Ninety percent of cultivated plants are flowering plants that reproduce by seed, but a few common ones are not. These *Pteridophytes* have a completely different reproductive system based on spores (see Chapter 3). Ferns are the most widely known Pteridophytes, and they belong to a branch of the plant kingdom that emerged early in plant evolution (Figure 2-1).

The seed plants are further divided into two groups. Gymnosperms are the smaller of these two groups, consisting primarily of evergreen cone-bearing plants such as pines, spruces, junipers, and yews. Their foliage is generally needlelike, and they do not produce showy flowers. Of the indoor plants, cycad, podocarpus, and Norfolk island pine are the common gymnosperms.

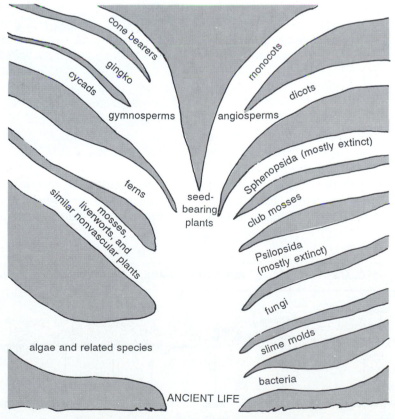

Figure 2-1 Evolutionary history of plants.

Angiosperms compose the majority of cultivated plants, including such diverse plants as rose, cabbage, and palm. It is within the angiosperm group that all ornamental flowering plants and nearly all food plants are found. The primary identifying characteristic of angiosperms is the growth of seeds completely within the plant ovary, which swells to become the fruit (see Chapter 3).

Monocots and Dicots

The angiosperms are further separated into the orders *Monocotyledoneae* and *Dicotyledoneae,* or "monocot" and "dicot" for short. The monocots are less numerous and are usually nonwoody plants that have a short stem and overlapping leaves arranged in a whorl, a form called a *rosette*. Their leaves are frequently long and narrow, with parallel veins running the length. Other identifying characteristics include flower petals fused at the base to form a bell and often a bulb or fleshy root system (see "Modified Roots"). Included among the monocots are all grasses, lilies, irises, onions, cattails, and most flowering bulbs.

Many dicots, on the other hand, grow to a large size. Their leaves have a branching vein pattern, and their flowers have a variety of shapes. Most trees and shrubs are dicots, as well as most fruits and garden vegetables.

Being able to recognize the difference between monocots and dicots is not of academic interest only. Whether a plant is a monocot or dicot can also determine its method of propagation and its susceptibility to weed killers.

Plant Families

The classification of plants leads ultimately to the smallest division, variety, or cultivar (Figure 2-2). For the most part the intermediate subdivisions are of interest only to botanists, with the exception of plant families, the division prior to genus. Each family groups a number of genera having like characteristics together. These families have both Latin and common names. Some families are listed in Table 2-2 with a sampling of the genera and each also includes key identifying characteristics.

Basic knowledge of plant families can help identify an unknown plant. For example, by remembering that all pea family members have pod-type fruits, you could deduce that such diverse plants as mimosa trees and snapbeans are both in that family.

In addition, plants in the same family are often susceptible to the same diseases. If a gardener has had a recurring problem with a disease, he could avoid buying plants with similar susceptibilities through familiarity with the family characteristics.

Horticultural Classification of Plants

Botanists are interested primarily in classifying plants by their evolutionary relationship to each other. Horticulturists frequently classify plants additionally according to use.

 Botanic gardens and arboreta (gardens specializing in trees) are public or private philanthropic gardens that offer garden access and educational programs to the public. Employment opportunities in a botanic garden might include supervising the display gardens and planning seasonal displays; giving talks and tours to the public; maintaining records and plant samples; or directing the entire garden, which would include budgeting, personnel, fund-raising, and so on.

Most cultivated plants are valued either for their ornamental or for their edible value, with a few multi-purpose plants such as herbs and ornamental vegeta-

Figure 2-2 A botanical classification of the tomato cultivar 'Big Boy.'

Table 2-2 Some Common Plant Families and Their Key Identifying Characteristics

Family name		Key identifying characteristics	Some included genera	Some included sample plants
Latin	*Common*			
Cactaceae	cactus family	fleshy growth habit and spines	*Mammillaria, Schlumbergera, Opuntia*	pincushion cactus, Christmas cactus prickly pear
Compositae (Asteraceae)	aster family	daisy or buttonlike flower heads composed of many tiny flowers which resemble petals	*Senecio Chrysanthemum, Rudbeckia*	florist cineraria, chrysanthemum, blackeyed susan
Cruciferae (Brassicaceae)	mustard family	flowers with four petals arranged crosswise	*Brassica, Alyssum*	cabbage, alyssum
Gramineae (Poaceae)	grass family	narrow, parallel-veined leaves encircling the stem at the base	*Poa Avena*	bluegrass, oats
Labiatae (Lamiaceae)	mint family	square stems; flowers have two lips	*Coleus, Mentha, Plectranthus*	coleus, mint, creeping charlie
Leguminosae (Fabaceae)	pea family	long, podlike fruits, characteristic flowers	*Lathyrus, Phaseolus*	sweet pea, green bean, honey locust
Liliaceae (Amaryllidaceae)	lily family	thin, grasslike leaves; bell-shaped flowers with six petals	*Chlorophytum, Lilium, Yucca, Allium*	spider plant, Easter lily, yucca, onion
Rosaceae	rose family	flower petals in multiples of five; flowers have many stamens	*Malus, Prunus, Pyracantha*	apple, flowering cherry, firethorn
Umbelliferae (Apiaceae)	carrot family	flowers in an umbrella like cluster	*Daucus, Anethum, Petroselinum*	carrot, dill, parsley

bles. Within the edible group, plants are commonly classified as the following.

> *Tree fruits:* apple, peach, plum, orange, avocado, and so on
>
> *Nuts:* pecans, walnuts, filberts, almonds, and so on
>
> *Small fruits:* blackberries, raspberries, strawberries, figs, grapes, and so on
>
> *Vegetables:* cauliflower, tomatoes, sweet corn, peppers, squash, and so on
>
> *Herbs:* rosemary, basil, dill, thyme, and so on
>
> *Grains.* wheat, rice, corn, oats, and so on.

The last category, grains, is generally not covered under horticulture. Grains are classified as field crops and fall under the study of crop science or agronomy.

Within ornamentals, the classification system links the plant to its form and use.

> *Trees:* oak, magnolia, pine, eucalyptus, and so on
>
> *Shrubs:* juniper, yew, lilac, azalea, and so on
>
> *Vines:* clematis, ivy, and so on
>
> *Groundcovers:* bugleweed, periwinkle, and so on

> *Garden flowers:* tulip, marigold, daylily, and so on
>
> *Turfgrasses and lawn substitutes:* bluegrass, St. Augustine grass, dichondra, and so on
>
> *Houseplants:* umbrella plant, African violet, spider plant, and so on.

Plant Identification

Determining the identity of a plant may be relatively easy if the plant is commonly grown, but it becomes difficult and time consuming if the plant is a wildflower or other uncultivated species. As a first step, cut off a stem with foliage and preferably flowers and/or fruits. Without flowers or fruits, identification may be difficult, unless the plant is a foliage houseplant. Enclose the sample in a plastic bag to lessen wilting, having sprinkled it previously with water if possible. Make note of the relative size of the plant, whether it is woody or nonwoody, and where it was growing (sun, shade, marsh, landscape, and any other relevant conditions). All this information can be helpful.

To have the unknown plant identified, begin by showing the sample to a knowledgeable local nurseryman. If he or she is unfamiliar with it, consult mail-order plant catalogs with color pictures or library books with photographs. If the plant was growing wild, botanical references may be necessary. However, these are usually complicated to use without a background in botany, because they include a considerable amount of technical terminology.

As a last resort, dry the sample by pressing it flat between several layers of newspaper and weighting it with a book. When the sample is completely dry (several weeks), it can be mounted on cardboard with tape and mailed to a university department of botany or horticulture along with the information on size, growing location, and so on, previously noted.

PLANT ANATOMY

Understanding how the parts of a plant form the whole, as well as the structure of those parts, is necessary to grow plants successfully.

 A plant anatomist generally has a Ph.D. and will be located at a university. His or her research will center on observing and identifying plant organs, generally at a microscopic level. Typical work might involve identifying and researching aspects of plant anatomy that could alter the plant's appearance, resistance to pests, keeping quality after harvest, or other useful traits. For example, a plant anatomist might research which variety of tomato is best suited to withstand shipping well by examining the parts of the fruit anatomy associated with shipability.

Vegetative Plant Organs and Their Function

Nearly all cultivated plants are composed of a limited number of basic parts, or organs: leaves, stems, buds, and roots (Figure 2-3). Collectively these are termed *vegetative organs* because they are not part of the sexual reproductive system of the plant. In its simplest form, a young plant would have the organs discussed in the next section.

Leaves

Leaves are the most obvious part of most plants. They are usually made up of two main parts: the wide section termed the *blade,* and the *petiole,* or leaf stem

Figure 2-3 Vegetative parts of a typical plant.

(Figure 2-4). The angle formed between the petiole and its supporting stem is the *leaf axil,* and in that axil will be a *bud.*

Figure 2-4 is a diagram of what is called a simple leaf. In contrast to this simple leaf, Figure 2-5 shows several compound leaves. A compound leaf is composed of individual *leaflets* but will only have a bud at the base of the entire leaf.

Leaves of the types shown in Figures 2-4 and 2-5 are common leaf forms on the angiosperm plants. But the leaves of cone-bearing gymnosperms are often scalelike or needlelike. Figure 2-6 shows sample leaves of gymnosperms.

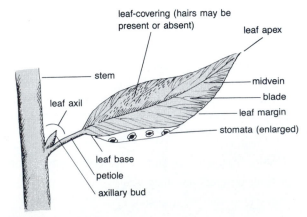

Figure 2-4 Parts of a simple leaf.

Figure 2-5 Types of compound leaves. *(Photo by Rick Smith)*

Parts of the Leaf Blade. Looking closely at a simple leaf, the visible parts are as follows.

Margin The leaf margin is its outside edge. The margin may be smooth, toothed, barbed, lobed, or in any other way different from smooth (Figure 2-7). Leaf margin characteristics are a means of identifying plants. Barbed or spined margins can also protect a plant from being eaten by grazing animals.

Veins The patterns of veins in leaves can be seen by looking closely or holding the leaf up to the light. Dicot plants have a central vein (midvein) with many branch veins. Monocots have parallel venation running the length of the leaf (Figure 2-8).

Leaf apex The apex is the tip of the leaf. The apex may be pointed, blunt, notched, or a number of other shapes.

Leaf base. The leaf base is the part of the blade that attaches to the petiole or directly to the stem (if the leaf has no petiole).

Leaf covering. Any hair, scales, or film on the leaf blade can be considered a leaf covering. Almost all leaves have an invisible wax layer, called the *cuticle,* which prevents water loss from the leaf surface. A leaf may have hair or scales in addition to this cuticle.

Stomata. Stomata are minute openings found primarily on the undersides of leaves (Figure 2-4). They are flanked on either side by *guard cells* that open and close the opening according to environmental conditions. Stomata regulate the flow of gases and water vapor in and out of the leaves.

Stipules. Stipules are small, often leaflike appendages found normally, at the base of a petiole. Not all plants have stipules, and in many plant species the stipules die, shrivel, and abscise relatively soon after the leaf has finished growing. One plant with particularly large stipules is sycamore (*Platanus* sp.), in which the stipule appears to surround the stem and is very leaflike.

Stipules may in prehistoric times have been protection for the leaf bud or emerging leaf. Now they may have a minor photosynthetic role or be modified into spines or tendris.

Figure 2-6 Leaves of gymnosperm plants. *(Photo by Kirk Zirion)*

Figure 2-7 Types of leaf margins. *(Photo by Kirk Zirion)*

(a)

(b)

(c)

Figure 2-8 Venation types: (a) elm leaf with midvein and branch veins, (b) maple leaf with three main veins and numerous branch veins, (c) amaryllis leaf with parallel veins.

Figure 2-9 Typical rosette-form plants.

Modified Leaves. Modified leaves have evolved to perform functions not customarily associated with leaves. The "petals" of poinsettia are modified leaves called *bracts.* Twining tendrils and spines are also modified leaves.

Stems

Growth of stems increases plant height or width. The stem transports water and other substances. It is also the site of leaf and flower attachment.

Modified Stems. Not all stems grow aboveground. Underground bulbs contain short stems and attached scales (onion layers, for example) which are modified leaves. The potato is an underground stem thickened to serve as a storage site and called a *tuber.* Frequent use is made of modified stems in plant reproduction, discussed in more detail in Chapter 4.

Shortened stems found in the aboveground portions of plants create the rosette plant form (Figure 2-9). Rosette plants are easily recognized because of the shortened stem; the leaves appear to grow from one central point and radiate outward like the overlapping petals of a rose. Many plants grow in rosette form, including African violets, cabbages, strawberries, daylilies, and all grasses.

Arrangement of Leaves on the Stem. Leaves can be arranged in an opposite or an alternate pattern on a stem. The pattern of leaf arrangement is important and is used in identifying plant species. As the names imply, leaves arranged oppositely grow in pairs, and those arranged alternately are attached stepwise down a stem (Figure 2-10). The site at which a leaf is or was once attached is called a *node,* and the section of stem between nodes is the *internode.* A bud will be present at each node, although it can be undeveloped and difficult to distinguish. Knowing the locations of nodes and internodes on a stem is important in pruning and propagation.

Internal Stem Structure. The life of a plant is dependent on the correct functioning of its vascular system, the network of pathways that moves carbohydrates, minerals, water, and other substances within the plant.

Figure 2-10 Types of leaf attachment and node locations.

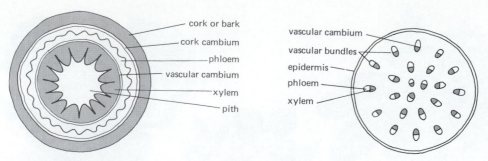

Figure 2-11 Cross sections showing arrangements of vascular tissue inside dicot and monocot stems.

These pathways, which can be thought of as a plant's circulatory system, are divided into xylem and phloem. Water and nutrients are moved up in the xylem from the roots, through the stem, and out to the leaves. Carbohydrate (the plant's "food") is manufactured in the leaves and moves down within the phloem to the roots or to other parts of the plant where it is needed.

Dicots have their xylem and phloem arranged in concentric rings inside the stem (Figure 2-11). The phloem is always found outside the xylem. Monocots, however, have xylem and phloem together in *vascular bundles* scattered throughout the stem. Both xylem and phloem extend not only through the stem of a plant but also up the petiole and into the leaf blade.

The vascular cambium is the third component of a dicot stem. The cambium manufactures new xylem and phloem cells to increase the transporting abilities of the vascular system. The cambium is a thin layer of rapidly dividing cells found between the xylem and phloem areas in the stem (Figures 2-4 and 2-11).

A second type of cambium is found in woody stems. This cambium, called the *cork cambium,* produces bark cells and is found directly beneath the bark surface.

Buds

A bud contains immature plant parts and can be of three types: *vegetative, flower,* or *mixed* bud. A vegetative bud contains a leaf or leaves and sometimes an embryonic shoot. A flower bud includes the rudiments of one or more flowers and is frequently larger than a vegetative bud. A mixed bud contains the potential to produce both a shoot and a flower.

The arrangement of buds on a stem places them in either *axillary* or *terminal* positions (Figures 2-4 and 2-12). Axillary buds are found in the leaf axils, usually directly above the point at which the petiole attaches to the stem. Terminal buds are located on the tips of stems.

In some cases buds form at other sites, such as along the veins of a leaf or at the petiole and leaf blade junction. Buds in these unusual places are called *adventitious* buds.

Roots

Roots are the fourth vegetative plant organ, and the point at which they attach to the aboveground portions of the plant is called the *crown*. Roots anchor the plant in its growing position. They also absorb and transport water and nutrients for use in photosynthesis. This absorption takes place mainly through the young tips of roots, rather than through the older larger parts of the root system. In addition, carbohydrates are stored in the older parts of the root system. This storage capacity of roots is very important because carbohydrates are the source of energy for generating new shoots if the top of the plant is damaged.

Root Types. In the development of a plant from seed, a single primary root is first produced. In many plants this primary root does not branch substantially and remains the primary site of anchorage and absorption. Such roots, called *taproots,* are found typically on citrus, dandelions, and carrot plants (Figure 2-13).

In most plants the primary root branches to form a netlike mass called a *fibrous root system* (Figure 2-13). Grasses, African violets, and squash are examples of plants with fibrous roots.

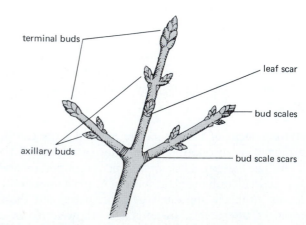

Figure 2-12 Stem with axillary and terminal buds.

Figure 2-13 Types of root systems.

A third type of root system has *fleshy roots* (Figure 2-13) which are thick like taproots but branch like fibrous roots.

Parts of the Root Tip. Regardless of whether a plant has tap-, fibrous, or fleshy roots, the structure of the root tips will be the same. At the tip of every root is the root cap. The cap is made up of a layer of cells that prevents damage to the rest of the root as it pushes through the soil (Figure 2-14). Directly behind is the root *meristem*, an area that produces new cells to replace cells scraped off as the cap pushes against the soil.

Beyond the meristem is the zone of elongation. Cells produced in the meristem are also deposited here, grow, and lengthen the root. Cell lengthening in this zone pushes the root through the soil.

Water and nutrients are absorbed in the next zone, the zone of maturation and absorption. Cells in this area form fragile root hairs (Figure 2-15) about 1\16 inch (1–2 millimeters) long through which water and nutrients enter the plant.

Root hairs often live only a few weeks, and are replaced by hairs developing on younger cells. In this way root hairs constantly contact fresh soil, assuring a steady supply of nutrients.

Modified Roots. Although most roots resemble those in Figure 2-13, modified roots exist, much the same as do modified leaves and stems. Roots fashioned for storing large quantities of carbohydrates are the most common modified roots. Sweet potatoes and beets are examples of this modification.

Adventitious roots grow from a stem instead of from other root tissues. The adventitious roots of ivy and orchids (Figure 2-16) attach the plants to trees.

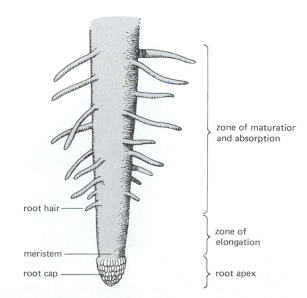

root hair

zone of maturation and absorption

zone of elongation

meristem

root cap

root apex

Figure 2-14 Parts of the root tip.

Figure 2-15 Root hairs of a radish seedling. *(Photo courtesy of Michael Knee, Ohio State University)*

Adventitious prop roots of corn (Figure 2-16) anchor it against uprooting by wind.

Sexual Plant Organs

Flowers and fruits are found on most cultivated plants at some time during the life cycle. They may appear continuously, seasonally, or only once, depending on the species and its growing conditions. Although the biological function of flowers is seed production, flowers are of-

(b)

(a)

(c)

Figure 2-16 Modified roots. *(a) photo by Kirk Zirion, San Luis Obispo, CA. (b) photo from Botany, 5th ed. by Carl D. Wilson and Walter E. Loomis, Copyright 1952, 1957, 1962, 1967, and 1971 by Holt, Rinehart and Winston. Reprinted by permission of Holt, Rinehart and Winston. (c) photo courtesy of Bob Hoffman, Huntington Beach Orchids, Huntington Beach, CA.)*

ten not essential for the continuation of the species because man reproduces plants from the vegetative parts.

Flower Types

Flowers are classified by their sexual parts. A flower with both male and female parts is called *perfect;* most species produce perfect flowers. However, a flower can be exclusively male or female, with blooms of opposite sexes needed for seed formation. Some species have male and female blooms on one plant and are called *monoecious.* Examples of monoecious species are cucumber, sweet corn, and pecan. Others have male flowers on one plant and female on another, with plants of opposite sexes needed to produce seed on the female. These *dioecious* species include holly, date palm, bittersweet, and ginkgo.

Occasionally flowers are sterile—that is, they have neither male nor female parts. Sterile-flowered plants are frequently prized by gardeners because they often have double sets of petals. The doubleness is caused by sexual parts mutating into petals.

Flower Structure

A typical flower is made up of a central female structure, the pistil, and numerous surrounding male parts called stamens (Figure 2-17).

The pistil contains three main parts: the knoblike top called the *stigma,* the thin vertical shaft called the *style,* and the bulblike base called the *ovary.* It is the ovary that contains the eggs which will develop into seeds.

Stamens are composed of the *anther* at the top, which produces pollen (functionally similar to sperm), and the supporting stalk, called the *filament.*

Surrounding or beneath the sexual parts are the petals and the *sepals* (the green, leaflike base beneath the petals). Under the sepals is the *receptacle,* the enlarged base on which the flower rests. The colored petals attract insects, which transfer the pollen to the female flower parts. They also shield the pistil and stamens from the weather. The sepals form the covering for the developing flower bud, protecting it from damage.

Figure 2-17 shows a typical form for a flower, but not all flowers have such easily distinguished parts. Instead the parts will be in various forms due to the divergent evolution of plant families and genera. Figure 2-18 illustrates other typical flower forms.

Moreover, not all flowers are produced singly. Often they occur in clusters, and even though each flower may be tiny, when grouped together they form a showy head. Flower clusters are named on the basis of the positioning of the blooms (Figure 2-19).

pea

snapdragon

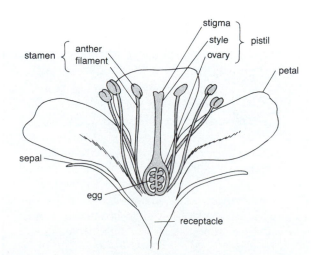

Figure 2-17 Parts of a perfect flower.

Figure 2-18 Two flower forms.

Fruit Production and Types

After flowering is complete, the flower parts shrivel and drop, with the exception of the ovary. The ovary continues developing, producing a swollen fruit with seeds.

(a)

(b)

(c)

Figure 2-19 Three types of inflorescences. (a) Head. (b) Spike. (c) Umbel.

The structure and characteristics of fruits vary widely among species of plants and are, as stated, one of the primary identifying features by which plants are botanically classified. Fruits can first be classified as either *fleshy* or *dry*. Fleshy fruits are those that contain appreciable amounts of water. Examples would be most eating fruits (plums, strawberries, oranges), most vegetables from which the "fruit" is eaten (cucumbers, squash, eggplant), and assorted garden plants such as crabapples, roses, and fuchsia.

Fruits are also classified by the position of the seeds relative to the fruit. *Drupe* fruits have a fleshy exterior and a single seed encased in a hard woody covering. Examples of drupe fruits are peaches and plums. *Pome* fruits are members of the apple family; they have multiple seeds inside and a swollen exterior ovary.

Pods are characterized by a row of seeds in a dry to fleshy case. Beans and peas are examples of this designation. A *berry* has individual seeds in separate fleshy sections that are attached to form the fruit. Of course, raspberries and blackberries meet this definition, but so does a pineapple, which is botanically classified as a berry.

A few fruits are classified as *achenes*. In an achene the seeds are borne outside of the fleshy part. A strawberry is an achene.

PHYSIOLOGICAL PROCESSES

Photosynthesis, respiration, translocation, absorption, and transpiration are complex processes that occur constantly in plants. Understanding these processes is essential to understand how plants function.

 A plant physiologist generally has a Ph.D. He or she may work at a university or for a large agricultural chemical company in the product development department, generally in the laboratory with excursions to growing areas to collect data for research projects. Physiologists research the internal processes of plant growth and development. Typical work might be developing and testing a new weed killer that kills specific botanically related weeds by disrupting their metabolic processes, but at the same time leaves the crop plant unaffected.

Photosynthesis

Photosynthesis is the process by which plants manufacture the carbohydrates they require to live and grow. The process is made possible by the green chemical chlorophyll. In the presence of water, light, and carbon dioxide, chlorophyll produces carbohydrates (starches and sugars) and, as a by-product, oxygen gas:

water + light + carbon dioxide gas

in the presence ↓ of chlorophyll

carbohydrates + oxygen gas

Light, the first essential for the photosynthetic process, generally comes from the sun. With the exception of plants that naturally grow in shady habitats, bright light intensity and long light duration maximize photosynthesis and increase growth.

Water, the second essential, is drawn from the soil by a process called *absorption*. Any lack of water will slow photosynthesis, and a severe deficiency will stop it completely.

Carbon dioxide, the third ingredient, is taken in from the air through leaf stomata. It is through this same pathway that the excess oxygen by-product is released into the air. Lack of carbon dioxide seldom causes insufficient photosynthesis outside. However, photosynthesis can be increased through the addition of supplemental carbon dioxide, which is occasionally used in commercial greenhouses (see Chapter 20).

Respiration

Respiration can be thought of as an opposite reaction to photosynthesis. Whereas photosynthesis makes carbohydrates out of the sun's energy, respiration breaks carbohydrates down into energy.

Respiration requires two ingredients—carbohydrates and oxygen—and yields three products—energy, water, and carbon dioxide gas:

carbohydrate + oxygen gas →

energy + water + carbon dioxide gas

The carbohydrates are already present in the plant. Oxygen is taken from the air primarily through the stomata on the leaves and also through the roots. Of the three products resulting from respiration, energy is the primary one used by the plant. It is needed for activities such as absorption of water and nutrients, chlorophyll formation, and development of flowers or fruits.

However, the carbon dioxide and water released in the reaction can also be reused in photosynthesis.

The photosynthetic process requires light but respiration does not. Respiration continues constantly in every cell of the plant, even at the same time photosynthesis is taking place. It does, however, generally occur at a slower rate than photosynthesis so that more carbohydrates will not be respired than are being photosynthesized, and the excess can be stored in the plant, for example in the roots or developing fruits.

The relative rates of respiration and photosynthesis govern the health and growth rate of plants. To grow a successful crop of vegetables, it is estimated that the photosynthesis rate should be 8 to 10 times higher than the respiration rate. This assures that the majority of the carbohydrates will accumulate in the plant or be used for new growth.

Correct environmental conditions are crucial for obtaining a high photosynthesis/low respiration ratio. Maximum photosynthesis can be encouraged by moderate to bright light intensity, sufficient water in the soil, and warm temperatures. Respiration, on the other hand, can be slowed by cooler temperatures. Because lowering the temperature also decreases photosynthesis, the best growing conditions are warm days with cool nights. Under these conditions the respiration rate is decreased at night, but daytime photosynthesis is not affected.

When photosynthesis and respiration rates are equal, there is no new plant growth. The carbohydrates being photosynthesized are entirely used up for the respiration of existing cells, and the plant merely subsists.

Occasionally a plant will have a higher rate of respiration than of photosynthesis. An example would be a small tree in a forest of larger trees. Shading from the larger trees reduces photosynthesis, while respiration remains moderate and steady. All the carbohydrate being produced, plus part of the stored reserve, is used for respiration. As a result, the tree stops growing and slowly dies because it is not producing enough carbohydrate even to keep its existing cells alive.

Translocation

The movement of carbohydrates, minerals, and water through the plant is called *translocation*. As discussed earlier, the main function of the xylem is water and mineral movement, and that of the phloem, sugar movement throughout the plant.

The forces that cause upward movement of water and minerals through the xylem are complex and beyond the scope of this text. It is sufficient to think of the movement as caused by upward pressure from the roots combined with pressure deficit when water vaporizes from the leaves.

The rate of water movement in the xylem is surprisingly rapid. This can be demonstrated by noting the time required to revive a wilted flower after it is placed in water. Complete revival of a flower on a 6-inch (15-centimeter) stem in 15 to 30 minutes is not unusual.

Movement of sugars in the phloem is called a *source to sink movement* because the sugars are taken from their site of manufacture (source) to a storage or utilization site (sink). One of the primary sinks for carbohydrate is the roots. Carbohydrate is stored there as starch compounds made up of many sugar molecules combined. The starches make compact, energy-concentrated packages that can be efficiently stored.

A second sink is developing flowers, fruits, and seeds. In order to survive, seeds must be well supplied with stored carbohydrate to provide energy for respiration after separation from the parent plant. The constant sugar flow to these sites also accounts for the continual increasing in size of fruits from bloom until maturity and explains why fruits are often sweet.

A third sink for photosynthesized sugars is the growing regions (called *meristems*) at the stem and root tips. They require the constant supply of sugars to grow and to manufacture compounds such as pigments for flowers and fruits, waxes for cuticle formation, and cellulose fibers to strengthen the stems.

Absorption

Water and mineral absorption is the fourth internal plant process. It is primarily a function of the roots; however, it can also occur through the leaves by a process called *foliar absorption*.

Absorption of materials through root cells is a complex topic, which is still being heavily researched today. Research has explained absorption as consisting of two processes: active absorption and passive absorption. Active absorption is an energy- and oxygen-requiring process, the primary process responsible for the absorption of mineral compounds. Passive transport through a process called *osmosis*, on the other hand, is a result of natural physical forces and is the main way in which water is absorbed.

Transpiration

Transpiration, the fifth process, is the opposite of absorption in the sense that it involves the loss of water from the plant. Water in liquid form in the plant is released into the air as vapor from the leaves, stems, and flowers. Of all the water taken from the soil by absorption, a relatively small amount is used in photosynthesis; the rest is lost through transpiration.

As mentioned earlier, the main parts of the leaf through which water is lost are the stomata (Figure 2-4). These are opened and closed by the kidney-shaped bordering cells called *guard cells*. An increase in the amount of water in the leaf and guard cells forces the stomata open and transpiration occurs. As the leaf loses water, guard cells go limp and the stomata close, decreasing transpiration.

Transpiration can be increased or decreased by environmental factors such as temperature, humidity, and wind. High temperature increases transpiration. High humidity decreases transpiration because humid air contains almost its capacity of water vapor, thus suppressing the transpiration of more water from the leaf. Wind usually increases transpiration by drawing newly transpired water vapor from the leaf. The dry air surrounding the leaf then causes rapid transpiration again.

Summary

These five processes—photosynthesis, respiration, translocation, absorption, and transpiration—are basic concepts in plant science. Knowledge of each and how each is related to the others is crucial to a thorough understanding of plant growth.

SELECTED REFERENCES FOR ADDITIONAL INFORMATION

Brako, Lois, David F. Farr, Amy Y. Rossman, and the U.S. National Fungus Collections Staff. *Scientific and Common Names of 7000 Vascular Plants in the United States.* Saint Paul: American Phytopathological Society, 1995.

Crocker, W., and L.V. Barton. *The Physiology of Seeds: An Introduction to the Study of Seed and Germination Problems.* Port Jervis, NY: Lubrecht and Cramer, 1990.

Els, David, ed. *The National Gardening Association Dictionary of Horticulture.* New York: Viking, 1996.

Esau, K. *Anatomy of Seed Plants.* 2nd ed. New York: Wiley, 1977.

Hopkins, Williams C. *Introduction to Plant Physiology.* 2nd ed. New York: John Wiley and Sons, 1998.

Radford, Albert E. *Fundamentals of Plant Systematics.* New York: Books on Demand, 1986.

Taiz, Lincoln. *Plant Physiology.* 2nd ed. Sunderland, MA: Sinauer Associates, 1998.

Vines, S.H. *Handbook of Plant Physiology.* New York: State Mutual Book and Periodical Service, 1985.

Wells, Diana. *100 Flowers and How They Got Their Names.* Chapel Hill, NC: Algonquin Books of Chapel Hill, 1997.

World Wide Web Addresses

http://www.botany.com/ A directory of plants by common and botanic names.

http://www.botany.com/. Botany.com Encyclopedia of Plants website. 2001.

http://www.botany.uwc.ac.za/Sci_ed/pupil/Angiosperms/. Anonymous. "Angiosperm Anatomy" website. Undated.

http://www.csdl.tamu.edu/FLORA/tfplab/labtabco.htm. *Herbarium.* Department of *Biology, Texas A&M University* "BOTANY 301—TAXONOMY OF FLOWERING PLANTS". 2001.

http://dallas.tamu.edu/weeds/anat.html. Texas Agriculture Experiment Station. "Plant Anatomy and Glossary". 1998.

http://gened.emc.maricopa.edu/bio/bio181/BIOBK/BioBookPLANTANAT.html. Farabee, M. J. "Plants and Their Structure". 2001.

http://www.hoflink.com/~house/pltphys.html. House, Ken. Biology Website References for Students and Teachers. 2001

http://www.inform.umd.edu/EdRes/Colleges/LFSC/life_sciences/.plant_biology/pb250/index.html. Revela, James. A. "Plant Biology 250 Lecture Notes" 1998.

http://www.inform.umd.edu/PBIO/pb250/index.html Plant taxonomy reference.

http://www.life.umd.edu/classroom/BSCI124/main.html. University of Maryland. BSCI 124 (Plant Biology) lecture notes website. 2001.

http://208.156.226.50/name.html National Gardening Association. "NGA Plant Name Finder". 2000.

http://www.uri.edu/artsci/bio/plant_anatomy/images.html. Roberts, Allison. "Plant Anatomy Bio 311". 1999.

http://web1.manhattan.edu/fcardill/plants/anatomy/homepage.html Cardillo, Francis M. "Plant Tissue Systems". Undated.

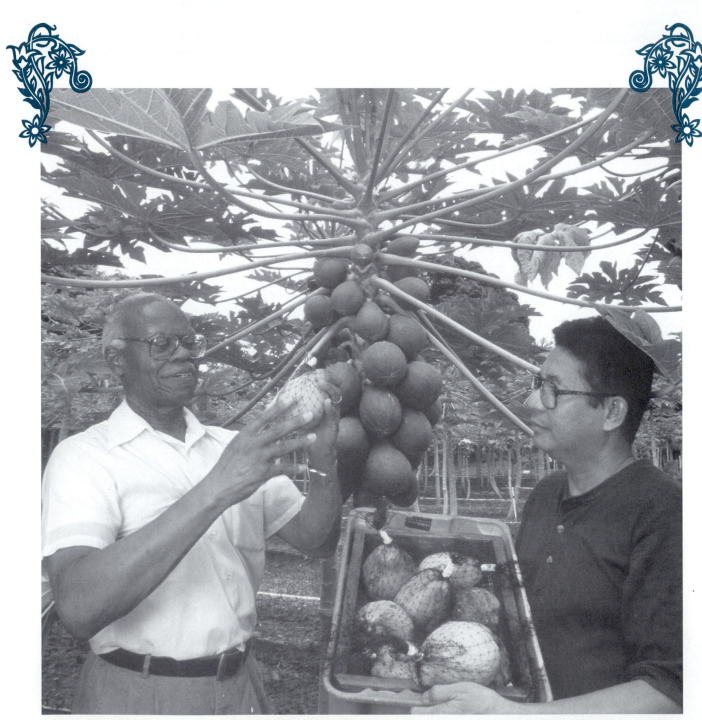

(Photo courtesy of Agricultural Research Service, USDA)

CHAPTER 3

PLANT GROWTH AND DEVELOPMENT

PLANT LIFE CYCLES

In a typical life cycle a plant will pass through two states, growth and rest. The growth phase occurs with suitable environmental conditions (temperature, rainfall, and the like) and is characterized by activities such as flowering, shoot lengthening, and leaf production. The rest phase, often called *dormancy*, also reflects the influence of environmental conditions such as cold, drought, or inadequate light. It is characterized by slowed or stopped growth, leaf drop, and/or death of the entire aboveground parts of the plant.

Dormancy should not be thought of solely as an outdoor plant phase. Many indoor plants also have dormant periods, induced by the shorter days of winter or by the plant adhering to its normal outdoor growth cycle although living indoors.

To understand the importance of growth and dormancy in the life of a plant, let us first classify plants according to the length of time they live (Figure 3-1). Plants fall into one of four botanical groups: *annual*, *biennial*, *perennial*, and *monocarp*.

Annual

An annual plant grows to maturity, flowers, produces seeds, and dies during one growing season. It experiences dormancy only as a seed.

To most gardeners an *annual* is a frost-tender flowering or vegetable plant grown in summer. In mild climates there are also *winter annuals* that are planted in fall to bloom through the winter. These common-usage definitions usually do not clash with the botanical definition of an annual. However, exception must be made for plants that would live more than one year in their warm native climate but are not able to survive winter in cold regions. These plants are not true annuals but function as annuals because of climate. Included in this group are vegetables such as tomatoes and flowers such as lantana and impatiens when grown in cold-winter areas.

Biennial

A biennial plant grows for one season, becomes dormant during winter, resumes growth in spring, flowers, and dies. The life cycle is completed in two years, hence the term "biennial." Biennials are relatively uncommon among cultivated plants. Some examples are hollyhocks, parsley, and sweet William.

Perennial

Perennials are plants that live for more than two growing seasons. They make up the third category of plants in terms of life cycle. Technically such plants as trees, shrubs, and bulbs are all perennial; however, in horticulture the term *perennials* often refers to perennial *flowers* such as daylilies and iris that grow back from the roots each spring.

Figure 3-1 The four types of plant life cycles.

Monocarp

A *monocarp* can live for many years but will flower only once in its lifetime and die afterward. Bromeliads and century plants (*Agave* spp.) are two examples of monocarpic plants. These plants bloom after several years of active growth. The tops then die, and new plants are produced by the root system of the old plant.

STAGES OF PLANT MATURATION

Regardless of how long a plant lives, it will usually pass through four stages of maturation: germination, juvenility, maturity, and senescence. These stages and the role each has in the life of the plant are discussed in the following sections.

Germination

Many plants begin life as germinating seeds. Germination starts when the seed absorbs water and ends when the primary root emerges (Figure 3-2). After germination, the seedling goes through a period called *estab-*

lishment, which lasts until the seedling is independent and photosynthesizing. Both these stages are exceedingly important in the life of a plant, because it is during these phases that plants are most susceptible to adverse environmental conditions and the greatest number of plants die.

Parts of the Seed

A seed must contain an *embryo* that develops into the new plant and a source of carbohydrates to supply energy to the embryo during germination and establishment. In addition, a covering called the *seed coat* helps prevent injury and drying (Figure 3-2).

In many seeds the carbohydrates required for the seed to germinate are stored in two *cotyledons*. These cotyledons resemble leaves and are attached to the stem when the seedling emerges from the ground. They are frequently called *seed leaves*, as opposed to the *true leaves* of the plant that are produced later (Figure 3-2).

In other species carbohydrates are stored in an *endosperm*. Whereas the endosperm is a single organ (Figure 3-3) used only for carbohydrate storage, cotyledons may serve not only as storage sites but also as photosynthesizing organs.

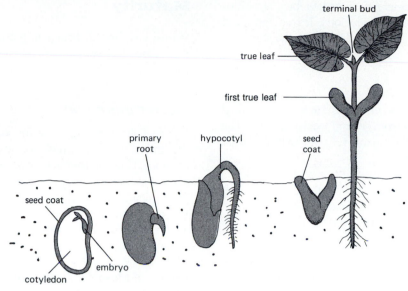

Figure 3-2 Germination of a bean seedling.

Environmental Requirements for Germination

For a seed to germinate and establish a new plant, correct environmental conditions are necessary. First, water must be present. Second, oxygen must be available. Third, the temperature must be within the range acceptable to that species.

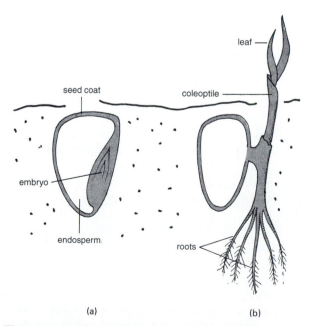

Figure 3-3 (a) A corn grain showing the enlarged endosperm. (b) The germinated corn grain. Note the absence of the cotyledon leaves found on the bean.

Uptake of water is the first process in germination. The uptake causes the seed to swell and triggers many chemical reactions, including an increase in the respiration rate. As carbohydrates are respired, energy is freed for the growth of the embryo.

Oxygen is necessary for the respiratory process; without it, respiration will not begin and the seed will not germinate. This fact can be demonstrated by the following. Seeds placed in a glass of water will absorb the water and swell, a sign that the germination process is beginning. If taken out of the water a day later and planted, germination will continue. But if the seeds are left in the water without access to the oxygen in the air, the process will stop after the swelling stage.

Suitable temperature is the third essential for germination. Temperatures either too high or too low will kill the embryo or prevent germination from proceeding. For most seeds, 70–80°F (21–27°C) temperatures promote vigorous germination, although plants native to cold climates will germinate at considerably cooler temperatures.

After sufficient swelling and respiration have taken place, the *radicle,* or primary root, will emerge. It grows down into the soil, beginning the absorption process, and soon afterward the shoot emerges.

Many times the shoot will arise from the soil as an arch, with the stem of the young plant forming a loop (Figure 3-2). The tip of the shoot is eventually pulled out of the soil by the growth of the stem, and the plant then straightens to an upright position. The pulling of the shoot tip from the soil prevents injury that could occur if it were pushed straight through the soil.

After the shoot emerges, the plant begins photosynthesis. The shoot tip develops leaves, and the cotyledons usually wither and drop from the stem.

Juvenility

After germination, most plants enter a period called *juvenility* that may last from several weeks to years. The plant will grow rapidly vegetatively but will not begin its reproductive process.

The exact causes of juvenility are not completely known. Pruning, temperature alteration, and applications of growth-regulating chemicals can change a plant from the juvenile to the mature state and back again.

Traits of Juvenility

In most cases the juvenile appearance of a plant resembles the mature form. However, some plants have recognizable traits that signal that the plant is in a juvenile state.

Leaf Form. Leaf form different from that found on the mature plant is a common juvenile trait. The leaves of some acacia species, for example, are frequently compound in the juvenile stage but simple on mature trees. Other plants including sassafras have deeply lobed and divided leaves when they are young, but predominately unlobed leaves when they mature.

Growth Form. Variation in form also distinguishes juvenile from mature plants. The most well-known example is ivy (*Hedera* spp.). Juvenile ivy forms are trailing or climbing, and the leaves are deeply lobed. Mature ivy, however, will produce upright shoots that grow without support and have leaves without lobes.

Juvenile branch form is sometimes found on young fruit trees as whiplike vertical shoots. Mature trees occasionally produce such juvenile branches. They usually arise from the base of the trunk and are called *suckers,* or *water sprouts* (see Chapter 11, "Pruning Trees").

Thorns. Thorns are a third common characteristic of juvenility. Citrus and locust trees are frequently thorny when young, but branches produced after the tree reaches maturity are thornless.

Leaf Retention. A tendency for the juvenile parts of a tree to hold leaves throughout the winter is the fourth characteristic of juvenility. Many trees demonstrate this phenomenon (Figure 3-4).

Plants may indicate juvenility other than by visible characteristics. In propagation, juvenile parts almost always root faster than mature portions.

Maturity

Maturity is the third stage in plant growth. It is during this stage that sexual reproduction takes place. The phases of this process are discussed here.

Flower Induction

Flower *induction* is the first indication of maturity and the first of the three phases of flowering. It is an initial chemical reaction that begins the flowering process and occurs before any physical changes take place in the plant. The factors that trigger blooming are well known for some species but unknown for others. Generally as a plant becomes older and larger, its likelihood of flowering increases.

Causes of Flower Induction. Induction of flowering is sometimes caused by environmental factors such as temperature or night length. In nature these factors program flowering into a seasonal pattern so that it occurs at the same time each year. Conditions that cause flowering can also be simulated by man to induce flowering at a preselected time. For example, controlling environmental conditions is a common practice in commercial greenhouses for the timing of flowering of such plants as poinsettias and Easter lilies for holiday sales.

Cool temperature is a crucial environmental factor that causes flowering of many plants. Some plants that grow and flower quite profusely in northern climates either die or grow only vegetatively in warmer locales. They must be *vernalized* (subjected to cold temperatures) before they can be successfully grown. It is because of insufficient vernalization that fruits such as apples, cherries, and pears are not grown in most parts of Florida.

In addition to certain fruit trees, biennial plants may also require vernalization for flowering. The cold received during the winter following the first season of growth induces flowering to complete the life cycle of the plant. Cabbages, carrots, and beets are all biennial plants that require vernalization. However, because they are raised for their roots and leaves and harvested the first year, the vernalization requirement is not important.

The vernalization needs of bulbs such as tulips, hyacinths, and crocuses are important as they are grown for their flowers. In warm climates the bulbs of these plants are sold *prechilled;* that is, they have already been vernalized by refrigerated storage. They will flower the following spring regardless of the climate they are grown in. However, they will not bloom more than one year unless they are growing in a climate where they will receive yearly natural vernalization.

Figure 3-4 An oak tree retaining its dead leaves in winter, a characteristic of juvenility. *(Photo courtesy of Dr. Claud L. Brown, University of Georgia)*

Daily night duration is a second environmental factor that can control flowering in plants. It is common knowledge that the nights are longest in winter, shortest in summer, and intermediate in spring and fall. The further north or south of the equator, the more extreme is the difference between the shortest night of summer and the longest of winter. Plants that respond to these changes in the night length are called *photoperiodic* and are common in all but tropical climates near the equator.

Two main types of photoperiodic plants have been studied intensively: *short-day* and *long-day* plants.[1] Short-day plants flower only when the daily light period is *less* than their critical photoperiod of, for example, 12 hours. Long-day plants will flower only if the light period is *more*

[1]The confusing emphasis on day length rather than night length stems from the first research on photoperiod. At that time it was mistakenly believed that the length of light rather than darkness caused flowering.

than their critical photoperiod. They can also flower under continuous light or during a long dark period interrupted at some point with a short light period of an hour or so. In an otherwise long night, a short period of light has the effect of cutting the dark period into two shorter periods. Provided each half is less than the critical period, the plant will respond as if it were exposed to only short nights, and flowering will take place (Figure 3-5).

The way in which night length triggers flowering involves a pigment called *phytochrome,* which alternates between two different forms, each with sensitivity to differing wavelengths of light in the red to far-red range (see Chapter 16 for a discussion of the wavelengths of light).

Light intensity can also control flowering. Plants that are *day neutral* and not affected by night length frequently respond to intensity. Generally, greater light intensity is required to induce flowering than is sufficient to keep a plant growing vegetatively. Consequently, many houseplants bought for their blooms never flower again after being moved from a greenhouse to indoors. The light intensity is simply too low to induce flowering.

Stress, such as from drought or the crowding of a plant's roots, can also stimulate flowering. One possible explanation is that the natural production of chemical compounds that trigger flowering of a mature plant is intensified when the plant is dying, and the remainder of the stored carbohydrates is used for flowering and continuation of the species through seed.

Flower Initiation

The second phase of flowering is *flower initiation.* Like induction, it is a phase of flowering invisible to the naked eye, but changes are taking place in the microscopic parts of the plant. It is during this phase that the vegetative meristems found at the stem tips or leaf axils change to flower meristems capable of developing into blossoms. This radical changeover from a vegetative to a reproductive structure takes place over several days or weeks. The center of the vegetative meristem, which was formerly mounded, becomes flattened. Small knobs of cells emerge in a spiral around the center (Figure 3-6); these are the beginnings of petals, stamens, and other flower parts. The meristem stops lengthening at this time but development continues.

Flower Development

The length of time from induction to bloom may vary from several weeks to six or eight months. Cold-climate woody plants usually have long developmental

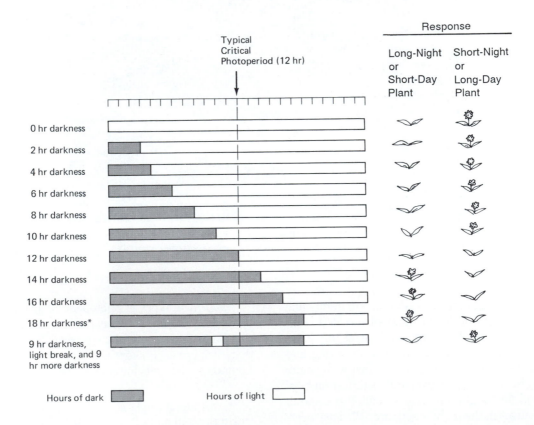

Figure 3-5 The flowering responses of long-day and short-day plants to varying dark/light ratios.

Figure 3-6 Electron microscope photograph of a developing flower meristem of an Easter lily. *(Photo courtesy of Dr. H. Paul Rasmussen, Logan, UT)*

Figure 3-7 Germinating pollen grains growing toward the ovary. These female parts of a flower are collectively called the *pistil*.

periods. The flower buds of these plants are partially formed in late summer, are vernalized over the winter, and finish developing and open in early spring. The opening of the flower is its final stage of development.

Pollination and Fertilization

After a flower opens, it is generally receptive to pollination. Whether it pollinates itself (*self-pollination*) or is *cross-pollinated* by another plant will affect the resulting seeds.

During pollination dust-size pollen grains found on the anthers are deposited by wind or insects on the stigma. Like seeds, they germinate and grow downward toward the ovary where the eggs are located (Figure 3-7). After reaching the ovary, the uniting of sperm (contained in the pollen) and eggs (in the ovary) occurs in the process of *fertilization.*

In some instances pollen produced by a flower may be unable to fertilize the eggs produced by that same flower. This *self-incompatibility* is found in about 40 percent of cultivated plants (see Chapter 7). In some cases the pollen grain simply never germinates. In others it germinates but grows poorly and never reaches the egg.

Cross-pollination is thus quite common. Separate fruiting and pollinating tree cultivars are frequently planted in fruit orchards to overcome self-incompatibility problems, because flowers that are not pollinated *and* fertilized usually will not develop fruit.

Cross-pollination can be a hindrance as well as an advantage. Sweet corn and field corn cross-pollinate if planted close together, making the sweet corn taste like corn raised for grain. Sweet and hot peppers will also cross-pollinate. The fruit, which develops from the ovary, will still be edible. But the seeds are the sexual union of two parents, and a mild pepper with hot-tasting seeds could be expected.

Although fertilization of the egg by pollen usually must occur before a fruit will develop, there are cases in which it is not necessary. For example, pollen does not fertilize the egg but the fruit still develops, resulting in a seedless fruit. Seedless fruits formed in this way are called *parthenocarpic*. Parthenocarpy can also be caused by applying growth-regulating chemicals and by specific environmental conditions. For example, excessively warm temperatures during pollination can cause seedless tomatoes to form.

Another way in which seedless fruits form is through the spontaneous death of the seed embryo in its early stages of development. Fertilization is needed to start fruit growth, but after the death of the embryo, the fruit will continue to mature to full size. Seedless fruits formed in this way are also parthenocarpic.

For the majority of cultivated fruits, living seeds are necessary for the fruit to develop normally. For example, if only half the eggs inside an apple are fertilized, only one side will grow; a misshapen fruit is thus produced. Cucumbers sometimes have this problem, with a portion staying small while the remainder develops normally (Figure 3-8). In cases where the fruit contains only one seed, the fruits will drop prematurely if the embryo dies. This happens in the case of apricots, peaches, and cherries.

Fruit Development and Enlargement

During fruit development sugars photosynthesized in the leaves constantly flow into the fruit. They supply energy for the developing ovary which will become the fruit.

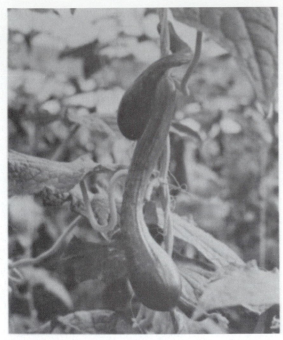

Figure 3-8 A greenhouse-grown cucumber that was not fully pollinated and failed to develop normally. *(Photo courtesy of the Cooperative Extension Service of the University of California)*

Ripening

When a fruit reaches the end of its enlargement period, ripening begins. The fruit may become soft or change color, and its flavor can change from sour to sweet.

Softening of fruits is the result of the breaking down of compounds called *pectic substances* which strengthen the cell walls and cement cells together. In overripening, mushiness occurs because too much of the pectic substance has been lost. Mealiness in apples results from such overripening. The cells are so easily separated from each other that they are chewed without breaking and releasing the flavor of the apple.

Color changes result from the depletion of chlorophyll and the accumulation of other pigments. The chlorophyll continually decreases while the other pigments increase in intensity until coloration is complete. *Carotene* is a pigment which gives orange color to fruits such as oranges and persimmons. *Anthocyanin* is another pigment, responsible for the red color of ripe strawberries and apples.

Senescence

Senescence is the aging past its prime of a plant or any of its parts. It occurs as a part of the natural life cycle of the plant or as a result of environmental factors.

The natural life span of the plant often determines when senescence begins. Annual plants begin senescence at the flowering stage and will die soon after seed formation is complete. Senescence after reproduction is also the rule for monocarpic plants which live several years before flowering.

Senescence of perennial plants is frequently called *decline*. Asparagus is a perennial vegetable that suffers from decline. The productive life of asparagus plants is generally 20 to 25 years. Once the plants reach this age, the yield drops off considerably, and the bed is replanted.

An entire plant need not die for the term *senescence* to be applied. The top portions of herbaceous perennials and bulbs senesce annually, but the root system remains alive. The lower branches of trees and leaves of houseplants such as red-margin dracaena (*Dracaena marginata*) and bamboo palm (*Chamaedorea erumpens*) frequently die while the top continues growing rapidly.

A colorful show of leaf senescence occurs during the fall when trees turn color. The coloring is due to the same pigment changes that occur with the ripening of fruit. However, environmental conditions trigger this senescence, in this case the shortening of day length in fall and the cooler temperature.

Abscission

The dropping of leaves, flowers, fruits, or other plant parts is called *abscission*. This process involves the manufacture of plant hormones and the formation of a zone of abscission. In simple leaves, this zone is formed at the point where the petiole connects to the stem (Figure 3-9). In compound leaves, the main abscission zone is also at the point where the petiole reaches the stem, although individual leaflets may also form abscission layers and drop.

 A plant illustrator produces drawings such as the ones in this book for books, catalogs, and magazines.

Flower abscission is often related to pollination, and again a hormone (auxin) is involved. The pollen contains auxin, which it carries to the pistil, triggering abscission of the flower parts. Because the petals, stamens, stigma, and style are no longer needed to attract pollinating insects, their function is complete and their abscission diverts more carbohydrate to the developing fruit.

Fruit abscission can occur at any point in development, but is most common after ripening. Dropping of ripe fruit is a type of seed dispersal. As with leaf drop, an abscission layer generally forms before a fruit drops.

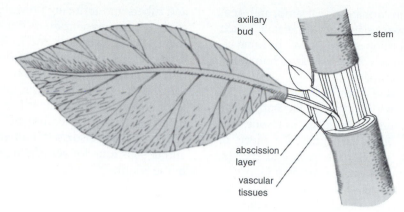

Figure 3-9 A leaf just prior to abscission.

Dormancy

Dormancy is a stage of plant development in which growth slows or stops. It affects all life stages from seed to maturity and plays an important part in adapting plants to their environment and ensuring their survival.

Dormancy occurs during periods not suitable for plant growth and is usually seasonally related. Winter dormancy is found among plants in cold-winter areas. Dry-season dormancy is found in other areas of the country that experience distinct wet and dry seasons, such as California.

Leaf abscission is a common indicator of dormancy, signaling the onset of winter dormancy in many deciduous (leaf-dropping) trees. Although photosynthesis can take place at temperatures below freezing (such as in evergreens like spruce and pine), the leaves of most plants are not adapted to withstand subfreezing temperatures. Instead, they abscise in autumn, and new leaves survive as buds.

The changeover from dormancy to active growth is called *breaking dormancy*. It normally results from changing environmental conditions—from those that induced the dormancy to those that encourage photosynthesis and growth.

Breaking Seed Dormancy

Seeds usually enter dormancy just prior to abscission of the fruit or senescence of the parent plant. The conditions required to break this dormancy and start germination are specifically geared to ensure the survival of the seedling.

For example, many seeds require winter vernalization before breaking dormancy and sprouting. This assures that germination will only take place in the spring, when the plant will have adequate time and suitable temperature to establish itself. If germination occurred as soon as the seed touched the ground in the fall, the young plant would be unlikely to survive the approaching winter. Simulated vernalization is sometimes used on seeds to break their dormancy in commercial nursery crop production. This process, called *stratification,* involves storing moist seeds at temperatures near freezing for one or more months. Peach and apple trees are examples of plants whose seeds are stratified before planting.

Another environmental condition that breaks dormancy in some seeds is heat. Forest fires destroy existing trees but also trigger the germination of dormant pine seeds. Because the seed coats of some pine seeds are woody and thick, only the intense heat from a forest fire will weaken them enough to allow water to be absorbed and the embryos to emerge.

Most seeds with hard seed coats require less drastic measures to break dormancy. Usually the seed coat erodes enough by weather to permit germination. It can, however, be weathered artificially by a process called *scarification.* This involves cutting, scraping, or otherwise injuring the seed coat enough to allow water absorption and subsequent germination.

Cold, heat, and weathering are all ways used to break seed dormancy but are seldom normal practices for the home gardener. The application of water is usually all that is required to start the germination process.

PLANT HORMONES AND GROWTH-REGULATING CHEMICALS

Plant hormones are chemicals made within the plant that produce changes in growth. They are effective in extremely small quantities and cause such responses as root formation, seedless fruit formation, leaf drop, and stem lengthening. Plant growth regulators are manufactured chemicals like plant hormones. They have the same effects and may be chemically quite similar to hormones, but they do not occur naturally.

At least nine groups of plant hormones have been identified, and it is likely that others will eventually be discovered. Scientists researching plant hormones encounter a number of difficulties. First, the effects of hormones are not the same from species to species. Second, slight changes in hormone concentrations can completely alter their effects. Third, two or more hormones frequently are found together, and it is difficult to determine which chemical is responsible for a particular effect. Table 3-1 lists some growth-regulating chemicals and the plants they are used on.

Auxins

Auxins were the first hormones discovered. In varying concentrations auxins have such diverse effects as promotion of rooting, formation of underground tubers and bulbs, prevention of fruit formation, defoliation, and prevention of abscission of leaves and fruits. Auxin manufactured in the terminal bud is also responsible for suppressing the sprouting of axillary buds further down the stem of a plant (see Chapter 14, "Pinching").

Synthetic auxins, which can be purchased in garden centers, have a number of practical home uses. Powders sold to encourage rooting of cuttings are composed of manufactured auxins in a talcum powder base. Commercially these products are known by the names IBA (indolebutyric acid), NAA (napthalenacytic acid), and IPA. 2,4-D sold to control lawn weeds is also an auxin. Applied at high concentrations, it kills many plants, but very low concentrations it is a growth enhancer.

High auxin concentrations can also be used to prevent fruit on ornamental trees when the fruit is an undesirable feature. Ginkgo and horse chestnut trees both have large bothersome fruits that can be prevented from forming by using an auxin spray in spring. In commercial agriculture, synthetic auxins are used to defoliate plants before harvest, prevent sprouting of potatoes in storage, and prevent premature fruit drop in orchards.

Gibberellins

Gibberellins are a second group of hormones. The first plant activity that was associated with gibberellins was stem elongation. Additional experiments showed other effects including breaking dormancy of seeds, buds,

and tubers and inducing flowering in plants that normally require vernalization or a specific photoperiod. Increases in flower, leaf, and fruit size were also found to be caused by gibberellins, leading to speculation that they would be wonder chemicals for agriculture, but this did not occur. However, gibberellins are used in greenhouses to form tall tree-form fuchsias and geraniums from cuttings because of their stem-elongation ability. They are also used to increase the size of grapes and to substitute for vernalizing azaleas and fruit trees in the South.

Cytokinins

Cytokinins are believed to work in conjunction with light to increase cell division and enlargement. They have also been shown to prevent chlorophyll degeneration and break axillary bud dormancy. Their only commercial horticulture use is in tissue culture, where they stimulate callus growth.

Growth Retardants

Synthetic and natural growth retardants make up the fourth classification of growth-regulating chemicals. They are sold under the trade names A-Rest®, B-Nine®, Bonzi®, Sumagic®, and Cyocel® for use on florist crops such as poinsettia and chrysanthemum. They slow elongation of the stems, making the plant sturdier and fuller. On fruit crops they are used to improve color and firmness and prolong storage life. Growth retardants can sometimes be used on hedges and lawns to slow growth and decrease maintenance (Figure 3-10). They can also be used to maintain bedding plants in a compact size, which gives them a neater appearance in formal landscapes.

Abscissic acid is a growth retardant that induces abscission and dormancy and inhibits seed germination. However, its action can be counteracted by growth-inducing chemicals like auxins, gibberellins, and cytokinins.

A number of growth retardants have also been used to control lawn growth and reduce mowing frequency. Maleic hydrazide and mefluidide are both sold commercially and radically reduce growth or stop flowering. EPTC and amidochlor reduce growth or flowering, but cannot stop it completely. These chemicals must be used carefully, and the results are not always predictable. Unacceptable color loss and injury can result, as can a lessened ability of the turf to recover from disease and pest problems.

Ethylene gas is another retardant, whose use dates back to the Chinese practice of ripening fruits in

Table 3-1 Selected List of Growth-Regulating Chemicals Used on Selected Plants

Purpose	Plants	Chemical
Control plant height	azalea	A-Rest®, Sumagic®
	bedding plants	A-Rest®, B-Nine®+Cycocel®, Bonzi®, Cycocel®
	bulbs	A-Rest®, Bonzi®
	carnation	Cycocel®, A-Rest®, Bonzi®
	chrysanthemums	A-Rest®, B-Nine®
	amaryllis	Bonzi®
	calla lily	Bonzi®
	dahlia	A-Rest®, B-Nine®, Bonzi®, Cycocel®, Sumagic®
	dracaena	A-Rest®
	gardenia	A-Rest®, B-Nine®
	geranium	Cycocel®, Sumagic®, Bonzi®
	grass	Bonzi®, Atrimmec®, Primo®
	holly	A-Rest®
	hydrangea	A-Rest®, B-Nine®
	lily	Bonzi®
	monstera	A-Rest®
	philodendron	A-Rest®
	Pilea spp.	Cycocel®
	poinsettia	A-Rest®, B-Nine®, Sumagic®, Bonzi®, B-Nine®+Cycocel®
	pothos	A-Rest®
	purple passion vine	Cycocel®
	roses (potted)	Cycocel®
	snapdragon	A-Rest®, Bonzi®, Sumagic®
	schefflera	Cycocel®, A-Rest®
	sunflower	Cycocel®
	wandering jew	A-Rest®
	tulip	A-Rest®, Bonzi®
	zinnia	A-Rest®, Bonzi®
Cause flowering	azalea	B-Nine®, Cycocel®, Bonzi®
	seedling geraniums	Cycocel®
	spathiphyllum	GibGro®
	bromeliad	Florel®
	Clerodendron	A-Rest®
Enhance flowering	camellia	gibberellin
	geranium	gibberellin
	cyclamen	gibberellin
	dutch iris	ethylene
Cause branching	azalea	Atrimmec®, Florel®
	bougainvillea	Atrimmec®
	chrysanthemum	Florel®
	fuchsia	Atrimmec®, Florel®
	gardenia	Atrimmec®
	geranium	Atrimmec®, Florel®
	grape ivy	Atrimmec®
	kalanchoe	Atrimmec®
	lantana	Atrimmec®, Florel®
	lipstick vine	Atrimmec®
	schefflera	Atrimmec®
	verbena	Atrimmec®, Florel®
Substitute for vernalization needed for flowering	azalea	GibGro®
Prevent flowering	azalea	GibGro®
	fuchsia	gibberellin
Enhance rooting	cuttings of plants	IBA, IPA, NAA

Source: Extracted in part from: Bailey, D.A. "Growth Regulators on Floricultural Crops in Commercial Flowers," *Growers Bulletin* 42(5):1–12, 1997.

Figure 3-10 A shrub treated (left) and untreated (right) with Atrimmec © growth retardant. *(Photo courtesy of PBI/Gordon Company, Kansas City, KS.)*

incense-filled rooms. It was later determined that it was the ethylene gas given off in the fumes that caused the accelerated ripening. Ethylene is also produced by the ripening fruits themselves and by cut flowers, or it can be manufactured.

In addition to ripening fruits, ethylene also ages plant parts such as flowers, causing them to abscise and induces flowering in a limited number of species. Pineapples and other members of the bromeliad family (many of which are used as houseplants) will bloom when treated with ethylene. The home practice is to enclose a ripening apple (a source of ethylene) in a plastic bag with the plant for about a week. If the plant is mature enough, flowers will appear in 1 or 2 months. In commercial greenhouses, the product used for ethylene generation is Florel® or ethephon.

Pinching Agents

Pinching agent chemicals are used commercially to kill terminal vegetative buds and therefore promote branching and a more bushy, attractive plant. Off-

Shoot-0® and Atrimmec® are the two chemicals mainly used for this purpose.

Vitamins

Vitamins, particularly B vitamins, are occasionally sold as stimulants for plant growth and for use after transplanting. They are the same vitamins sold for human use but act more as hormones in plants.

The effectiveness of vitamins on improving plant growth has not been fully determined. When vitamins are used on bean seeds, limited experiments have shown that they improve germination rates and decrease the time from seed sowing to harvest. Increases in yield have also been reported on some species. When vitamins are used with auxins, improved rooting of cuttings has been found, although it is questionable whether this is due to the vitamin or to the auxin. Overall, it is very possible that vitamins do improve plant growth in some cases. However, they should never be used in place of fertilizers or proven-effective chemicals sold for use on plants.

SELECTED REFERENCES FOR ADDITIONAL INFORMATION

Galston, A.W. *Life of the Green Plant.* 3rd ed. Englewood Cliffs, NJ: Prentice-Hall, 1997.

Kozlowski, Theodore T., and Stephen G. Pollardy, ed. *Growth Control in Woody Plants.* Morgan Kaufmann Publishers, 1996.

Nickell, L.G. *Plant Growth Regulators: Agricultural Uses.* New York: Springer Verlag, 1981.

Owens, S. *Higher Plant Reproduction.* New York: Chapman and Hall, 1999.

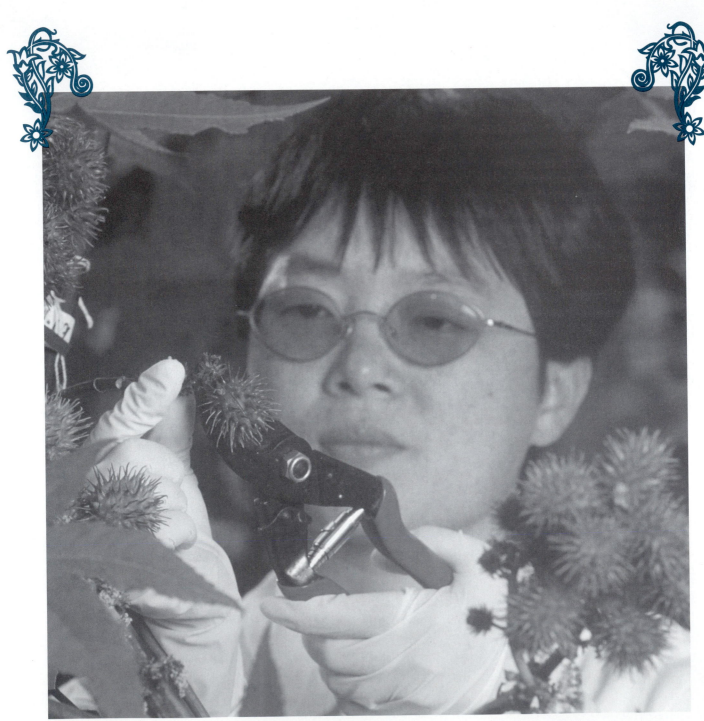

(Photo courtesy of Agricultural Research Service, USDA)

CHAPTER 4

PLANT PROPAGATION

SEXUAL AND VEGETATIVE PROPAGATION

Plants are reproduced, or *propagated*, either sexually or vegetatively. Sexual propagation involves seeds or spores that result from the union of male and female cells. Vegetative propagation is based on using parts of an existing stock to generate new plants. These *daughter plants* will have traits identical to those of the plants from which they were derived.

Propagation by seed is possible in almost all cultivated plants, and is used extensively. However, vegetative propagation is equally, if not more important in horticulture for several reasons. First, it is often faster than seed propagation. Most importantly, exact genetic duplicates of the parent plant are obtained. These duplicates are called *clones* and are essential for the preservation of desirable characteristics (such as fast growth or high yield) that could be lost in sexual reproduction. Seedlings may resemble the parent plant on which they were produced, but they will seldom be exactly like it. This is a result of genetic recombination of male- and female-carried genes. Just as brothers and sisters produced by the same parents are not all alike, seeds produced by a plant will not grow into identical plants. The gene mixing will result in offspring with a variety of different traits.

Whether a plant is reproduced sexually or vegetatively depends on many factors: the ease of germinating the seed, the number of plants to be grown, and the importance of preserving a special trait possessed by a parent plant. Table 4-1 lists plant categories according to use and tells the ways in which they are normally propagated.

Seed Formation

Seeds can be formed as a result of a plant fertilizing itself (self-pollination) or fertilization by another plant (cross-pollination). Offspring resulting from cross-pollination are called *hybrids* and carry traits of both parents. Although the term *hybrid* can apply to any seed resulting from cross-pollination, a seed package bearing the word *hybrid* indicates that the included seeds are the result of special breeding and are generally superior to other cultivars. Selected hybrid seeds produce healthier, faster-growing plants, a phenomenon called *hybrid vigor*.

Hybrid seed results from the controlled crossing of two groups of plants, one designated to be the *female line* and the other designated the *male line*, of known genetic makeup. The crossing of the two *genetically pure lines* produces seed that will have the best traits of both parents (Table 4-2).

Seed saved from hybrid plants will not grow good plants the following year. Only the original hybrid seed bears the desirable traits. The seeds included in fruit produced from hybrid seed result from random cross- or self-pollination. They will not produce the same superior plants.

Table 4-1 Primary Propagation Methods for Cultivated Plants

Plant category	Primary method of propagation
Annual flowers	Seed
Annual vegetables	Seed
Fruit trees	Seed for rootstock, followed by grafting of vegetatively propagated scion
Groundcovers	Vegetatively by cuttings
Houseplants	Vegetatively by cuttings
Ornamental trees	Seed
Perennial flowers and bulbs	Vegetatively by cuttings, underground storage organs, or division
Perennial vegetables	Vegetatively by division
Turfgrass	Seed
Vines	Vegetatively by cuttings

Figure 4-1 Basic Mendelian pea cross for seed shape.

Mendelian Genetics

Our knowledge of modern genetic science such as the above comes originally from the work done in the 1800s by a Czech monk named Gregor Mendel. Mendel's work started with the observation that, in both the plant and animal kingdoms, offspring resemble their parents. Using pea plants, he began a series of plant-breeding experiments that lasted for many years and on which he took detailed scientific notes. He studied seven different traits: plant height, flower color, flower position on the plant, seed color, seed shape, pod color, and pod shape.

Mendel suspected that the characteristics of an offspring are dependent on contributions of "traits" from both parents, and that specific characteristics from each parent are passed on directly rather than being blended together to give an offspring that looks like a combination of its parents. In other words, an offspring

may look exactly like the female parent, even though it is a product of both parents. This manifesting of the traits in a visible way is called the *phenotype* of the offspring as opposed to the actual hereditary content of the offspring, which is the *genotype*.

As Mendel suspected these traits are in pairs in each parent, and they later came to be called *genes* and classified initially as either "recessive" or "dominant" depending on their ability to physically manifest themselves (phenotype) in the offspring. It is now known that *partial* dominance exists in genes, that it is not a winner-loser situation when they are combined.

Figure 4-1 shows an example of the most basic type of plant breeding that Mendel did—in this case the crossing of a pea with a shrunken-shaped seed with one with a full round seed. In each case the *s* represented the recessive gene for seed shape (shrunken-shaped seeds) and the S the dominant gene (full seeds). It is customary to write recessive genes as a lowercase letter and dominant genes as a capital letter.

The basic cross involves a pea parent that is *homozygous* for the gene of shrunken-seed shape, meaning that both of its genes are recessive "s," expressed as *ss* because each parent has the genes in pairs. This is crossed with a homozygous parent that is homozygous for the

Table 4-2 Sample Cucumber Hybridization

Inbred parents	Traits of parents	Traits of resulting hybrid offspring
Cucumber A line (female line)	Long, thin shape (desirable trait)	Long, thin shape
	Pale color (undesirable trait)	Deep green color
	Matures early in season (desirable trait)	Matures early
	Large, tough seeds (undesirable trait)	Small, tender seeds
Cucumber B line (male line)	Short, thick shape (undesirable trait)	
	Deep green color (desirable trait)	
	Matures late in season (undesirable trait)	
	Small, tender seeds (desirable trait)	

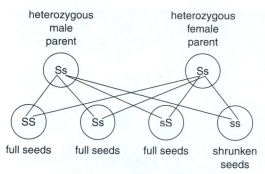

possible combination of offspring:
two homozygous and two heterozygous

Figure 4-2 Crossing of heterozygous offspring.

dominant full-seed shape: S, expressed as *SS*. When combined, these would potentially yield, in a pool of four offspring all "heterozygous" (Ss) offspring that contain one gene from each parent. But the offspring will all have full seeds (the phenotype), even though they contain the gene for shrunken seeds.

Figure 4-2 shows what would most likely happen in regard to seed shape if two of these offspring were crossed and yielded four more offspring. The genes recombine to give two heterozygous offspring (Ss and sS) and two homozygous offspring (SS and ss). In regard to the phenotype (visual expression of the genes), one of the three offspring would have full seeds because the S is dominant, and one would have shrunken seeds.

Techniques of Plant Breeding

Amateur plant breeding is an interesting activity, but it seldom yields very spectacular plants. Desirable characteristics are the result of luck in most cases. For example, a person who crosses red and white petunias may assume that pink offspring will be produced due to the mixing of red and white. But plant genetics is a complex science, so the result is more likely to be a combination of reds, whites, and shades between.

When breeding plants, a few basic rules of genetics must be followed. First, the two plants to be crossed nearly always must belong to the same genus and often to the same species. Cultivars within a species will cross-pollinate in most cases.

Only in rare instances do plants cross between genera. One example was the cross of English ivy (*Hedera helix*) and Japanese fatsia (*Fatsia japonica*) to produce the popular plant *Fatshedera*, a semivining shrub with ivy-like leaves intermediate in size between the parents.

In nature, seeds from cross-pollination make up only 4 percent of all seeds produced. Understandably so, because the pollen-bearing anthers within a flower are close to the female style, making self-pollination most likely.

Among plants that never self-pollinate are species that produce male and female flowers on separate plants (holly and bittersweet, for example). Clearly, self-pollination of these species is impossible.

The mechanics of plant breeding are not difficult. Self-pollination can be achieved by sealing a flower in a paper bag just before it opens (usually in the early morning). The flower will open within the bag and, having no other source of pollen, will self-pollinate. Tapping the bag will aid in releasing the pollen from the anthers so that it can land on the stigma.

For cross-pollination, flowers that are ready to open should be selected from two parent plants. The petals and anthers of the female parent should then be removed with small scissors and the blossom enclosed in a paper bag. Proceeding to the male parent, the flower should be opened by hand and a small paintbrush inserted to pick up some of the pollen. The pollen should then be taken to the female parent, brushed lightly over the style (Figure 4-3), and the bag replaced.

In several days the pollen will fertilize the eggs, and the style will shrivel and fall. The bag can then be removed and the flower tagged with the names of the parents.

A plant breeder who has developed an outstanding new cultivar is able to patent it for 17 years. During this period it is illegal for anyone else to propagate the cultivar without permission.

When seed is mature and ready to be harvested, usually either the seed pod splits or the fruit drops to the ground. The color of the mature seeds will darken as they ripen.

Cleaning removes the ovary tissues that surround the seed. Dry fruits such as the pods of beans or fruits of many trees usually split open when mature, and the seeds can be shelled out with the fingers. With fleshy fruits such as peach or squash, the fruit should be cut apart and the seeds extracted and dried on a paper towel for several days.

The best conditions for storing seed depend on the species. However, for general storage of vegetable and flower seeds, a sealed container kept in the refrigerator is sufficient. Glass jars, plastic containers, and freezer bags all are suitable because they maintain constant humidity. The average refrigerator temperature of about 40°F (4°C) slows respiration and keeps the seed viable for the greatest possible time.

Figure 4-3 Pollination of a hibiscus bloom by brushing the pollen on with a paintbrush. *(USDA photo)*

 A contracted seed grower is a farmer who multiplies the limited amount of seed provided by the seed breeder or seed company into a quantity large enough for commercial sale. A seed grower works outdoors and must be diligent in his or her farming practices to prevent unwanted cross-pollination or contamination of the seed with other plants. The seed that the seed grower harvests is generally sold back to the breeder from which the original genetic material was obtained, or to a seed distributor specified by the company that bred the seed.

An allied employment opportunity is for a fieldman who works with the contracted growers. The fieldman advises the farmers on how to produce a good-quality crop, reports to the company on expected yield, and arranges for delivery of the seed to the processing plant.

The life span of seeds in storage varies from days to years among species. Excess vegetable seeds are commonly stored for use the following year. Chapter 6 gives the maximum time seeds of common vegetables can be stored under recommended storage conditions and still produce satisfactory plants.

 A horticulturist can also work in seed processing for a large commercial seed company. People are needed to monitor and oversee seed drying, cleaning, packaging, and shipment.

Even under ideal storage conditions, the longer the seeds are stored, the fewer seeds will germinate. The seedlings that do emerge will be weaker. Therefore when planting seed that has been stored, it is advisable to sow the seed thickly to compensate for decreased germination.

Commercial Seed Production

Many flowers and vegetables are grown from seed because it is the cheapest source plant material. Vegetable seed varieties are chosen for their shipability, disease resistance, herbicide resistance, ability to withstand mechanical harvesting, and characteristics that are preferred by consumers (size, color, taste, and so on). Flowers are selected for growth habit, size, flower color, disease resistance, and other qualities.

Commercial seed production involves:

1. *Breeders.* The breeders are in charge of developing new cultivars that are in demand by growers. A breeder designs breeding programs, evaluates the results, does statistical analysis of the crossings, and decides which cultivars to make available for commercial growing.

2. *Field staff.* The field staff contracts with outside growers to produce the seed of the new cultivars, monitors the growing, advises the growers about growing techniques, and monitors to some extent the quality and quantity of seed that is produced.

3. *Warehouse staff.* The warehouse staff takes delivery of the seed produced by contracted growers, cleans it, treats it with chemicals if necessary, packages it, and organizes it within the warehouse.

4. *Stock seed controllers.* Stock seed controllers maintain the purity of the produced seed in regard to the quality of plants that grow from it. They rogue out mutated plants and also remove any seed stock whose quality has begun to deteriorate. They maintain the breeding plant stocks for the breeder.

5. *Seed salespeople.* The seed salespeople visit growers to advise them of the newest cultivars available for growing. They discuss the merits of new cultivars, take orders for supplies of seed to be used in production, and arrange for delivery of the seed from the warehouse.

 There are many jobs for seed sellers, either in retail (to the public) or wholesale (to retailers such as nurseries that sell to the public). The seller must be knowledgeable about the seeds and give presentations to growers about how to best grow them. Seed sellers use advertising, sales programs, and trial and demonstration plantings to promote sales.

GROWING PLANTS FROM SEED

Seed Growing Outdoors

Sowing seed outdoors is called *direct seeding* and is used for many commercially grown vegetables, flowers, and herbs (See Chapter 6) as well as in home gardening. Because the plants are seeded where they will mature, it is important to choose the site carefully. It should provide the correct amount of light for the species and should also be well drained (water should not stand after a rain). The soil must be loose and crumble easily. Soils that crust or pack should be conditioned as described in Chapter 5 and spaded and worked to a depth of about 1 foot (30 centimeters) to assure suitable soil conditions for root growth. Clods should be broken and any rocks or debris removed.

 A seed analyst works in a laboratory determining the purity and percentage of seeds that will germinate from a given seed sample. Large quantities of seed are frequently contaminated with weed seeds, disease, or inert matter such as soil or leaf particles. Some seeds might be broken and dead and would not sprout. Seed that is tested can be "certified" to assure the grower of its quality. Both seed producers and universities employ seed analysts.

The seeds should be sown at a depth and spacing recommended on the package. If sowing directions are not given, the general rule is that seeds should be planted one and one-half times as deep as their diameter. Spacing between seeds can be estimated from the mature size of the plant. Thicker sowing is permissible and even recommended, with the understanding that weaker seedlings will be pulled out to achieve proper spacing for mature plants.

Soil used to cover seeds should be free of even small clods to enable the emerging shoots to reach the soil surface easily. Rubbing the soil between the palms of the hands will pulverize it to the proper degree.

Timing is important when planting seeds outdoors. If planted too early in spring, the seeds will fail to germinate or germinate slowly because of cold weather. Planted too late, they may not mature within the growing season. Package directions are the best source of planting information.

Seeds planted in moist soil do not require immediate watering. They are able to absorb moisture from the soil to start germination. In dry soil, watering will supply moisture and settle the soil around the seed so that the emerging root will have immediate contact with a source of water and nutrients.

Maintaining adequate soil moisture during the germination and establishment parts of a plant's life is essential. A drying of the soil at these times will be fatal to the seedlings.

Soil crusting can occur after rain or watering; it can be recognized by the smooth, glazed appearance of the soil surface. Air and water entry into the soil is inhibited and germination reduced. If soil begins to crust, the surface should be broken carefully with a light rake or hand tool so emerging seedlings are not injured.

A liquid fertilizer *starter solution* (Chapter 5) can be applied to seedlings after germination is complete. Although this is not essential, research has shown that a starter fertilizer will increase the growth rate of seedlings substantially. Seedlings are considered established after true leaves appear (see Chapter 3). True leaves indicate that the root system is absorbing water and nutrients from the soil, and the plant is no longer dependent on carbohydrate stored in the seed.

Natural Reseeding of Outdoor Plants

A number of cultivated plants produce seed that germinates naturally the following season. Among these plants are tomatoes, corn, and petunias. Seedlings that grow untended are called *volunteers* and generally do not produce plants of the same quality as the parents.

Seed Growing Indoors for Home Horticulture

Seedlings of vegetables, flowers, and herbs can be raised indoors. They should be started six to eight weeks ahead of when they will be required for the garden.

A container for raising seedlings need not be elaborate. The bottom of a milk carton or plastic dairy product container is sufficient, provided holes are made at the base for draining of excess water. For large numbers of seedlings, plastic, styrofoam, or wood nursery flats can be purchased. Compressed peat moss disks (Figure 4-4) that swell with water and form a plantable container for one seedling are also available in garden centers. Cells of plastic egg cartons can also be used for individual seedlings. Plastic pony or pot packs from plants bought in previous years can also be used.

Figure 4-4 A compressed peat moss disk (Jiffy 7 peat pellet) before water is added to it, after water is added to it, and with a seedling. *(Photo courtesy of Jiffy Products of America, Inc., Batavia, IL)*

The material in which seeds are planted is called the *growing medium*, and it must be fast-draining and not prone to packing. Commercial potting soils sold for houseplants fit these requirements and are convenient for raising seedlings. They are not actually soils at all but mixtures of natural treated ingredients. Pure garden soil should not be used, because it packs when restricted in a container and becomes unsuitable for seedling growth.

Sterilization of the growing medium is an important precaution when growing seed indoors because it kills fungi in the medium that cause plant diseases. But it is only necessary if outdoor soil is a part of the medium. To sterilize the medium, it should be dampened, placed in a baking dish, and covered with foil or a lid. The oven temperature should be set at 160 to 180°F (71°–82°C) and the medium baked for about an hour or until it is heated throughout. If all the growing medium is not used, the excess should be kept in a sealed plastic bag to prevent recontamination.

 A transplant producer germinates the seeds produced by the seed grower and grows them to an adequate size for transplanting into the field or into the greenhouse. This is a hands-on job involving planting, watering, fertilizing, and otherwise caring for very large numbers of plants in the most vulnerable stage in their development. The transplant producer generally sells the transplants to distributors or commercial growers.

Houseplant potting soil seldom requires any further preparation than sterilization. After it has cooled,

Figure 4-5 A commercially produced seed-starting system for homeowners. *(Photo courtesy of A.M. Leonard Co., Piqua, OH, (800) 543-8955)*

Figure 4-6 Stringy tomato seedlings raised in insufficient light. *(Photo by Kirk Zirion)*

it can be poured into a clean seeding container and the seeds planted immediately. Spacing between seeds can be close provided the young plants will be transplanted soon after they emerge.

Water can be supplied to container-grown seedlings by sprinkling or misting whenever the growing medium begins to dry. An easier way is to water the seeds after planting and then enclose the seed container in a clear plastic bag. The plastic prevents moisture from evaporating. No further watering will be required until after the seeds have germinated and the plastic is removed. If the container is made of plastic or other waterproof material, it is not necessary to enclose the entire container. A sheet of plastic food wrap can be laid across the top of the container to hold in moisture. A home seed starting unit can also be used (Figure 4-5).

For most rapid germination, seeded containers should be placed at a temperature of 70 to 80°F (21–27°C). A sunny area is not required for germination and, in fact, should not be used for containers covered or enclosed in plastic. The heat from the sunlight will build up under the plastic and kill the seeds.

Though not essential for germination, bright light or direct sunlight is needed after seedling emergence. The limited light entering through a window, even a bright south window, is often not sufficient to raise healthy transplants. If the seedlings become pale and "stringy" (Figure 4-6) instead of compact, poor garden plants will result. For raising most transplants, artificial light (Chapter 16), a greenhouse, or a greenhouse window is advisable. Exception can be made for seedlings of houseplants (such as palms and Norfolk Island pines) and for bright-light climates listed in Chapter 16.

Commercial Production of Seedlings in Greenhouses

Bedding plants are a commerical classification of plants that are sold in spring and in some climates in fall. They are usually seedlings of annual flowers, annual vegetables, or perennials, and as a business they are worth $1.2 billion (wholesale) annually in the United States. The seedlings are sold in several types of containers. Four- or six-plant *pony packs* (Figure 4-7) or 4-inch (20-cm)-diameter pots are the standard for retail sale to homeowners. For commercial growers selling to commercial producers of cut flowers or vegetables, plastic flats or *plug flats* (also known by the commercial name thermoform flats) (Figure 4-8) are the norm.

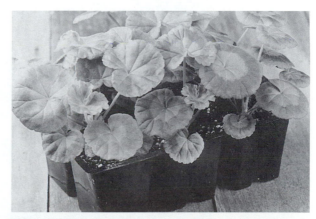

Figure 4-7 A pony pack of geraniums. *(Photo courtesy of National Garden Bureau, Downer's Grove, IL)*

Figure 4-8 A thermoform plug flat of seedlings. *(Photo courtesy of Blackmore Co., Belleville, MI)*

These flats are composed of from 100 to 800 small cells, each of which holds one plant. The system of production is highly mechanized and sometimes automated. The trays are mechanically seeded using a commercial seeder (Figure 4-9), and then germinated in growth rooms that precisely control temperature. They are then moved to a greenhouse or outdoors for several weeks, during which time they are irrigated and fertilized with overhead irrigation systems (see Chapter 20). The seedlings are sold to commercial growers such as vegetable growers who transplant them directly into the field, or to commercial flower growers who transplant them into containers such as pony or cell packs for further growing and eventual retail sale.

Damping Off

The period from seedling emergence until several true leaves have formed is the primary time that seedlings are susceptible to damping-off disease. If a fungus is present in unsterilized growing media, the organism rots the seedling stem, which discolors and shrivels at soil level. Within a few days the plant collapses and dies (Figure 4-10).

Once damping-off disease is detected, it can spread rapidly and kill the majority of the seedlings. Fungicidal soil drenches (see Chapter 18) are moderately effective in stopping the spread, but the best prevention is sterilizing the growing medium before planting. Use of an inorganic material such as vermiculite or perlite to cover the seeds is also helpful.

Transplanting Seedlings at Home

It is advisable to transplant seedlings to larger growing containers as they begin to crowd each other. As leaves of neighboring plants overlap, less light will be avail-

Figure 4-9 A small-scale commercial seeder for plug production. *(Photo courtesy of Blackmore Co., Inc., Belleville, MI)*

Figure 4-10 A damping-off infection in seedlings. *(Photo courtesy of Dr. R.E. Partyka, Columbus, OH)*

able to each seedling, and growth will slow accordingly. The competition of crowded roots for nutrients also causes stunting. At this point, usually corresponding with emergence of true leaves, the seedlings should be transplanted into separate growing containers. Seedlings are very delicate, and transplanting must be done carefully to avoid injury. A pencil or knife can be used to lift up groups of seedlings, after which they can be gently separated with the fingers. Seedlings should always be picked up by a leaf rather than by the stem, because any injury to the stem will kill the seedling.

When transplanting, each plant should be watered to assure contact between the growing medium and the roots. The seedlings should be kept in a shaded location for two to three days to minimize transpiration water loss while the roots damaged during transplanting resume water absorption.

After this period of reestablishment, the seedlings can be replaced in bright light and fertilized weekly with diluted liquid fertilizer (Chapter 5).

Hardening Off

Seedlings raised in a controlled environment grow under less strenuous conditions than they will be exposed to in a garden. To acclimatize them to outdoor conditions before transplanting, they should be subjected to a *hardening-off* period. When the weather is suitable, the transplants should be moved outdoors for several hours each day. Gradually, the time can be increased until they are brought indoors only if night temperatures are predicted to drop below the range acceptable for the species. After about a week the transplants will be hardened and less succulent appearing. Hardened plants are less likely to wilt or shock at transplanting.

Transplanting Outdoors

If possible, an overcast day should be chosen for transplanting outdoors to minimize wilting. If this is not possible, transplanting should be done late in the afternoon to allow the roots to recuperate overnight before the plant is exposed to bright sunlight and other drying conditions the following day. The roots should be disturbed as little as possible and the plant watered afterward. Some wilting should be expected, but if it is severe or lasts more than one to two days, the plant will need protection during reestablishment. Shade should be provided by a cardboard tent or newspapers. The soil should be moist at all times so the roots can easily replace water lost through transpiration.

GROWING PLANTS FROM SPORES

Of all the commonly cultivated plants, only one type, ferns, bears spores. In seed plants, the egg (ovule) and sperm (pollen) unite and mature into seeds on the parent plant. In ferns, a spore drops on the ground and

Figure 4-11 Fern prothalli. *(From Botany, 5th ed. by Carl E. Wilson and Walter E. Loomis. Copyright 1952, 1957, 1962, 1967, 1971 by Holt, Rinehart and Winston. Reprinted by permission of Holt, Rinehart and Winston)*

grows into a flat plant called a *prothallus* (Figure 4-11). The prothallus then develops the sperm and egg, which unite and form the new plant.

Mature, healthy ferns (with the exception of highly ruffled types that are sterile) develop hundreds of organs called *spore cases* on the undersides of their fronds (Figure 4-12). The spore cases resemble small brown dots or lines, and protect and release the spores.

Several months are required for a fern to develop and ripen spores. When the spores are mature, the spore cases open and release them into the air. The spores themselves look very much like a brown powder.

To capture spores before they are released, the maturity of the spore cases can be tested by lightly tapping the frond over a piece of white paper. When specks appear on the paper, the frond should be cut off and enclosed in an envelope for several days. The spore cases

will then open in the envelope, yielding a tiny amount of spores in the bottom.

Spores have the same growing requirements as seeds: warmth, moisture, air, and a growing medium. A small pot or flat should be filled with damp houseplant soil, then misted, and the spores scattered over the surface. The pot should be enclosed in plastic and set in a shaded location. In several weeks the spores will begin to germinate and grow into flat, green *prothalli* (plural of prothallus). As they grow, the prothalli should be misted daily because the egg and sperm they form can only unite in water. Eventually fronds will grow through the center of the prothalli, which will slowly die. The young ferns can then be transplanted to individual pots as soon as they are large enough to handle.

VEGETATIVE PROPAGATION

Propagating plants vegetatively involves the use of nonsexual plant organs such as leaves, stems, and roots. Whereas a plant propagated from seed may not develop identically to its parent, vegetatively propagated plants almost always do. Vegetative propagation thereby preserves the work of plant breeders and natural mutations that have yielded valuable plants.

 The term *plant propagator* can encompass those who produce seedling transplants, but generally it refers to those who propagate vegetatively. A full-time propagator is generally employed by a large nursery, but propagation is done part-time as one of the duties of a general nursery employee.

A second reason is to reproduce plants that seldom flower or produce only sterile flowers. Foliage plants and navel oranges fall in this group.

A third reason is the relatively short time vegetative propagation requires to produce a mature plant. A cutting can be rooted in as little as two weeks. To raise the same size plant from seed would require a month or more.

Cuttings

Cuttings are the most widespread vegetative propagation method. Cuttings are vegetative plant parts such as leaves, stems, and roots that regenerate their missing parts to form new plants. They are cut from parent plants called *stock plants*. The environment required for

Figure 4-12 Spore cases on the underside of a holly fern leaf. *(Photo by Rick Smith)*

growing cuttings is the same as for germinating seeds: warmth, moisture, and a growing medium.

Rooting Media

Most parts used for vegetative propagation are taken from aboveground portions of the plant and must regenerate roots. The growing medium that supports and surrounds the rooting area will determine whether roots will form and their quality. The main requirement is that it should drain quickly to admit air to the rooting area yet retain some moisture.

There is not one superior rooting medium. Many combinations of materials are used, including the following five: two parts (by volume) sand and one part peat moss, one part perlite and one part vermiculite, pure vermiculite, pure perlite, and pure sand. The ingredients listed in these media are discussed in more detail in Chapter 15. As with seed media, media used for rooting should always be sterilized.

Cuttings from Outdoor Plants

Shrubs, vines, herbs, groundcovers, and a few perennial flowers are propagated by cuttings taken at different times of the year. Cuttings from woody outdoor plants are classified as hardwood, semihardwood, or softwood, depending on the degree of woodiness of the cutting. A fourth classification, herbaceous, includes cuttings from nonwoody plants such as many groundcovers.

Hardwood Cuttings.   Hardwood cuttings (Figure 4-13) should be taken in late fall through early spring from the matured wood of the past season's growth. They are generally about 6 to 10 inches (15–25 centimeters) long and

Figure 4-13 Hardwood cuttings of juniper (left), sycamore (middle), and boxwood (right). *(Photo by Kirk Zirion)*

may come from either deciduous or evergreen plants such as juniper, viburnum, and apple.

The specific procedures for rooting cuttings vary among species, but generally the cuttings are wrapped in plastic and stored over winter in the refrigerator. In spring they are unwrapped, and the lower 1 to 2 inches (2.5–5 centimeters) is pushed into damp rooting medium. The cuttings should be enclosed in a plastic bag to prevent moisture loss and placed in a 70 to 80°F (21–27°C) location.

Roots should form in less than 6 weeks, and dormant buds should begin growing shortly thereafter. The cuttings can be hardened off and transplanted several weeks later.

Semihardwood Cuttings.   Semihardwood cuttings are taken in summer from the partially matured new growth of woody plants. The cuttings are made 3 to 6 inches (8–15 centimeters) long and, because they lose water through their leaves very rapidly, should be placed as they are cut in a premoistened plastic bag.

The procedure for rooting is the same as for hardwood cuttings, but rooting is often more rapid.

Softwood Cuttings.   Softwood cuttings are taken in late spring from succulent new growth produced that season. They generally root more consistently and quickly than either hardwood or semihardwood cuttings. Picking the cuttings at the right stage of maturity is crucial to success. Shoots that are growing rapidly are too tender and prone to rotting. Ideally, the shoots should be slightly flexible but should snap when bent to a 90° angle. They should be harvested into wet plastic bags and rooted using the same directions as for hard- and softwood cuttings.

Because the younger parts of plants generally root more easily than more mature parts, softwood cuttings are the most reliable way to vegetatively propagate most woody outdoor plants.

Herbaceous Cuttings.   Herbaceous cuttings are the equivalent of softwood cuttings but are taken from herbaceous plants such as coleus and impatiens that never become woody. Most houseplants are also herbaceous and are discussed in the next section.

Herbaceous cuttings can be taken and rooted at any time during the growing season. They are generally 2 to 4 inches (5–10 centimeters) long and taken from the tip of a stem. As with all the types of cuttings previously discussed, the bases should be planted in damp rooting medium and the cuttings enclosed in a plastic bag to increase humidity. Keeping the cuttings wet from the time they are cut until they are stuck in the medium will reduce wilting.

Root Cuttings. Root cuttings are sections of thickened roots without leaves or stems attached. Whereas most outdoor plants can be propagated by one or more types of top cuttings, only a few will grow from root cuttings. These species have the ability to generate an adventitious bud on the root section.

Root cuttings of most outdoor plants are best taken in early spring just before new growth begins. At this time carbohydrate reserves are maximum, and the plants are ready to emerge from dormancy. Like other cuttings, root cuttings should be protected from drying by a plastic bag and can be harvested by simply digging out medium-sized roots from around the base of an established plant. The sections can be cut into 3-inch (8-centimeter) lengths and planted horizontally 1/2 inch (1–2 centimeters) deep in rooting medium. The medium should be kept damp, and shoots should appear in four to six weeks.

Indoor Plant Cuttings

Cuttings of indoor plants can be of the following four types, depending on the part of the plant from which they are taken. The procedure for rooting the cuttings is the same as for cuttings of outdoor plants.

Stem Tip Cuttings. Stem tip cuttings are the most common type (Figure 4-14) and consist of the top 2 to 4 inches (5–10 centimeters) of a growing stem.

Leaf Bud Cuttings. Leaf bud cuttings (Figure 4-14 and Figure 4-15) are taken from a plant after the stem tip has been used. Each cutting includes a short section of stem, one leaf, and its corresponding axillary bud. When the base of the cutting roots, the axillary bud breaks dormancy and forms the new stem of the plant. Any plant that can be propagated by stem tip cuttings can be propagated by leaf bud cuttings, although the latter method is slower.

Stem Section Cuttings. Stem sections or cane cuttings (Figure 4-14 and Figure 4-15) are short pieces of thickened, leafless stem. They are used primarily for obtaining large numbers of offspring from cane-forming plants such as dumbcanes (*Dieffenbachia* spp.) and Hawaiian ti plants (*Cordyline* spp.). First, the top of the plant is removed for a stem tip cutting. The remaining stem is then cut into pieces containing two to three nodes each. The sections are laid horizontally half-buried in the rooting medium. A plastic covering is not essential pro-

stem tip

leaf bud

stem section

Figure 4-14 A typical dicot plant, showing where stem tip, leaf bud, and stem section cuttings would be found.

vided the medium remains moist, and roots and shoots will be generated in one to three months.

Leaf Cuttings. Leaf cuttings are a single leaf and sometimes its petiole. Most indoor plants cannot be propagated using only a leaf; only certain species generate buds where the leaf blade joins the petiole or along the leaf veins or margins (Table 4-3). The propagation procedures vary among genera. Leaves of African violet and peperomia are picked with the petiole attached. The leaf

Figure 4-15 Rooting stem tip (left), stem section (middle), and leaf bud (right) cuttings. *(Photo by Rick Smith)*

is buried in the rooting medium up to the blade, and new plants form at the soil line (Figure 4-16).

With *Bryophyllum* and other succulents such as jade plant, burro tail, and echeveria, leaves are picked directly off the plant and laid on the surface of the medium. Because they are succulents, a medium largely composed of sand is best; enclosing them in a plastic bag is not advisable. With *Bryophyllum*, up to 20 new plants will form around the leaf margin. With others, single plants will appear where the leaf was broken from the plant.

Fibrous-rooted begonias such as rex begonia produce new plants along the veins of mature leaves and also where the petiole connects to the leaf blade. Whole

leaves or pieces of leaves should be laid on a damp rooting medium so that the veins make contact. Cutting the veins with a knife stimulates bud formation, and three to five cuts can be made on each leaf.

Another houseplant grown from a leaf is the snake plant (*Sansevieria* spp.). A mature leaf can be cut crosswise in 1-inch (3-centimeter)-long sections and the base of each stuck in rooting medium (Figure 4-16). Remember that the orientation of the cuttings is critical. The end of the section that was originally on top must be up, and the part that was lower then planted in the rooting medium. Sections inserted upside down do not root. After the leaf section roots, a new plant will be produced at the base and grow up beside the leaf section (Figure 4-16).

It should be noted that variegated or yellow-banded sansevierias propagated by this method produce green offspring. To maintain variegation, the variety must be propagated by division.

Propagation from leaf cuttings is slow. The generation of an adventitious bud and roots is an energy-consuming process. Several months will be required before the young plants are large enough to be transplanted.

Table 4-3 Selected Plants Propagated by Leaf Cuttings

Common name	Botanical name
Rex begonia	Begonia × rex-cultorum
Jade plant	Crassula argentea
Echeveria	Echeveria spp.
Mother of millions, Bryophyllum	Kalanchoe beauverdii, laxiflora, and pinnata
Peperomia	Peperomia spp.
African violet	Saintpaulia ionantha
Snake plant	Sansevieria spp.
Burro tail	Sedum morganianum
Christmas cheer	Sedum × rubrotinctum

Suggestions for Successful Rooting of Cuttings under Home Conditions

Attempts to propagate plants by cuttings meet with varying degrees of success, depending on the care one takes and the readiness with which the species generates new

Figure 4-16 Leaf cuttings of a snake plant (Sansevieria) (left), begonia (middle), and peperomia (right). Note the locations of the new plants that formed. *(Photo by Rick Smith)*

roots. Following the suggestions given here will maximize chances of success with most species. But for plants that are difficult to propagate, consulting one of the references listed at the end of the chapter may be necessary.

First, select cuttings from shoots without flowers or flower buds, or if this is not possible, pick off the flowers. A cutting with blossoms will channel energy into the reproductive parts and will be less likely or slower to root.

Also note the positions of nodes on the stem of the cutting. Roots are often generated first at these sites, so cuttings should be made with at least one node near the base.

Also, leafy cuttings wilt easily and once severely wilted are less likely to root. Therefore cuttings should always be kept moist after cutting and before being stuck in the medium to slow water loss from the leaves. A humid chamber that will minimize transpiration can be constructed by draping a piece of polyethylene film tent fashion over wire supports (Figure 4-17).

Any leaves that will be covered after the cutting is stuck into the rooting medium should be removed. If left on, they will rot and provide a breeding ground for disease organisms.

Likewise, during the rooting period, leaves that die and drop from the cuttings should be removed along with whole cuttings that appear dead.

Finally, use of a rooting hormone can increase rooting speed and success. The auxins discussed in Chapter 3 are available in powder form in a talc base, as dissolvable powders for mixing with water, or as liquids, and are applied to the stems of cuttings before insertion into the rooting medium.

Gentle tugging of rooting cuttings about once per week is the standard test for determining whether rooting has occurred. If the cutting slips out easily, no anchoring roots have formed. The cutting should be inspected for signs of rotting and, if still healthy, can be reinserted into the medium. If the cutting does not pull out with gentle tugging, it already has roots. The plastic bag can then be opened to accustom the plants to normal humidity and be removed entirely after several days. After one week the cuttings can be transplanted to pots.

A few outdoor plants, such as willow and pussy willow, and many tropical indoor plants (Table 4-4) can be rooted and grown in water. The main reason more species will not root in water is lack of oxygen in the water that is, in this case, the rooting medium. If air can be bubbled in (for example by an aquarium pump), many more species can be rooted this way.

Stem tip cuttings are usually used, with the lower leaves removed and the cutting absorbing water through the cut end until roots are formed. One of the easiest

Figure 4-17 A rooting chamber made of plastic sheeting draped over wire supports. *(Photo by Rick Smith)*

Table 4-4 Selected Cuttings for Rooting In Water

Common name	Botanical name
Chinese evergreen	*Aglaonema* spp.
wax begonia	*Begonia semperflorens*
angel wing begonia	*B.* spp.
grape ivy	*Cissus rhombifolia*
coleus	*Coleus* × *hybridus*
dumbcane	*Dieffenbachia* spp.
dracaena	*Dracaena* spp.
pothos	*Epipremnum aureum*
botanical wonder	× *Fatshedera lizei*
rubber plant	*Ficus elastica*
nerve plant	*Fittonia vershaffeltii*
Tahitian bridal veil	*Gibasis geniculata*
English ivy	*Hedera helix*
polka dot plant	*Hypoestes phyllostachya*
impatiens	*Impatiens wallerana*
bloodleaf	*Iresine herbstii*
prayer plant	*Maranta leuconeura*
philodendron	*Philodendron* spp.
Swedish ivy	*Plectranthus* spp.
African violet	*Saintpaulia ionantha*
German ivy	*Senecio macroglossus*
arrowhead	*Syngonium podophyllum*
wandering jew	*Zebrina* spp. and *Tradescantia* spp.

methods is to fill a glass with water and cover the mouth with foil. Holes can then be punched in the foil and the cuttings inserted and held upright by the foil. To minimize transpiration the cuttings should be kept out of direct sun.

Commercial Rooting of Cuttings

Commercial greenhouses produce large numbers of cuttings for use in producing florists' plants such as chrysanthemums, poinsettias, and foliage plants. Shrubs are generally grown from softwood or stem-tip cuttings because these root the fastest, therefore decreasing the time required to produce a plant for sale.

A greenhouse generally has one or more benches exclusively used for propagation and equipped with an *intermittent mist system* (Figure 4-18). This system is designed to keep transpiration at a minimum until the cuttings regenerate roots and can take up water and nutrients. To accomplish this, the leaves of the cuttings are kept barely moist at all times by light mistings of water that last about 2 to 10 seconds, spaced at intervals of 2 to 10 minutes. The interval of time between mistings will depend on the rate of evaporation in the

Figure 4-18 A mist nozzle. *(Photo courtesy of Vocational Education Productions, California Polytechnic State University, San Luis Obispo, CA)*

greenhouse. The goal is to remist just before the leaves are fully dry, but not to wet them excessively. Excessive wetting leaches nutrients from the leaves and can also increase the occurrence of disease infection.

The intermittent mist system consists of a time clock that times the interval of misting, an electric solenoid valve that turns the mist on and off, and piping and mist nozzles over the plants.

In some cases an electronic leaf (Figure 4-19) can be substituted for a time clock. This is a small metal sensing plate attached to an electronic unit that controls the solenoid valve. The sensing plate is wet each time the mist turns on, and its increased weight causes it to drop, completing an electrical circuit. When water evaporates from it, the electrical circuit is broken, triggering the mist to turn back on. An electronic leaf is preferable to a time clock for misting in that it compensates for environmental changes, such as change from cloudy to sunny, and avoids overmisting that might occur in a situation with a time clock.

A root-zone heating system is often also used (Figure 4-20) to warm only the bases of the cuttings, but not increase transpiration from the leaves. The heat encourages faster and more reliable rooting of cuttings.

Crown Division

Crown division is probably the most common and reliable home propagation method. It is used for many herbaceous perennials, a few shrubs, and many houseplants such as ferns, asparagus ferns, African violets, and spider plants. In division, one plant is separated into two or more pieces, each containing a portion of roots and crown (Figure 4-21).

The division can be done at any time in the growing season; however, plants in active growth must be

Figure 4-19 An electronic leaf to control a mist system. *(Photo courtesy of E.C. Geiger, Inc., Harleysville, PA)*

Figure 4-20 A root-zone heating system suitable for use with either pots or flats. *(Photo courtesy of E. C. Geiger, Inc., Harleysville, PA)*

treated carefully to minimize water loss through the leaves. Therefore division of dormant plants in early spring is common in cold-winter areas of the country.

Shrubs to be divided must be multistemmed, that is, have a number of stems growing out of the ground instead of branches arising from one single short trunk. They should be divided when dormant if possible, by cutting through the crown with a spade so that it is bro-

Figure 4-21 Division of a plantain lily into several smaller crowns. *(Photo by George Taloumis)*

ken down into several sections. Each section is then dug, keeping as many roots as possible, transplanted to the new location, and pruned to lessen shock. A part of the parent crown can be left to rejuvenate the shrub. With herbaceous perennials, the entire plant is dug up and cut apart. The sections are then replanted and watered.

When dividing houseplants, the parent can be removed from its pot, cut or pulled apart, and the sections repotted. Pruning is normally unnecessary. If wilting occurs, the newly potted sections can be left in plastic bags for several weeks until the roots are reestablished.

Layering

Layering is another home method for propagating vines and also shrubs that have a trailing growth habit or flexible branches. A low, flexible shoot is bent to the ground during any time the plant is in active growth and held in place by a bent wire or heavy stone. It is then mounded with soil a short distance back from the tip. Within a few months, roots should have formed on the covered portion of stem, and the layer can be cut from the parent plant and transplanted.

Runners, Rhizomes, and Stolons

These three structures are types of horizontal stems produced as a natural means of vegetative reproduction. *Runners* (Figure 4-22) grow aboveground and are found on such plants as strawberries, ferns, spider plants, and strawberry begonias. *Rhizomes* and stolons grow at ground level or below ground and are typically found on bamboo, grasses, and some irises. There are subtle botanical differences among the three, but basically they are handled the same in propagation.

As modified stems, these organs have buds, nodes, and internodes. A runner will have only one bud at the tip, whereas stolons and rhizomes may have several. The nodes are the sites where new plants will form. With a runner on an outdoor plant, the stem can be positioned as it begins to form the new plant. This is a common practice to avoid overcrowding in strawberries (Chapter 8). The runner will root without additional attention and can be severed from the parent and moved after rooting if desired.

Because houseplants are grown in pots, the runner plants may never make contact with a rooting medium without help and can live attached to the parent indefinitely. When a new plant is wanted, a small pot of growing medium can be placed under the runner plant. Roots will form in less than a month, and the runner can then be detached. When this method is impractical

Figure 4-22 Rooting a runner of a spider plant. *(Photo by Kirk Zirion)*

Figure 4-23 A snake plant with two young offsets. *(Photo by Kirk Zirion)*

(such as with hanging plants), the runner can be detached and treated as an unrooted stem-tip cutting.

Propagating plants from rhizomes and stolons is done much the same way. The stem connecting the young plant to its parent is cut after a moderate amount of roots has formed, and the young plant is transplanted to the new location.

Suckers and Offsets

Suckers and *offsets* (Figure 4-23 and Figure 4-24) are young shoots that grow from the roots or stems of mature plants. They are functionally similar to rhizomes and stolons and are found in many shrubs and houseplants such as bromeliads, succulents, and cacti. Offsets on cacti (Figure 4-24) are frequently produced on top of the plant and can be broken off and rooted without difficulty. Suckers from the bases of plants may or may not have developed root systems independent from the parent. If so, they can be transplanted directly. If not, they are treated as cuttings.

Storage Organs: Bulbs, Corms, and Tubers

Underground organs, are produced by some herbaceous perennials such as lilies, gladiolas, and amaryllis as an underground repository of stored carbohydrate. Botani-

Figure 4-24 A pincushion cactus with offsets. *(Photo by Rick Smith)*

cally, these are modified stems with nodes, buds, and modified leaves. Their natural means of vegetative reproduction is the formation of clones of themselves (called *bulbils, cormels,* or *tubers*) around the base of the parent (Figure 4-25). These can be broken off (preferably while the plant is dormant) and planted in new locations. Blooming of storage organs may take two to three years after the year they are produced, because a minimum size must be reached before flowering will occur.

Air Layering

Air layering is a technique used primarily on large houseplants such as rubber plants. It involves inducing stem rooting of a plant that still has its own root system. It can be used to obtain new plants from a large branched spec-

Figure 4-25 Removing a daughter bulb from the mother.

imen, to shorten a plant that has grown too tall, or to make a plant that has become tall and leafless short and bushy again. New plants formed by air layering can be very large, up to 3 feet (1 meter) tall if desired.

Many plants having woody stems can be air-layered (Figure 4-26). As the first step, choose the section of the stem where the new root system of the plant is desired. Then a strip of bark about 1 inch (2 centimeters) wide should be cut around the stem and the bark pulled off. This girdling removes the phloem and cambium but not the xylem, which will still translocate water to the top of the plant.

Next, place a handful or two of damp sphagnum moss over the girdled area and wrap with plastic (Figure 4-26). Twist-ties or tape secure the moss at both ends and seal in moisture.

Within two to three months, roots should be growing in the sphagnum. During the waiting period the sphagnum should be checked every one to two weeks for dampness and rewet if necessary because roots will not form if the moss is dry. When several roots 2 to 3 inches (5–8 centimeters) long have formed, the air layer can be cut from the parent and transplanted to its own pot. If it wilts severely, humid conditions should be provided for about a week.

Budding and Grafting

Budding and grafting are fairly complex methods of propagation used for reproducing valuable fruit and ornamental cultivars in nurseries; they are discussed only briefly here. An amateur who wishes to try budding or grafting should plan the project ahead of time and consult several of the reference books listed at the end of the chapter for more in-depth information.

 Grafting is a specialized form of propagation that requires precise techniques and years of experience to be successful. A grafter would typically work in a fruit tree nursery, and the quality of his or her work is determined by the percentage of "takes" (successful grafts) obtained from the plants joined together.

Essentially, budding and grafting unite two genetically different plants so that they heal together and function as a single plant. Budding transfers a bud of one plant to another plant that will function as the root system, whereas grafting attaches a small branch to another plant.

Most frequently, budding or grafting combines two cultivars of a species into one plant that exhibits the

Figure 4-26 Steps in air layering: (1) girdling the branch, (2) applying damp peat moss to the wound, (3) wrapping the peat moss in plastic and sealing the ends, (4) root formation is evident. *(Photo from Plant Propagation: Principles and Practices by Hartmann and Kester, 1990. Reprinted by permission of Prentice Hall)*

best features of each. For example, a plant with a vigorous root system but little ornamental value (called the *rootstock*) could be grafted onto a plant having good ornamental qualities but weak roots (called the *scion*).

Grafting can also serve other purposes: (1) repair of girdled trees that would otherwise die; (2) creation of unusual plant forms such as tree roses or trees with weeping heads atop strong, straight trunks; and (3) changeover of fruit trees in an old orchard to a new cultivar.

Grafting and budding rely on the activity of cambium cells and must be done when these cells are actively dividing and will heal the grafted area quickly. To determine the correct stage for grafting, bark is examined for *slippage*. If it can be separated easily from the underlying wood, the cambium cells are active and the grafting or budding is feasible using the most common techniques. This is usually in early spring before new growth starts,

because this greatly reduces water loss through transpiration and improves chances that the cambia of the two plants will grow and unite them. Other grafting techniques can be used even on dormant plants.

Grafting

There are many styles of grafting. The style used will depend on the species and the reason the graft is being made. General rules for grafting are:

1. The rootstock diameter must be equal to or larger than that of the scion, but the scion wood is usually a pencil thickness or slightly larger.
2. The cambium of the stock and scion must be in contact, and preferably over as great an area as possible. If cambium contact is not made, the graft will not heal, and the scion will die.

(a)

(b)

Figure 4-27 Steps in performing a whip graft: (a) stock and scion cut to equal size and ready for grafting, (b) stock and scion joined and held together with a rubber strip. *(Photo courtesy of Vocational Education Productions, California Polytechnic State University, San Luis Obispo, CA)*

3. The stock and scion must fit tightly together, and the joint must be protected from drying. A tight graft union is achieved by wrapping the area with special rubber ties or waxed string. Drying is then prevented by a coating of wax.

Figure 4-27 shows the basic steps involved in a style of graft called whip grafting.

Budding

Budding (Figure 4-28) involves removing a patch of bark with a bud from a scion and laying it directly against the cambium of the stock. It is a less risky process than grafting because the stock is only slightly damaged if the union does not heal. As with grafting, the area of the union must be protected from drying with rubber ties.

Tissue Culture

Tissue culture, also called *micropropagation*, is the propagation of plants from nearly microscopic portions of parent plants (Figure 4-29). The technique has undergone a growth explosion in the past decade and progressed from being primarily a university laboratory procedure to a commercial propagation method of significant economic

Figure 4-28 Bud patches to be used in budding (above) and a bud patch in place on a branch (bottom). *(Photo courtesy of Dr. Robert F. Carlson)*

Figure 4-29 Tissue-cultured strawberries. *(Photo courtesy of Barbara M. Reed, National Clonal Germplasm Repository, Corrallis, OR)*

consequence. The technique has two distinct advantages over traditional propagation: (1) It enables mass production of a cultivar from an extremely limited amount of *parent stock* and in a relatively small area. (2) It enables the propagator to eliminate disease-causing viruses from the parent material, an achievement unattainable through the use of pesticides, and to propagate numerous virus-free offspring that are healthy and vigorous. This importance of propagation from virus-free parent stock has only recently come to be appreciated as the detrimental effects of unrecognized virus infection have become known.

 A tissue culture lab technician will find work in the labs of companies that specialize in propagating plants by this method. College coursework in biology, botany, or horticulture and some experience with tissue culture techniques are entry-level requirements.

In its simplest form, tissue culture involves the following steps:

1. Removal of the previously disinfected meristem of a shoot tip using a surgical knife and tweezers under sterile conditions.

2. Placing of this *explant* in a test tube of gel-type growing medium containing plant hormones, again under sterile conditions. The test tube is plugged with sterile cotton to prevent entry of disease-causing organisms.

3. Waiting during a growth and establishment period of several weeks or months during which the explant multiplies many times in size.

4. Division under sterile conditions of the cell mass of *callus* into numerous near-microscopic pieces

that are placed in more gel-filled test tubes containing hormones and plugged with cotton.

5. Waiting for the development of roots and shoots induced by the hormones in the medium.

6. Transferring of the small plantlets from test tubes to pots under greenhouse growing conditions for establishment in normal growing medium.

7. Transferring of plants to outdoor growing conditions.

Tissue culture is not an amateur activity, because it is nearly impossible to achieve the sterile conditions necessary. However tissue-cultured plants still in test tubes are sold in nurseries occasionally as novelty items, particularly orchids, which were the main plants tissue cultured for many years. The test tube is left sealed and treated as a miniature terrarium. When the plant outgrows the tube it can sometimes be transplanted to a pot, although the process is not always successful.

GENETIC ENGINEERING[1]

Genetic engineering of plants is a branch of biotechnology, the harnessing of the biological machinery of bacteria to manufacture otherwise hard-to-obtain products, combat genetically caused diseases, improve the tolerance of plants to adverse environmental conditions, and attain other similar laudable goals. Genetic engineering as a means of plant improvement changes the genetic makeup of plants without either breeding or selection, the conventional routes of improvement. Its main advantage is that while plant breeding and selection normally involve only the genes within a species, genetic engineering makes possible the transfer of genes between completely unrelated plants or bacteria, and, in rare cases, even from animals to plants. As an example of the usefulness of this ability, most flowers are not available in blue cultivars, and it is impossible to find or breed blue cultivars because no members of the species carry the genes for this color. However, using genetic engineering, it is possible for blueness from, for example, bachelor buttons to be transferred to roses, and in fact blue roses have already been developed by genetic engineering in Australia.

The obvious advantage then is that when the possibilities for plant improvement through breeding and selection have been exhausted because of the limited *gene pool* (the genes the plants in the species naturally carry), a species can continue to be improved by bringing in outside genes.

[1]The author gratefully expresses appreciation to Dr. Rick Ward of Michigan State University for his help with this topic. This section could not have been written without his counsel.

Genetic engineering is a profession that requires a Ph.D. although lab technicians who work under the main research scientist may have only bachelor's degrees. Genetic engineers may work for large commercial companies researching, for example, a particular crop such as tomatoes for Campbell's soup. They also work in universities.

Procedure for Genetic Engineering

The procedure is, of course, complicated, but a simplified version is outlined as follows:

1. The first step is to identify the characteristic that is to be incorporated, for example, disease resistance, frost resistance, or color.

2. The second step is then to either create or isolate from a donor plant the genetic material that will give this characteristic. To isolate the material the donor organism is ground up and then all extraneous cell contents are eliminated chemically until only the genes remain.

3. The third step is to take this extract of genetic material and incorporate it into a *vector*, which will convey the genes to the target plant to be genetically engineered. The first vector is a *plasmid*, a microscopic ring of genetic material found in bacterial cells and available in quantity from scientific supply companies. *Enzymes* (chemicals that cause or accelerate chemical reactions without being affected themselves) are used to open the plasmid rings to admit the new genetic material and then to close them. The treated plasmids, not all of which will have accepted the new genetic material, are then taken up naturally by specialized bacteria in a lab procedure.

4. The next step is to check that *transformation* has taken place, that the plasmids are in the bacteria. This is done by growing the bacteria individually into colonies on a gel medium that has been specifically designed to react through color or another visible sign when in the presence of a transformed bacterium.

5. Those bacteria that are seen to have been transformed are then transferred to new gel medium, grown, and then checked for the presence of the specific genes wanted. This is necessary because although it will have been ascertained that genetic material was trans-

ferred, it is uncertain without further checking whether the specifically desired genes are there. One way to accomplish such checking is through chemical analysis. If the trait to be transferred were disease resistance (conveyed by a gene that triggers a plant to produce poison that kills the disease organism, for example), one could check through chemical analysis for the presence of the poison in the bacteria.

6. The next step is to transfer the genetic material in the bacteria into a second plasmid vector within an organism that will put the genes into the target plant. The organism normally selected for this in genetic engineering of plants is the bacterium that causes the disease crown gall (see Chapter 13). This bacterium has the astounding and rare natural ability to transfer genes to its host plant as a part of its life cycle, and this is the property that makes it invaluable because it can transfer other genes as well.

    Crown gall naturally produces tumors in plants by producing cytokinen and indole acetic acid hormone. To eliminate the tumor-causing characteristic and take advantage only of its infective potential, the bacterium is first *disarmed*, that is, rendered incapable of causing galls.

7. The final step is to infect the target plant with the crown gall bacteria that now contain the genetic material to be incorporated. Pieces of root, leaf sections or other plant parts are *inoculated* (touched with the bacteria), and those that become infected (again checked by chemicals in the medium that show a reaction) can then be grown as tissue-cultured plants (see previous section) to regenerate a whole plant. The new plants are now called *transgenic* because they contain genes from outside their own genus. The new genetic material has become an integral part of the plants' genetic makeup and will reproduce itself in every new cell produced.

Although the above-described method is the most common, others are used as well. One uses a device called a particle gun that actually fires 22 caliber cartridges containing tungsten particles coated with the plasmids carrying the foreign genes. The plant to be genetically altered is literally shot in a vacuum, and the genetic material is propelled through the cell walls and becomes a part of the genetic makeup of the plant.

Another method is *protoplast* transformation. Cells from the target plant have their cell walls removed chemically (thus becoming protoplasts), and are then soaked in plasmids that have had the new genetic material inserted. The new plasmids incorporate into the protoplasts, and the protoplasts are then regenerated by tissue culture into whole plants.

Commercial Applications of Genetic Engineering

In commercial horticultural production, research has mainly centered on vegetable and fruit crop genetic engineering with a limited amount on flower crops. Although genetically engineered crops are not in widespread cultivation, many are being field-tested to see if the genetic enhancement imparted in the laboratory will translate to benefits in field cultivation. Generally, genetic engineering efforts have focused on imparting disease and pest resistance, with a limited amount directed toward extending the length of time the crop remains fresh for the consumer, altering color (in flowers), and imparting cold-temperature resistance (in strawberries).

One of the best known genetically engineered horticultural crops is the 'Flavr Savr' tomato. This tomato was genetically engineered to retain a firm texture longer than normal. Tomatoes destined for fresh eating must be hard-harvested to prevent bruising, unlike canning tomatoes that can be mechanically harvested—a less expensive process. Fresh tomatoes must also be transported quickly and with careful packaging to assure that they arrive at the supermarket in an attractive condition that will be appealing to the buyer.

'Flavr Savr' contains genetic material that inhibits the expression of the genetic material that causes the fruit to soften when it ripens. The softening part of ripening is thus slowed, although the flavor continues to develop. This allows mechanical harvesting, increased transport time, and longer fresh shelf life in the supermarket.

A second genetically engineered crop receiving widespread attention is a Thompson Seedless grape variety engineered to be virus resistant. Thompson Seedless is one of the most commonly cultivated table grapes and is also a component of blended wines. Scientists hope that genetically engineered virus resistance will reduce both the expense of chemicals and their entry into the environment because it will no longer be necessary to spray to prevent the disease.

Table 4-5 lists a number of horticultural crops currently in research projects involving genetic engineering.

Opponents of genetic engineering object on several bases. Some fear that something as unnatural as genetic engineering could upset the ecosystem in unknown ways. They feel that the accelerated pace of genetic change (as compared to traditional breeding methods) at which highly specifically designed plant cultivars can be developed could inundate the environment with bizarre plants, causing an unstable ecological situation.

Another objection is based on the fact that a bacterium, *Bacillus thuringiensis*, is the source of genetic material put into plants to cause their cells to produce

Table 4-5 Selected Genetically Engineered Crops

apples	cranberries	raspberries
asparagus	eggplants	squash
broccoli	gladioli	strawberries
cabbage	grapes	sweet potatoes
carnations	lettuce	tomatoes
carrots	peas	walnuts
chrysanthemums	peppers	watermelons
corn	poplar	
cucumbers	potatoes	

an insect poison, thus causing them to be very resistant to insect attack. Some organic farmers fear that this biological pesticide, which they use to control infestations of worms, will no longer be effective due to insect resistance as a result of widespread incorporation in many crops. This will deprive the organic farmers of one of their most important natural insecticides, which they apply judiciously as needed.

The Environmental Protection Agency (EPA) has approved a number of genetically engineered plants. Over three million acres of genetically engineered corn, cotton, and potatoes were planted in the United States in 1997. However, a class action suit has been filed against the Environmental Protection Agency by 31 groups who charge that the EPA has been negligent in its approval of genetically engineered crops.

SELECTED REFERENCES FOR ADDITIONAL INFORMATION

Arbury, Jim. *The Complete Book of Plant Propagation.* Newtown, CT: Taunton Press, 1997.

Bajaj, Y.P., ed. *High-Tech and Micropropagation V.* New York: Springer-Verlag, 1992.

Ball, Vic, ed. *Ball RedBook,* 16th ed. Batavia, IL: Ball Publishing, 1997.

Coastline Community College. *Plant Reproduction.* Fountain Valley, CA: Coastline Community College.

Dirr, Michael, Charles W. Heuser, and Bonnie L. Dirr. *The Reference Manual of Woody Plant Propagation: From Seed to Tissue Culture.* Athens, GA: Varsity Press, 1987.

Garner, R.J. *The Grafter's Handbook.* London: Cassell, 1993.

Hartmann, H., D. Kester, F. T. Davies, and R. L. Geneve. *Hartmann and Kester's Plant Propagation: Principles and Practices.* 7th ed. Upper Saddle River, NJ: Prentice Hall, 2002.

Klock, Peter. *Plant Propagation: Seeds, Cuttings, Division, Layering, Grafting.* London: Ward Lock, 1997.

Kyte, Lydiane, and John Kleyn. *Plants from Test Tubes: An Introduction to Micropropagation,* Portland, OR: Timber Press, 1996.

MacDonald, Bruce. *Practical Woody Plant Propagation for Nursery Growers.* Portland, OR: Timber Press, 1987.

Nau, Jim. *Ball Culture Guide, the Encyclopaedia of Seed Germination.* Batavia, IL: Ball Publishing, 1996.

Nau, Jim. *Ball Perennial Manual: Propagation and Production.* Batavia, IL: Ball Publishing, 1996.

Styer, Roger C., and David S. Koranski. *Plug and Transplant Production: A Grower's Guide.* Batavia, IL: Ball Publishing, 1997.

Toogood, Alan R. *Plant Propagation: The Fully Illustrated Plant-by-Plant Manual of Practical Techniques.* New York: Dorling, Kindersley Publishing, Inc., 1999.

Van Patten, George F., and Alyssa F. Bust. *Gardening Indoors with Cuttings.* Washougal, WA: Van Patten Publishing, 1997.

World Wide Web Addresses

http://aggiehorticulture.tamu.edu/propagation/propagation.html   Texas Agricultural and Mechanical University Cooperative Extension Service, "An Overview of Asexual Propagation" *Plant Propagation website.*

http://www.ghg.net/beyer/propag.htm.   Beyer, Bob and Lana, "Introduction to Plant Propagation", Ornamental Horticulture for the Houston and Gulf Coast Area website.

http://www.hcs.ohio-state.edu/mg/manual/prop.htm Kerrigan, Jack and Margaret Nagel. Ohio State University Cooperative Extension Service: "Plant Propagation" *Master Gardener Training Manual website.*

http://wwwusers.imaginet.fr/~jmh/SeedsOfLife/intro.html Brenckmann, Françoise: "Introduction: What is a Seed?" *Seeds of Life.*

PART TWO

GROWING PLANTS OUTDOORS

(Photo courtesy of Agricultural Research Service, USDA)

CHAPTER 5

OUTDOOR SOILS AND FERTILITY

The soil is a blanket of loose material that covers the majority of the Earth's land surface. It is a dynamic mixture of minerals, decaying plants and animals (organic matter), microorganisms, water, and air. Relative to plants, the soil serves two primary purposes: first, it provides mechanical support for the plant by permitting roots to grow through it; second, it provides a reservoir for water, air, and nutrients essential for healthy plant growth. How well a particular soil performs each of these functions largely determines the fertility of the soil, that is, how well plants will grow in it.

 A soil scientist is different from a horticulturist, but all graduates in horticulture take one or more soil science courses. A university-trained horticulturist with a minor in soil science might work in a lab testing soil samples or be employed by a large horticultural producer monitoring and supervising soil management programs.

CLASSIFICATION

Mineral soils are classified primarily by the size and quantity of mineral particles present in the soil. Soil particles range from clay particles, which are so small that they require an electron microscope to be seen, to silt, and then to sand, which can be seen readily by the unaided eye. Each type of soil particle possesses unique properties that influence plant growth. However, the properties of silt are in many respects intermediate between those of clay and sand, and silt is present in relatively small quantities in many soils. Therefore, a basic understanding of soils can be obtained by considering mainly the properties of clay and sand particles. Awareness of these properties will enable the gardener to adapt cultural practices so that the soil environment will be the most conducive to healthy plant growth.

Organic soils (including muck and peat) contain predominately decaying organic matter with smaller amounts of mineral components. They are generally found in small pockets in the north central and northeastern sections of the United States and Canada as well as in the Everglades region of Florida. Organic soils, although very fertile, require careful management; they do not, in many respects, respond in the same manner as mineral soils. Because organic soils are not important as home garden soils, they are not discussed in this text. Persons gardening in organic soils should consult a soils text or their county extension agent for further information.

Soils seldom contain only one type of particle. Instead they contain several types in varying percentages, and the soil is classified according to the type of particle that predominates (Table 5-1). For example, if a soil contains mostly clay particles but also a small amount of sand, it will be classified as a clay soil. When the percentage of sand increases to a point where it nearly equals the percentage of clay, the soil is classified as a loam. When the percentage of sand further increases to the point where it greatly surpasses the percentage of clay, the soil is designated as a sandy soil. Thus the term

Table 5-1 Sample Classifications of Soil by Particle Types

Soil	Average particle size	Suitability as garden soil
Clay	Smallest	Fair
Loamy clay	Small	Fair
Clay loam	Medium	Good
Loam	Medium	Good
Sandy loam	Medium	Good
Loamy sand	Large	Good
Sand	Largest	Fair

(a)

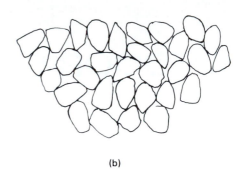

(b)

Figure 5-1 (a) Clay particles lie flat and have smaller pore spaces between them than (b) sand particles.

loam, which is often used on seed packages to describe the ideal soil in which seeds should be planted, refers to a natural soil having nearly equal percentages of sand and clay particles.

As mentioned previously, clay and sand particles each possess unique characteristics. If a soil is classified as a loam, its properties will be intermediate between those of clay and sand particles. The higher the percentage of one type of particle, the more closely the soil properties will resemble those of that particle group. A loamy clay soil, for example, mostly possesses the characteristics of clay particles but is modified somewhat by the sand it contains. A clay loam, on the other hand, is basically a loam with a slight extra amount of clay.

Characteristics of Clay Soils

Clay soils (Table 5-2) are often referred to as *heavy soils,* composed primarily of platelike clay particles. These particles are flat and disk-shaped, and their small size and flattened shape are responsible to a large extent for the properties of clay soils. Because the particles are flat, they are able to lie in close contact with one another. This closeness makes the spaces between individual clay

particles relatively small compared to the spaces between more rounded particles such as sand. This might be likened to pennies (clay particles) being stacked in a jar as compared to marbles (sand particles). The spaces between the pennies are much smaller than the spaces between the marbles (Figure 5-1). Conversely, the jar filled with marbles, which is likened to a sandy soil, contains fewer particles but larger spaces. These small spaces between clay particles (called *pore spaces*) account for one of the main properties of clay soils: slow water and air movement into and out of the soil.

Water movement through pore spaces causes a soil to be either fast- or slow-draining and determines whether its drainage is good (fast) or poor (slow). Thus

Table 5-2 Properties of Sand and Clay Soils

Clay	Sand
Ability to hold a large amount of water	Inability to hold water
Slow movement of air and water in the soil (poor drainage)	Rapid movement of air and water in the soil (good drainage)
Small, numerous pore spaces	Large and fewer pore spaces
Tendency to expand when wet and contract and crack when dry	Does not expand or contract as a result of water
Good ability to attract and hold nutrients for plant growth (high cation-exchange capacity)	Poor ability to attract and hold nutrients for plant growth (low cation-exchange capacity)
Tendency to compact under pressure when wet	No tendency toward compaction
Plastic when wet and highly moldable	Not plastic when wet

Figure 5-2 The plasticity of clay allows it to be formed into ribbons when wet. *(Photo courtesy of Vocational Education Productions, California Polytechnic State University, San Luis Obispo, CA)*

soils composed mostly of clay are considered poor-draining. They hold water for long periods of time, and this can be detrimental to plant growth.

Another property of clay soils attributable to the flattened shape of the particles is *plasticity*. This is the property of clay responsible for its being sticky and moldable when wet (Figure 5-2) and hard and cloddy when dry. Because clays are highly plastic when wet, they are also subject to compaction, a packing down of the soil particles that practically closes up all the pore spaces (Figure 5-3). Compaction is the result of pressure such as walking or driving on the soil when it is wet. It makes water penetration and drainage poor, resulting in poor plant growth due to a lack of air in the soil. Another result of the plastic properties of clay is that it expands when water is added and con-

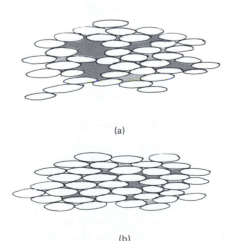

(a)

(b)

Figure 5-3 (a) Aggregated clay soil, (b) compacted clay soil. Note the decreased pore space in the compacted soil.

tracts when it dries out. This tendency to expand and contract results in the surface cracking of the soil that is often evident when clay soils become very dry.

Clay soils also have a great capacity to attract and hold nutrients. This property (called *cation-exchange capacity*) will be discussed in more detail later. Basically, it means that nutrients tend to be stored in clay soils rather than washed away by rainfall or irrigation.

Characteristics of Sandy Soils

Sandy soils (Table 5-2), or *light soils,* can be thought of as possessing the opposite properties of clay soils. Whereas water moves through clay soils slowly via the small pore spaces, in sandy soils it moves freely and rapidly through large pore spaces. Sandy soils do not retain moisture well; nor do they expand and contract, become sticky or cloddy, or compact readily. The cation-exchange capacity of sandy soils is quite low, so nutrients wash away readily when water drains through.

The difference between the properties of clay and sand particles is due mainly to the larger size of sand particles and their rounded or irregular shape. Being irregularly shaped, they do not pack tightly (compact). Nor do they slide easily past one another when wet, as clay particles do. Consequently they lack the plastic properties inherent in clay soils.

Characteristics of Loam Soils

Loam soils possess qualities of both sand and clay proportional to the relative amount of clay or sand particles in the loam soil. They are considered to be ideal for most plants because they do not have the extreme properties of either sands or clays. For example, clay soils may become waterlogged and unworkable for many days after rain, and sandy soils hold so little water that irrigation must be applied frequently. However, a loam soil will absorb a large quantity of water and not require frequent irrigation, yet it will dry enough to be workable in a much shorter time than will a clay soil. Loams also hold more nutrients than do sandy soils; however, they are not as prone to becoming cloddy or compacted as clay soils.

The Soil Texture Triangle

The *soil texture triangle* (Figure 5-4) is a graphic device that allows a person to classify a soil by weight (heavy or light) and by class name (sand, clay, loam, sandy clay, and so

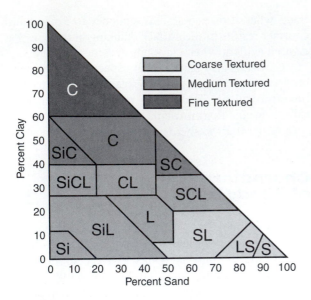

Key: C-Clay
L-Loam
Si-Silt
S-Sand

Figure 5-4 The Soil Texture Triangle.

on). It combines and converts the relative percentages of each major component (as measured by its weight) into a textural class or name such as loam, clay, silty clay loam, and so on. Silt is not an axis of the triangle because it is automatically factored in as the remaining component not accounted for when sand and clay are combined.

For example, a soil that is 45 percent sand, 25 percent clay and 30 percent silt is a loam (L), generally the most preferred soil type. One that is 50 percent clay and 30 percent sand is a clay (C). One that is 15 percent clay and 60 percent sand is a sandy loam (SL).

The determination of the percentages of each component by weight is normally done commercially in a lab, but an internet reference at the end of the chapter will explain how it can be done without lab facilities.

ORGANIC MATTER IN THE SOIL

In addition to the mineral components of soil such as sand and clay, *organic matter* is also a valuable component. In fact plants can be grown in pure organic matter quite satisfactorily without any inclusion of sand or clay. Organic matter in the soil is the result of vegetative matter such as leaves and roots and animal products such as manure returning to the soil and decom-

posing. Though organic matter is present in most soils in relatively small quantities (a few percent), it greatly influences the properties of the soil. It commonly, for example, is responsible for half of the cation-exchange capacity of a soil. In sandy soils organic matter increases water- and nutrient-holding capacity; in clay soils it improves drainage and air movement and decreases compaction and plasticity. Its net effect, then, is improvement of any soil to which it is added.

One of the most important effects of organic matter on clay soils is to cause *aggregation*. Clay particles have a tendency to lay very close together and exclude air. When organic matter decomposes, the resulting *humic acid* acts as a glue and causes the clay particles to clump together loosely (Figure 5-3). Because the small clay particles are then clumped together, they form much larger particles that function more like sand. The effect of this aggregation is that because the soil has larger pore spaces, water and air movement is improved, resulting in better plant growth. When clay soils are worked while wet, these aggregates are broken apart, resulting in a poorer soil. For this reason, clay soils should never be disturbed until they are dry enough to let the soil crumble easily in the hand.

Another beneficial effect of soil organic matter is that it decreases the tendency of soils to crust. *Crusting* refers to the formation of a thin, hard soil layer at the soil surface that impedes the movement of water and air into and out of the soil and may physically impede the germination of seed. Crusting is most common in soils that are deficient in organic matter and is seldom a problem where careful soil management has regularly returned organic matter to the soil.

CATION-EXCHANGE CAPACITY

The *cation-exchange capacity* (CEC) of a soil is the relative capacity of the soil to attract and hold nutrients (cations) on the surface of the soil particles. Soil and organic matter particles have negative charges; because particles with opposite charges attract each other, they are able to attract positively charged ions from the soil water and hold them against the surface of the soil particle (Figure 5-5). This process is called *adsorption*. In soils with a poor CEC such as sand, the number of cations each sand particle can attract is quite low, due largely to the small surface area of each particle. Thus, when nutrients are added to the soil, only a small amount can be held by the soil particles. The rest remain dissolved in the soil water and are washed away readily with rainfall or irrigation.

Figure 5-5 Negatively charged clay particles attract and hold cations, providing a reservoir of nutrients.

Figure 5-6 Adhesion and cohesion of water molecules around a marble show how water is held in the soil.

In soils with a high cation-exchange capacity such as clays or those high in organic matter, a very large quantity of nutrients can be held by each soil particle. Thus, when a fertilizer such as potassium is added, a large amount of the nutrient will be held in the soil. These adsorbed nutrients cannot be leached away by watering; instead, they are stored in the soil for future use by plants. This *buffering capacity* makes the soil easier to manage and means that chemicals added to the soil will affect the soil less readily than in soils with a poor CEC.

Understanding cation-exchange capacity is important in fertilizer application. Fertilizers need to be applied less frequently to soils high in cation-exchange capacity because of their nutrient storage capabilities. A high cation-exchange capacity is therefore a desirable soil quality.

Of the three commonly applied fertilizer elements—nitrogen, phosphorus, and potassium—only nitrogen and potassium leach readily in soils that are low in CEC and thus must be applied frequently. Fertilization can be reduced in high CEC soils if fertilizers containing nitrogen in an ammonia form, such as urea or ammonium nitrate, are used.

AIR AND WATER IN THE SOIL

Varying amounts of air and water are contained in any soil. Most higher plants require that their roots be in contact with adequate amounts of both water and air. Thus a good soil will provide a balance between water and air. Whenever either water or air is lacking in the soil, the plant will die. In a sandy soil, which tends to provide abundant quantities of air but does not retain water well, lack of water commonly restricts plant growth. On the other hand, in a heavy clay soil where water drainage is poor, plant growth may be restricted by a lack of adequate air for root respiration.

In any soil the pore spaces are filled with either air or water. At any given time the percentage of pores filled by water or air is determined by the size of these pores and the relative abundance of either air or water. There-

fore the balance between air and water is partially determined by the soil itself (pore size and number) and partially determined by the quantity of water supplied.

Pore size affects the air/water ratio in the soil because small pores hold water better than do large pores. When water is added to the soil, it has a tendency to move downward in response to gravity. It is a force called *adhesion,* the attraction of solid materials (soil particles) to liquids, which holds water in the soil rather than allowing it to flow through. Adhesion can be demonstrated by dipping a marble in water and observing the thin film of water that adheres to it. At the same time water is adhering to soil particles, it is also *cohering* to other water molecules. *Cohesion* is illustrated by the droplet of water that forms at the bottom of the marble after it is removed from the water (Figure 5-6). This droplet is composed of a layer of water molecules adhering directly to the marble and another layer cohering to the first layer of molecules. When too many layers of water molecules cohere to the first layer, the force holding them together is not strong enough to support the weight of the droplet, and the droplet will fall from the marble. The same phenomenon occurs in soil. If a pore space is small, a greater quantity of water will remain, due to the combined forces of adhesion and cohesion.

Using this information and knowledge of the size of pore spaces found in clay and sand, it is easily seen why clay has a higher water-holding capacity than sand. The presence of a large number of small pores in clay, as compared to the fewer but larger pores in sandy soil, explains the difference.

Water, in addition to moving downward in the soil, may also move sideways or upward. Water moving in these directions is due to the combined forces of adhesion and cohesion and is called *capillary water.* Capillary water is analogous to water that moves upward in a straw placed in a glass of water (Figure 5-7). In this case the water adheres to the sides of the straw, and then the water molecules cohere to each other, thus creating a column of water that extends above the

Figure 5-7 Capillary action moves water upward in this straw in the same way that water moves upward in the soil.

level of water in the glass. In the soil, water moves upward in a column through the pores to a root through which the plant absorbs it or to the surface, where it evaporates. Through the process of capillary water movement, water that has not been utilized by the plant or has evaporated is constantly replaced by water

from below. Capillary action illustrates the importance of watering deeply rather than applying only a small quantity of water to the soil surface.

If water is added to the soil through rainfall or irrigation the soil will eventually become completely saturated to the point that it can hold no more water. When this point is reached, and after any excess (called *gravitational water*) has drained away, the soil is said to be at *field capacity*. As time passes and water evaporates or plants absorb the water in the soil, the water content drops until there is no longer any water left that plants can use. At this point, called the *permanent wilting point,* the only water left in the soil is *hygroscopic water,* which is too tightly held to the soil particles for plants to remove. Thus, available water for plant growth is that water between field capacity and the permanent wilting point.

Figure 5-8 illustrates the energy that is required to extract water from a field of soil. To extract water from the soil, a plant must overcome the *matric potential* of the soil. The matric potential is the potential energy of water in its attraction to soil. In an unsaturated soil, matric potential comes from capillary and adhesion forces previously discussed. Plants must use more energy through their roots than the energy with which the water is held, and the amount of energy required changes as the soil

Figure 5-8 Soil moisture retention curve.

Table 5-3 Capacity of Typical Soils for Water Retention

Soil type	Field capacity (% of vol.)	Permanent wilting* point (% by vol.)	Crop available (% by vol.)
Sandy Loam	17	9	8
Loam	24	11	13
Clay	36	20	16
Heavy Clay	57	28	29

*Permanent wilting point is an average for crop plants. Even at permanent wilting point, many drought-tolerant plants can extract water from the soil.

dries. At field capacity, very little energy is required, but as permanent wilting point approaches, more is required, until the plant eventually is unable to remove water from the soil.

Table 5-3 shows the relative amounts of the soil water that are available to a plant: the difference between field capacity and permanent wilting point. Because soils that are predominantly clay and "heavy" are capable of holding more water, it is not surprising that their available crop water is greater (29 percent in a heavy clay) than that of sandy loam (8 percent).

SOIL LAYERS

Most soils possess distinct layers that affect plant growth. Soil scientists divide the soil into layers called *horizons*. The topmost horizon is called the A horizon which also corresponds with the *topsoil*. In undisturbed soil the topsoil may extend to a depth of several feet or may be as shallow as a couple of inches. The A horizon is characterized by relatively high levels of organic matter, a darker color than the deeper horizons, and generally higher fertility. The topsoil layer is the one that is usually involved in soil preparation activities and that provides the best substrate for plant growth.

Below the A horizon lie the B and then the C horizons, which are collectively referred to as the *subsoil*. These horizons are low in organic matter and relatively infertile. They are progressively lighter in color than the topsoil. Roots may extend into the B horizon but, except for large woody plants, rarely extend into the C horizon.

Unfortunately, home gardeners are often in the position of trying to garden in areas where developers have removed the topsoil during construction, leaving only subsoil. Because of the inherently infertile nature of subsoil, the gardener will have to make extensive use of soil amendments and careful soil management practices in order to garden successfully. With time and

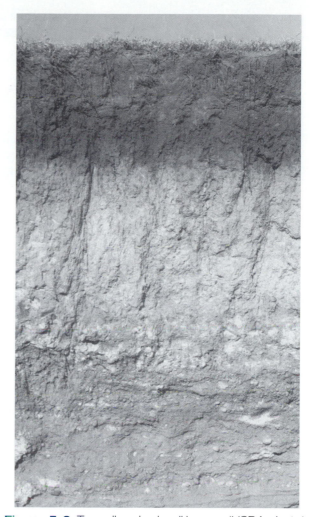

Figure 5-9 Topsoil and subsoil layers. *(USDA photo)*

careful management even the most infertile subsoil can produce a beautiful garden.

For the gardener, the most important layers are the topsoil and the subsoil (Figure 5-9). In addition to the natural topsoil and subsoil layers, another layer called a *hardpan* (Figure 5-10) may occasionally be present. A hardpan is a layer of compacted soil that slows or stops the movement of water and air and the growth of roots into lower layers. This results in the total displacement of air in both large and small pores, with consequent damage to plant roots due to the absence of air. When air is lacking, roots become susceptible to several disease organisms present in the soil and will be weakened or killed.

Hardpans may occur naturally or as a result of heavy equipment moving over the soil. If a hardpan is present, the best solution is to dig through it with a posthole digger or, if this fails, to construct raised planters. The presence of a hardpan and accompanying poor drainage can

Figure 5-10 A soil hardpan.

be identified by filling a planting hole with water and waiting for it to drain away. If the water remains after a half-hour, the hardpan is still present.

EROSION

Erosion is the movement of soil particles from one place to another and is a process that is detrimental to the fertility of the soil. Water or wind are usually the agents of the movement with the soil either being washed away by heavy rain or improper irrigation or blown away by wind.

Though erosion is a more serious problem for commercial agriculture (see Soil Conservation section of this chapter), the home gardener should also be aware of possible erosion problems. Erosion is most likely to be a problem when gardening on steep banks where water erosion will result in movement of soil from the top of the bank to the bottom. To minimize erosion, the gardener should be sure that appropriate plant materials are used to cover the bank. Plants with extensive, fibrous root systems and spreading growth habits are useful for controlling erosion. When planting a bank, the gardener should disturb the soil as little as possible. Preparing individual planting holes for shrubs or groundcovers is, for example, preferable to rototilling the entire bank. Where preparation of the entire bank is necessary, as in the case of seeding a grass, planting should proceed immediately after soil preparation, and the soil surface should be mulched to stabilize the soil until the grass is sufficiently established to hold it.

Even where banks are minimally disturbed during planting, mulches or commercially available materials such as jute netting should be used to minimize erosion. Irrigation systems used on banks should be carefully matched to the water infiltration rate of the soil so that runoff does not occur. This means that systems will have to deliver water slowly so that the water will soak into the soil rather than run down the slope carrying soil with it.

Where extensive bank areas are to be planted, commercial hydroseeding is often employed. *Hydroseeding* is a process in which the seed is mixed with water, fertilizer, and a soil stabilizer and then sprayed on the bank. Hydroseeding is an effective method of establishing erosion-preventing plants on steep banks.

SOIL BIOLOGY

The soil is inhabited by a vast and ever-changing population of small, often microscopic, plants and animals. These organisms affect the soil and the higher plants growing in it in a number of ways. They improve drainage and aeration, assist in the decomposition of organic matter, aid in plant nutrient and water uptake, cause diseases, or feed directly upon plants, insects, or pathogens. Although some are harmful, causing plant injuries or diseases, most are beneficial and form an essential part of the soil complex. Included among soil-inhabiting organisms are such animals as earthworms, insects, and nematodes and such microscopic plants as mycorrhizae, algae, fungi, and bacteria.

The study of soil organisms, called *soil biology,* has in recent years revealed an interdependent population of organisms in the soil that make up an interrelated ecosystem or *food web* about which little was previsously known.

Broken down in terms of size and type, the soil organisms are classified as follows:

Fauna (Animals)

Macrofauna ("large" + "animals") include mammals like gophers and mice, plus burrowing insects such as crickets and beetles, grubs, spiders, ants, sowbugs/pillbugs, centipedes, earthworms, slugs, snails, and the like. Their tunneling brings subsoil to the surface on a continual basis. These are the parts of plants that you see in the soil, primarily roots, but also tubers, tuberous roots, and bulbs.

Perhaps the most interesting animal in the garden soil is the common earthworm. These animals feed by passing soil containing organic matter through their bodies, where it is subjected to digestive enzymes and a grinding action. The material that has passed through the earthworm's body is called a *casting*. Castings are much higher in organic matter and nutrients than the surrounding soil. In addition, the earthworm burrows increase soil aeration and drainage.

Earthworms are most abundant in moist clay or loam soils rich in organic matter. In cold climates they are least abundant in the spring, with populations increasing all summer. Because earthworms are killed by early fall freezes in bare soil, they should be protected by surface mulches to maintain high populations in the home garden. After the first cold snap of the year, earthworms begin to acquire cold tolerance and migrate deeper into the soil where the temperature is more favorable.

Mesofauna ("medium-size" + "animals") include tiny soil worms, root mealy bugs, and mites, which can be small but are not microscopic. Often they eat dead plants to start the decomposition process. But some mites eat fungi or prey on tiny insects, eggs, and larvae.

Microfauna ("microscopic" + "animals") include organisms such as mites, springtails, nematodes, and amoebas. Some of these are beneficial, but many are parasites that attack plants. Protozoa are microorganisms that are not strictly animals, but prey on the microflora.

Microflora (microscopic + plants)

Microflora are the bacteria, fungi, and their relatives the Actinomycetes, and algae; these add a significant amount of organic matter to soils. They are extremely numerous. Although individually very small, their total weight can approach half a percent of the weight of a given amount of soil, and their importance in the ecosystem is very significant. They decompose dead plants and animals, and make humus. The by-products of their activity bind soil particles into aggregates, which both improves aeration and decreases erosion potential by giving the soil a rough texture rather than a smooth surface over which rainwater can run unchecked.

The numbers of each organism depend on soil conditions such as moisture, aeration, organic matter content, temperature, and the species of plants present. Soil acidity and alkalinity also affect organism populations; thus different microfloral ecosystems and balances are created in different soils.

Bacteria

Of special note are the bacteria called *Rhizobia*, which relate in a *symbiotic relationship* (a relationship within which both organisms benefit) with the plant roots of members of the pea family (legumes). The bacteria supply nitrogen present in the air to the plant.

Cyanobacteria (which used to be called blue-green algae) are a special type of bacteria with chlorophyll. They photosynthesize like higher plants and are important in rice paddies and wetlands because they too fix nitrogen.

Fungi

Likewise, *Mycorrhizae* fungi also have a symbiotic relationship with the root cells of many plant species. They live inside the cells but protrude into the nearby soil, effectively increasing the nutrient and water absorption capacity of the root system up to tenfold. In exchange for this they use about 5 to 10 percent of the carbohydrates being translocated down to the roots.

Fungi normally include a threadlike system of mycelia for exploration of the soil and nutrient absorption. Although most fungi are *saprophytic* (that is, they eat only dead materials), one type has a sticky system of mycelial hoops that entangle roundworms. A fine filament of the fungus penetrates and fills the body of the worm, then absorbs its contents. Another predatory fungus has threadlike rings. When a roundworm accidentally passes through the ring, the friction of its body causes the ring cells to swell and trap the worm, often in as little as one-tenth of a second, and the worm is then devoured.

Actinomycetes

Actinomycetes are the source of many human antibiotics such as streptomycin, actinomyacin, neomycin. They are found mostly in soils with a neutral pH and are the smell of "fresh earth" after a rain. In terms of soil ecology, they are particularly effective at breaking down hard-to-decompose soil materials such as cellulose (wood fiber) and chitin (a primary ingredient in the hard shells of certain insects).

SOIL PH

The pH of a soil is a measure of its acidity or alkalinity. It is measured on a logarithmic scale that runs from 0 to 14, although a soil pH below 4 or above 9 is uncommon (Figure 5-11). A pH of 7 is neutral, whereas a pH above 7 is alkaline and below 7 is acid. The optimum range for the majority of plants is 6.5 to 7.0. Many plants adapted to a lower pH are called *acid-loving* while other plants are tolerant of alkaline soils.

A soil pH that is not optimum may be corrected by adding materials to the soil to change the pH. Because the pH scale is a logarithmic one, a pH of 6.0 is 10 times more acidic than a pH of 7.0; a pH of 5.0 is therefore 100 times more acidic than a soil with a pH of 7.0. Thus relatively large quantities of acid- or alkaline-forming materials must be added to cause even a small change in pH, especially in clay soils. Table 5-4 lists some soil amendments used for adjusting soil pH.

Lime is one material utilized to raise soil pH. It is available in several different grades and types, all of which will make the soil more alkaline. Ground limestone or calcitic limestone ($CaCO_3$) is the most common and the least expensive type available. Dolomitic limestone ($Ca \cdot MgCO_3$), although slightly more expensive than calcitic limestone, performs the added function of providing magnesium, a necessary nutrient lacking in some soils. Whichever type of lime is used, it will take between three and six months to accomplish the desired pH change. This is due to the slow speed with which lime dissolves. To speed up the pH change as much as possible, buy finely ground lime, and mix it thoroughly into the soil. Surface applications of lime are only minimally effective in raising the pH below the soil surface.

Acidification of the soil is generally accomplished through the use of the chemical sulfur or aluminum sulfate or by using organic amendments such as oak leaves, pine needles, or sphagnum moss that are acid-forming. (Tables 5-4 and 5-5) When using organic amendments, large quantities are needed, and the change is very slow. Many fertilizers such as ammonium sulfate and potassium sulfate are also acidic and can be used to acidify soils, though large amounts cannot be used due to the danger of damage to plants from excess fertilization. Sulfur or aluminum sulfate also changes soil pH but only after the use of relatively large quantities. Because of the high cost of acidifying materials, large-scale soil acidification is generally not done. Instead the soil immediately surrounding an acid-loving plant is adjusted. Agricultural sulfur is the most commonly used acidifying agent and is less costly than aluminum sulfate; however, a greater quantity is required than of the latter, and the effect is slower. Aluminum sulfate is more acid-forming than sulfur but must be used carefully because, if overused, the aluminum can become toxic to plants. Soils with a high buffering capacity will require larger quantities of pH-adjusting materials than those with a low buffering capacity.

SOIL IMPROVEMENT

Soil Testing

The best way to determine the pH and nutrient content of a soil is to have it professionally tested. The results can then be used in the soil improvement program. In most of the United States a soil test will be performed for a small charge by the Cooperative Extension Service. The usual procedure is to submit a soil sample to

Figure 5-11 The pH scale of soil.

Table 5-4 Common Soil Amendments

Amendment	Uses	Nitrogen requirement	Comments
INORGANIC AMENDMENTS			
Perlite	Work into heavy soils to improve drainage and aeration and reduce compaction	None	Expensive; lasts a long time; best for planters; lightweight; floats
Sand (coarse)	Same as perlite	None	Less expensive than perlite but required in large quantities for substantial improvement of clay soils; heavy; do not use beach sand because of salts
Topsoil (loam)	Work into very sandy soils to improve water and nutrient holding	None	Inexpensive but heavy; may be contaminated with weed seed, nematodes, fungi, and other undesirable organisms
Vermiculite	Improves water and nutrient holding in sandy soils	None	Expensive; lightweight; if soil is worked frequently, the structure of the vermiculite will be damaged, reducing its effectiveness
Scoria (lava rock, cinder rock)	Same as perlite	None	Less expensive than perlite; medium weight
ORGANIC AMENDMENTS			
Bark	Mix with heavy soils to improve drainage and aeration and decrease compaction; some improvement of water and nutrient-holding capacity in sandy soils	Add about 1 lb (0.5 kg) of actual nitrogen per cubic yard (cu. meter)	Usually inexpensive; breaks down fairly slowly; finely ground bark is best
Compost (also called humus)	Improves both clay and sandy soils; contains some nutrients	None	Variable; may have a high ash content
Leaf mold	Similar to compost but different in that it is derived primarily from leaves	None	May acidify soil slightly
Manure (composted)	Improves both clay and sandy soils; contains some nutrients	None	May be high in both salts and ash; may have a strong odor; inexpensive
Michigan peat (Hypnum peat, muck)	Improves both clay and sandy soils if used in large quantities	None	Inexpensive, but large quantities are needed; high ash content; not readily available west of the Rockies
Redwood compost	Best for clay soils, with some improvement of sandy soils	Add about ¼ lb (0.1 kg) of actual nitrogen per cubic yard	Inexpensive, however, does not hold water or nutrients well
Sawdust	Improves clay soils	Add about 1 lb (0.5 kg) of actual nitrogen per cubic yard (cu. meter)	Very inexpensive. Breaks down so rapidly that nitrogen deficiencies may develop despite the addition of nitrogen; improves soil for a relatively short time. If used, mix with a better amendment. Some toxicity may occur if used fresh.
Sphagnum peat moss	Excellent for clay or sandy soils; high water and nutrient capacity	None	Expensive, but less is required; hard to wet when fully dry; acidifies soil
Sewage sludge (Milorganite)	Improves both sandy and clay soils; adds some nutrients	None	Inexpensive; may have a high ash content and a strong odor; do not use in the vegetable garden if root vegetables are grown
Rice hulls	Improves both clay and sandy soils	Add about ¼ lb (0.1 kg) of actual nitrogen per cubic yard (cu. meter)	Inexpensive, holds water well

Table 5-5 Chemical Soil Amendments for pH Adjustment and Micronutrient Addition

Name	Chemical formula	Speed of effectiveness	Use
Limestone	$CaCO_3$	Slow	Raise soil pH
Hydrated lime	$Ca(OH)_2$	Rapid	Raise soil pH
Gypsum or calcium sulfate	$CaSO_4$	Medium	Add calcium, which displaces sodium in soils and improves drainage
Sulfur	S	Slow	Add sulfur and lower soil pH
Epsom salts	$MgSO_4\ 7H_2O$	Rapid	Add magnesium
Aluminum sulfate	$Al_2(SO_4)_3$	Rapid	Add sulfur and lower soil pH
Calcium nitrate	$Ca(NO_3)_2\ 2H_2O$ (analysis 15-0-0)	Rapid	Add calcium and nitrogen and raise soil pH
Ammonium sulfate	$(NH_4)\ 2SO_4$ (analysis 20-0-0)	Rapid	Lower pH sharply, add sulfur and nitrogen
Magnesium sulfate	$MnSO_4$		Lower pH and add magnesium
Iron sulfate	$FeSO_4$		Lower pH and add iron
Chelated iron	9–12% Iron		Add iron in absorbable form
Borax	$Na_2B_4O_7\ 10H_2O$		Add boron
Copper sulfate	$CuSO_4$		Add copper and lower pH
General micronutrient fertilizers	Contain a combination of most essential micronutrients		Add deficient micronutrients

the county agent, who will forward it to the soil testing laboratory at the university. In other areas commercial soil testing laboratories must be used.

The most important step in soil testing is taking the sample. The sample must be representative of the area being tested or it will be meaningless. To obtain a representative sample, a combined sample is taken from the area. The best technique is to establish a grid pattern over the area being sampled (Figure 5-12) and then take a small sample from each grid intersection. Each individual sample can best be taken by digging a hole 6 to 8 inches (15–20 centimeters) deep and then skimming off a small slice of soil (Figure 5-13) from the top to the bottom of the hole. At least a dozen samples should be combined in a plastic bucket and mixed thoroughly, and then a 1-pint (0.5 liter) sample should be taken from this.

10'- 20'
(3-6 m)

Figure 5-12 Following a grid pattern for soil testing will assure a representative soil sample.

Figure 5-13 Slicing off an undisturbed section of soil for testing.

Improvement of Garden Soil through Addition of Amendments

A *soil amendment* is a material added to soil to improve physical properties such as drainage, CEC, aeration, and water retention. It may or may not supply nutrients, depending on the amendment. The addition of amendments for soil improvement is a fundamental gardening practice and will greatly influence gardening success.

The best amendment is chosen by considering several factors including the problem to be corrected and its seriousness, properties of the amendments available, and cost. Often the best amendment will not be practical because of cost, and a less expensive amendment will have to be used.

The most important rule to remember about the use of any soil amendment is that amendments must be used in large quantities to achieve the desired benefits. The addition of a small amount will not make any appreciable difference in soil properties. The amount of amendment required will depend on the amendment chosen and the severity of the soil problem. As a general rule, an amount of amendment equal to 25 percent of the volume of the soil should be applied.

The easiest way to apply an amendment is to spread it over the surface of the soil to a depth of several inches (8–10 centimeters) and then spade or rototill it into the soil. The thickness of the amendment layer needed can be determined by comparing the soil depth to be spaded or rototilled to the amendment depth. For example, if a 2-inch (5-centimeter) amendment layer is used and the soil is spaded 8 inches (20 centimeters) deep, the amendment added equals approximately 25 percent of

the soil volume. If large quantities of amendments are added, the process is best repeated several times, with about a 2-inch (5-centimeter) layer of amendment incorporated each time.

Each type of soil amendment has properties that must be evaluated when determining its suitability. An important consideration is the quantity of nitrogen it contains and how rapidly it decomposes. If the amendment decomposes rapidly and contains little nitrogen, nitrogen may actually be taken from the soil to supply the nitrogen needed for the process of decomposing the amendment. This nitrogen will be released when decomposition is completed, but in the meantime nitrogen deficiency symptoms will often occur in plants growing in the area. This effect will be intensified if the amendment is finely ground and the soil is warm and moist, because these conditions are conducive to the most rapid decomposition of the amendment and thus the greatest nitrogen deficit. To counteract this, a fertilizer containing nitrogen should be added with the amendment. Amendments that take up nitrogen excessively include sawdust and wheat straw.

Other considerations are salinity and ash content. Certain amendments are high in chemicals such as sodium and ammonium salts. This may not be a serious problem in the eastern United States and Canada where abundant rainfall occurs; however, in the drier western parts of the hemisphere, rainfall is often insufficient to wash away the salts, resulting in salt toxicity to plants. Manure often contains high levels of ammonia salts and therefore will "burn" plants if used in large quantities.

Ash content refers to the amounts of mineral matter present in the amendment. In organic amendments the presence of large quantities of ash is usually undesirable, as ash does not improve the soil. Leaf molds, manures, and compost often contain large amounts of ash.

Inorganic amendments (not of plant or animal origin) such as vermiculite and perlite are usually too expensive to be used in soils, although they are commonly used in potting soils. Sand is sometimes used but must be used in large quantities to be effective. When they are used, the effects of inorganic amendments are often permanent, whereas organic amendments must be added periodically. Table 5-4 lists soil amendments, their characteristics, and uses.

Composting

Compost (Figure 5-14) is composed of partially decomposed plants and is made from rotted garden refuse such as weeds, grass clippings, fallen leaves, and plant clippings. Many cities now have "greenwaste" recycling. Garden waste is collected and brought to a central recycling

Figure 5-14 Compost. *(USDA photo)*

area where it is chopped and allowed to decompose. The resulting compost is then sold back to the public or for commercial use.

A home compost pile can be relatively simple or quite complicated. In its most basic form, the plant refuse can be simply piled in an out-of-the-way spot and allowed to decompose slowly without any further attention. With this method, compost will be made very slowly, and three to four months may pass before appreciable compost forms.

To form compost at a faster rate, a compost heap must be built. A wire or wooden bin should be constructed as shown in Figure 5-15. A layer of refuse should then be spread in the bottom 6 to 9 inches (15–23 centimeters) deep, sprinkled with a handful of high-nitrogen fertilizer, covered with a thin layer of soil to supply microorganisms, and wet slightly. The layering and wetting process is repeated until the bin is full. The composting material should be kept damp by sprinkling with water as necessary and turned and mixed with a pitchfork every four weeks to ensure even decomposition of the entire pile. If not turned, the pile will generate heat and decompose mainly in the center, with the

soil and fertilizer

organic refuse

Figure 5-15 Cross section of a compost pile built in layers for rapid decomposition. Keeping the pile moist speeds the process.

outer layers rotting only very slowly. Compost can be considered ready for use when the material is evenly dark-colored and decomposed to the point where the materials are not recognizable. It is then used for soil amending or mulching.

Among the other materials that can be composted are vegetable peelings and rotten vegetables, coffee grounds, and other kitchen refuse other than meat scraps. Diseased plants should not be composted because there is a chance the disease-causing organisms will survive despite the heat of decomposition (often up to 150°F) (66°C). Woody materials such as twigs and small branches can be used but will break down very slowly. Their decomposition can be hastened by chopping them into small pieces before adding them to the pile.

SOIL CONSERVATION

Soil deterioration can be due to erosion from wind or water. To preserve the soil, several techniques are used:

Contour Planting

Contour planting is planting in curving rows to follow the natural changes in elevation of the land (Figure 5-16). The aim is to counteract the natural tendency of water to drain down the slope by the steepest and therefore quickest route, forming erosion gullies. Contour planting is especially important in areas where rainfall is rapid and intense, as this rainfall pattern is the most apt to cause rapid runoff and gully formation.

Shelterbelts

Shelterbelts (Figure 5-17) are areas of trees planted to act as wind barriers. They are a useful soil conservation technique in areas where wind erosion is a problem and can also increase crop yield by 25 percent. Choosing the right species is important and the right species will depend on the climate, so the Cooperative Extension Service is the best source of information.

Filter Strips

Filter strips are areas of either planted or native vegetation between a potential, pollutant-source area (such as a crop area where pesticides and fertilizers are regularly applied) and a surface-water body that receives runoff. The term *buffer strip* is sometimes used interchangeably. The purpose of a filter strip is to trap water runoff that may contain sediment and organic matter, and plant nutrients and pesticides that are either chemically bound to the sediment or dissolved in the water. A filter strip protects the purity of any water downhill from it by stopping or reducing the movement of sediment, organic matter, and some nutrients and pesticides before the runoff enters a stream, lake, or pond.

Grass species are preferred to broadleaf plants for filter strips because they form dense sod, have a fibrous root system that holds the soil tightly in place, and provide more complete soil cover. Solid cover prevents *splash erosion*, the dislodging and washing away of soil particles due to the impact of heavy raindrops. Splash erosion is also prevented by planting in rows close enough to produce a plant canopy of leaves that completely shades the soil when the crop is full grown.

Normally filter strips are 10 to 25 feet (3–6 meters) wide: 10 feet (3 meters) for a slope of less than 1 percent and up to 25 feet (6 meters) for a land slope of up to 30 percent. For maximum trapping efficiency, the runoff must pass over the filter in the form of a shallow, even sheet of water. Therefore, the filter strip must be placed to make sure that they intercept runoff before it forms

Figure 5-16 Contour planting or "strip cropping" for erosion control. *Photo used with permission of the USDA.*

Figure 5-17 Shelterbelt trees.

natural drainage channels in the field. Also, filters should be planted on the contour to provide the best conditions for a shallow, uniform flow into the filter area.

Sustainable Agriculture

Sustainable agriculture, also known as *alternative agriculture* or *sustainable farming,* is a reduced-chemical approach to farming. It is a modified form of standard agriculture that currently can involve continuous cropping. In continuous cropping commercial agricultural production, the soil is essentially "mined" for its ability to produce crops. It is planted year after year, often with the same annual crop, and relatively large quantities of chemical fertilizers and pesticides are used as needed. Continuous cropping results in depressed overall soil fertility, depressed productivity of the land due to soil compaction and reduced soil microbe activity, and reduced numbers of natural predators of plant pests.

Sustainable agriculture should not be confused with organic gardening as addressed in Chapter 13. Sustainable agriculture is characterized by substituting increased labor (either in the form of actual manual labor or mental labor) for fertilizers and pesticides. It also includes other soil-management practices such as crop rotation, nonstandard tillage practices (such as strip cropping), and nonstandard planting practices (such as intercropping).

Sustainable agriculture, which may integrate livestock as well as plants, began with research in the 1970s and 1980s. The research aimed at integrating key components—crop rotation, nonstandard tillage techniques, careful soil fertility management, water conservation and water-quality protection, and methods of pest control that emphasize limited use of chemicals and integrated pest management (see Chapter 13). A 1990 act of Congress called the Sustainable Agriculture Research and Education Program (SAREP) provided funds for research on integrated crop/livestock farming and for Cooperative Extension Services to disseminate knowledge of sustainable agriculture techniques.

Sustainable agriculture is now encouraged by local and national government allotments of money for research into "best management practices" such as cover crops, green manure crops, and strip cropping (farming only strips of soil with fallow area between). In addition, farmers can recoup up to $3,500 per year for implementing practices such as rotations, biological pest control, and soil testing and for purchasing specialized equipment. These techniques preserve the soil and the ecosystem of microbes and insects that live in it.

The principal concern about sustainable agriculture is whether it can be adequately productive and profitable to compete successfully with standard agricultural production techniques such as monoculture (the planting of only one species per field). The period during which a farmer changes over from standard to sustainable agriculture can be financially perilous. Weed control is particularly difficult without large inputs of labor, which are expensive in comparison to weed-controlling chemicals. In general, a sustainable farm will be smaller than a conventional farm.

PLANT NUTRITION

All plants obtain a large number of mineral nutrients from the soil. Because many nutrients are known to be utilized by plants, they are classified for ease of discussion into two groups based on the relative quantities utilized by the plant. The first group, called *macronutrients,* includes those nutrients used in fairly large quantities by plants. The macronutrients include nitrogen (N), phosphorus (P), potassium (K), calcium (Ca), magnesium (Mg), and sulfur (S). The second group, called *micronutrients,* or trace elements, is equally important but is used by plants in small quantities. The most important micronutrients are iron (Fe), copper (Cu), zinc (Z), boron (B), molybdenum (Mo), chlorine (Cl), and cobalt (Co). An example of the difference in quantity of nutrients used by a plant is that an acre (0.4 hectare) of corn might easily remove 150 pounds (60 kilograms) of nitrogen from the soil while removing less than an ounce (28 grams) of cobalt.

For optimum plant growth to occur, all of the nutrients needed by that plant not only must be present in the soil but must be present in a form the plant can utilize.

 A horticulturist may specialize in plant nutrition and be responsible for the fertilizing and monitoring of the fertility status of large acreages of plants in a nursery, orchard, vineyard, or vegetable production fields.

Macronutrients

Because macronutrients are those nutrients used by plants in the largest quantities, they are the primary ingredients in most garden fertilizers. Of the six macronutrients plants obtain from the soil, calcium, magnesium, and sulfur are generally present in large enough quantities to be adequate for many years and seldom need to be added when fertilizing. For cases where these nutrients are lacking, a list of fertilizer sources is given in Table 5-6. Phosphorus, potassium, and particularly ni-

Table 5-6 Fertilizers and Their Characteristics

Name	Analysis	Speed of nutrient release	Effect on pH
INORGANIC AND CHEMICALLY MANUFACTURED			
Ammonium sulfate	20-0-0	Fast	Very acid
Sodium nitrate	15-0-0	Fast	Alkaline
Potassium nitrate	13-0-44	Fast	Neutral
Ammonium nitrate	33-0-0	Fast	Acid
Urea	46-0-0	Fast	Slightly acid
Monoammonium phosphate	11-48-0	Fast	Acid
Diammonium phosphate	18-46-0	Fast	Acid
Treble superphosphate	0-40-0	Medium	Neutral
Superphosphate	0-20-0	Medium	Neutral
Potassium chloride	0-0-60	Fast	Neutral
General balanced granular	10-10-10, 5-10-10, and so on	Various	Varies
ORGANIC			
Bone meal, steamed	2-27-0	Slow	Alkaline
Bone meal, raw	4-22-0	Slow	Alkaline
Hoof and horn meal	13-0-0	Slow	
Dried blood	12-0-0	Medium	Acid
Sewage sludge	2-1-1	Medium	Acid
Activated sewage sludge (microorganisms added)	6-5-0	Medium	Slightly acid
Cattle manure (fresh)	0.5-0.3-0.5	Slow	Slightly acid
Chicken manure (fresh)	0.9-0.5-0.8	Slow	Slightly acid
Horse manure (fresh)	0.6-0.3-0.6	Slow	Slightly acid
Sheep manure (fresh)	0.9-0.5-0.8	Slow	Slightly acid
Swine manure (fresh)	0.6-0.5-0.4	Slow	Slightly acid
Wood ashes	0-2-6	Medium	Alkaline
Rabbit manure (dried)	2.25-1-1	Slow (unless steamed)	Slightly acid
Cattle manure (dried)	2-3-3	Slow (unless steamed)	Slightly acid
Swine manure (dried)[a]	2.25-2-1	Slow (unless steamed)	Slightly acid
Seaweed	1.7-0.75-5	Slow	
Cottonseed meal	6-2-1	Medium	Acid

[a]The nutrient content of the dried, processed manures not listed can be approximated as four times that of the fresh manure.

trogen are frequently not present in sufficient quantities for optimum growth; they are therefore the main macronutrients supplied in fertilizers.

Nitrogen

When nitrogen is lacking, plants grow slowly and lose their deep green color. The application of nitrogen will often result in a flush of leafy growth and the return of a deep green color within a week or two. Anyone who has applied nitrogen fertilizers to a lawn can vouch for the resulting fast growth and increased necessity of frequent mowing.

Because nitrogen stimulates rapid vegetative growth, it should be applied frequently to plants grown for their foliage such as turfgrass, leafy vegetables, and indoor foliage plants. It should also be regularly applied to all young plants to encourage the rapid development of leaves that photosynthesize sugars needed to fuel subsequent flowering and fruiting. High levels of nitrogen fertilization do, however, have the disadvantage of stimulating leafy growth—often instead of flowers and fruit. A common example is the vegetable gardener who uses too much high-nitrogen lawn fertilizer on tomatoes. The result is a tomato plant of astounding size with healthy, dark green leaves but that never produces many flowers or fruit. In this case a lower percentage N fertilizer should have been used once the plants had produced sufficient vegetative growth.

Another disadvantage of high nitrogen fertility levels is that succulent new growth of shrubs and trees may continue too late in the summer, resulting in cold injury when winter arrives. Thus, when nitrogen fertilizers are used, the consequences must be assessed and the quantity and timing of each application regulated to maintain adequate, but not excessive, nitrogen in the soil.

A characteristic of nitrogen that greatly affects fertilizing rate and frequency is that nitrate nitrogen is not bound to soil particles as are many other nutrients. Virtually all nitrogen added to soil in the form of nitrate will remain dissolved in the soil water. As a consequence, it readily leaches from the soil. This means that whenever heavy rainfall follows nitrogen applications, much of the nitrogen will be leached away and lost. The drainage characteristics of the soil should thus be assessed when deciding how often to fertilize with nitrogen.

Especially in a sandy soil, nitrogen will be leached when rainfall occurs, so that *slow-release fertilizers,* or frequent small applications of quick-release (nitrate) fertilizers rather than a single large application are best. Slow- and *fast-release fertilizers* and the forms of nitrogen contained in fertilizers and used by plants are discussed under "Fertilizers." In slower-draining clay soil, loss of nitrogen is slower and the benefits from slow-release fertilizers are less valuable.

Another consequence of the lack of nitrate adsorption is that virtually all nitrates remain dissolved in the water in the soil. If too much fertilizer is applied, severe root damage may result. With nutrients that are adsorbed, burning is less likely because some will be adsorbed to soil particles.

Nitrogen can be added to the soil in many different forms, each having characteristics that determine its suitability for a particular usage. Consult Table 5-6 for characteristics of a number of common nitrogen sources.

Phosphorus

Phosphorus, the second macronutrient, influences many plant functions including flowering and fruiting, root development, disease resistance, and maturation.

Phosphorus behaves very differently from nitrogen in the soil. It is insoluble in the soil water and, when added to the soil, reacts readily with aluminum, iron, calcium, and other elements, forming various compounds unusable by plants. Consequently, relatively little phosphorus is ever in the soil water at any one time, and there is negligible leaching and little danger of plant damage from overapplication. The problem, then, with phosphorus is to maintain enough of it in the soil water for plant use. This can best be done by maintaining the pH between 6.0 and 7.0, maintaining high amounts of organic matter, and wherever possible, placing the fertilizer in bands next to plants to minimize soil contact.

Because phosphorus is insoluble, it will remain where it is placed and will not leach through the soil. Therefore, it must be supplied in the root zone of the plant. Surface applications of phosphorus will have little effect on deeply rooted plants.

An important, rather specialized use of phosphorus is its use as a transplant or starter fertilizer. (Starter fertilizers are characteristically low in nitrogen and potassium but high in phosphorus.) Starter fertilizers are applied as a liquid at the time of transplanting and can bring about reduced transplanting shock and a resulting quick resumption of growth. It is not uncommon for tomatoes that received a starter fertilizer to produce fruit several weeks sooner than a similar plant that did not receive starter fertilizer.

Potassium

The third main fertilizer element is potassium, essential for starch formation, movement of sugars in the plant, formation of chlorophyll, and flower and fruit coloring. It also plays a role in the opening and closing of leaf stomates for gas exchange.

Potassium (or potash) is abundant in many soils. However, most of the potassium is not available for plant use, necessitating the addition of potassium in usable forms. Potassium, unlike phosphorus, leaches fairly readily and may be wasted if heavy rainfall occurs after application. The problem of leaching is less severe in clay soils because potassium will be absorbed or held by soil particles and thus prevented from leaching. This potassium can then be drawn upon and used by plants at a later time.

Heavy applications of this nutrient are undesirable because the plants will absorb more potassium than they need if it is available. This results in fertilizer waste with no benefits in plant growth or appearance, or what is called luxury consumption. Moreover, overdoses of potassium will injure or "burn" plants (overdoses of nitrogen are even more likely to do so) by killing the root tips.

Potassium fertilizers can be used most efficiently by being added frequently in small amounts rather than in a single, larger application.

Micronutrients

Although micronutrients are just as important to plant growth as macronutrients, they are needed in such small quantities that soils generally contain enough to supply most plants. Occasionally micronutrients are lacking. Frequently, however, a micronutrient deficiency occurs even though large quantities of the nutrient are in the soil. At certain pH levels, nutrients react chemically with other elements in the soil, forming compounds that cannot be utilized by the plant. Each nutrient has a definite pH range within which it is most available to plants.

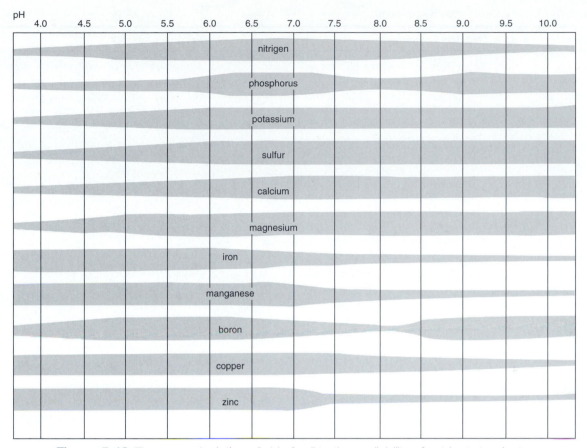

Figure 5-18 The general relation of pH of soil to the availability of nutrients to plants.

In most soils, as can be seen in (Figure 5-18), the majority of nutrients is readily available between pH 6.5 and 7.0. If pH shifts out of this range, certain elements are likely to become unavailable. If micronutrient deficiencies occur consistently, check soil pH.

Correcting micronutrient deficiencies involves correcting soil pH, if practical, and applying the nutrient to foliage and roots. When applying micronutrients to the soil, it is best to purchase the nutrient in *chelated* form. Although more expensive, it is formulated to be unreactive with other elements in the soil and will remain available to the plant. Foliage applications of micronutrients (as discussed under "Fertilizer Application") will speed the elimination of deficiency symptoms. Optimum soil moisture and warm temperature also decrease chances of microelement deficiencies.

NUTRIENT DEFICIENCIES

Either a lack or an unavailability of any macro- or micronutrient will impede normal plant growth and development. Symptoms vary according to the species, and it is often difficult to diagnose less common defi-

ciencies by sight alone. For example, lack of calcium in tomatoes shows as a brown rotting patch at the blossom end (called *blossom end rot*) rather than on the leaves. Table 5-7 lists the most common visual symptoms of nutrient deficiencies.

The only sure determiner of deficiency is a professional tissue analysis done in a plant laboratory. In tissue analysis, the actual amounts of elemental components in the plant leaves are measured and can then be compared to ideal contents for that species. Periodic tissue analysis during the growing season can be a great aid for large growers to maximize fertilizer application efficiency, minimize fertilizer cost, and consequently maximize plant growth rate for the best cost/benefits ratio and lowest production costs.

FERTILIZERS

Fertilizers are compounds that, when properly used, provide the necessary nutrients for plant growth. In common usage, they are divided into organic and inorganic types.

Organic fertilizers are derived from the decomposition of plant and animal products and include blood

Table 5-7 Nutrient Deficiency Symptoms*

Nutrient	Deficiency symptoms	Fertilizer sources
calcium (Ca)	New leaves (top of plant) are distorted or irregularly shaped.	Anything with the word "calcium"; also gypsum.
nitrogen (N)	General yellowing of older leaves (bottom of plant). The rest of the plant is often light green.	Anything with the words "ammonium," "nitrate," or "urea." Also manures.
magnesium (Mg)	Older leaves turn yellow at edge leaving a green arrowhead shape in the center of the leaf.	Anything with the word "magnesium"; also Epsom salts (magnesium sulfate).
phosphorus (P)	Leaf tips look burnt, followed by older leaves turning a dark green or reddish-purple.	Anything with the words "phosphate" or "bone." Also greensand.
potassium (K)	Older leaves may wilt, look scorched. Interveinal chlorosis begins at the base, scorching inward from leaf margins.	Anything with the words "potassium" or "potash."
sulfur (S)	Younger leaves turn yellow first, sometimes followed by older leaves.	Anything with the word "sulfate."
Micronutrients		
boron (B)	Terminal buds die, witches' brooms form.	Anything with the words "borax" or "borate."
copper (Cu)	Leaves are dark green, plant is stunted.	Anything with the words "copper," "cupric," or "cuprous."
iron (Fe)	Yellowing occurs between the veins of young leaves.	Anything with the word "iron chelate."
manganese (Mn)	Yellowing occurs between the veins of young leaves. Pattern is not as distinct as with iron. Reduction in size of plant parts (leaves, shoots, fruit) generally. Dead spots or patches.	Anything with the words "manganese" or "manganous." Often required with zinc application.
molybdenum (Mo)	General yellowing of older leaves (bottom of plant). The rest of the plant is often light green.	Anything with the words "molybdate" or "molybdic."
zinc (Zn)	Terminal leaves may be rosetted, and yellowing occurs between the veins of the new leaves.	Anything with the word "zinc."

*(Information used courtesy of The University of Arizona Cooperative Extension)

meal, bone meal, manure, and sewage sludge. Organic fertilizers have the advantage of acting as soil amendments in addition to supplying nutrients. Because nutrients from organic fertilizers are released slowly during decomposition, burning of plants caused by too much fertilizer dissolved in the soil solution at one time is less likely. In general, organic fertilizers are much more expensive than chemical fertilizers on the basis of the amount of nutrients supplied. Additionally, because decomposition is slow in cold soils, organic fertilizers are not reliable fertilizers in most areas during winter and early spring.

Inorganic fertilizers are manufactured using raw materials such as natural gas and phosphate rock. They are much more concentrated than organic fertilizers. Some inorganic fertilizers are formulated to release nutrients rapidly so that rapid plant response to the fertilizer occurs. Others release nutrients slowly over a period of time. These two types of fertilizers are described as *fast release* and *slow release,* respectively. Synthetic organic fertilizers such as urea generally react the same as inorganic fertilizers though they are more concentrated.

Relatively recently, new products that seek to combine the soil-building attributes of organic fertilizers with the relatively concentrated fertilizer analysis of chemical fertilizers have become available. These products combine a concentrated organic base, rich in humic acid, with chemical fertilizers, often in slow-release form. Because they are rich in humic acid, they promote aggregation of clay soils without having to add the quantity of organic matter necessary with other organic materials. Their concentrated fertilizer analysis means that sufficient fertilizer is available to the plants. Thus these products overcome the disadvantages of chemical fertilizers by building the soil and of organic fertilizers by containing sufficient fertilizer.

To understand the difference between organic and chemical fertilizers, it is helpful to understand the *nitrogen cycle,* defined as the series of transformations that nitrogen undergoes in the environment (Figure 5-19).

Nitrogen comprises about 70 percent of the air. However, plants, with the exception of those in the legume family and a few grasses, cannot use this atmospheric nitrogen. Legumes (beans, peas, peanuts, and others), with the aid of bacteria that live on their roots, are able to enrich the soil by "fixing" atmospheric nitrogen found in the air spaces in the soil.

Most plants can use nitrogen only when it is present in the soil in either the nitrate (NO_3-) or the am-

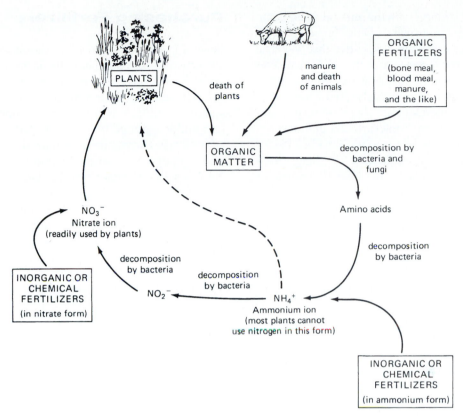

Figure 5-19 The nitrogen cycle, showing the breakdown of nitrogen-containing compounds from organic matter so that nitrogen can be used by plants.

monium ($NH4^+$) chemical form. Though some plants use nitrogen in the ammonium form effectively, most garden plants use nitrogen primarily in the nitrate form. This means that a considerable amount of decomposition must occur before nitrogen (which is added in the form of leaves or blood meal, for example) moves through the decomposition cycle to the point that nitrates are available. This accounts for the slow-release characteristic of organic fertilizers.

If a chemical fertilizer such as ammonium nitrate is added to the soil, part of the nitrogen is already in the nitrate form and part is in the ammonium form. The portion in the nitrate form can be used immediately by the plant. Depending on the plant, the ammonium portion can be either used immediately or converted into nitrates through the decomposition process. This accounts for the rapid plant response to ammonium nitrate and the possibility of burning plants due to high concentrations of usable nitrogen.

It is important to realize that whether the fertilizer is added in an organic form, such as animal manure, or in the form of a chemical fertilizer, such as ammonium nitrate, the plant absorbs the nitrogen primarily as nitrates. The only difference is that the nitrogen is added

at a different point in the nitrogen cycle. The main disadvantage of chemical fertilizers is that if organic matter is not added to the soil (by compost, natural plant residues, or other means), the amount of organic matter in the soil will decline, causing a worsening of physical characteristics of the soil.

Fertilizer Forms

Fertilizers are usually available in several different forms—liquids, soluble powders, granules, and tablets.

Liquids are popular primarily because of their ease of application (usually with a hose-end sprayer) and the quick response obtained. These fertilizers often contain micronutrients so that through foliar absorption, micronutrient deficiencies can be alleviated. Because all the nutrients are in a water-soluble form, it is easy to burn plants if too much fertilizer is used. In addition, much of the nitrogen will leach away if heavy rainfall or irrigation occurs after fertilization, and the benefits of the fertilizer will be short-lived.

Soluble powder fertilizers are popular because they dissolve readily in water and can be applied in the same manner as a liquid fertilizer with the same advantages and disadvantages. They are usually a better buy than liquid fertilizers. They often contain a blue dye that will stain hands and clothes but serves to alert the applier who uses a hose-end sprayer that the sprayer is working properly.

The most common and widely used form of fertilizer for both commercial and home use is the granular form. During the manufacture of a granular fertilizer, the nutrients are compressed into bead-sized particles that are heavy enough to eliminate problems of blowing. Moreover, because the nutrients need not be immediately soluble in water (as in liquid fertilizers), slow-release formulations are available to prolong the release of nutrients and decrease the risk of burning.

Tablet and spike-shaped fertilizers, which have long been available for house plants, are becoming increasingly popular for outdoor use by gardeners. Though more expensive than granular forms, they do not have to be measured; one or two tablets are simply dropped in the planting hole. Additionally, tablets release their nutrients slowly over a period of several months up to a year, thus minimizing waste and burning of the root system. They are designed for use in planting holes or pots and do not work well when applied to the soil surface.

Purchasing Fertilizers

Several factors are involved in determining which fertilizer is the best buy: analysis, form, organic or inorganic, slow or fast release, and price.

The analysis of a fertilizer is a statement of the quantity and type of nutrients contained in the fertilizer and is required by law to be printed on every bag of fertilizer. The standard form for the statement of fertilizer analysis is a set of three numbers separated by hyphens, for example, 5-10-20. Each number refers to the percentage of a particular nutrient contained in the fertilizer (Figure 5-20). The first number always refers to the percentage of nitrogen, the second number to the percentage of phosphorus in a chemical form called *phosphorus pentoxide* (P_2O_5), and the third number to the percentage of potassium in a chemical form called *potash* (K_2O). Thus, for the sample analysis above (5-10-20), the fertilizer contains 5 percent nitrogen, 10 percent phosphorus compound, and 20 percent potassium compound. Another way of stating this is that the fertilizer contains 5 pounds of nitrogen, 10 pounds of phosphorus compound, and 20 pounds of potassium compound in every 100 pounds of fertilizer.

Fertilizers are often grouped into three categories based on the relative amounts of nitrogen, phosphorus,

Figure 5-20 Three bags of fertilizer showing the analysis in large numbers. *(Photo courtesy of Vocational Education Productions, California Polytechnic State University, San Luis Obispo, CA)*

and potassium they contain. The first group is called *balanced fertilizers* because all the numbers in the analysis are identical (8-8-8). The second group, called *complete fertilizers,* contains nitrogen, phosphorus, and potassium but in varying amounts (5-10-15). The last category, called *single element fertilizers,* contains only one of the three major nutrients such as urea (46-0-0).

The analysis is the first item to look at in buying any fertilizer, because nutrients are the important part of the fertilizer. Other ingredients may be present that make the volume of fertilizer larger and make it appear to be a better bargain. These filler materials may be useful as soil amendments but cannot be considered a good buy when purchased at the price of fertilizer.

Consider both the total quantity of nutrients (by adding the numbers) and the relative amounts of nitrogen, phosphorus, and potassium. If you are buying a lawn fertilizer where the primary nutrient needed is nitrogen, the nitrogen analysis is a prime consideration. For most, a balanced fertilizer that contains about equal proportions of nitrogen, phosphorus, and potassium should be chosen.

Next look at the form of the fertilizer. It is difficult to compare liquid pricewise with either granular or insoluble powder forms because of the weight of the water; however, the latter two can be compared. Powdered fertilizers should be distinctly less expensive than granular forms. If a substantial difference in price does not exist, there is little justification for purchasing a powdered fertilizer.

Whether a fertilizer is organically derived may be an important consideration if you are committed to the exclusive use of organic materials. If an organic fertilizer is selected, you should be prepared to pay a much higher price for the same quantity of nutrients and should not expect a high fertilizer analysis. This is because, by definition, an organic fertilizer will contain bulky organic material in which the nutrients are contained, and shipping costs will often be a substantial component of the price. Either fast- or slow-release fertilizers can be purchased in inorganic form, but the slow-release types will be more expensive. Whether or not they are worth the added expense depends on how they are to be used.

Price often becomes the primary consideration in the purchase of fertilizers for home use. In order to determine the best buy, all of the factors discussed above should be weighed relative to the price.

Fertilizer Application

Topdressing

The method of fertilizer application will vary according to the type of plant being fertilized, but the basic rule is that the fertilizer must reach the root zone. For shallow-rooted plants such as turfgrasses, groundcovers, annual flowers, and some shrubs, surface applications, called *topdressings,* followed by rainfall or irrigation will be sufficient to supply nutrients to the root zone. When surface applications are used, you should broadcast the fertilizer around the plant, being careful to remove fertilizer that collects around the stem or upon the leaves.

Preplanting Incorporation

Fertilizers can also be incorporated into the soil along with soil amendments. This *preplant incorporation* method of application has the advantage of getting phosphorus into the root zone of the plants.

Sidedressing

Sidedressing is a form of surface fertilizer application useful on rows of plants such as those found in vegetable gardens and some landscape plantings. Using this method, a narrow band of fertilizer is applied along one or both sides of the row (Figure 5-21) over the root area of the plants but away from the crown, where it could cause burning injury by contact. This banding makes fertilizing long rows of plants relatively easy and avoids the sloppiness and fertilizer waste that can occur with topdressing. It also minimizes contact of the fertilizer granules with the soil and thus helps prevent susceptible nutrients such as phosphorus from reacting with soil components.

For more deeply rooted plants such as large shrubs and trees, surface applications are not effective, because

Figure 5-21 Sidedressing a row of peas. *(Photo by Rick Smith)*

Figure 5-22 A liquid needle feeder. *(Photo courtesy of A. M. Leonard Inc., Piqua, OH. (800) 543-8955)*

Figure 5-23 The drill hole method for fertilizing trees.

the roots are located 18 inches (45 centimeters) or more below the soil surface; most of the phosphorus and a large amount of the potassium will never leach down to the roots. In addition, if surface fertilizer applications are used routinely, surface roots will be encouraged and these will reduce the plant's ability to survive drought.

Needle Feeding

Two methods are commonly employed to place fertilizer in the root zone of trees. The first is the use of a feeding needle, often called a *root feeder*. This apparatus (Figure 5-22) attaches to a garden hose and contains a fertilizer chamber and a pipe with a pointed tip. When the water is turned on, it passes through the fertilizer chamber, dissolving the fertilizer, through the pipe, and out the tip. The water passing through the tip aids the operator in pushing the pipe to the desired depth in the soil. This method is easy and provides both water and available nutrients to the root zone when injected at several points over the root zone of the tree.

Drill Hole

Another effective technique that does not involve the purchase of any special equipment is the *drill hole method*, whereby holes are pounded 12 to 18 inches

(30–45 centimeters) deep into the ground throughout the root zone of the tree (Figure 5-23). These holes are then filled with a granular fertilizer according to the recommended rate and covered with peat moss or topsoil. To be effective, 10 to 20 holes for each inch (2.5 centimeters) of trunk diameter must be made throughout the root zone of the tree. Holes may be made with a crow bar, reinforcing bar, pipe, or auger.

Fertilizer spikes are a modification of the drill hole method. The same results can be obtained using the drill hole method and at less expense.

Foliar Feeding

Another method of supplying small quantities of nutrients to plants is by means of *foliar feeding*. Diluted water-soluble fertilizers are sprayed on plant foliage, and the plant then absorbs small quantities through the leaves.

Because leaves absorb only small quantities of nutrients, this method is best reserved for micronutrients. The advantage of foliar fertilization is that if deficiency symptoms are obvious, improvement will be evident in a matter of days. In addition, where the micronutrient is likely to become unavailable in the soil, soil application can be supplemented. As a general rule, foliar fertilization does not supply enough nutrients to completely satisfy requirements and should be used in addition to regular fertilizer that is added to the soil, or sufficient liquid fertilizer should be used to wet the soil.

SELECTED REFERENCES FOR ADDITIONAL INFORMATION

Biondo, Ronald J., and Charles B. Schroeder. *Introduction to Plant and Soil Science and Technology*. Danville, IL: Interstate, 2000.

Black, Charles A., and Skip DeWall. *Soil Fertility Evaluation and Control*. Elkins Park, PA: Franklin Book Co., 1993.

Bush-Brown, J., and L. Bush-Brown. *America's Garden Book*. Rev. ed. New York: Hungry Minds, Inc. 1996.

Cameron, K., and R. G. McClaren. *Soil Science: Sustainable Production and Environmental Protection*. 2nd ed. New York: Oxford University Press, 1996.

Foth, H. D., and Boyd G. Ellis. *Soil Fertility*, 2nd ed. Boca Raton, FL: CRC Press, 1997.

Harpstead, Milo I., Thomas J. Sauer, and William F. Bennett. *Soil Science Simplified*. 4th ed. Ames, IA: Iowa State University Press, 2001.

Hausenbuiller, R. L. *Soil Sciences Principles and Practices*, 3rd ed. Dubuque, IA: W. C. Brown, 1985.

Leeper, G. W., and N. C. Uren. *Soil Science: An Introduction*. Melbourne: Melbourne Univ. Press. 1993.

Plaster, Edward J. *Soil Science*. Albany: Delmar Thomson, 1997.

Singer, Michael J., and Donald N. Munns. *Soils: An Introduction*. Upper Saddle River, NJ: Prentice-Hall. 2001.

World Wide Web Addresses

http://www.ar.wroc.pl/~weber/humic.htm   Weber, Jerzy. "Soil Humic Substances."

http://www.compostinfo.com/ University of Florida. Florida's Online Composting Center. 2001.

http://dir.yahoo.com/Science/Agriculture/Sustainable_ Agriculture/. Yahoo.   "Directory of Sustainable Agriculture Links." 2001.

http://ext.usu.edu/wsare/docs/basics.htm   This is a reference on sustainable agriculture.

http://www.eosc.osshe.edu/peers/lessons/soils.html. Peers Site Map. "Soils."

http://www.epa.gov/compost/   Environmental Protection Agency. "Composting". 2001.

http://www.esf.edu/pubprog/brochure/soilph/soilph.htm. Bickelhaupt, Donald. "Soil pH: What It Means". 1993.

http://www.extension.uiuc.edu./~robsond/solutions/ horticulture/soils.html   University of Illinois. "Horticulture Solutions Series. Soils and Fertilizers".

http://koning.ecsu.ctstateu.edu/plant_physiology/minerals. html   Koning, Ross E. 1994. "Mineral Requirements". *Plant Physiology Website*.

http://www.metla.fi/info/vlib/soils/old.htm   Heiskanen, Juha, "Soils and Substrates." 2000.

http://www.rain.org/~sals/Petrick.html   2001. Petrik Laboratories, "Understanding the Soil Process"

http://www.sc2000.net~czaremba/aplabs/plantpig.html Zaremba, Charles. Lab exercise for photosynthesis.

http://www.sc2000.net/~czaremba/explanations/photo. html   Zarwamba, Charles, "PHOTOSYNTHESIS".

http://www.sc2000.net/~czaremba/explanations/Plantnotes. html   Zaremba, Charles, Plant Structure and Growth. 2001.

http://www.soil.ncsu.edu/extension/pubs.php.   North Carolina State University, "NC State University Soil Science" (A list of North Carolina Extension Publications dealing with soils)

http://www.soils.wisc.edu/barak/soilscience326/listofel. htm.   Barak, Philip. Essential Elements for Plant Growth. 1997.

http://www.statlab.iastate.edu/soils/nssc/educ/Edpage.html. USDA - Natural Resources Conservation Service, "National Soil Survey Center Soil Science Education Website" 2001.

http://www.urbanext.uiuc.edu/ecosystems/teacherguide4. html   University of Illinois Extension. "Soils and Ecosystems Teacher's Guide".

http://www.urbanext.uiuc.edu/ecosystems/teacherguide5. html   University of Illinois Extension. "Sustainable agriculture".

http://www.uog.edu/cals/site/users/soil/soil/fertft1a.html Motavalli, Peter, Thomas Marler, Frank Cruz and James McConnell Fertilizer Facts - Essential Plant Nutrients. 2001.

(Photo by the author)

CHAPTER 6

VEGETABLE GARDENING

Vegetable gardening is one of the most rewarding activities in home horticulture. It combines the enjoyment of cultivating plants and watching them grow with the pleasure of harvesting fresh produce at the peak of flavor and maturity. Most people will argue that "home grown" tastes much better than "store bought," and there are several reasons why. Vegetable cultivars grown on a commercial scale must have certain characteristics including the property of shipping well without bruising or becoming overripe. Therefore many commercial vegetable cultivars are bred to be less juicy. They must be picked and shipped while not completely ripe in order to reach the consumer undamaged, and this frequently diminishes the ultimate flavor quality. Secondly, from the moment a ripe vegetable is picked, its quality will normally begin to slowly deteriorate. In some cultivars of sweet corn, for example, sugar begins to turn to starch almost immediately, and the flavor of the vegetable can be completely altered in less than two hours. Although not always the case, the general rule of "the fresher, the better" is fairly reliable.

The economic benefits of vegetable gardening for saving on food bills have been widely acclaimed. Studies quote that a vegetable plot of a given size will supply all the vegetable needs of a family for a year. Although this is true with intensive, systematic cultivation and conscientious use of all vegetables as they become available, for most families it will not hold true. It is more reasonable to regard a vegetable garden as a money saver with recreational value that provides superior quality vegetables harvested and eaten at their prime.

TYPES OF VEGETABLES

Just like all other plants, vegetables are either annual, biennial, or perennial. Most fall in the annual group including peppers, squash, and beans. Others are technically biennials but are treated as annuals because only the vegetative parts are eaten. These include vegetables such as beets. Still fewer are perennials, including asparagus and rhubarb. Surprising to many people is the fact that the tomato is a perennial, but it is not frost-tolerant and is therefore treated as an annual. Table 6-1 lists the common garden vegetables and their classifications as annuals or perennials.

More important than whether a vegetable is annual or perennial is its classification as either *warm season* or *cool season* (Table 6-1). Warm-season vegetables thrive at daytime temperatures ranging from 65 to 90°F (18–32°C) and at nighttime lows not less than about 55°F (13°C). At lower temperatures they grow slowly. Some, like tomatoes, fail to develop fruit, whereas others, like peppers, develop fruit but produce only small fruits which are not fleshy or full. These vegetables will not live through a frost.

The opposite is true with cool-season vegetables. They will tolerate light frost and grow best at daytime temperatures from 50 to 65°F (10–18°C). In warmer weather, quality is often poor, for example, lettuce may become bitter. Some cool-season vegetables will flower when leaves alone are wanted, and this too is undesirable. The flowering, called *bolting*, is a response to the shortening of the nights and the warmer weather during the summer.

Table 6-1 Garden Vegetable Cultivation

Vegetable	Cool/ warm season	Grown as annual, perennial	Edible portion	Moderate planting for family of four	Distance between plants	Distance between rows
Artichoke, globe	Cool	Perennial in zones 9-10, annual in others	Flower buds	3 or 4 plants	4' (1.2 m)	5' (1.5 m)
Artichoke, Jerusalem	N/A	Perennial	Tubers	10-15 plants	1' (0.3 m)	5' (1.5 m)
Asparagus	N/A	Perennial	New shoots	30-40 plants	1' (0.3 m)	5' (1.5 m)
Bean, lima	Warm	Annual	Fruit	15-25' (5-8 m) row	6" (15 cm) (bush) 2' (0.6 m) (pole)	30" (0.5 m)
Bean, snap or green	Warm	Annual	Fruit	15-25' (5-8 m) row	3" (8 cm) (bush) 2' (0.6 m) (pole)	2' (0.6 m) 2' (0.6 m)
Beet	Cool	Annual	Leaves, root	10-15′ (3-4.5 m) row	2" (5 cm)	18" (0.5 m)
Broccoli	Cool	Annual	Flower buds	7-10 plants	2' (0.6 m)	3' (1 m)
Brussels sprout	Cool	Annual	Axillary buds	7-10 plants	2' (0.6 m)	3' (1 m)
Cabbage	Cool	Annual	Leaves	10-15 plants	2' (0.6 m)	3' (1 m)
Carrot	Cool	Annual	Root	20-30' (6-9 m) row	2" (5 cm)	18" (0.5 m)
Cauliflower	Cool	Annual	Flower head	10-15 plants	2' (0.6 m)	3' (1 m)
Celery	Cool	Annual	Petioles	20-30' (6-9 m) row	5" (13 cm)	18" (0.5 m)
Chinese cabbage	Cool	Annual	Leaves	10-15' (3-5 m) row	1' (0.3 m)	30" (0.75 m)
Collard	Cool	Annual/perennial	Leaves	10-15' (3-5 m) row	18" (0.5 m)	3' (1 m)
Corn	Warm	Annual	Seeds	4 rows of 20-30' (6-9 m)	6" (15 cm)	3' (1 m)
Cucumber	Warm	Annual	Fruit	6 plants	2' (0.6 m)	4' (1.2 m)
Eggplant	Warm	Annual	Fruit	4-6 plants	2' (0.6 m)	3' (1 m)
Endive	Cool	Annual	Leaves	10-15' (3-5 m) row	10" (25 cm)	18" (0.5 m)
Kale	Cool	Annual	Leaves	10-15' (3-5 m) row	10" (25 cm)	2' (0.6 m)
Kohlrabi	Cool	Annual	Enlarged stem base	10-15' (3-5 m) row	3" (8 cm)	2' (0.6 m)
Leek	Cool	Annual/perennial	Enlarged stem base	10' (3 m) row	2" (5 cm)	2' (0.6 m)
Lettuce	Cool	Annual	Leaves	10-15' (3-5 m) row	12" (0.3 m) (head) 6" (15 cm) (leaf)	2' (0.6 m) 2' (0.6 m)
Melon	Warm	Annual	Fruit	5-6 hills	6" (15 cm)	6' (2 m)
Mustard green	Cool	Annual	Leaves	10' (3 m) row	8" (20 cm)	18" (0.45 m)
Okra	Warm	Annual	Fruit	10-20' (3-6 m) row	18" (0.5 m)	3' (1 m)
Onion, green	Cool	Annual	Leaves	10-15' (3-5 m) row	1" (2-3 cm)	18" (0.5 m)
Onion, bulb	Cool	Annual	Bulb	30-40' (9-12 m) row	3" (8 cm)	18" (0.5 m)
Parsnip	Cool	Annual	Root	10-15' (3-5 m) row	3" (8 cm)	18" (0.5 m)
Pea, English	Cool	Annual	Seeds	30-40' (9-12 m) row	2" (5 cm)	3' (1 m)
Pea, edible pod or Chinese	Cool	Annual	Fruit	30-40' (9-12 m) row	2" (5 cm)	4' (1.2 m)
Pepper	Warm	Annual	Fruit	5-10 plants	2' (0.6 m)	3' (1 m)
Potato	Warm	Annual	Tubers	100' (30 m) row	1' (0.3 m)	30" (0.75 m)
Pumpkin	Warm	Annual	Fruit	1-3 plants	4' (1.2 m)	6' (2 m)
Radish	Cool	Annual	Root	4' (1.2 m) row	1" (2-3 cm)	18" (0.5 m)
Rhubarb	N/A	Perennial	Petioles	2-3 plants	3' (1 m)	4' (1.2 m)
Rutabaga	Cool	Annual	Root	10-15' (3-5 m) row	3" (8 cm)	18" (0.5 m)
Spinach	Cool	Annual	Leaves	10-20' (3-6 m) row	3" (8 cm)	18" (0.5 m)
Squash, summer and zucchini	Warm	Annual	Immature fruit	2-4 plants	2' (0.6 m)	4' (1.2 m)
Squash, winter	Warm	Annual	Mature fruit	2-4 plants	4' (1.2 m)	6' (2 m)
Sweet potato	Warm	Annual	Root	50' (15 m) row	1' (0.3 m)	3' (1 m)
Swiss chard	Cool	Annual	Leaves	3-4 plants	1' (0.3 m)	30" (0.75 m)
Tomato	Warm	Annual	Fruit	10-20 plants	2' (0.6 m) (bush) 1' (0.3 m) (staked)	3' (1 m)
Turnip	Cool	Annual	Root	10-15' (3-5 m) row	2" (5 cm)	18" (0.5 m)

Another factor affecting the culture of vegetables is whether they are grown for roots, leaves, or fruit (Table 6-1). First, fertilizer selection must be considered. Leaf vegetables produce best with nitrogen-rich fertilizer that encourages vegetative growth. Root and fruit vegetables, on the other hand, are best with a fertilizer formula containing less nitrogen. In fact an excess of nitrogen will prevent some vegetables from bearing at all. Second, leaf and many root vegetables can be grown successfully in semishady areas. Fruit-producing vegetables generally require full sun because they are unable to photosynthesize the necessary carbohydrate for flowering and fruiting without bright sunlight and may produce only foliage in shady situations. This fact is important for city dwellers whose gardens are shaded by tall buildings or for other gardeners who find themselves restricted to a shady area. Undue time and effort need not be wasted on fruit crops that will produce no reward. The space can be reallocated to vegetables that will grow more successfully.

CLIMATIC FACTORS

Vegetable gardening is considered a spring and summer activity by most people, with seed sowing in spring and harvest through summer ending at the first killing frost. But in mild-winter areas of the country, vegetable gardening continues all year. In most other areas, it can be extended far beyond the traditional season by choosing the right crops and cultural practices.

Vegetable Gardening in Mild-Winter Areas

The mild-winter areas of North America (see Figure 6-1) are its vegetable bowl. Here most of the fresh vegetables are grown for winter consumption. Gardeners located in one of these areas are fortunate, for they can grow fresh vegetables the entire year and can easily supply all their vegetables from the garden.

 A vegetable crop grower or farm manager is responsible for planting, schedules, monitoring crop growth, supervising pest and disease control, and supervising harvest of one or more crops under his or her supervision. A manager may be paid a share of the profits of the season's crops in addition to a salary.

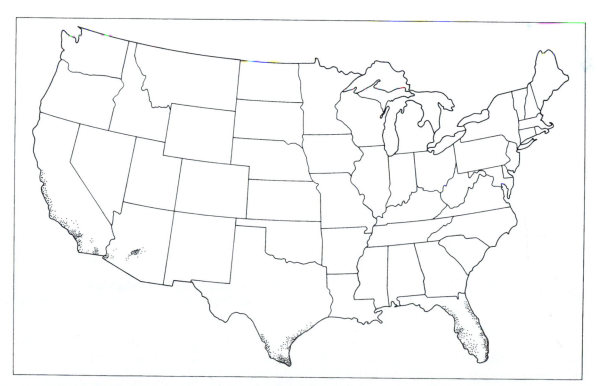

Figure 6-1 Areas of the continental United States where vegetables can be grown throughout the winter.

The climate characteristics of the locale will determine which vegetables can be raised in each season. In the Deep South, for example, winter will be cool and nearly frostless and summer will be very hot. The cool-season vegetables can be raised in winter and the warm-season vegetables in summer when the temperatures exceed the acceptable range for the former. In the Florida Keys, the only tropical area of North America north of Mexico, winter temperatures are moderate (50–75°F, 10–24°C), high enough for warm-season vegetables but not too high for cool-season vegetables. In the summer only warm-season crops could be grown. Along much of the coast of California temperatures are moderated by the Pacific Ocean; there is a minimal difference between winter and summer temperatures. They often fall just short of the acceptable range for warm-season vegetables, and gardeners in this area grow primarily cool-season crops all year.

Vegetable Gardening in Temperate Climates

Most of the United States and Canada have widely varying winter and summer temperatures. Temperatures rise slowly in spring, reach a maximum in summer, and cool slowly as winter approaches. In these climates spring and fall are the best seasons for cultivating cool-season crops, whereas summer is most conducive for raising warm-season crops.

The number of days from the last spring frost until the earliest fall frost defines the period of growing warm-season crops. This *frost-free period*, an important concept in vegetable gardening, varies widely with latitude, ranging from 250 days in the Deep South to 60 days in parts of North Dakota.

The importance of knowing the number of frost-free days lies in its relevance to cultivar selection of warm-season crops. Modern plant breeding offers the home gardener a wide selection of vegetable cultivars bred for taste, canning suitability, and rate of maturity. The *days to maturity* is the number of days from either seed or transplant (depending on the vegetable) to the time when the crop is ready for harvest. For example, there are tomato cultivars that mature in as little as 54 days and others that require 80 or more frost-free days.

The benefits of fast-maturing vegetables to gardeners in the far North are obvious. Even gardeners in more moderate climates value fast-maturing vegetables because they are ready for harvest earlier in the season.

Cool-season vegetables are not restricted to the frost-free period. Not only are they tolerant of light frost, but their flavor may actually be improved by it. Extending the vegetable gardening season in temperate climates is largely a matter of taking advantage of the spring and fall growing periods, which have occasional frosts but are generally favorable for the growth of cool-season vegetables.

Early spring, for example, is fine for sowing peas, lettuce, carrots, and other cool-season vegetables grown from seed. It is also ideal for setting out transplants of broccoli, cabbage, or cauliflower. Although they may appear to grow slowly at first, they will establish their roots and grow faster later in the season. Planting time for cool-season crops starts as soon as the ground is thawed and dry enough to be workable.

Fall Vegetable Gardening

Fall vegetable gardening is a second way to extend the growing season. It involves delaying the planting of cool-season vegetables so that they mature after the rest of the crops have been harvested. Although many of the cool-season vegetables planted in spring are started as transplants, they are usually seeded when grown for a fall crop. This adds six to eight weeks to the "days to maturity" designation of the cultivar and must be factored in when deciding the date to sow seed. Crops such as leaf lettuce and radishes may be started in late summer or even fall, because they mature quickly, but slower crops should be seeded in early summer. The following vegetables are suitable for summer sowing and fall harvest:

Beets	Kohlrabi
Broccoli	Leeks (sow in spring or over wintered seedlings)
Cabbage	Lettuce (leaf and semihead)
Carrots	Mustard greens
Cauliflower	Parsnips (sow in spring)
Chard	Peas, English and edible pod
Chinese cabbage	Radishes
Collards	Spinach
Endive	Turnips
Kale	

Overwintering Vegetables in the Ground for Continuous Harvest

The cold tolerance of some root vegetables comes as a surprise to many gardeners. It is not unusual to discover these vegetables growing through the snow in spring and still good to eat. Occasionally even leaf vegetables such as lettuce and parsley will remain green under the snow cover and renew growth in the spring.

This overwintering phenomenon is due not only to the cold hardiness of the vegetable itself but also to the insulating effects of snow cover and the latent heat of earth (Chapter 20). When these two conditions combine, the soil may remain unfrozen or just at freezing all winter, despite air temperatures reaching far below 32°F (0°C).

It is both possible and practical to make use of these conditions to "store" root vegetables in the ground and extend their harvest through winter. The technique is fairly simple. A section of garden should be set aside for the winter harvest area and planted with root vegetables timed to mature in late summer or fall. These include beets, carrots, Jerusalem artichokes, leeks, parsnips, large radishes (Chinese or icicle-type cultivars), rutabagas, and turnips. In fall the area should be mulched with compost, newspapers, leaves, or any other suitable material. The depth of the mulch needed to prevent freezing will depend on the expected winter temperatures. If winter temperatures only reach 20 to 25°F (−7 to −4°C), a layer of mulch 6 to 8 inches (15–20 centimeters) thick would be sufficient. The depth should be increased, ranging up to 2 feet (60 centimeters), if the temperature will drop below 0°F (−18°C). Continual snow cover on the mulch will provide additional insulation. As a precaution, the winter garden can be located near the house in a protected area to preserve the ground heat.

Harvest of the vegetables can begin any time. The mulch should be removed, the roots harvested, and the mulch replaced each time.

PLANNING A VEGETABLE GARDEN

Choosing the Area

A potential vegetable area should have both fast-draining soil for vigorous root growth and full sunlight to encourage maximum growth rate and flowering. Accordingly, the garden should be situated away from trees and buildings. A previously gardened area will probably possess better soil than one that has been planted in grass, but any soil could be improved to a suitable quality by adding amendments, as discussed in Chapter 5, or by green manuring, as described later in this chapter.

Determining the Size of the Garden

For the inexperienced gardener, a small plot is preferable to a large one. After the first gardening season, the amount of work involved and the number of vegetables produced will be known. The gardener can expand or decrease the size accordingly. A 25 × 25-foot (8 × 8-meter) plot will enable a first-time gardener to grow most popular vegetables except for large space consumers like melons and potatoes.

Garden Layout

The first tasks are to decide which vegetables will be grown and to lay out on paper where they will be located in the garden. Rows should be planted running east and west, if possible, to prevent shading. Taller or trellised crops should be placed north with shorter crops to the south. Figure 6-2 shows a typical layout for a plot producing many different kinds of vegetables.

It is often difficult to estimate how much of each vegetable to plant; Table 6-1 gives an average planting for a family of four for each vegetable. When using these estimates, it is best to divide the amount to be planted into two or three parts, and stagger the sowing of the seed so that the harvest is continuous throughout the season. This practice, called *succession cropping*, will assure a steady supply of vegetables instead of an overabundance at one time.

For example, two leaf lettuce sowings made two weeks apart in spring will provide lettuce into summer. A third sowing in late summer will yield a fall crop, giving a total of three successive crops of lettuce. Succession cropping requires careful management of the garden space. It entails more work than does the straight "plant and harvest" method, but the crop rewards are greater throughout the season. Almost all vegetables with the exception of the slow-maturing warm-season crops can be succession cropped.

Intensive or Block Gardening

In intensive or block gardening (Figures 6-3, 6-4), vegetables are grown in blocks rather than rows. It was the original form of gardening practiced before the plow was invented and is still popular in countries where land is scarce. In the United States and Canada block gardening is useful in cities because it gives maximum yield from minimum space.

For block planting, a garden should consist of squares or rectangles no more than 4 feet (1.2 meters) across to keep the plants within reach of the path. Seed is sown by scattering it evenly over the area or in row fashion.

A variation of the original block garden involves growing vegetables in raised beds. The soil in these beds is thoroughly conditioned and fertilized to grow superior-quality crops.

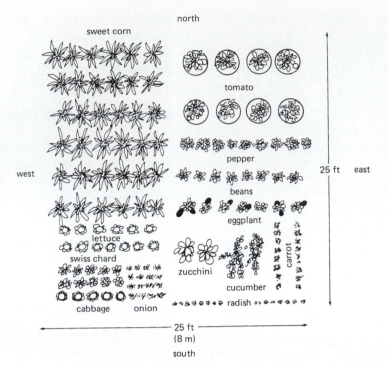

Figure 6-2 A typical vegetable garden plan.

Figure 6-3 A block garden made up of four blocks.

118

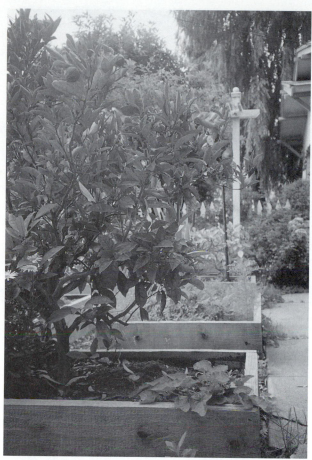

Figure 6-4 A block garden made with raised beds. *(Photo by the author)*

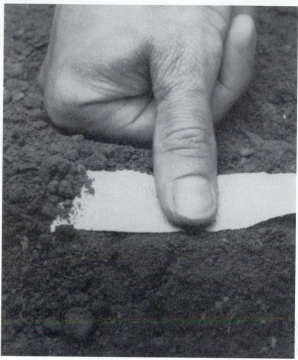

Figure 6-5 Using seed tape to plant a row of vegetables. *(Photo courtesy of W. Atlee Burpee and Co., Warminster, PA)*

Vegetable Cultivar Selection and Seed Purchasing

The Cooperative Extension Service of most states yearly evaluates new and existing vegetable cultivars to determine the best ones for the climate of that state. A list of the recommended cultivars is then compiled, published, and made available to the public. This list is probably the single most important information source on vegetable cultivars a gardener can have.

Seed catalogs list numerous cultivars and their merits but say little about their drawbacks. The gardener must evaluate the suitability of each cultivar to his or her needs and climate, noting days to maturity, resistance to disease, and other characteristics. It is wise to use catalogs from companies located in a climate similar to one's own so that the cultivars that are offered are geared to similar climates. For example, southern gardeners should avoid most catalogs from northern companies. It would be un-

fortunate to pick a cultivar by the picture, only to find it a poor performer under local growing conditions.

Many people buy their seeds from racks in groceries or nurseries. If so, the packages should be examined for the year in which they were packaged for use. If the stamped date is not the current year, the seed is old and may germinate poorly. Moreover, a seed rack may have a limited selection of cultivars, all or none of which may be recommended for the area.

Seed tape (Figure 6-5) is an expensive way to purchase seeds but will save labor. The seeds are sandwiched between layers of plasticlike material that dissolves when the row is irrigated.

Overbuying vegetable seed is a common mistake and results in leftover seed at the end of the growing season. Most vegetable seeds can be kept for the following season by storing them in the refrigerator (see "Seed Storage," Chapter 4). Table 6-2 lists the relative keeping quality of various vegetable seeds.

Ornamental Vegetable Cultivars

Vegetables can be used not only for eating but also as ornamentals (Figures 6-6 and 6-7). Leaf lettuce, for example, makes an attractive flower border. Cultivars of

Table 6-2 Approximate Life of Vegetable Seeds Stored Under Cool Conditions

Vegetable	Years
Asparagus	3
Bean	3
Beet	4
Broccoli	5
Brussels sprout	5
Cabbage	5
Carrot	3
Cauliflower	5
Celery	5
Chard, Swiss	4
Chinese cabbage	5
Collard	5
Corn	1-2
Cucumber	5
Eggplant	5
Endive	5
Kale	5
Kohlrabi	5
Leek	3
Lettuce	5
Muskmelon	5
Mustard	4
Okra	2
Onion	1-2
Parsley	2
Parsnip	1-2
Pea	3
Pepper	4
Pumpkin	4
Radish	5
Rutabaga	5
Spinach	5
Squash	5
Tomato	4
Turnip	5
Watermelon	5

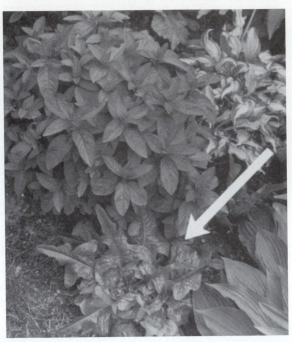

Figure 6-6 Dark green oakleaf lettuce used as a border in a perennial flower bed. *(Photo by George Taloumis)*

Figure 6-7 A potted pepper (right) mixes in with pots of marigolds. *(Photo courtesy of National Garden Bureau, Downer's Grove, IL)*

vegetables selected for their colorful flowers, foliage, or fruits are available through seed catalogs. A few are listed in Table 6-3.

Patio pot gardens are another ornamental use of vegetables. Ornamental peppers and chard grow well in containers. Cherry tomatoes can be planted in hanging baskets, and runner beans trained on strings to screen a porch or balcony. By using vegetables as ornamental plants, the gardener with space restrictions can also make better use of a limited area of land.

Soil Preparation

The time spent in soil preparation will depend on whether the area was previously gardened or not. If it was and the soil condition is good (not packed down, for example), little preparation will be needed except for pulling or killing weeds and mixing in fertilizer before planting, called *preplanting incorporation*. Preincorporation replaces nutrients used by the previous season's crop and positions a fresh supply in the root zone. Chapter 5 gives directions on preincorporating fertilizers.

Table 6-3 Ornamental Vegetable Varieties

Vegetable	Ornamental species or cultivars	Remarks
Artichoke	*Cardunculus* species (common name, cardoon)	Large gray leaves and thistlelike purple flowers; petioles and roots are edible
	Scolymus species (common artichoke)	Purple flower heads, gray foliage; smaller plant than cardoon
Artichoke, Jerusalem	Any	Mature plants are 6 ft (2 m) or more tall with yellow sunflowers
Asparagus	Any	Feathery foliage and red berries in fall
Bean, bush	'Royalty'	Purple pods
Bean, pole	'Scarlet Runner'	Red flowers; shade- and cool-temperature-tolerant
Cauliflower	'Purple Head' and 'Royal Purple'	Heads tinged lavender at maturity
	'Greenball'	Head green at maturity
Chard	'Rhubarb'	Red petioles
Corn	Indian corn, 'Rainbow'	Red, yellow, and white kernels
	Strawberry corn	Small, strawberry-shaped ear with red kernels
	'Gracillima'	Dwarf size
	'Japonica'	Leaves striped white and pink
Cabbage, ornamental	Park's Super Cole Hybrids	Ruffled leaves are lavender, pink, and green; does not head
Kale, flowering	Park's Super Cole Hybrids	Ruffled leaves, red- or green-tinged
Lettuce, leaf	'Oakleaf'	Dark green lobed leaves
	'Red Salad Bowl'	Red-tinged leaves
	'Ruby'	Red leaves
Pepper, bell	'Golden Calwonder'	Yellow fruits
	'Bell Boy Hybrid'	Red fruits
	'Midway'	Red fruits
Pepper, hot	'Fiesta'	Red, yellow, and green fruits
Eggplant	'Morden Midget'	Deep purple, miniature fruit
Sunflower	No cultivars suggested	Enormous yellow flowers on 6- to 12-ft (1.8- to 3.6-m) stems; edible seeds excellent when toasted
Tomato	'Patio'	Dwarf plant to 1 ft (0.3 m) tall with golf-ball-sized tomatoes
	'Small Fry'	Vining plant producing large quantities of cherry tomatoes; good in a hanging basket
	'Tiny Tim'	6-in. (15-cm)-high plant with tiny tomatoes
	'Goldie'	14-in. (35-cm) tall plant produces 1 in. (3 cm) plum-shaped yellow fruit
	'Minibel'	A cascading growth habit for hanging basket culture and 2-in.- (5-cm-) diameter red fruit

Annual weed seedlings can be controlled by simply *scraping* the soil surface with a hoe, provided the seedlings are still small enough to be cut off at soil level. The area should not be dug to eliminate annual weeds, as this perpetuates the problem by exposing new seeds and encouraging them to germinate. Perennial weeds that have come back from the roots can be dug out individually, if they are few, or treated with a weed killer if they are numerous. A local nursery can recommend a chemical to control existing perennial weeds, but weed samples should be brought to the nursery to be sure of selecting an effective chemical. The nursery also needs to know that the area will be planted as a vegetable garden so that a chemical with a short effective period (called its *residual life*) can be selected and not affect the crops. If the area was previously a lawn, a typical situation, the sod must be removed, although as garden preparation is begun in fall it can be simply turned under and allowed to rot in place. In either case eliminating the grass with a weed killer will be necessary if the lawn contains vigorous grass species that spread by rhizomes, because it is nearly impossible to remove all rhizomes by hand and those left in the soil will resprout continually and create a weed problem.

After sod removal it may be necessary to mix in organic matter to improve the structure of the soil, to "loosen it up." Nurseries sell rotted manure, bark products, and various other soil conditioners for this purpose.

PLANTING THE GARDEN

Sowing Seed

With the exception of block gardening, vegetables grown from seed are planted in rows or hills. When sowing in rows, each row should be marked with a labeled stake and the seed sown thickly and in accordance with package directions. Heavy sowing assures a sufficient stand of plants in case of poor germination.

 Another career available to those interested in growing vegetable crops is self-employment in a growing and roadside marketing business.

With slow-germinating seeds such as carrots, a fast-germinating seed is sometimes planted in the same row to mark its location until the other seedlings emerge. In soils that tend to crust, the rapidly germinating seeds also break the surface to ease the emergence of less vigorous crops. Radishes germinate in 3 days and are often used for row marking. They can be pulled as soon as the main crop emerges but are usually left in place. Because radishes mature in as little as 21 days, they can be harvested in time to prevent crowding of the main crop. This technique, called *interplanting*, saves space and is frequently done with small, fast-maturing crops such as radishes and leaf lettuce. A full crop of leaf lettuce, for example, can be raised by seeding between transplants of broccoli or cauliflower (Figure 6-8).

Directions on packages of squash and vining vegetables like cucumbers normally specify planting in *hills*. *Hill planting* refers to grouping several plants together, not planting seeds in a mound of soil.

After seeding it is important to keep the surface of the soil moist. Drying of the soil surrounding the germinating seeds will kill the emerging seedlings and necessitate resowing.

Buying Transplants

Selecting well-grown transplants contributes to the success of the garden, and quality should not be sacrificed in favor of price. The plants should be short, sturdy, and with foliage to the base. Those with yellowed foliage or bare stems should be avoided because they have been growing too long in the transplant container and are likely to be root-bound and slow to begin growth. In some cases they will flower prematurely and there will be no crop because the plants are not large enough to bear fruit.

Setting Out Transplants

If possible, an overcast day or early evening should be selected for transplanting to lessen shock. Watering the plants before removing them from the pack is advisable.

If the plants were grown in individual cells of a plastic pack, there will be very little root shock during transplanting. However, if after removing a transplant, it is found to be root-bound, the root ball should be cut shallowly on each side (Figure 6-9). The cutting will force branching of the roots into the soil. If uncut, the roots may continue growing within the original soil ball, and establishment will be slow.

Transplants grown in pots of compressed peat moss can be transplanted with the pot intact, but the top of the pot should be broken off down to the soil. If the rim is not broken and is exposed to the air, it will wick water from the peat moss pot, restricting the penetration by the roots and slowing growth.

Figure 6-8 A crop of leaf lettuce planted between broccoli plants.

Figure 6-9 Cutting through the wrapped roots of a pot-bound transplant. *(Photo by Rick Smith)*

Figure 6-11 A row of beets before thinning (on the left) and after thinning (on the right). The thinnings are in the foreground. *(Photo courtesy of National Garden Bureau)*

Vegetable transplants should be planted slightly deeper than they were originally grown. Tomatoes should be set with much of the stem below the soil surface (Figure 6-10) because they will form adventitious roots along the submerged stem. These stem roots create a larger root system and a more vigorous plant.

After planting, the transplants should be watered with starter fertilizer as discussed in Chapter 5. The watering assures full soil contact with the roots and provides maximum soil moisture to prevent shock.

VEGETABLE GARDEN MAINTENANCE

Thinning

Thinning is the removal of excess seedlings that are spaced too closely for best growth (Figure 6-11). Thinning can be done either once or twice for each crop.

One-time thinning should be done as soon as the leaves of neighboring plants touch and should remove as many seedlings as necessary to achieve the recommended final spacing for the vegetable.

Twice-over thinning is practical for vegetables grown for their leaves, such as chard, lettuce, and Chinese cabbage. The first thinning is done while the plants are seedlings, but it removes only enough plants to prevent severe overcrowding. The second thinning takes place 2 to 4 weeks later and thins the remaining plants to the final spacing. By the second thinning, plants are large enough to be harvested and eaten.

Weeding

Weed control is very important to a successful vegetable garden. The detrimental effect of weeds on plant growth is masked, showing up as a slowed plant growth rate due to the competition of the crop with the weeds for water, nutrients, and light. Many gardeners incorrectly attribute this sluggish growth to poor soil or adverse weather, not knowing the real cause.

Because seedlings have limited root systems, they are especially vulnerable to competition from vigorous weeds within the row. If unchecked, the weeds will eventually crowd out the desirable plants altogether.

Figure 6-10 Deep planting of a tomato encourages stem rooting.

If seedling plants are invaded by weeds, pulling is probably the only control method. Later, hoeing or mulching is effective. Because the roots of vegetable plants are shallow, hoeing for weeds should be a scraping rather than a digging operation. The scraping will cut weed seedlings at the soil line, and if they are annuals, they will not resprout.

Mulching

Mulch is a layer of plant-derived or synthetic material laid on the soil surface over the roots of plants. Mulching can considerably reduce the time spent weeding and will conserve soil moisture. In hot areas of the South and Southwest it helps keep soil from overheating and consequently is beneficial to root growth. Because vegetable gardens are tilled every year, many people find it easier to use plant-derived mulches that will decay in the soil. Compost, leaves, and grass clippings from lawns that have not been recently treated with weed killer are satisfactory.

Mulch should be applied 2 to 3 inches (5–8 centimeters) thick as a general rule. Transplants should be mulched sparingly about 1 inch (3 centimeters) deep when young, gradually increasing the depth to 3 inches (8 centimeters) as the plants grow taller. For low-growing plants such as lettuce and radishes, mulch should be applied on either side of the row at a depth of 1.5 to 2 inches (4–5 centimeters). Vining crops such as melons should be mulched over the entire area where the vine will grow. Covering the area will help keep the fruits clean and reduce losses from rot diseases that occur when the vegetables lie on the soil.

Newspapers held down with soil are another suitable vegetable garden mulch. Although not attractive, they are readily available and inexpensive.

Plastic mulches (Figure 6-12) are very useful in northern vegetable gardens because they absorb heat and radiate it into the soil underneath. They will speed up the growth of warm-season crops by heating the soil and can mature the crop a week or more ahead of normal. Either clear or black plastic mulch may be used. The former will raise the temperature more than the latter; however, weeds will grow beneath it.

For especially early harvests, a plastic mulch can be used with transplants of cucumbers and melons. The time gained by using transplants combined with the warmth provided by the plastic will encourage rapid growth and fruit set.

Plastic mulches can be applied in two ways. For row crops, strips of plastic are laid on either side of the row, with the plants between. For vegetables in hills, the area can be covered and a flap can be cut in the plastic so the shoots can emerge.

Hot Caps and Frost Protection Devices

The use of *hot caps* (Figure 6-13), paper or plastic domes set over plants in early spring, is another method of hastening growth. The caps admit light but insulate against heat loss, keeping the air surrounding the transplants several degrees warmer than the outside air. This slight rise in temperature is often enough to speed growth considerably during early spring. As an added feature, hot caps offer frost protection. They enable gardeners to set out transplants of warm-season vegetables and sow seeds several weeks ahead of the last predicted frost date.

Frost protection of warm-season crops in fall is more difficult because of the size of the plants. If only light frost is expected, a layer of newspapers over the plants will often give adequate protection.

Figure 6-12 Laying a plastic mulch in a commercial field. *(Photo from the USDA)*

Figure 6-13 Hot caps being used to cover transplants in the garden. *(USDA photo)*

Watering

Watering the vegetable garden should begin as soon as seeds are planted and be done as often as necessary to prevent wilting. Do not assume, however, that every time a plant wilts, the garden needs watering. Many times large-leaved squash or cucumber plants wilt repeatedly during the hottest part of the day due to an inability to absorb enough moisture to compensate for the enormous water loss through the foliage, a phenomenon called *diurnal wilting*. However, if they do not recover at night, water is definitely needed. If, after watering, they still do not recover, the plants may be infected by a wilt disease or nematodes.

Watering the vegetable garden should soak the soil to a depth of about 18 inches (0.5 meter). Perennial vegetables such as asparagus and artichokes will need to be irrigated more deeply.

Fertilizing

If fertilizer is worked into the soil of the vegetable garden during preparation, it will often be sufficient to supply the needs of the crops to maturity. On sandy soils, on soils of low fertility, or if deficiency symptoms are noted, a midsummer sidedressing of balanced granular fertilizer may spur crop growth. Nitrogen fertilizers can stimulate leaf production at the expense of the fruit and should be used sparingly after plants have reached mature size.

Training

Peas, runner beans, tomatoes, and cucumbers are vining plants that can be trained on stakes, strings, or wire. Training keeps the fruits from becoming dirty and can lessen the chances of fruit rot by preventing contact of the fruits with the soil. It also optimizes space usage in the garden and makes harvesting easier.

Teepee Training

Teepee training (Figure 6-14) can be used for a group of several cucumber or bean plants. Ideally the teepee should be built first and then seeds planted at the base of each stake. The vines are tied to the stakes at 1-foot (0.3-meter) intervals.

Staking

A single sturdy stake (Figure 6-15) will support one tomato plant or two to three runner beans. Beans will climb without help, but tomatoes should be tied loosely with pieces of cloth at intervals along the stem.

Figure 6-14 Teepee training of cucumbers.

Figure 6-15 Staking is one of the most common ways to train tomatoes.

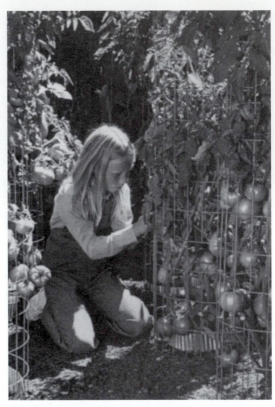

Figure 6-17 Tomato cages. *(USDA photo)*

String Trellises

Runner beans and peas are trained easily on a trellis made of twine strung between poles (Figure 6-16). The poles should be pounded in deeply not more than 10 feet (3 meters) apart to provide a sturdy support for the trellis and vines.

Tomato Cages

Tomato cages using a wire-mesh cylinder either purchased or made at home (Figure 6-17) are an easy way to support tomatoes. Concrete reinforcing wire has an ideal strength and a good mesh size for tomato cages; many types of galvanized fencing (not chain link) are also suitable. It is only necessary that the mesh be sufficiently large to reach through to pick ripe fruit. Tying the vines is not required.

CROP ROTATION

Crop rotation is the planting of crops in different areas of the garden every year. It involves, for example, switching the locations of the beans and cabbage each

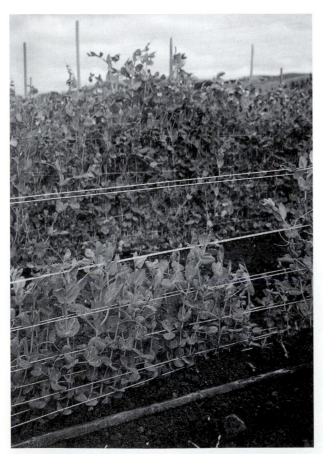

Figure 6-16 A string trellis used to support peas. *(Photo by the author)*

Figure 6-18 Suggested crop rotation for a home vegetable garden.

year. Crop rotation deters the buildup of soil-carried disease organisms associated with particular crops. Although crop rotation is not essential, it is a sound practice that does not require much effort.

The easiest form of crop rotation is to simply reverse the vegetable garden plan each year, being careful to avoid shading problems due to placement of tall plants. The progression shown in Figure 6-18 is another rotation sequence.

COVER AND GREEN MANURE CROPS

Cover cropping is used to maintain fertility in established vegetable gardens, but green manure crops are used to improve poor soil before a garden is planted. In most areas of the country, cover crops are planted in fall over the vegetable garden. They grow during fall and winter and are turned under in spring, supplying organic matter and increasing the fertility of the soil as they decay. Green manure crops perform similarly but are planted at any time of year, grown halfway to maturity, and then turned under to decay.

Green manuring can be used in place of adding organic soil amendments and is much less expensive. The general practice is to sow the green manure crop, wait until it is 8 to 10 inches (20–25 centimeters) high (or until flowering if it is of the pea family), turn it under, wait 10 to 14 days, and then resow another green manure crop within three weeks. The plowing and resowing are repeated as many times as necessary until the soil is of acceptable quality. Members of the pea family such as alfalfa, clover, cowpeas, soybeans, and some cultivars of vetch are among the best green manure and cover crops. They absorb nitrogen from the air and, when turned under to decay, will release that nitrogen into the soil. Winter-growing cover crops include wheat, rye, ryegrass, and buckwheat. Buckwheat is very tolerant of adverse soil conditions and is especially useful in very poor soils.

COMMERCIAL VEGETABLE PRODUCTION IN THE FIELD

Large-scale commercial vegetable production for both fresh market and processing is spread throughout the United States during the summer months, although some states focus on particular crops well suited to their soils and climates. For example, celery requires cool temperatures throughout its growing period. But tomatoes will thrive in even the hottest summer temperatures. Therefore celery is commonly produced in the cooler parts of California near the Pacific Ocean, but tomatoes are grown in the South in summer, among other growing sites.

During the fall, production of vegetables for fresh market for winter shifts to the South and Far West and in the winter, to Florida, and California, with some in Arizona, Texas, and parts of the South. In Canada the main provinces that produce vegetables are Ontario, Quebec, and British Columbia. Greenhouse production of tomatoes and cucumbers is common in many states and in Canada, but few other vegetables are grown in greenhouses.

Major vegetable crops grown in North America in terms of acreage are lettuce, tomatoes, and broccoli.

Vegetables that are to be sold fresh have a higher value than those that will be processed—for example, canned, frozen, dried, or made into products such as soup. This is because the quality demanded by consumers for fresh vegetables is very high. When vegetables are to be processed, there can be small irregularities, mixed sizes, or even small amounts of insect damage because these will not be seen in the finished product. However, fresh vegetables must be completely insect- and disease-free to be accepted by most consumers.

The main vegetables that are sold fresh are lettuce, tomatoes, onions, celery, cabbage, sweet corn, and carrots. The main processed vegetables are peas, potatoes, tomatoes, corn, spinach, and beets. The trend in recent years has been toward more production of fresh vegetables and fewer destined for processing. This reflects emphasis on healthy eating.

Field Growing of Vegetable Crops

A person who decides to become a commercial vegetable producer must first evaluate the land that has been selected to decide if it is suitable for growing vegetables, and which vegetables. Most vegetables give best yields in high-quality rich soils. A soil test followed by adjusting and preparing the soil to optimum quality for a particular crop is essential. The state extension service can assist with soil testing, site selection, and land preparation advice.

As with other commercial enterprises, maximizing production while minimizing cost inputs are the major strategies in economical production. Inputs include seed/seedlings, fertilizers, pesticides, water, and the labor for all these and harvest (unless mechanical harvesting is practiced).

Seedbed Preparation

Seedbed preparation comes next, and this is done just before planting the crop. The goal is to leave the soil level and with numerous pore spaces that will allow oxygen and water to enter freely. How this is done will depend on the type of soil. Leveling will come first. Leveling eliminates uneven drainage problems that can cause irregular growth in the field. If the field is not leveled, there may be areas where water forms a pool after rain or irrigation. Plants that are left standing in water often have damaged roots because the roots were deprived of oxygen while submerged underwater.

Many times the crops are grown on raised beds about 3 feet (1 meter) wide. At this width, workers can access all the plants for hand weeding and harvesting. Figure 6-19 shows the preparation of a raised bed.

Whenever possible the field is plowed when the moisture content is correct, neither too dry nor too wet. This is so that it will have a fine surface texture through which seeds can germinate.

If the soil has been producing crops for several seasons under irrigation and machine traffic for spraying and similar maintenance, it may have developed a *hardpan* (See Chapter 5). The hardpan under the soil must be broken up every few years by *chiseling or subsoiling* at the end of the growing season. If not, drainage will be poor.

Crusting is damaging of the surface soil structure so that there is no aggregation, and the surface becomes smooth, dry, and sometimes water repellent. It is caused by the impact of rain or irrigation water on soil that contains insufficient organic matter to sustain aggregation. The surface aggregates break down. The formation of soil crust is the main reason that soil bed preparation is done only immediately before planting. But after a crust is formed, it is difficult for seedlings to emerge. Soil crusting can be prevented by anticrustants such as dilute sulfuric or phosphoric acids. Organic mulch such as chopped straw can also prevent crusting by breaking the force of the water drops at the same time that it holds moisture in the top layer of soil.

Figure 6-19 A machine for making raised beds for vegetables. *(Photo courtesy of Sustainable Agricultural Machinery, Wodonga, Australia)*

Seeding

Most commercial vegetable seeding is done by machine. This reduces labor costs not only in the seed planting stage but also later, because less hand thinning of the seedlings will be needed for them to be at the optimum spacing for growth. Machines called *rotovators* deliver precise, although slow, planting seeds while in some cases shaping the seedbed, fertilizing, and applying herbicide.

Growing from Transplants

Some vegetables are field grown from seeded transplants, including broccoli, cauliflower, eggplants, and cabbage (Figure 6-20). This allows the growers more control in the critical germination and establishment stages of the crop because the transplants can be grown under the best temperature and irrigation conditions for the specific crop, and not subjected as seedlings to the changes in weather that would occur in the field (See Chapter 4, "Plant Propagation").

Transplants can be put in the field by hand labor, or in some cases by machine transplanters designed to water and fertilize the transplants at the same time as they are planted (Figure 6-21).

Irrigation may be necessary if natural rainfall is not sufficient. Even a slight water stress without visible wilting can reduce growth. Sometimes a high salt content in the soil will make a plant unable to absorb water from moist soil. The green tissues of the plants may become thick but the plant will not wilt. Nonetheless, quality will be reduced (Figure 6-23).

Figure 6-20 A worker loads plug flats of vegetable seedlings. *(Photo by the author)*

Figure 6-22 A commercial vegetable weeder. *(Photo courtesy of Sustainable Agriculture Machinery, Wodonga, Australia)*

Figure 6-21 A commercial planter for vegetable seedlings. *(Photo courtesy of Laanen Plant Systems)*

Maintenance of the Crop in the Field

Several growing activities must be performed while the crop is in production. Cultivation may be necessary four-week control to break soil crust. Depending on the crop, this can be done by hand hoeing or with an implement attached to a tractor (Figure 6-22).

Figure 6-23 A boom system for irrigation passes over plug flats of vegetable seedlings. *(Photo by the author)*

Sometimes the plant roots are damaged because soil is soggy. In wet soil the roots are deprived of oxygen and die. In this case, they die because they are unable to absorb water that the plant needs, and the effect is the same as for a water deficiency due to drought.

In the case of a simple water deficiency due to drought, wilting will be the first visible sign to occur. The cell will lose turgor. Turgor is the rigidity of a leaf that has the maximum amount of water it can contain. When turgor is lost the leaves will wilt, and eventually cell destruction will occur. But even before that happens, lack of water will cause the stomates to stay closed. They do not absorb carbon dioxide, photosynthesis is reduced or stops, and growth stops as well.

The grower must irrigate according to the type of root system that the crop has: shallow, medium, or deep. Table 6-4 lists the rooting depths of some vegetable crops.

The most critical period for water stress will depend on the crop. Foremost is during the period of initial establishment, flowering, and fruit enlargement. With onions and radishes it is during bulb formation. Cole crops can be particularly stressed during head formation.

Irrigation techniques depend on the area, finances, the crop, and many other factors. Use of high- or low-pressure sprinklers is the most common irrigation technique. Sprinkler pipes used to be moved by hand throughout the field, but now they move on wheels down the rows or pivot in a circle.

For water conservation, probes measure the water content of the soil down to 3 to 5 feet (1–1.5 m) below the surface. A probe is calibrated for soil type. With the aid of specially designed computer programs, a report including soil-moisture graphs and recommended next irrigation amounts and hours of operation can be generated, based on the probe data. Precise soil-moisture measurement has led to documented water and energy savings of 10 percent to 20 percent.

Spraying for insects and diseases is essential (except for organically produced vegetables) to assure a flawless crop acceptable to the consumer. See "Pest Control," (Chapter 13).

Table 6-4 Vegetable Crops and Their Rooting Depths

Rooting depth		
Shallow	**Medium**	**Deep**
cole crops	beans	asparagus
celery	peppers	watermelon
corn	summer squash	muskmelon
potatoes	cucumber	pumpkins
radishes	root crops other	winter squash
onions and their relatives	than onions	sweet potato

Harvesting and Post-harvest Handling

Harvesting can be done by hand or by machine. Many vegetables for fresh market must be hand harvested to avoid damage that would make the crop unsalable. But in some cases a combination of machinery and hand harvesting can be used. A "mule train" is a machine that follows workers down the rows as they harvest. The workers place the lettuce or other crop on a conveyor belt where it moves to a receiving unit for packaging in the field.

All potatoes, and all tomatoes and sweet potatoes that will be processed are harvested by machine, as well as some peas and beans. Most carrots, radishes, and beets are also machine harvested, along with pickling cucumbers, some corn, and onions.

Post-harvest handling is the treatment of the crop after it leaves the field but before it is shipped for sale. Post-harvest handling of vegetables includes washing, sometimes spraying with *antidessicant waxes* to prevent wilting before sale, sorting (which may be mechanized), and cooling.

Removal of the "field heat" is an important part of post-harvest handling. Field heat is the warmth that is left in the vegetables from the heat of the sun in the field. This excess warmth must be removed so that the vegetable will reach the correct storage temperature for maximum life. Cooling can be done by a process known as *hydrocooling*, in which blocks of ice are floated in the wash water of the vegetables. *Vacuum cooling* takes place in a refrigerated storage. Cold air is pulled through the boxed vegetables, absorbing the heat and quickly bringing vegetables to proper storage temperature. Humidity and ventilation are also controlled and brought to a level ideal for storing the crop.

CULTURE OF COMMON GARDEN VEGETABLES

Asparagus

Asparagus (*Asparagus officinalis*) is a perennial vegetable grown for its tender sprouts. Some gardeners grow asparagus for its tall, fernlike foliage and ornamental red berries. Sprouts appear early in the spring in temperate climates and throughout the year in tropical areas.

Sandy soils high in organic matter are best, but this vegetable will succeed in any soil. If grown in loam or clay soils, crowns should be planted 5 to 6 inches (13–15 centimeters) deep; in lighter soils, crowns should be planted 6 to 8 inches (15–20 centimeters) deep.

Asparagus can be grown from either seed or 1- to 2-year-old crowns. Seed will not produce an appreciable crop until the third year except in zones 10 and 11 where harvesting may begin the second year. Consequently, growing from crowns is the preferred method.

Selection of an asparagus cultivar should, like other vegetables, be based on the adaptability of the cultivar to local growing conditions. In addition to this basic essential, it is useful to know that recently *all-male cultivars* of asparagus have been developed and can be grown from seed. Normally asparagus is dioecious, and a cultivar is a mixture of male and female plants. However, female plants normally produce a lower yield of asparagus, diverting a part of their energy instead to fruit and seed production. Male plants flower, but don't fruit, so that their surplus carbohydrate can be stored instead in the root system to foster additional vegetative yield in the form of spears. Examples of male hybrid cultivars are 'Jersey Gem,' 'Jersey General,' 'Jersey King,' 'Jersey Knight,' and 'Greenwich.'

Asparagus crowns should be planted 8 to 12 inches (20–30 centimeters) apart in rows in the spring or fall. Harvesting should be delayed until the second year to allow the plants to become established. During this time, plants should be watered regularly and fertilized with a balanced fertilizer. During the second year, spears should be harvested for a two-week period and then allowed to leaf out. In the third year, spears may be harvested until the plants show weakening by producing small-diameter spears or ones that branch close to ground level. Care should be taken not to harvest for too long a period, because this will weaken the crowns and cause a lower yield the following year.

Asparagus spears should be harvested when they are 6 to 8 inches tall (15–20 centimeters)—before they begin to branch. The best way to harvest is to snap the tips by hand (Figure 6-24). The shoots will break at the

point where they become fibrous and inedible. During warm weather, beds can often be harvested daily or even twice daily. If a frost threatens, all emerged sprouts should be harvested regardless of length because they are not frost-tolerant.

Asparagus should be refrigerated immediately after harvest because its quality diminishes rapidly in warm temperatures. Flavor will be retained best under high humidity and at a temperature just above freezing.

 A plant photographer produces the photos seen in gardening magazines, on seed packets, in catalogs, and in books. Magazines that regularly publish articles on gardening frequently have plant photographers on staff.

Beans

The common bean (*Phaseolus vulgaris*) includes many different garden beans known as "snapbeans," "stringless" or "string beans," "wax beans," and "green beans." Beans may be either climbing vines called *pole types* or *runner beans,* or low-growing, bushlike plants called *bush* or *dwarf types.* The edible pod is eaten while immature and should be picked before the seeds inside enlarge. Because quality declines rapidly after harvesting, beans should be refrigerated and kept in high humidity. For best quality they should be eaten within 24 hours.

Beans are warm-season annuals and will not tolerate cool temperatures or frost. Therefore beans should be planted only after the weather has warmed in the spring. Any well-drained soil is acceptable for growing beans. A pH of between 6.5 and 7.0 is best.

Beans should be planted 1 inch (2–3 centimeters) deep and 2 to 3 inches (5–8 centimeters) apart and do not require thinning. If bush beans are grown, staking is not necessary; however, if pole types are grown, stakes should be provided, preferably at planting time. Beans should be sidedressed with a balanced fertilizer prior to flowering if leaves are not a dark green color.

Summer irrigation is required in much of the South and Southwest to prevent blossom drop. This malady is aggravated by hot, dry winds.

Bush beans normally yield sooner than pole beans, but the harvest period will be shorter. Succession plantings should be made two to three weeks apart to prolong the harvest. Although pole beans take longer to begin producing, they will often bear until killed by frost if all pods are kept picked.

Figure 6-24 Picking asparagus by the snap method. (*USDA photo*)

Beets

Beets (*Beta vulgaris* var. *crassa*) are one of the easiest vegetable crops to grow. They can be grown in all climate zones in the summer and in warmer zones as a winter crop. The plants are only about 1 foot (30 centimeters) tall, with red-tinged leaves originating from a large red root. The root is the primary edible portion of the plant. The leaves are also edible; in fact they are more nutritious than the roots.

Beet roots can be harvested and eaten at any time. However, the quality is best before they reach 3 inches (8 centimeters) in diameter. Leaves can be harvested either when the plants are thinned or by picking individual leaves from the best plants. No more than two leaves should be removed from a single plant, or root formation will be retarded. Roots can be stored for three to five months in the coldest part of the refrigerator if tops have been removed.

A light, well-drained soil with a pH above 6.5 is preferred; however, beets will grow in most garden soils. Because beets tolerate light freezes, they should be planted as early as possible in the spring or year-round in mild-winter areas. For most cultivars a "seed" will contain several embryos, resulting in several plants at each location. Thinning is thus necessary to space the plants 2 inches (5 centimeters) apart. If greens are desirable, thinning can be delayed until several 3- to 4-inch (8- to 10-centimeter) leaves have formed. The tenderest beets are produced when plants are growing quickly, so a sidedressing of a balanced fertilizer is desirable four to six weeks after planting. Irrigation should also be provided if rainfall is inadequate. When harvesting roots, pull beets starting when they reach one inch (3 centimeters) so that more room is provided for those remaining.

Broccoli

See "Cole Crops."

Brussels Sprouts

See "Cole Crops."

Cabbage

See "Cole Crops."

Cantaloupe

See "Cucurbits."

Carrots

Carrots (*Daucus carota* var. *sativus*) are an easy-to-grow root crop for all climate zones. They are a cool-season crop that can be grown as a summer crop in all zones and as a winter crop in warm-winter regions.

The best quality carrots are produced in deep loose soils with good drainage. In heavy soils roots will often be crooked or forked (Figure 6-25); short-rooted cultivars are likely to produce the best results in such soils.

Seeds should be planted shallowly in soil that has been well prepared to eliminate clods. Where crusting of the soil surface is a problem, organic matter should be mixed with the soil to cover the seeds.

Carrot seeds are small and slow to germinate. To aid emergence and to mark where the carrots are planted, radish seed can be mixed with the carrot seed. The radish seed will germinate rapidly and break the surface crust as well as mark the location of the carrot row. The radishes can then be removed after the carrots germinate.

It is important to thin carrots to provide adequate room for the roots to develop. Thinning can be done several times to ultimately leave 1½ to 2 inches (3–5

Figure 6-25 Carrots deformed like these are caused by rocky, compacted soil and too much moisture. *(Photo courtesy of Stoke's Seeds, Inc., Buffalo, NY)*

centimeters) between carrots. The last thinning can usually be delayed until the thinnings are pencil-size and can be eaten. Harvesting can then continue until they are between 1 and 2 inches (3–5 centimeters) in diameter. After harvest, carrots should be refrigerated in high humidity. If fully mature, they will keep for four to five months under these conditions. Younger carrots can generally be stored only four to six weeks.

If carrot leaves are a light green, a sidedressing with a balanced fertilizer will improve yield. As with most root crops, fast growth will produce a more tender and tasty root.

Cauliflower

See "Cole Crops."

Chard

Chard (*Beta vulgaris* var. *cicla*) (Figure 6-26) is probably the easiest to grow and the most useful leaf vegetable for the home garden. As the botanical name indicates, chard is actually a beet but is grown for its spinachlike leaves rather than for its roots. The culture

Figure 6-26 Swiss chard. *(USDA photo)*

is the same as for beets. Leaves are harvested individually by cutting the outer ones off at the base as they reach edible size. Chard will produce leaves year-round in mild areas and from early spring until late fall in colder areas. It is a vegetable that is practically pest-free and is highly recommended for the home garden.

Cole Crops

The term *cole crop* refers to several cool-season crops related to cabbage and having similar cultural requirements. Included are broccoli, Brussels sprouts, cabbage, cauliflower, collards, and kale, as well as several less common vegetables.

All cole crops thrive under cool temperatures and will survive light frosts. In hot-summer areas they should be grown as a spring or fall crop; where winters are warm, they should be grown as a winter crop. Collards alone are tolerant of very hot summer weather.

Any well-drained soil is acceptable. A pH between 5.5 and 6.5 is ideal, but a wide range is satisfactory in practice. All cole crops germinate readily from seed planted ¼ inch (6 millimeters) deep or may be started from transplants. Abundant moisture and a sidedressing with a high-nitrogen fertilizer are beneficial. The most serious problem with growing cole crops is the control of caterpillars and aphids.

Individual Requirements

Broccoli. (*Brassica oleracea*, Italica Group). "Sprouting" cultivars that form a large number of small heads rather than one large head should be selected for planting. The edible part is the flower head, which should be harvested before the buds begin to open. If all heads are kept picked, broccoli will continue to produce for many months (Figure 6-27).

Figure 6-27 The main head of this broccoli has been harvested, but it is continuing to produce edible side shoots. *(Photo courtesy of National Garden Bureau)*

Brussels Sprouts. (*Brassica oleracea*, Gemmifera Group). The edible parts are the small axillary buds that form along the main stem. As the plant grows taller, the lower leaves can be removed until leaves remain only at the top of the plant. At this point the plant will somewhat resemble a palm tree with side growths looking like small cabbages.

Sprouts should be harvested when about 1 inch (2–3 centimeters) across but before they begin to open. If the weather is warm, they will often be soft and leafy rather than hard. Flavor and quality will improve as cool weather arrives in fall. If long storage of sprouts is desired, the entire plant should be pulled up in fall and replanted in damp sand in a cool cellar. Sprouts will store for several months in this manner.

Cabbage. (*Brassica oleracea*, Capitata Group). Cabbage should be harvested when the heads are hard. If allowed to remain in the garden for too long, the heads will split, especially after a heavy rain. Because cabbage heads tend to mature at the same time, succession planting or the use of several different cultivars with varying harvest dates is desirable. Cabbage will store for a month or more under cold temperatures and high humidity.

Cauliflower. (*Brassica oleracea*, Botrytis Group). This is one of the more difficult cole crops to grow. For best results, the plants should be kept growing vigorously by watering and fertilizing; any slowdown in growth will cause the production of a small head of poor quality. Transplants that are not vigorous should be avoided because these will often produce a head only 1 inch (2.5 centimeters) or so in diameter immediately after transplanting, a phenomenon called *bolting*.

Like broccoli, the edible parts of cauliflower are the immature flower buds. The head is harvested while the buds are still pressed closely together. Quality declines rapidly once the buds begin to separate. To produce white heads similar to those available in markets, the heads must be *blanched*. This is accomplished by tying several of the upper leaves over the head as soon as it begins to form in order to shade it (Figure 6-28). Several self-blanching cultivars are also available. These will naturally cover the developing flower buds with leaves, but only if the weather is cool. Failure to blanch cauliflower will result in heads light green in color that, although more nutritious, have a stronger flavor. Cauliflower plants produce only one head and should be pulled out after the head is harvested.

Collards. (*Brassica oleracea*, Acephala Group). This nonheading cabbage is tolerant of both very cold weather and the heat of summer; as a result it is one of the most reliable plants for the production of greens. In-

Figure 6-28 Blanching cauliflower. *(USDA photo)*

dividual leaves can be harvested at any time, although during hot weather the flavor is likely to be strong.

Kale. (*Brassica oleracea*, Acephala Group). Kale is another easy-to-grow plant for greens, but it will not survive hot summers. It will, however, often survive the coldest of winters, producing tender greens as soon as the snow melts. To harvest, remove the outer leaves individually. Kale is less prone to insect problems than are other cole crops.

Collards

See "Cole Crops."

Corn

Sweet corn (*Zea mays* var. *saccharata*) is one of the most popular home-garden vegetables. It is a warm-season crop and should be planted only after danger of frost is past.

Corn will thrive in a wide variety of well-drained soils. A pH between 6.0 and 6.5 is best. It should be planted 1 to 2 inches (2.5–5 centimeters) deep and 8 to 10 inches (20–25 centimeters) apart. Several short rows rather than one long row should be planned to ensure pollination. After corn has germinated, it is necessary to control weeds, provide water, and give one or two sidedressings of high-nitrogen fertilizer. Because corn tends to reach harvesting stage all at one time, several succession plantings should be made.

Because hundreds of cultivars of sweet corn are available, the gardener can pick from midgets less than 3 feet (1 meter) tall to giants 6 to 7 feet (2 meters) tall. However, regardless which cultivar is selected, popcorn or field corn should not be grown within 100 feet (30 meters) of sweet corn to prevent cross-pollination and poor kernel quality.

Corn is harvested when the silk (female flowers) on the ears turns dark brown and the kernels are plump and filled with milky sap. Because (with the exception of "super sweet" cultivars) the sugar turns to starch quickly after picking, sweet corn should be cooked immediately after harvest.

Cucumbers

See "Cucurbits."

Cucurbits

The group of vegetables called *cucurbits* includes many warm-season vine crops with similar growth requirements. Members of the group include cucumbers, squash, melons, and pumpkins. In all cases the edible portion of the plant is the fruit.

Cucurbits require a warm climate for optimum growth and fruit development. The best quality of fruits is obtained when summers are long and hot. This does not preclude the growing of these plants in northern areas; however, plantings should be made only after all danger of frost is past and night temperatures are consistently above 45°F (7°C). Because melons require the highest temperatures and the longest growing season, melon cultivars should be selected carefully for adaptability to the local climate. Winter growing of cucurbits is restricted to the warmest areas of California, Florida, and Texas.

Cucurbits can be started either from seed sown ½ inch (1–2 centimeters) deep in late spring or from transplants. Because they do not transplant easily, seedlings should be purchased only if they are grown in peat pots or other plantable containers.

Cucurbits can be planted in rows or hills. If only a few plants are to be grown, the hill method is best because of better space utilization. Vines can be either allowed to run along the ground or trellised. Trellising works very well for cucumbers but is generally not used for melons and squash because of the heavy fruits. Gardeners are often disappointed by the lack of fruit set early in the year on cucurbit plantings. However, with most cultivars this is natural because cucurbits produce separate male and female flowers, and the normal pattern is for male flowers to be produced first, followed later by female flowers and corresponding fruit set.

Individual Requirements

Cucumbers. Compared with melons, cucumbers (*Cucumis sativus*) require a shorter growing season and less heat and therefore are adapted to a wider range of climates. Two types of cucumbers are commonly grown in the home garden, depending on their proposed use. These are *pickling cucumbers* and *slicers*. Pickling cucumbers are bred for their qualities as pickles and are harvested while immature. Slicers are bred for eating fresh, are larger, and have thicker skins and a more attractive, less warty appearance. For the home gardener who desires cucumbers for both pickling and fresh uses, pickling cultivars are best because they are of a good quality for fresh use as well as for pickles.

 An international horticulturist works abroad in advising or supervising production of vegetables, fruits, or flowers for export. The U.S. government employs horticulturists in development projects, as do multinational companies. University training or extensive experience as a grower are usually required. Previous experience abroad (such as in the Peace Corps) and knowledge of the language of the country are sometimes sought-after requirements.

Bitterness in cucumbers is due to a number of factors including genetic composition, excessively warm temperatures, and lack of water. Though bitterness is difficult to prevent, adequate water will decrease the chance of bitter cucumbers.

Melons. The most commonly grown melons are cantaloupes (more correctly, muskmelons, *Cucumis melo*, Reticulatus Group), watermelons (*Citrullus lanatus*), and honeydew melons (*Cucumis melo*, Inodorus Group). All require a long growing season, hot days, and plenty of moisture. The best areas for melon growing are warm regions of the country, although cultivars are available that can be grown in most of the United States and Canada. Honeydews require the most heat, followed by watermelons and muskmelons. Because the climate requirements are rather stringent, selection of cultivars adapted to local conditions is imperative. As a general rule, cultivars that produce small or midget fruit perform best where summers are cool.

Often one of the most difficult tasks the home melon grower faces is determining when the fruit is ripe. Watermelons should be harvested when the stem of the melon has shriveled, whereas muskmelons should be picked when the stem slips easily from the fruit with slight pressure from the thumb. Honeydew melons are at their best when they develop a yellowish skin color and sweet smell.

Squash and Pumpkins. Squash and pumpkins (*Cucurbita maxima, C. pepo, C. mixta, or C. moschata*) are the easiest

Figure 6-29 A selection of winter squashes in storage. *(Photo courtesy of National Garden Bureau)*

of the cucurbits to grow, and they produce an abundance of fruit. They are divided into two types based on whether the immature or the mature fruit is eaten. Those squash eaten when the fruit is mature and possessing a hard skin are *winter squash* (Figure 6-29). The name is derived from the fact that these types ripen in late fall and can be stored in a cool place through much of the winter. Pumpkins are essentially a type of winter squash. *Summer squash*, eaten while immature, include zucchini, yellow crookneck, and yellow straightneck squash (Figure 6-30). They are perishable and must be eaten soon after harvest.

Winter squash should be harvested when the rind becomes hard, evident by an inability to puncture it with a thumbnail. Usually a change in color will ac-

Figure 6-30 A selection of summer squash: patty pan on the left, yellow crookneck at bottom right, zucchini in the basket on the left and yellow straightneck in the basket on the right. *(Photo courtesy of the National Garden Bureau)*

company the ripening process. Summer squash are picked before the skin begins to harden but not necessarily when they are very small. If summer squash become large, they are still edible by scooping the seeds from the center prior to cooking.

When selecting cultivars of squash and pumpkins, consider selecting bush types instead of vine types, because the former require much less room and produce the same quality of fruit.

Eggplant

Eggplant (*Solanum melongena* var. *esculentum*) is a warm-season vegetable related to tomatoes and peppers. The plant is bushy with heart-shaped leaves and purple, star-shaped flowers. The edible portion is the egg-shaped fruit, which is generally large and purple in color and is harvested when it has reached the size appropriate for the cultivar but before the skin has lost its natural sheen. Storage of eggplant should ideally be at 50 to 55°F (10–13°C). It can be damaged by storage in the refrigerator.

Eggplants require a long period of warm weather to set and mature fruit properly. When growing eggplant in northern climates, early maturing cultivars should be

planted. They are grown from transplants because the plants are very slow to develop from seed. A starter fertilizer should be used at transplanting, followed by a sidedressing of a balanced fertilizer after the first fruit has set. Because eggplants are subject to several soil-borne diseases, they should be rotated in the garden to avoid being grown in the same area where related plants (tomatoes, peppers, potatoes) were planted.

Honeydew Melons

See "Cucurbits."

Kale

See "Cole Crops."

Lettuce

Lettuce (*Lactuca sativa*) is a cool-season vegetable. It is available in a number of types (Figure 6-31), those that form a firm head (called *head lettuce* or *iceberg*), semi-heading types such as bibb and cos (romaine), and a

Figure 6-31 Four types of lettuce, clockwise from upper left: head or iceberg type, a curly leaf lettuce, romaine or cos type, and butterhead type. *(Photo courtesy of W. Atlee Burpee and Co., Warminster, PA)*

type that produces no head at all, called *leaf lettuce*. Heading and semiheading lettuce is harvested by cutting the head from the roots when it is full grown and, in the case of head lettuce, firm. Leaf lettuce can be harvested by removing leaves individually or by cutting the plant at ground level. Head lettuce and romaine will keep one to two weeks in the coldest part of the refrigerator, but leaf and bibb lettuce should not be stored more than one or two days.

Lettuce will grow in most soils; however, emergence of the seedlings in heavy soils may be hindered by soil crusting. To remedy this, mix sieved organic matter into the soil used for covering the seeds. The crop requires cool temperatures and is therefore best grown as either a spring or a fall crop. It can also be used as a winter crop in mild-winter areas, provided no hard freezes occur. During hot weather lettuce will become bitter and tend to bolt. As a preventive measure, plant only during cool weather and choose heat-resistant cultivars.

Leaf and semiheading cultivars are generally seeded shallowly and later thinned to a 10-inch (25-centimeter) spacing. Head and romaine lettuce may be either seeded or transplanted. Transplants should be spaced at least 10 inches (25 centimeters) apart; seedlings should be thinned to the same spacing as soon as they begin to crowd. If not thinned, the lettuce will be prone to rot and may fail to head.

Difficulty in germinating may be due to overly warm soil. If this is suspected, place the seed in water in the refrigerator overnight prior to planting. Lettuce should be kept well watered and should be sidedressed with nitrogen-rich fertilizer once during its growing period.

Melons

See "Cucurbits."

Muskmelons

See "Cucurbits."

Onions

Onions (*Allium cepa*) are a popular, easy-to-grow vegetable crop. They grow best in a sandy soil rich in humus. When onions are growing in heavy soils, large amounts of organic matter should be added before planting.

Onions are classified as a cool-season crop, but are tolerant of a wide range of temperatures including frost. The best-quality bulbs will be produced when temperatures are cool during the early part of the growing season and moisture is abundant.

Onions can be grown from seed, transplants, or small bulbs called *sets*. Green onions (sometimes mistakenly called "shallots") can be easily grown from seed, but bulb types will be ready for harvest sooner if sets or transplants are used. They may be planted 3 to 4 inches (8–10 centimeters) apart or closer together with every other plant pulled for use as a green onion. Regular watering and one or two sidedressings of balanced fertilizer during the growing season are recommended.

The most important factor affecting onion growing is photoperiod. Onions form bulbs only when the night is shorter than the critical photoperiod for that cultivar. If the critical photoperiod is reached too soon after seeding, the onions will bulb before the plants are large enough, and the resulting bulbs will be very small. If the critical photoperiod is never reached, the onions will never form bulbs.

To avoid bulbing problems, select only cultivars recommended for the area. In areas where both winter and summer crops can be raised, a different cultivar must be selected for each, because of the difference in night length between the seasons.

Onions can be harvested at almost any stage of maturity, depending on the proposed use. Young onions, regardless of cultivar, can be harvested before bulbing for use as green or "bunching" onions. In this stage they are perishable and should be refrigerated. If mature onions with bulbs are desired, a cultivar bred for its storage characteristics should be selected, and the onions should be allowed to remain in the garden until the tops fall over. This signals that the bulb is mature. After digging out the bulb, cut the leaves off about 1 inch (2–3 centimeters) above the bulbs. The bulbs should be dried outside if the weather is warm or in slatted crates or mesh bags in a warm indoor location. After several weeks of drying, the onions can be stored in a cool, dark location.

Peas

Peas (*Pisum sativum*) are a cool-season garden crop that shows a spectacular improvement in quality over purchased produce when home grown and picked immediately prior to eating. Peas require cool temperatures and abundant moisture to grow well. They will tolerate moderate freezes but not heat. They are best grown as an early spring or late fall crop or, in mild-winter areas, as a winter crop. Where warm weather makes growing peas difficult, a heat-resistant cultivar such as 'Wando' should be chosen.

Pea seeds are sown about 1 inch (2–3 centimeters) apart and are not thinned. A common practice is to sow peas in double rows about 8 inches (20 centimeters) apart and train two rows on a single string trellis.

It is important to supply water regularly to peas and to pick the pods as soon as they are mature to lengthen the harvest. A sidedressing with balanced fertilizer is recommended when the plants are 4 to 6 inches (10–15 centimeters) tall.

The peas should be harvested when they begin to fill out the pod but before becoming tough. Storage of peas is not recommended, but if storage is unavoidable, they should be kept in the coolest part of the refrigerator.

Peppers

Peppers (*Capsicum annuum*) are known as either "sweet" or "hot." The sweet peppers include bell, pimento, Mexican stuffing ('Anaheim') and sweet banana types. The hot peppers usually have fruits that taper to a point and can be green, red, or yellow. Sweet peppers may be picked any time after the fruit has attained full size and the color characteristic of the particular cultivar. Bell peppers that are not picked soon after they reach full size will eventually turn red; they are still edible and not hot, and some people prefer them at this stage because they are less bitter. Hot peppers can be picked when the full size and proper color have been reached or can be allowed to dry on the plant for winter use. However, picking the fruits as they ripen will increase the total fruit production.

A common problem encountered by home gardeners raising both hot and sweet peppers is cross-pollination. Cultivars of peppers cross freely, and although the fruits will each retain their proper taste, the seeds will bear the characteristics of both parents. The seeds of sweet peppers grown near hot peppers should therefore be removed before eating.

Another problem is poor fruit set, which is usually weather-related. Both excessively warm temperatures and cloudy, wet weather will prevent fruit set.

Fresh peppers should be stored in the refrigerator and will retain good quality there for a week or more.

The soil, climate requirements, and culture are the same as for eggplant.

Pumpkins

See "Cucurbits."

Radishes

Radishes (*Raphanus sativus*) are the fastest-growing garden vegetable and will produce an edible root in 21 to 30 days. These quick-maturing radishes are the red globe-shaped type. The second type, called *Chinese* or

Figure 6-32 An icicle winter radish, also known as daikon. *(Photo courtesy of W. Atlee Burpee and Co., Warminster, PA)*

icicle radishes, takes longer to mature, but produces larger roots that are white. Globe radishes are suitable for planting in the spring and should be picked as soon as they reach edible size, or they become tough and pithy. Icicle cultivars (Figure 6-32) are preferred as a fall crop where the weather is warm.

Both are cool-season crops and will grow best in temperatures below 70°F (21°C). Under warm temperature conditions they tend to be hot and to bolt. Seed should be sown in early spring to avoid the warm temperatures that slow growth and cause bolting. Plants should be thinned to 1 inch (2–3 centimeters) apart soon after germination. Several sowings made one week apart are recommended to assure a steady supply. Radishes can also be seeded along with carrots or leeks or between transplants of other crops because they mature quickly and can be harvested before crowding occurs.

Squash, Summer

Summer squash includes yellow straightneck, yellow crookneck, and zucchini. See "Cucurbits."

Squash, Winter

Winter squash includes acorn, butternut, spaghetti, and other thick-rind types. See "Cucurbits."

Swiss Chard

See "Chard."

Tomatoes

Tomatoes (*Lycopersicon lycopersicum*) are the most popular of all garden vegetables. Although they are actually herbaceous perennials, they are grown as annuals in the vegetable garden and develop into branched bushes or vines with compound leaves and yellow flowers.

 A cannery field manager is a liaison between the canners and the growers of the produce to be canned, such as tomatoes. The field manager works with the contracted growers to make sure the vegetable is grown correctly to reach the quality standards required by the cannery, including those for size, ripeness, color, and freedom from pests and diseases.

Botanically, tomatoes are related to eggplants and peppers but are easier to grow than either. The fruit will be most flavorful when allowed to ripen fully on the plant. Green tomatoes can be eaten pickled or fried.

Tomatoes will succeed in a wide range of soil types provided the drainage is adequate. However, they should not be planted in the same place year after year due to the increased danger of wilt diseases and nematodes. A pH between 6.0 and 6.8 is best.

Like peppers and eggplants, tomatoes are a warm-season crop. For fruit to set, temperatures must be warm, because pollen is usually not shed when night temperatures are below 59°F (15°C) or above 80°F (27°C). In areas with cool summers or short frost-free growing seasons, early cultivars should be selected because they not only mature more quickly but grow better at cool temperatures. Cultivars that produce small fruits such as cherry tomatoes tolerate a wide range of temperatures.

Although tomatoes can be raised easily from seed planted directly in the garden, they are usually raised from transplants for an earlier harvest. Spacing of the plants will depend on whether the cultivar to be grown is "determinate" or "indeterminate." Determinate tomato cultivars grow into bushy plants that can be either staked, caged, or allowed to sprawl along the ground. Indeterminate types grow long and vinelike and definitely require support.

A balanced fertilizer should be applied prior to planting and a nitrogen sidedressing one month later after which nitrogen applications should be restricted. When planting tomatoes from transplants, a starter fertilizer is beneficial.

Although pruning is sometimes advocated for home-grown tomatoes, most experts now agree that pruning only slows fruit production and increases sun scald of the fruit as well.

Turnips

Turnips (*Brassica rapa*) are a cool-season root crop related to cole crops. Though the root is the primary edible portion, the leaves are also edible; in fact some cultivars are grown mainly for their leaves. Turnip leaves do not store well and should be picked immediately prior to eating. Roots, on the other hand, store for long periods if the tops are removed and if kept in a cool, humid environment. They should be harvested when they are 2 to 3 inches (5–8 centimeters) in diameter.

Turnips grow best in a light- to medium-weight soil rich in organic matter but will produce large crops in almost any soil as long as the drainage is adequate. They are planted as a spring or fall crop in most areas or as a winter crop in mild-winter areas. In locales with hot summers, fall plantings are most successful.

Turnips are grown from seed that germinates rapidly in the garden. The young plants grow vigorously and will need to be thinned so that they are about 3 inches (8 centimeters) apart when they are 3 inches (8 centimeters) tall. Thinning will not be required if the turnips are grown for greens only.

During the growth period the plants should be kept well watered and sidedressed once or twice with a light application of balanced fertilizer. This will help the plants to grow rapidly and will assure tender roots.

Watermelon

See "Cucurbits."

Zucchini

See "Cucurbits."

MINOR VEGETABLES

Black-Eyed Peas

Black-eyed peas or cowpeas (*Vigna unguiculata* or *V. sinensis*) is a warm-season crop seeded 2 to 4 inches (5–10 centimeters) apart in rows 6 feet (2 meters) apart. A slightly acid soil is best. The plants are beanlike, erect, and normally do not require training, although some cultivars of *V. s. sesquipedalis*, also called *cowpea*, are vining, require support, and produce yard-long (one-meter) pods. The crop is harvested when the pods, which are normally 3 to 8 inches (8–20 centimeters) long, are slightly yellow but before the seeds inside are dry.

Artichokes

Two very different plants are called "artichokes." The first is the globe artichoke (*Cynara scolymus*), a tender perennial generally grown only in areas where the ground does not freeze. The second is the Jerusalem artichoke (*Helianthus tuberosus*), a hardy perennial that can be grown throughout the United States and Canada.

Globe Artichokes

The globe artichoke is a large, thistlelike plant up to 6 feet (1.8 meters) in diameter. The large flower buds it produces in late spring and early summer are the edible portion of the plant (Figure 6-33).

The globe artichoke grows best in areas where winter temperatures are mild and summers are cool. Areas along the California coast, where summer fogs are common, are ideal. In hot-summer regions artichokes often produce buds that open rapidly and have a leathery texture. In zones 1 through 7, artichokes can be grown in containers in summer and moved indoors in the winter, or they can be dug after cold has killed the leaves and the crowns stored in damp sawdust until spring. A heavy mulch will generally provide sufficient protection in zones 8 and 9 to overwinter the crowns.

Artichokes are usually grown from divisions that consist of a woody stalk, a root, and leaves. This division should be planted vertically, with the base of the

leaves slightly below the soil surface. In most areas late winter or early spring is the best time to plant. Allow 3 to 4 feet (1–1.2 meters) between plants. A starter fertilizer is beneficial at planting.

After the plants begin to grow, plenty of water and nitrogen fertilizer should be provided. When buds appear, they should be harvested before they begin to open by cutting the stem ½ to 1 inch (1–3 centimeters) below the base and then refrigerated until ready to eat. An artichoke will keep two to three weeks if it is kept in the coldest part of the refrigerator in a plastic bag.

After the buds are cut, the supporting stem will shrivel and should then be cut back to healthy growth. New shoots will sprout from the base to provide next year's harvest. In cold-weather regions nitrogen fertilizer should be withheld after midsummer to enhance cold tolerance.

Jerusalem Artichokes

The Jerusalem artichoke is a sunflower with coarse, hairy leaves that grows to a height of 6 feet (1.8 meters). The edible portions of the plant are the underground tubers that are produced abundantly in the fall. The tubers are harvested after frost or may be left in the ground all winter and dug as needed. After harvest the tubers can be stored from two to five months under high humidity and at temperatures as close as possible to freezing.

The Jerusalem artichoke can be grown in zones 2 through 9. However, the best quality and highest yields occur in the colder zones. Jerusalem artichokes thrive in any well-drained soil. Highest yields are obtained in soils with poor nitrogen content, so Jerusalem artichokes may be grown in the poorest section of the garden.

Propagation is from tubers that may be purchased from nurseries or seed companies or the produce department of a grocery. Tubers are planted 3 to 5 inches (8–13 centimeters) deep (depth is not critical) and 5 to 10 inches (13–25 centimeters) apart in spring. The plants are very vigorous and generally do not require any fertilization or care until harvest. When harvesting, remove all tubers except a few to start the next season's crop. If many tubers are left in the ground, the plant may become a weed problem.

Celery

Celery (*Apium graveolens*) is one of the most difficult vegetables to grow in the home garden because, in addition to its preference for soil rich in organic matter, celery requires a long growing season with cool temperatures (65–70°F, 18–21°C). If temperatures are too warm, it will become tough. If prolonged periods with

Figure 6-33 A field of globe artichokes. *(USDA photo)*

temperatures below 50°F (10°C) occur, it will bolt before leaf and stalk development take place.

Celery is generally grown from transplants 3 to 4 inches (8–10 centimeters) tall planted 10 inches (25 centimeters) apart. Seeds require three weeks to germinate and another month to reach transplanting size. Regular watering and frequent fertilizing with a high-nitrogen fertilizer will maximize growth. Plan on approximately a five-month period from seeding to harvest.

Blanching excludes light from the stalks and decreases their chlorophyll content. It was a common practice in the past and can be accomplished by mounding soil around the lower stalks or otherwise shading them. Most gardeners today do not blanch celery because it results in a lower nutritional value and many cultivars are "self-blanching."

It is harvested by cutting the plant at ground level or by removing individual stalks as needed. Celery will keep for several weeks if refrigerated and kept in a humid condition.

Chinese Cabbage

Chinese cabbage (*Brassica rapa*, Pekinensis Group), sometimes called *celery cabbage*, is a cool-season annual grown for its leaves (Figure 6-34). Seeds should be sown early in spring or as fall approaches to avoid hot weather and flowering. Seedlings should be thinned to 8 inches (20 centimeters) apart when they begin to crowd; the heads are later harvested individually. This is a very fast-growing leaf vegetable prized for Chinese dishes, and several distinct types with varying leaf shapes are grown.

Curly Endive

Endive (*Cichorium endivia*) (Figure 6-35) is grown for its slightly bitter leaves, which are used fresh as a salad green. Seed should be planted in early spring, and leaves can be harvested whenever they are large enough to eat. For better flavor the mature outer leaves can be tied over the inner leaves to blanch them.

Kohlrabi

Kohlrabi (*Brassica oleracea*, Gongylodes Group) (Figure 6-36) is a relative of cabbage grown for its enlarged stem. Plants are harvested when the enlarged portion is 2 to 4 inches (5–10 centimeters) in diameter. See "Cole Crops" for cultural information.

Figure 6-34 Chinese cabbage of the Napa type. *(Photo courtesy of W. Atlee Burpee and Co., Warminster, PA)*

Figure 6-35 Curly endive. *(Photo courtesy of W. Atlee Burpee and Co., Warminster, PA)*

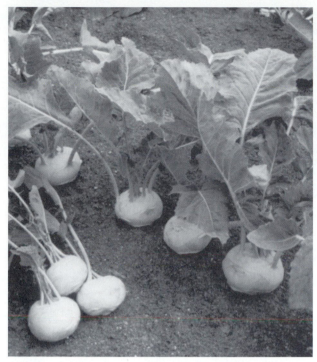

Figure 6-36 Kohlrabi 'Grand Duke.' *(Photo courtesy of W. Atlee Burpee and Co., Warminster, PA)*

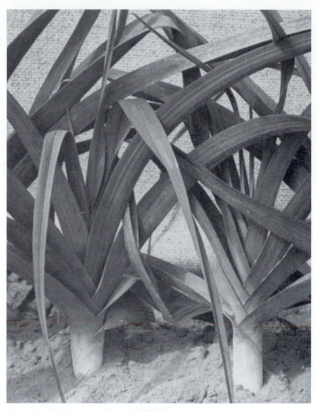

Figure 6-37 Leeks. *(USDA photo)*

Leeks

Leeks (*Allium ampeloprasum*) (Figure 6-37) are a non-bulbing relative of the onion with a mild flavor. They are generally grown from seed and require a long, cool growing season. The white underground stem is eaten at any size. Soil can be mounded around the plants to blanch more of the stem, thus producing more edible portion per plant.

Mustard

Mustard (*Brassica juncea*) is a fast-growing, cool-season, vegetable grown for its strong-flavored greens. It is seeded directly in the garden in early spring, and individual leaves are picked when they reach 3 to 4 inches (8–10 centimeters). Planting early is important because mustard will flower and die as soon as the weather becomes warm.

Okra

Okra (*Abelmoschus esculentus*) (Figure 6-38) is a large, fast-growing plant that produces edible seed pods. To grow well, okra requires hot weather, well-drained soil, and a relatively long growing season. The pods should be picked before they reach 3 inches (7.6 centimeters) in length.

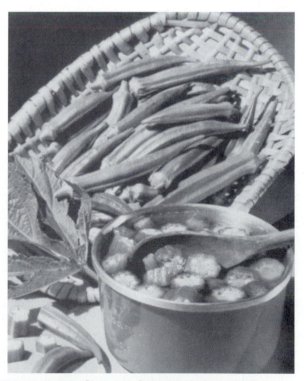

Figure 6-38 'Clemson Spineless' okra. *(Photo courtesy of W. Atlee Burpee and Co., Warminster, PA)*

Figure 6-39 Parsnips. *(Photo courtesy of W. Atlee Burpee and Co., Warminster, PA)*

Parsnips

Parsnips (*Pastinaca sativa*) (Figure 6-39) produce white, carrot-shaped roots with a sweet, nutlike flavor. Cool-season vegetables, they should be seeded in deeply worked soil very early in the spring and grown like carrots. The roots can be harvested at any size and can be allowed to remain in the ground for winter and spring eating.

Parsnip seeds are among the shortest-lived of all vegetable seeds, so storage of old seed is not recommended.

Potatoes

Potatoes (*Solanum tuberosum*) are grown for their underground tubers and will grow in most climate zones. However, quality will be best in a climate with cool nights.

Potatoes are generally grown from seed potatoes, small pieces of a potato that contain at least one "eye" or bud per section. Only certified seed potatoes that are guaranteed to be disease-free should be grown. A cultivar that can be grown from seed started in flats about eight weeks prior to transplanting is also available.

New potatoes are thin-skinned, immature potatoes harvested when the plants begin to flower and the potatoes are still small. They should be eaten immediately after digging in summer. To have potatoes for storing for winter use, delay harvest until the tops of the plants have died in fall.

Unless a large garden space is available, potatoes are not considered worth raising in the home garden. The quality will not differ substantially from that of supermarket potatoes, and monetary savings are usually min-

imal. In addition, home-grown potatoes require careful attention to control insects and diseases.

Rhubarb

Rhubarb (*Rheum rhabarbarum*) is a perennial grown for its leaf stalks, which have an acid, fruitlike flavor. It will grow in all USDA zones. One- or two-year-old crowns should be planted in spring and light harvesting started the following spring by removing one or two leaves per plant. The leaves should not be eaten, as they contain oxalic acid and are poisonous.

Flower stalks should be removed as soon as they appear to divert the strength of the plant into vegetative growth. Fertilizing once or twice per year is sufficient, and clumps should be divided about once every four years.

Rutabagas

Rutabagas (*Brassica napus*, Napobrassica Group) are grown for their large turniplike roots, which have a flavor similar to that of a strong turnip. Their culture is the same as for turnips.

Snow Peas, Chinese Pea Pods, Edible-Podded Peas (*Pisum sativum* var. *macrocarpon*)

Culture the same as for peas. See Figure 6-40.

Spinach

Spinach (*Spinacia oleracea*) is a nutritious cool-season vegetable grown for its strongly flavored leaves. It should be seeded in early spring directly in the garden and harvested before the weather becomes hot because the plants flower in warm weather.

Sweet Potatoes

Sweet potatoes (*Impomoea batatus*) can be grown in warm climates with a frost-free growing period of 120 days or longer. They are grown from transplants spaced 24 to 30 inches (0.6–0.7 meter) apart and can be harvested as soon as they reach edible size. If plans include storing the sweet potatoes for winter use, harvest should be delayed until frost kills the tops. The roots are then dug, cured at 85°F (30°C) for three weeks, and stored at 55°F (13°C). Care should be taken that the roots are not bruised and that curing is performed properly, otherwise sweet potatoes will not store well.

Figure 6-40 Snow peas or Chinese pea pods. *(Photo courtesy of W. Atlee Burpee and Co., Warminster, PA)*

 The position of an agricultural librarian could be filled by someone with training in horticulture or another agricultural specialization, knowledge of computers, and organizational skills. Larger libraries in universities and in botanic gardens normally have an agricultural librarian on staff to make decisions on the purchasing of new texts, direct patrons to sources of information, and organize the material available in the library.

SELECTED REFERENCES FOR ADDITIONAL INFORMATION

Abraham, G. *The Green Thumb Garden Handbook.* Guilford, CT: Globe Pequot Press, 1992.

Bush-Brown, J., and L. Bush-Brown. *America's Garden Book.* Rev. ed. NY: Hungry Minds, Inc., 1996.

Crockett, J. U. *Crockett's Victory Garden.* New York: Little, 1977.

Doty, Walter L. *All About Vegetables.* San Ramon, CA: Ortho Books, 1996.

Fell, Derek. *Vegetable Gardening with Derek Fell.* NY: Barnes and Noble, 1996.

Maynard, Donald E. and George J. Hochmuth. *Knott's Handbook for Vegetable Growers.* 3rd ed. New York: Wiley, 1997.

Pleasant, Barbara. *Vegetables.* Des Moines, IA: Meredith Books, 1998.

Swiader, John M. and George W. Ware. *Producing Vegetable Crops.* Danville, IL: Interstate Publishers, Incorporated, 2001.

Time-Life Publications. *Summer Vegetables.* Alexandria, VA: Time-Life Books, 1988.

World Wide Web Addresses

http://www.aces.uicu.edu/ipm/news/fvnews.html   University of Illinois Extension. "Illinois Fruit and Vegetable News." (Oct 12, 2000)

http://www.actahort.org/books/4/   International Society for Horticultural Science. "Symposium on Vegetable Growing Under Glass." ISHS Acta Horticulturae 4.

http://edis.ifas.ufl.edu/MENU CV:CV   University of Florida Extension Service. "Commercial Vegetable Production Guide for Florida."

http://www.eas.asu.edu/~nfapp/html/links.htm   National Food and Agricultural Policy Project NFAPP's Fruit and Vegetable Links.

http://www.gardenguides.com/Vegetables/vegetabl.htm   Garden Guides. "Vegetable Guide" 2001.

http://www.lane-ag.org/scarl/prodsys/prodsys.htm   Russo, Vincent. Vegetable Production Systems. August 10, 2001.

http://www.lus.edu/horticulture/4083/class3b/   Motsenbocker, Carl. Vegetable Growing Principles: Environmental Factors Influencing the Growth of Vegetables (Ch 6). Sept 25, 1998.

http://www.msucares.com/pubs/pub 2036.htm.   Harris, Pat; James H. Jarratt; Frank Killebrew; John D. Byrd, Jr.; Rick Snyder. "Organic Vegetable IPM Guide." MSU Cares website of Mississippi State University Extension Service. July 19, 2001.

http://ohioline.osu.edu/b672/   Ohio State University Cooperative Extension Service. "2001 Ohio Vegetable Production Guide" website, Bulletin 672-01. 2001.

http://www.thevegetablepatch.com/   Anonymous. "The Vegetable Patch" October 2001. This is an organic gardening site.

http://vric.ucdavis.edu/veginfo/veginfor.htm   University of California Vegetable Research and Information Center, "Vegetable Information" Sept. 10, 2001.

(Photo courtesy of USDA)

CHAPTER 7

GROWING TREE FRUITS AND NUTS

Tree crops are raised for their edible fruits or seeds. Most fall into the designations of fruits or nuts, but a few, such as the avocado and olive, grow on trees but fit neither category.

When a gardener decides to grow fruit or nut trees, the species and the number selected will depend on many factors, among them the following:

1. *Climate.* Minimum winter temperature as well as average summer temperature and length will determine which species and cultivars can be grown. A few tree crops such as avocado and citrus are semitropical and will tolerate temperatures only to the mid-20°F (-2°C) range. Others such as sour cherries are adapted to climates with widely varying winter and summer temperatures. They will not survive without a cool winter period and will fail to produce fruit and even die in a warm-winter area.

Unexpected spring frosts in temperate areas can also cause problems with some fruits that flower early. Apricots are notably vulnerable because of their early bloom period.

Summer temperature and length can also affect ability to grow some fruit trees successfully. During some years, the plant will not receive enough warmth for proper fruit ripening and sugar formation, or the growing season will not last long enough for the fruit to develop completely.

2. *Property size.* The size of the property will also determine how many fruit and nut trees can be raised. Owners of small properties frequently incorporate fruit and nut trees into the landscaped areas in lieu of orna-mental trees. Others use dwarf cultivars that grow only about 8 feet (2.4 meters) high. Although this is a solution with most fruit trees, no nut trees are available in dwarf form.

3. *Expected maintenance time.* Although some species have little problem with pests and diseases, others, such as the apple, will require spraying every two weeks in addition to yearly pruning and fertilization. The considerable amount of time that will need to be devoted to maintenance should be taken into consideration.

SITE SELECTION

A property may have several possible sites for planting fruit and nut trees. If so, the following criteria (in order from most to least important) should be taken into account when the spot is selected.

Soil Drainage

Few plants will grow in an area with poor drainage, and trees are no exception. Water must drain away rapidly to allow air to penetrate to the roots. If it does not and the roots are in soggy soil for an appreciable period, the tree will die. *Stone fruits* (those with a single, hard pit such as cherry, plum, peach, and nectarine) are the most susceptible to poor drainage.

Amount of Sunlight

Fruit trees should have as much sun as possible to bear heavily and ripen their crops. Six hours of direct sun per day is considered minimum for average fruit production. Although a sunny spot is also desirable for nut trees and will speed their growth considerably, it is not as essential as for fruit trees. The nut trees, being ultimately larger, will compete for sun, whereas dwarf fruit trees will not.

Land Slope and Exposure

In all but subtropical and tropical climates, situating fruit and nut trees on the upper part of a slope is ideal because it partially protects against frost injury. As discussed in Chapter 1, cool air flows downhill, leaving the higher area of the slope several degrees warmer. The few degrees difference may prevent a frost that kills the season's crop. Unless it is unavoidable, tree crops should not be planted in a low spot or at the bottom of a slope. Flat land is preferable to those areas.

The direction in which the slope should ideally face will depend on the weather conditions in the area. South-facing slopes warm earlier in the spring, and trees will start growth sooner. However, this also means that they may be in a tender state and susceptible to late spring frosts. Consequently, a north-facing slope is suggested in climate areas with spring frost problems, but southern slopes are best in areas that are completely or nearly frost-free. South slopes are also advisable where summer temperatures are likely to be too cool for complete fruit maturation, because a south-facing slope receives direct sun the entire day and will be several degrees warmer.

INCORPORATING TREE CROPS IN THE LANDSCAPE

On small properties, a homeowner may have to work fruit and nut trees into the landscape design in lieu of setting aside a separate orchard area. How the trees can be incorporated will depend on the unique features of the property, but the following suggestions can aid in planning.

1. *Substitute nut trees for deciduous shade trees.* Because nut trees are large, they work well as substitutes for shade trees. Many have a pleasing fall color, although the flowers are inconspicuous. Large fruit trees also make handsome shade trees. However, if spraying is a necessary part of the maintenance program, the job will be harder with a large tree.

2. *Use dwarf trees as a screen.* Three or more dwarf trees planted close together (about 6 to 8 feet or 1.8 to 2.4 meters apart) will eventually grow to form a screen. Staggering the plants in two rows (Figure 7-1) will create the screen effect faster without crowding the trees.

3. *Use dwarf fruit trees as patio trees.* Fruit trees are very ornamental, with lovely spring blossoms and heavy sets of ripe fruit. Dwarf types can be conversation pieces, but semidwarfs should be planted if shade is desired.

4. *Espalier fruit trees.* The flat training method called *espalier* (Figure 7-2) can fit several dwarf fruit trees into the smallest yard. The tree will still bear a worthwhile amount of fruit, although probably not as much as it would growing in a normal form.

Figure 7-1 Fruit trees used as a screen in a landscape.

Figure 7-2 Pear trees espaliered diagonally. *(Courtesy of the American Floral Marketing Council)*

TREE SELECTION

In selecting an ornamental plant, the cultivar may be picked for its outstanding flowers or vivid fall color, but neither is crucial to its success in the landscape. But choosing a proper cultivar is very important in fruit and nut growing. Choice of the wrong cultivar can lead to failure, so it must be chosen carefully. Read not only the nursery catalog description but also state Agricultural Extension Service publications for information on the best cultivars for the area. Look for the following information in making a decision.

Unless you live near a nursery specializing in fruit and nut trees, the best selection is available from mail-order nurseries. Several firms specialize exclusively in fruits and nuts and offer many cultivars of each species. Some are listed under "Selected References" at the end of this chapter.

Winter Hardiness

Some fruit cultivars are bred for growing in the South, whereas others of the same species are bred for the North. A particular southern peach variety may not survive a northern winter, much less produce the expected size and quality of fruit.

Required Fruit Maturity Period

Cultivars of fruits bred for cold climates ripen more quickly than those bred for warm climates where the summer is longer. A cultivar should be chosen that will mature and ripen within the limitations of the frost-free growing season (Chapter 6). Moreover, fruits advertised as maturing extremely early in the season will often be inferior in quality to later-maturing cultivars.

Disease Resistance

Selecting disease-resistant cultivars can eliminate much of the labor of spraying against fungus- and bacterium-caused diseases. Apple varieties, for example, are available that are resistant to scab, fire blight, and cedar rust, three common disease problems. Chinese chestnuts are immune to chestnut blight, whereas the American chestnut is so susceptible that only a few trees are alive in the United States today.

Fruit Use

Some fruit cultivars are primarily for fresh eating, whereas others (originally bred for commercial orchard use) are best for drying or canning. Prune versus fresh-eating plum cultivars are examples. Any fruit bred for processing can, of course, be eaten fresh, but its quality may be poorer than that of a fresh-eating cultivar.

Fruit Appearance

Size, coloring, and shape are included under fruit appearance. Many gardeners have a preference for yellow over red apples, for example, or a cultivar that produces mammoth fruits. But appearance should not be allowed to outweigh the important factors of hardiness, disease resistance, maturity period, and yield.

Pollination Requirements

Many fruits and nuts require cross-pollination with another cultivar or, with nuts, a tree of the opposite sex to bear a crop. Many homeowners have grown healthy trees for years without a crop, due to failure to recognize the pollination requirement of the variety.

DWARF FRUIT TREES

Nut trees are available only in *standard* (normal) sizes. However, most fruit trees come in three sizes: standard, semidwarf, and dwarf. Standard trees grow to the normal size for the species and are large in most cases. A standard apple tree can grow 30 feet (10 meters) across and to an equal height, although 20 to 25 feet (6–8 meters) is more typical.

Semidwarf trees range from 10 to 15 feet (3–5 meters) and dwarfs from 5 to 12 feet (1.5–4 meters). Most fruit trees sold to home gardeners are semidwarf or dwarf, for various reasons. First, because they are shorter, picking, pruning, and spraying are easier, and a ladder is not usually required. Dwarfed trees also bear sooner. An average of five to six years from planting is not unusual for a standard tree to bear its fruit crop, but a dwarf will bear the second year after planting and may even produce a few fruits the first year. Dwarf and semidwarf trees occupy less space, and a homeowner with limited land can grow more species and cultivars of fruit in an area than with standard trees.

Dwarf trees are produced in one of two ways. The most common is by grafting onto a *dwarfing rootstock* that prevents the cultivar from growing to full size. The grafting combines the dwarf characteristic of one cultivar with the exceptional fruit of another to produce a plant with the desirable characteristics of each.

Sometimes three cultivars are involved, a rootstock, a scion, and an *interstock* (Figure 7-3). An interstock would be used, for example, when the stock and scion are *incompatible* (will not graft together), but each is compatible with the interstock. Grafting with an interstock might also be used to combine superior fruiting (in the scion), dwarfing (in the interstock), and a vigorous root system (in the rootstock). Many nurseries use this technique commonly to avoid the poor root systems inherent with many dwarfing rootstocks.

A second type of tree dwarfing is called *genetic dwarfing,* whereby the fruiting portion of the tree remains small due to its genetic makeup rather than because of a dwarfing rootstock. However, genetically dwarf trees are also frequently grafted. The genetically dwarf top is grafted onto a particularly vigorous or disease-resistant root system for the qualities the rootstock can contribute to the tree.

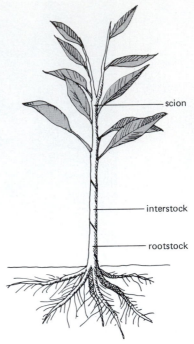

Figure 7-3 An interstock grafted between the rootstock and scion on a fruit tree.

PLANTING FRUIT AND NUT TREES

Planting fruit and nut trees is the same as planting an ornamental tree. Deciduous trees are normally sold leafless and bare-rooted for planting in the fall or spring. Evergreens such as olive, citrus, and avocado are sold in plastic sleeves or containers and can be planted at any time of the year.

The hole should be dug large and deep enough to accommodate the root system without crowding. After the plant is settled in the hole, it should be filled around with the removed topsoil. Organic matter such as peat moss or compost can be added but is not necessary. Watering thoroughly afterward is essential.

 A commercial tree fruit or nut grower often specializes in one particular fruit, such as apples, and may either own his or her own land or work as a manager for a large fruit-growing company. A grower is outside for much of the day supervising all aspects of growing including planting, training, pruning, pest and disease control, and harvesting.

Figure 7-4 The graft union of a fruit tree. *(Photo by Rick Smith, Los Osos, CA)*

Most fruit trees and many nut trees are grafted, and the placement of the graft union above ground at planting is important (Figure 7-4). This placement is necessary to retain the special qualities of the rootstock. For example, it may have disease resistance or cause the top to be dwarfed instead of full size. If the scion (top portion) comes in contact with soil, it will send out roots of its own, and the desirable rootstock characteristic will be lost.

Staking for support may be necessary for fruit and nut trees for the first year or two until the root system grows large enough to anchor the plant. Many dwarf cultivars are very shallow and fibrous-rooted, so permanent staking or trellising to prevent blowing over may be advisable. A short stake attached to the truck is usually sufficient. Trees should be tied to the stakes using a wire covered with a section of rubber hose or pieces of rag. Bare wire or string may injure the bark by rubbing or girdling the trunk.

Normally fruit and nut trees are pruned before shipping, but if not, they will have to be pruned after planting. This is discussed under "Pruning."

MAINTENANCE OF TREE CROPS

Watering

Watering of a fruit or nut tree varies with the species. Some, such as the olive, are very drought-resistant and adapted to arid climates, whereas most others will produce their best crops only with regular rainfall or irrigation.

As a general rule, a newly planted tree should be watered deeply every one to two weeks during its first growing season. Decrease the frequency in subsequent years to about every three weeks, depending on the soil water-holding capacity and natural rainfall.

Watering with a drip irrigation system (see Chapter 11) conserves both water and labor. Many commercial orchards use drip irrigation systems, and these systems can be highly recommended for home use.

Fertilization

Tree crops are normally fertilized yearly on a precautionary basis rather than because the tree has shown any distinct nutrient deficiency. Without seasonal soil tests it is impossible to apply the exact amount of fertilizer required, and yearly fertilizing is used to prevent possible deficiencies.

Methods for applying fertilizer include the drill hole method explained in Chapter 5 and the surface topdressing method. The drill hole method is used when the area under the tree is covered with grass or groundcover plants that would be injured by the fertilizer application.

General recommendations specify applying granular-type fertilizers in spring before or just as new growth starts. About ½ cup (120 milliliters) of nitrogen-rich fertilizer per inch (2.5 centimeters) of trunk diameter will supply an adequate amount of nutrition. A second application should be given in midsummer.

Micronutrient fertilizers are frequently used to prevent iron and zinc deficiency on citrus and pecans growing in alkaline soil. Either foliar or soil applications should be made repeatedly until the foliage regains its healthy green color.

Pruning

Some fruit trees—citrus, olive, and avocado—for example, require no pruning to change their shapes. Most nuts also do not require pruning, and indeed many are so tall that it is impractical to do so after the first few years. Most deciduous fruits should be pruned, however.

The three main pruning styles are the central leader, modified leader, and vase shape. Espalier training styles are less commonly used. The pruning style selected depends mostly on the species of tree; some adapt more easily to one style than another. In all cases the object of the training is to produce a tree with well-placed, strong branches capable of supporting a heavy weight of fruit without breaking.

Central Leader Form

Central leader form (Figure 7-5a) is used on most nut trees, on sweet cherries, and occasionally for apples and pears. The pruning starts at planting, if necessary, with the selection of the *central leader* and several *scaffold*

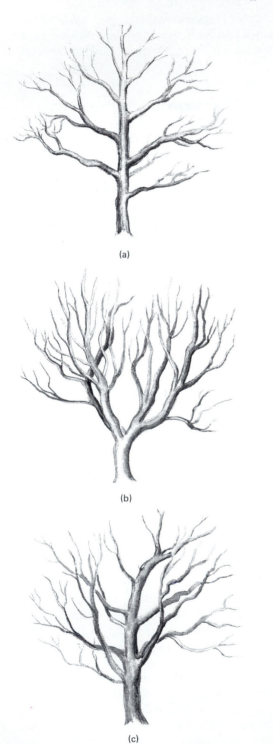

(a)

(b)

(c)

Figure 7-5 Fruit trees pruned to central leader (a), vase (b), and modified central leader (c) forms.

branches spaced widely around and down the trunk. Figure 7-5a shows a young tree pruned to a central leader style. Pruning after the first several years should consist of removing rootstock suckers (Figure 7-6) and competing branches arising from the trunk.

Vase Form

Vase form, called *open-center form*, allows two to five scaffold branches to grow from a short trunk (about 1 to 2 feet, or 0.3 to 0.6 meter). The branches originate close together on the trunk, and the central leader is removed to a side branch.

A vase-shaped tree (Figure 7-5b) is pruned to a final form directly after planting by selecting the lowest, strongest branches on the tree. Vase pruning opens the center of the tree to admit light and encourage fruiting and keeps the bearing branches of the tree low for easy picking. Vase pruning is common on peach, apricot, and nectarine.

Modified Central Leader Form

In a modified central leader form (Figure 7-5c), the tree is first trained to a central leader form for two to three years. The next year the leader is cut back to a strong

Figure 7-6 Pruning suckers from the base of a tree. *(Photo by the author)*

side branch, leaving three to five scaffold branches on the tree. With the leader gone, the side branches grow stronger and longer, and the height of the tree is kept in check. Modified central leader form is suggested for apples, pears, plums, and sour cherries.

Espalier Forms

Espaliers can take many forms (Figure 7-7). Training must start as soon as the tree is planted; it consists of pruning, bending, and tying the branches as they mature to achieve a symmetrical shape.

Although espaliered fruit trees do not produce as large a crop as conventionally pruned trees, the amount is acceptable considering the small space they use. Apple and pear trees are among the best for espalier training. Peach and plum trees are more difficult but possible. Cherries and nut trees should not be espaliered because their production will be seriously decreased.

Training

Whereas pruning is the primary method for training a tree, several other supplementary techniques help create the desired form. Many fruit trees have branches that grow upward instead of outward, with narrow branch angles called *crotches*. As the branches enlarge, they exert pressure at the crotch, causing it to become weak with a tendency toward splitting under a snow or fruit load. However, when young and flexible, a branch can be trained to develop a wide crotch by two methods.

Tying (Figure 7-8) involves bending a flexible young branch down and securing it to the ground with a stake. The branch should be tied for one or two seasons and frequently checked to make sure it is not being injured. After the rope is removed, the branch will grow outward instead of upward.

Figure 7-8 Tying a fruit tree branch to the ground to encourage a wide branch crotch to form.

Spreading (Figure 7-9) uses forked or nail-tipped pieces of wood wedged tightly between a young branch and the trunk of a tree to widen the crotch angle. As with tying, the spreader is left in place for one or two seasons to encourage the crotch to remain open.

Fruit Thinning

Fruit thinning is the removal of a portion of the fruits on a tree while they are still small. Many deciduous fruit trees (nut trees are excluded) produce more fruit than can fully mature. If unthinned, the fruit will be

Figure 7-7 Sample espalier forms for fruit trees.

Figure 7-9 A branch spreader in place in a tree. *(Photo by Rick Smith)*

undersized and frequently poorly colored. In addition, the excessive fruit load may weaken the tree. Bearing in alternate years may result, with the fruitless year devoted to regaining the vigor of the tree. Because most gardeners want fruit every year, the advantages of thinning are obvious.

Fruits should be thinned when they are 1/2 to 3/4 inch (13–20 millimeters) in diameter, and after the natural drop of young fruit has occurred. Twisting fruits will remove them quickly and easily.

The number of fruits that should be left on a tree depends on the species. Table 7-1 lists approximate distances that should be left between fruits along a branch for moderate fruit load. In practice, fruits are not spaced evenly but are close together on strong branches and far apart on weak ones. The spacing guidelines should be taken only as a guide and not an absolute. The largest, healthiest fruits should always be left on the tree, regardless of their relative spacing on a branch.

If a gardener has a large orchard of, for example, apples, hand thinning may be impractical. In that case the trees may be thinned chemically with a hormone application shortly after bloom. The gardener should consult his county Cooperative Extension Agent for specific recommendations.

Table 7-1 Suggested Fruit Load after Thinning for Common Fruit Species

Species	Distance between fruits
apples	6–8 in. (15–20 cm)
apricots	1½–2 in. (4–5 cm)
avocados	Thinning not necessary
cherries	Thinning not necessary
citrus, figs, mulberries	Thinning not necessary
nectarines, olives	4–5 in. (10–13 cm)
peaches	6–8 in. (15–20 cm) for early varieties, 4–5 in. (10–13 cm) for later ones
pears	Thinning not necessary but can be thinned lightly to increase size
persimmons	Thinning not necessary
plums, European	Thinning not necessary
plums, Japanese	2–4 in. (5–10 cm)
quinces	Thinning not necessary

Renovating Neglected Fruit Trees

When buying an older home, it is common to inherit at the same time at least one neglected fruit tree. The tree frequently will be bearing sparsely, and the fruit produced will be small and of poor quality. It may have dead limbs or be diseased.

The question of whether or not to save the tree is a personal choice. It will probably never bear the quality of fruit of a newer cultivar, but fruit quality can be improved to an acceptable grade if the tree is given some attention. The tree may be valuable for shade and flowers and thus be worth renovating from this standpoint.

Renovation pruning should be spaced over three years to avoid stimulating excess sucker growth. The first pruning should remove dead wood, suckers, and some of the crossing branches. (See Chapter 11 for pruning specifics.) Leaves and fruits that drop in the fall serve as a source of disease reinfection. They should be removed, and a disease and pest control program should be started.

During the next two seasons excess branches can be gradually removed to open up the head, admitting light and improving fruit production. Fertilizer can be withheld until after pruning is complete but should be applied immediately if deficiency symptoms are present.

FAILURE TO BEAR

The failure of a tree to bear may be caused by one or a combination of the following factors.

Age

The age at which a tree can be expected to produce its first crop varies with the species. Most dwarf and semidwarf types bear by the third or fourth year after planting.

Pollination Problems

In most fruit and all nut trees, there must be pollination and fertilization to set a crop. Exceptions to this rule are rare but include fruits such as navel oranges, persimmons, and some figs and pears. These fruits are able to set *parthenocarpically*, that is, without fertilization.

The remaining tree crops require pollination, either self or cross. Self-pollination is the rule on most nut trees, the majority of which bear separate male and female flowers on the same tree and are pollinated by wind. Self-pollination is also found in a few fruit trees, but more often fruit trees are incapable of pollinating themselves and are called *self-sterile*. Pollination with another *compatible* cultivar flowering at the same time is necessary if a crop is to be produced.

The causes of self-sterility are primarily the following:

1. *Incompatibility of the pollen with the female flower parts.* This can be compared to an allergic reaction in which the pollen grains do not function correctly when in contact with the female flower parts of the same plant. Pollen may fail to germinate or grow too slowly to the egg, even though it will perform normally with other cultivars.

2. *Nonviable pollen.* The pollen does not germinate.

3. *Failure of the pollen and female flower parts to attain maturity at the same time.* Pollen is shed after the female flower parts have passed their receptive stage or before they have attained it.

4. *Dioecious plants.* In the case of a few tree crops such as the date, male and female flowers are produced on separate plants. There is simply no pollen produced with which the tree can self-pollinate.

In summary, cross-pollination with a compatible cultivar flowering at the same time is a prerequisite for successful growing of many tree crops and should be carefully taken into account when purchasing trees.

Insufficient Winter Chilling

In climates with mild winters the winter temperatures may not be sufficiently low to provide the necessary chilling period to produce flowers. Deciduous fruit and nut trees are most susceptible.

Insufficient Sunlight

A tree receiving insufficient sunlight will produce lush vegetative growth but few flowers and fruits.

Frost or Rain during the Pollination Period

These two weather conditions can damage flowers or restrict bee activity. Thus, no crop will set.

COMMERCIAL FRUIT PRODUCTION

Planting

High-density planting of fruit trees to maximize yield per acre or hectare is the most modern method of commercial production of fruits such as pears, plums, sweet cherries, and particularly apples. Formerly, full-size trees were planted at wide spacings that allowed the trees to develop to full size before their canopies touched. However, this left the bulk of the land uncovered by foliage for years while the trees matured. Thus, production cost *per acre over time in years* was high because of this downtime. Recently cultivars on dwarfing rootstocks previously used only for home production in small spaces have been used for commercial production.

The main concept behind this planting technique is the relationship between orchard production (measured by the amount of fruit harvested) as compared to the area of the fruiting canopy (foliage area that is occupied by fruiting branches) per acre (hectare) or canopy acre. The speed of canopy establishment is closely related to the volume of fruit produced and, hence, the dollar return. In other words, the faster the trees grow to make a canopy that makes maximum use of the photosynthetic area of the land, the sooner profit will be made by the grower. In most cases, taking into account the increased cost for increased numbers of trees to plant the area, it is still financially advantageous to growers to use the intensive technique.

Intensive orchards for fruits such as pears and apples use tree densities of up to 2,000 trees per hectare or greater (typically 11–12 feet or 3.5 meters between rows and 5 feet or 1.5 meters in the row) using the recently available dwarfing rootstocks. The trees are rarely over 8 feet (2.5 meters) in height, so that picking-labor cost is reduced substantially.

The trees are often characterized by short internodes and a dense foliage head. Sometimes they are columnar in form, and there are even experiments using weeping forms of peaches for high-density planting.

One drawback of the intensive planting method is that in some cases the trees are incapable of supporting the heavy fruit load without the branches breaking. Trellising or another training system must be put into the orchard to support the branches. This is an additional cost, but it is not enough to offset the increased profit from early, heavy production.

Irrigation

Selecting an irrigation system depends on knowledge of the crop's water requirements as well as a basic understanding of the relationships between water, plants, and soil, as explained in Chapter 5.

The amount of water required by the crop changes during the growing season because more water is required during fruit development, ripening, and flowering. Water trees during any of these crucial periods of "peak water demand" or a less than an optimum amount of water in the soil will result in reduced yield or crop quality.

For fruit trees, microirrigation, or drip irrigation, is the preferred method of water application. It is efficient both in terms of water conservation and energy conservation because water is applied at low pressure and directly to the soil so that there is no evaporative loss. In a soil with a lot of clay, water will move sideways faster than it will move downward into the soil. In a loamy soil, the water will move sideways and downward at about the same rate, wetting the soil in an almost circular pattern. But, in a sandy soil, the water moves downward faster than it moves sideways. Thus, the pattern of wetting must be taken into account to determine how long the trees must be irrigated to wet the entire root volume.

To set up a drip irrigation system, the grower will need many components:

- *Source emitters,* such as drip tape, soaker hose, or a heavy-wall drip line to which point-source emitters are attached, are crucial. Flow rate is pressure dependent and therefore varies by the thickness of the plastic. Plastic thickness also affects the emitters durability and longevity in the orchard. A rating of 25 mil (for up to 25 years of use) is normal in an orchard. If a heavy-wall drip line is used with point-source emitters, the tubing diameter will depend on the distance the water must flow. Smaller diameters are suitable for orchards in which the water does not need to run very far to reach all the trees, and larger diameter tubes are suitable for longer runs.

- *Point-source emitters* are small, water-emitting plastic devices that actually distribute the water to the plant (Figure 7-10). These are normally spaced 12 inches (30 centimeters) apart, but in sandy soils they may need to be placed closer together. Likewise, with large perennial crops, one may need more than one emitter per plant.

- *Water filters,* are used to make sure that no soil or other particles enter the water line. Not all water sources are suitable for drip irrigation. The solids content of the water will affect the functioning of the system, because a high solids content can plug the emitters. A 150-mesh stainless-steel screen is adequate to remove particles that would clog the system when using a commercial water supply or groundwater.

  There are also sand filters. Sand filters are the best choice when irrigating with slow-moving or stagnant surface water such as from a pond. A sand filter will need to be followed by a screen filter. Regardless of the filter type chosen, the filter will need to be cleaned periodically by flushing.

- *Pumps, pressure regulators, and valves.* A qualified irrigation designer will select the correct pump size and other components based on field slope, acreage, and soil type, and decide on the spacing of the source emitters based on their flow rates.

Fertilizing

See Chapter 5.

Figure 7-10 A microsprinkler head used with a drip irrigation system. *(Photo courtesy of Chapin Watermatics, Watertown, NY)*

Disease and Pest Control

See Chapter 13.

INDIVIDUAL CROPS AND THEIR GROWING REQUIREMENTS

Almonds (*Prunus amygdalus* var. *dulcis*)

Almond growing is restricted mainly to USDA zones 7 to 9 and to the western states. The primary problem of growing in colder zones is not minimum winter temperature but an early bloom period that hinders pollination and sometimes frosts the blossoms. The spring rains and high humidity of the east are also a hindrance to pollination and nut production in that climate. Lack of winter chilling excludes almonds from zones 10 and 11.

Almond trees are available in only one size, and they grow 20 to 30 feet (6–9 meters) high. They should be pruned to a modified leader or vase form and can be expected to bear the third or fourth year after planting. Cross-pollination with another cultivar is required in almost all cases.

The almond is produced inside a shell, which is in turn encased in a peachlike hull. In fall the hulls split, dropping the ripe nuts in their shells to the ground.

Apples (*Malus pumila* descendants)

Apples are grown in zones 4 through 9. Special tropical cultivars must be selected for zone 9 because apples have a chilling requirement. Most apples are self-fertile and are available in dwarf, semidwarf, and standard sizes.

For greatest strength, apples should be pruned to a modified leader system. Dwarfs can be expected to bear the first or second year after planting.

Apples are highly susceptible to disease and insect problems, and regular spraying will usually be required for good fruit. However, disease problems can be minimized by selection of disease-resistant cultivars.

Apricots (*Prunus armeniaca*)

Like apples, apricots require winter chilling. They can be grown in zones 5 through 8 and some parts of zone 9. Apricots do not have a cross-pollination requirement and are available in standard and semidwarf sizes.

Apricots bear one to two years after planting and are among the most ornamental of fruit trees. However, their early bloom period exposes them to frost damage. Training can be done in either a modified leader or a vase shape, and pruning should be done yearly to restrict height and increase spread.

Avocados (*Persea americana*)

Avocados are strictly subtropical and tropical (zones 9–11) and are hardy only at 20–24°F (−7 to −5°C). Their deep green, evergreen foliage and round forms make them valued as ornamentals.

Avocados divide into three races: the Guatemalan, with large fruit; the West Indian, which has large fruit but is more tropical; and the Mexican, which is hardier but has smaller fruit. Most cultivars are available only in standard size, though the cultivar 'Little Cado' is dwarf. When mature they range from 30 to 60 feet (10–20 meters) tall (depending on the cultivar) and may be of equal width. Avocados tend to biennial bearing, and although cross-pollination is not essential, it will result in a better fruit set. Avocado cultivars are divided into two groups, A and B, based on their pollination requirements. For best results choose one tree from each group.

Avocados require little pruning. Flowers appear from winter through spring, depending on the cultivar, and the fruit will mature 7 to 12 months later. About three years is average from planting until the first set of fruit.

Cherries (*Prunus avium* and *P. cerasus*)

The sweet cherry (*Prunus avium*) is known for fresh eating, whereas the tart cherry (*Prunus cerasus*) is most useful for pies and other cooking uses. Both have a chilling requirement. The sweet cherry can be grown in zones 5 to 7; the tart cherry is more cold-hardy and will survive to zone 4.

Sweet and tart cherries are available in dwarf, semidwarf, and standard sizes. Although tart cherries do not require cross-pollination to produce fruit, sweet cherries have very particular pollination requirements. Careful attention must be paid to selecting compatible cultivars from among the many sweet cherry cultivars available.

Pruning of cherries should be modified leader for sweet types and modified leader or vase for sour cultivars.

Chinese Chestnuts (*Castanea mollissima*)

The chestnut that was originally grown for eating in North America was the American sweet chestnut. But in the early 1900s a blight disease killed nearly all, and now the Chinese chestnut is the primary source of nuts. This blight-resistant tree will grow in zones 4 through 8 and averages 40 to 50 feet (12–15 meters).

The chestnut will set nuts without cross-pollination, but a heavier set will result if another tree is situated nearby for pollination. Bearing will begin after five to eight years if the plant is from seed, or after two years with a grafted nursery plant. The nuts are perishable and should be stored in the refrigerator in a plastic bag to prevent drying.

Citrus (*Citrus* spp., *Fortunella* spp.)

All citrus are evergreen shrubs or trees adaptable to growing in zones 9 and 10. A few such as the kumquat will survive in zone 8. The major citrus grown in North America are grapefruit (*Citrus × paradisi*), kumquat (*Fortunella margarita, F. japonica*), orange (*Citrus sinensis*), lemon (*Citrus limon*), lime (*Citrus aurantifolia*), and tangerine or mandarin orange (*Citrus reticulata*).

Most are available in standard and dwarf sizes and do not require cross-pollination. Citrus require little pruning. Suckers and crossing limbs may have to be removed occasionally.

Because of their year-round growth cycle, citrus should be fertilized twice per year instead of once. They bear when very young, and most species retain their fruits in excellent condition on the plant for many months, the exception being the tangerine.

Figs (*Ficus carica*)

The fig is a deciduous tree or large shrub grown in zones 7 through 10. It can even be grown as far north as zone 5 with winter protection. Although the top will die back to the ground, it will resprout each spring from the roots and produce a crop.

Depending on the cultivar, the fig can grow from 30 to 80 feet (10–25 meters) at maturity. No dwarf forms are available. The fruit sets parthenocarpically (without pollination) when the plant is about four years old. In zones 9 and 10 two crops per year (one in June and one

Figure 7-11 Hazelnuts are excellent nuts for cultivation on small properties. *(USDA photo)*

Figure 7-12 Hickory nuts. *(Photo courtesy of Stark Bros. Nurseries, Louisiana, MO)*

in August to November) can be expected. In colder zones only one will set and is matured in late summer.

Because fig trees can grow very large, pruning is necessary to keep them low enough to harvest the fruit. Young trees should be trained to a low, flat, vase shape, but after achieving this form, no seasonal pruning is recommended.

Hazelnuts (*Corylus avellana*)

The hazelnut (Figure 7-11) is a small tree or shrub that grows 10 to 15 feet (3–5 meters) in height and equal that in width. It is hardy in zones 4 through 9. Oregon's Willamette Valley produces 90 percent of U.S. hazelnuts.

Hazelnuts bear separate male and female flowers on the same plant and are pollinated by wind. Although they will set some nuts without cross-pollination, a better set can be expected with it.

Little pruning is required for hazelnuts except for removal of suckers and thinning to permit light to enter the center of the plant. They bear when two to three years old, and a harvest of 5 to 10 pounds (2–5 kilograms) of nuts is average for a mature 10- to 15-year-old plant with cross-pollination. The nuts are harvested after drop.

Hickory Nuts (*Carya ovata* and *C. lacinosa*)

Hard-shell hickories (Figure 7-12) are related to pecans; in fact there are many hybrids between them called *hicans*. They are grown in zones 4 through 8. Hickory trees are rugged looking, with taproots and a strong central leader form. They bear at 3 to 4 years of age if grafted but may take 10 to 15 years from seed.

A mature hickory is quite tall (up to 70 feet; 20 meters), and pruning is practical only when the plant is

young. Many cultivars are self-sterile and require cross-pollination. The nuts are harvested from the ground.

Mulberries (*Morus alba, M. rubra, M. nigra*)

Mulberry trees of one of the above species grow in any zone of the United States. Many are used as ornamentals; fruitless and/or grafted weeping types are sold for this purpose.

The species planted will largely determine its ultimate height. The white mulberry (*Morus alba*) will grow to 80 feet (25 meters) and produce white, pinkish, or purple fruits. *Morus nigra* is the shortest at 30 feet (10 meters), and the southern red mulberry (*Morus rubra*) is intermediate at 60 feet (20 meters).

Mulberries are monoecious, with flowers of separate sexes. Pruning should aim to maintain the naturally spreading branches low so that the fruits will be within reach when ripe.

Nectarines (*Prunus persica* var. *nucipersica*)

Nectarines are a mutation of peaches and not a hybrid between a peach and plum, as is popularly believed. They have been cultivated since the days of the Roman empire.

Culture of nectarines is the same as for peaches, except that they are more susceptible to fruit rots and may require more spraying. They do not require cross-pollination and should be trained to a vase shape. Both dwarf and standard sizes are available. Standard trees are moderately small at maturity, 12 to 15 feet (4–5 meters). USDA zones 5 through 10 and sometimes 11 are suitable for growing nectarines.

Olives (*Olea europaea*)

Olives will grow in zones 9 and 10 or wherever winter temperatures do not fall below 12°F (−11°C). They require hot summer temperatures to mature their fruits, but they are grown for their ornamental value alone in coastal climates where excess humidity decreases the fruit production.

Olive trees are small, 25 to 30 feet (8 to 9 meters), single- or multitrunked trees with silver-gray foliage. They transplant easily (even at a very old age) and are moderate to fast growing. Pruning does not seek to create any particular form, but it should develop strong scaffold branches and remove suckers that would eventually cause a bushlike form. Dead wood and twigs are common and should be removed yearly.

Cross-pollination is not a prerequisite for olive production. Biennial bearing is common but can be minimized by fruit thinning to leave two to three olives per foot of branch. Olives are picked green and processed in lye water and brine to make Spanish or black olives. Picked ripe, they are soaked in brine and spices to make Greek-style olives.

Peaches (*Prunus persica*)

Peaches grow in zones 5 through 10, although care must be exercised in selecting a cultivar with a low winter chilling requirement for zones 8 through 10. They are generally trained to a vase shape atop a short 18-inch (0.5 meter) trunk. Pruning should be done every year to thin the head and keep the top low. Bearing begins at about the third year and will peak during the eighth to twelfth years, with no cross-pollination required. Compared to other fruit trees, peaches are short-lived; trees tend to decline rapidly after 10 to 12 years as a result of winter injury and wood rot infection.

Pears (*Pyrus communis*)

By choosing cultivars carefully, pears can be raised in zones 4 through 9, although zones 5 through 8 provide the best climate. Most pear cultivars are vulnerable to the bacterial disease "fire blight," and highly susceptible cultivars such as 'Bartlett' should not be planted east of the Rocky Mountains.

The trees are long-lived, and although some are self-fertile or set fruit parthenocarpically, most need cross-pollination. Pears are available in dwarf, semi-dwarf, and standard sizes and should be trained to a modified central leader. They naturally grow very upright, and spreading of the branches may be required.

Figure 7-13 Pecan nuts. *(Photo courtesy of Stark Bros. Nursery, Louisiana, MO)*

Pecans (*Carya illinoinensis*)

The pecan (Figure 7-13), like many other nut trees, has a taproot and a strong central leader growth habit. It is available in only one size and will easily grow to 60 or more feet (18 meters) tall when mature.

Zones 6 through 8 are best for pecan growing. The tree is hardy even further north, but the growing season is usually not long enough to mature the nuts.

Pecans begin bearing at four to seven years. Trees are monoecious, but more and larger nuts will be borne if cross-pollination is provided. A biennial bearing tendency may develop in older trees. Compared to other nut trees, pecans require relatively high maintenance. In alkaline soils, zinc deficiency is common.

Persimmons (*Diospyros kaki*)

The persimmon (Figure 7-14) is a deciduous tree useful as both an ornamental and a fruit tree in zones 6 through 10 and occasionally in zone 5. The fruits, which are apple size and brilliant orange, mature after the leaves drop. Some people question the edibility of persimmon fruits because of their unpleasant astringency when they look and feel ripe. Astringency is partly varietal, but it can be avoided by ripening the fruits until they are extremely soft or by freezing them before eating.

Persimmons are available only in standard or semi-dwarf size, with standards growing into round-headed trees 20 to 60 feet (12–18 meters) tall at maturity. The species has brittle wood and should be pruned to modified leader form for strength. Annual pruning may be used to maintain a small size. Most bear fruit within two to four years after planting, and most do not require cross-pollination.

Figure 7-14 Persimmon fruits. *(Photo courtesy of W. Atlee Burpee and Co., Warminster, PA)*

Plums (*Prunus domestica and P. salicina*)

Plums are of two types, the European (*Prunus domestica*) and the Japanese (*P. salicina*). The European group includes the blue and prune plums including the well-known 'Damson' cultivar. Japanese plums encompass the larger red and yellow plums popular for fresh eating. They are very early bloomers and should be rejected in favor of Europeans in climates where frosts frequently kill fruit tree flowers.

Plum trees are available in both semidwarf and standard sizes. Their training method and pollination requirements vary by type. Europeans are most often trained to a modified central leader and will usually set fruit without a cross-pollinating cultivar. Japanese usually require crossing with another Japanese cultivar and are normally trained to be vase-shaped.

Zones 5 through 10 are suitable for plum growing. The trees are long-lived, produce outstanding spring flowers, and can be expected to bear the first crop three to four years after planting.

Quinces (*Cydonia oblonga*)

The quince was a much more popular fruit a few years ago than it is now. It is used today primarily for preserves and for baking whole like an apple. The fruit is greenish-yellow when mature and about the size and shape of an apple. It is covered with a peachlike fuzz that is wiped off before the fruit is used. The fruit is picked while hard, preferably after a light frost, and should be treated gently to avoid bruising.

Quinces grow in zones 5 to 10 on spreading trees that grow to 15 feet (5 meters) in height. They are slow growing, and little pruning is required beyond the initial vase-shape training. The fruit is produced on the current season's growth.

Bearing age for quinces is three to four years, and about one bushel of fruit per tree can be expected at peak bearing. Cross-pollination is not required, but spraying for diseases and insects is usually necessary because fire blight is a common problem.

Figure 7-15 English walnuts. *(USDA photo)*

Walnuts (*Juglans regia* and *J. nigra*)

Walnuts are of two types: English (Carpathian) (*Juglans regia*) (Figure 7-15) and black (*Juglans nigra*) (Figure 7-16). The English type is the smooth-shelled walnut raised commercially and sold in nut baskets. The black walnut is a rough-shelled, native American tree more commonly grown for shade than nuts. Both are hardy in zones 5 through 10 and reach an average height of 50 feet (15 meters) or more when full grown. However, the summer in the colder zones may be insufficient to develop kernels of English walnuts, resulting in empty nuts.

Figure 7-16 Black walnuts. *(USDA photo)*

Normally, walnut trees are not pruned except to remove crossing and dead branches; they naturally grow in a vase or modified central leader shape. Cross-pollination is not necessary, and the nuts are harvested after they drop. Small crops can be expected starting two years after planting.

Care should be taken to plant walnut trees away from ornamental plantings and vegetable gardens. This is necessary because the roots secrete a chemical that is toxic to many other plants.

SELECTED REFERENCES FOR ADDITIONAL INFORMATION

Bountiful Ridge Nurseries. *Retail Catalog.* Princess Anne, MD 21853.

California Nursery Company. *Retail Catalog.* Niles District, Fremont, CA 94536.

Childers, N. F. *Modern Fruit Science: Orchard and Small Fruit Culture.* 9th ed. Gainesville, FL: Norman F, Childers Publications, 1995.

Cumberland Valley Nurseries. *Retail Catalog.* McMinnville, TN 37110.

Downing, Andrew J. *Fruits and Fruit Trees of America.* Temecula, CA: Reprint Services Corporation, 1993.

Fitzgerald, Toni J., and Rhonda M. Hart. *Gardening in the Inland Northwest: Growing Vegetables, Berries, Grapes and Fruit Trees.* Chehalis, WA: Painted Sky, 1997.

Hansen's New Plants. *Retail Catalog.* Fremont, NB 68025.

Hilltop Orchard and Nurseries. *Retail Catalog.* Rt. 2, Hartford, MI 49057.

Hayden, R. A., M. A. Ellis, and R. E. Foster, eds. *Midwest Tree Fruit Handbook.* Upland, PA: DIANE, 1998.

Jackson, David and N. E. Looney, eds. *Temperate & Subtropical Fruit Production,* NY: Oxford University Press, 1999.

Lundy's Nursery. *Retail Catalog.* Rt. 3, Box 35, Live Oak, FL 32060.

Maib, Karen M., Preston K. Andrews, Gregory A. Lang, and Kent Mullinix, eds. "Tree Fruit Physiology: Growth & Development: A comprehensive manual for regulating deciduous treefruit growth & development." *Good Fruit Grower.* Dec. 1996.

Otto, Stella B. *The Back Yard Orchardist.* Maple City, MI: Otto Graphics, 1996.

Otto, Stella B. *The Backyard Orchardist: The Complete Guide to Growing Fruit Trees in the Home Garden.* 2nd ed. Maple City, MI: OttoGraphics, 1995.

Salunkhe, D. K. and S. S. Kadam. *Handbook of Fruit Science & Technology: Production, Composition, Storage, & Processing.* Boulder, CO: netLibrary, Incorporated, 1999.

Somerville, Warren. *Pruning & Training Fruit Trees.* Woburn, MA: Butterworth-Heinemann, 1996.

Stark Bros. Nurseries. *Retail Catalog.* Louisiana, MO 63353.

Swenson, Allan A. *Fruit Trees for the Home Gardener.* New York: Lyons Press, 1994.

Westwood, Melvin N. *Temperate-Zone Pomology.* 3rd ed. Portland, OR: Timber Press, 1993.

World Wide Web Addresses

http://www.caf.wvu.edu/kearneysville/wvufarm1.html University of West Virginia. "Kearneysville Tree Fruit Research and Education Center" 2001.

http://www.ext.vt.edu/pubs/envirohort/426-841/426-841.html Relf, Diane. "Tree Fruit in the Home Garden" August 2000.

http://www.tfrec.wsu.edu/ Washington State University. "Tree Fruit Research and Extension Center" Oct 19, 2001.

http://tfpg.cas.psu.edu/ Penn State College of Agricultural Science. "Pennsylvania Tree Fruit Production Guide 2000-2001" 2001.

(Photo by the author)

CHAPTER 8

BUSH AND OTHER SMALL FRUITS

The species classified as small fruits are the small perennial plants such as strawberries, grapes, raspberries and blackberries (called *brambles*), and blueberries. Currants, gooseberries, and cranberries are less well-known members.

Because of their size, small fruits can be grown on almost any property. On a large property, enough fruit can be grown for canning and freezing, as well as fresh eating. The small fruits are easier to grow than many tree crops. Cross-pollination is seldom required, and spraying for disease and insects is uncommon. Only one or two species are regionally limited, and most bear within one or two years after planting. By carefully choosing cultivars to mature serially throughout the growing season, a harvest of small fruits can be assured from early summer until frost.

PLANNING THE SMALL FRUIT GARDEN

Site Selection

Rapid soil drainage and full sunlight are the two site prerequisites for successful small fruit growing. An area where water stands after a rain is unsatisfactory and will kill plantings quickly. Without full or nearly full sun, the plants will grow only vegetatively and will fail to bloom or ripen fruit. Beyond these considerations, virtually any site is acceptable. Areas with good garden soil are ideal, but even poor-soil areas can be amended to a satisfactory condition with compost or other organic soil additives.

Many gardeners prefer to plant all their fruits in one area. If this is the case, care should be taken to space and arrange the plants to prevent the taller species from shading the smaller species (strawberries). Because the sun in the northern hemisphere is always slightly in the southern sky, this requires placing the taller plants north of the intermediate and short ones. Figure 8-1 gives two sample layouts for fruit garden areas.

Small fruits can also be incorporated in a landscape. Currants, gooseberries, and brambles can be used as hedge plantings that will grow to 4 to 5 feet (1.2–1.5 meters) tall when mature. Highbush and rabbiteye blueberries grow even taller, reaching 10 feet (3 meters). Low-growing strawberries produce lovely white blossoms in spring and make an excellent border for a perennial flower bed. They can also be used as groundcover plants in small sunny areas or in planters around a patio or deck area. Finally, vining grape plants can be trained to an overhead arbor or espaliered against a wall or fence.

Plant Selection

Of the small fruits discussed in this chapter, all but the cranberry are highly climatically adaptable. The choice is therefore a matter of personal preference and, most important, selecting a cultivar adapted to the climate. The county or state Cooperative Extension Service will furnish literature on the small fruit cultivars most successful in a particular state.

Figure 8-1 Two sample layouts for small fruit gardens.

GENERAL GROWING REQUIREMENTS FOR SPECIFIC SMALL FRUITS

Blackberries (*Rubus* spp.)

Blackberries are one of the bramble fruits. They look essentially like raspberries except the fruits are larger, and when picked, the center core remains in the berry (with a raspberry it pulls out at picking). They withstand more heat and drought but less cold than do raspberries.

 Blackberries and raspberries are particularly good crop choices for those who wish to run their own U-Pick farms. A person who likes to work with the public and also enjoys plants would be a good candidate to own his or her own U-Pick farm.

Blackberries fall into two categories based on growth habit, height, and climate adaptation. Erect blackberries grow 3 to 5 feet (1–1.5 meters) high and grow without supports. The canes of trailing blackberries reach 6 to 8 feet (2–2.4 meters), and the plants must be grown on a trellis or staked for support. Erect blackberry types are most often grown in the East, Southwest, and Midwest and are the more cold-hardy of the two types, surviving winter temperatures of down to −20°F (−29°C). Trailing types are the traditional blackberries of the Pacific Coast states and southern states, although they are occasionally grown as far north as Michigan.

There is often confusion in blackberries when the names *dewberries*, *boysenberries*, *loganberries*, and *youngberries* are used. Dewberries can be one of any species or cultivar of trailing blackberries. Loganberries, actually the cultivar 'Logan,' are a cross between a blackberry and a raspberry. Another blackberry/raspberry cross cultivar is 'Phenomenal,' which was later crossed with a dewberry to produce the cultivar 'Young,' hence *youngberries.* The origin of boysenberries is similar to that of youngberries.

Blueberries (*Vaccinium* spp.)

Blueberries will grow from northern Minnesota and Canada to Florida and to the West Coast; however, they have a soil pH requirement of between 4.2 and 5.5. Because soils that are naturally this acidic are uncommon, pH modification is essential for blueberry growing in most cases. Chapter 5 discusses the use of ammonium sulfate and other compounds for lowering soil pH. Blueberries also grow best in soils high in organic matter. The addition of organic matter such as compost or peat moss will satisfy this need and help to keep the soil pH low.

The species of blueberry that should be raised depends on the regional climate. Table 8-1 lists the three main cultivated blueberry species and their identifying characteristics and geographic range. Two cultivars of blueberries should be planted for cross-pollination, as some are nearly self-sterile, and both more and larger fruits will be produced if a pollenizer is provided.

Besides the cultivated species, additional species such as the dryland blueberry (*V. pallidum*), mountain blueberry (*V. membranaceum*), and lowbush blueberry (*V. angustifolium*) grow in the United States and Canada. However, they are normally wild, and cultivars selected for home growing should be chosen from the three cultivated species.

Cranberries (*Vaccinium macrocarpon*)

Cranberries have exacting growth requirements: moist, peat-bog conditions and a pH of 4.2 to 5.0. Because these conditions are hard to create, cranberries cannot be recommended as a home fruit.

Figure 8-2 White currants. *(USDA photo)*

Currants (*Ribes sativum* and *R. nigrum*)

Currants (Figures 8-2 and 8-3) are very resistant to cold temperature and will survive in even the coldest areas. Their growing area spans the northern United States and Canada wherever moist, cool conditions exist. In the South they grow less satisfactorily due to prolonged summer heat.

Table 8-1 Cultivated Blueberry Species for Home Gardens

Species	Characteristics	Range
Rabbiteye blueberry (*Vaccinium ashei*)	6–8′ (2–2.4 m) with large fruits; moderately heat- and drought-resistant	Southern states and other areas having a frost-free growing period of at least 160 days
Highbush blueberry (*Vaccinium corymbosum*)	10–15′ (3–4.6 m) tall (unpruned); not drought-resistant; sand and peat soil best; hardy to −20°F (−29°C), requires winter chilling and should only be grown north of zone 8	Zones 5–7, East Coast, Pacific Northwest, Midwest
Evergreen blueberry or evergreen huckleberry (*Vaccinium ovatum*)	Evergreen with ornamental foliage used by florists; grows to 15′ (4.6 m) tall, fruits are small, shiny, black, and strong-flavored	Pacific Coast from central California to British Columbia

Figure 8-3 Currants. *(Photo courtesy of Dr. Danny L. Barney)*

 Cranberries are a very specific commercial crop whose production is restricted to certain areas of Massachusetts, Michigan, Canada, New Jersey, Washington, Wisconsin, and Oregon. Their production is unique, and a person interested in becoming a specialist in cranberries would need to focus his or her education in this field and then secure work in that area. Individual growers work cooperatively under the popular "Ocean Spray" label.

The plants are small, attractive shrubs ranging in height from 3 to 5 feet (1–1.5 meters). The plants should be set 4 to 5 feet (1–1.5 meters) apart when grown in the garden or about 3 feet (1 meter) apart when used as an ornamental hedge.

The most popular currants are the red varieties (*Ribes sativum*), which produce large clusters of fruits and are easy to grow. White (*Ribes sativum*) and black (*Ribes nigrum*) currants are relatively rare, but the black varieties are popular in Europe, where the fruits are made into juice.

Gooseberries (*Ribes hirtellum* [American] and *R. uva-crispa* [European] cultivars)

The growing requirements of gooseberries are similar to those of currants: moist soil and average-to-cool summers such as are found in the Midwest, North, and northern Pacific states. These species are very cold tolerant but will not thrive in a hot summer climate. In warm summer areas planting in partial shade will reduce summer heat and may allow the crop to be raised successfully.

Gooseberries are unknown to most home gardeners. They are shrubs about 3 feet (1 meter) high and densely covered with thorns (with the exception of one or two cultivars). The fruits are the size of large grapes and are green to red when mature, depending on the cultivar. They can be eaten fresh, as well as made into pies, jam, or jelly.

Grapes (*Vitis* spp. and hybrids)

Grapes can be grown for fresh eating, wine making, or making into jellies and juice. When a cultivar is selected, the choice should be based on climate and the main use for which the fruit will be raised. Native American grape cultivars (*Vitis labrusca*) are the easiest to raise and very productive. The familiar Concord grape is one American type used commonly for jellies and juice, but others are available that are of fresh-eating or wine-making quality.

The European grape (*Vitus vinifera*) is grown in warmer areas and is the primary species for wine making. Excellent cultivars for fresh eating such as 'Thompson's seedless' are also included in this group.

 A horticulturist interested in grape production will find many opportunities in production of table grapes for fresh eating and especially in production of wine grapes. A grape specialist is in charge of planting, pruning, pest and disease control, and harvesting. Winemakers (enologists) are frequently horticulturists, and this specialization requires particular knowledge of cultivars, precise time of harvest, and food science to make a high-quality wine.

The French-American hybrid grapes combine the vigor of American cultivars with the wine-making quality of European grapes. The resulting varieties are excellent for wine use and are raised throughout the country for that purpose.

Finally, muscadine grapes (*Vitis rotundifolia*) are the least common, being confined to the southern states where temperatures do not fall below 10°F (−23°C). The fruits are large and are borne in small clusters.

Raspberries (*Rubus* spp.)

Raspberries (Figure 8-4) are the second bramble fruit, growing wild in some areas of the country. Red (*Rubus indaeus*) and black (*Rubus occidentalis*) raspberry culti-

Figure 8-4 Raspberries in spring, showing the previous season's canes above and the current year's canes growing from below. *(Photo courtesy of Dr. Bernadine C. Strik, Oregon State University)*

vars are the most common, with the red having a slightly tarter taste than the black, more cold tolerance, less susceptibility to disease, and greater yield potential. Novelty cultivars in yellow and purple are also available.

Raspberry cultivars (along with strawberries) can be *spring-bearing* or *everbearing* (also called *primocane*). Everbearing plants produce two harvests per year: one in early summer and a second in fall.

Strawberries (*Fragaria × ananassa* cultivars)

Strawberries are one of the most popular small fruits. Although they are perennial, the productive life of a strawberry planting is limited to about two to four years. The first year following planting, a heavy crop will be produced. The second year, the plants will bear only two-thirds as much, and the third year, only one-third as much. The decline is due to invasion of the plants by viruses that lessen their productivity, hybrid vigor decline, and competition from weeds. The bed should be dug out and replanted.

 A strawberry producer can run his or her own U-Pick farm, grow berries for seasonal fresh sale, or grow for the frozen fruit market. Commercial strawberries are replanted yearly. Those interested in strawberry propagation could also secure employment in select production nurseries, which supply large numbers of virus-free plants to growers on a yearly basis.

Like raspberries, strawberries are available in *spring-bearing* (also called *June-bearing*) and *everbearing* (called *day-neutral*) cultivars. The everbearers produce lightly throughout the summer until fall, but the quality of the berries is sometimes inferior to that of spring-bearing plants.

PLANNING, SOIL PREPARATION, AND PLANTING

Unlike tree crops, in which one or two plants assure a plentiful harvest, small fruit plants must be ordered in larger quantities. Table 8-2 lists the small fruits, the estimated yield from each plant at maturity, and the suggested number of plants for a family of five.

The amount of soil improvement necessary will depend on the present condition of the soil. If the area selected has been a garden, little soil preparation may be necessary. However, if an unworked area is chosen, considerable effort may be needed to create a favorable soil environment. If practical, a soil test should be taken, and fertilizer and pH-altering materials added to the area based on the results of the soil analysis. Organic matter should be incorporated deeply to loosen a clay soil or to improve the water-holding capacity of a sandy one. If a sodded area is selected, the sod should be turned under the fall prior to planting so that it will decay and can be worked into the soil in spring. The success or failure of the crops will depend greatly on the soil conditioning prior to planting.

Most small fruits are planted in spring and begin producing fruit the following season. The ground is cultivated in early spring, and the plants are transplanted while dormant, using the spacings listed in Table 8-2, and then watered.

Small fruit cultivars are commonly ordered from mail-order nurseries and will arrive without any soil around the roots to decrease shipping cost. Planting is best done immediately, but if it must be delayed, the shipping box should be opened and the plants moistened if

Table 8-2 Small-Fruit Cultivation Information

Fruit	Minimum planting distance between		Estimated yield per mature plant	Suggested number of plants per family of 5
	plants	rows		
blackberry, erect	4–5′ (1.2–1.5 m)	6–8′ (2–2.5 m)	1 qt (1l)	15–20
blackberry, trailing or semierect	6–10′ (2–3 m)	6–8′ (2–2.5 m)	4–10 qt (4–10 l)	8–10
currant, black	4–5′ (1.2–1.5 m)	6–8′ (2–2.5 m)	2–5 qt (2–5 l)	4–6
currant, red	3–4′ (1–1.2 m)	6–8′ (2–2.5 m)	5–10 qt (5–10 l)	2–3
blueberry	6–8′ (2–2.5 m)	8–10′ (2.4–3 m)	3–4 qt (3–4 l)	8–10
gooseberry	3–4′ (1–1.2 m)	6–8′ (2–2.5 m)	4–5 qt (4–5 l)	4–6
grape	3–4′ (1–1.2 m)	8–10′ (2.5–3 m)	¼–½ bushel (9–18 l)	5–10
raspberry, spring-bearing	3–4′ (1–1.2 m)	6–8′ (2–2.5 m)	1–1.5 qt (1–1.5 l)	20–25
raspberry, ever-bearing	3′ (1 m)	8′ (2.5 m)	1 qt (1 l) in spring and ½ qt (0.5 l) in fall	15–20
strawberry, spring-bearing	1–2′ (30–60 cm)	4′ (1.2 m)	½–1 qt/ft of row* (0.5–3 l/m of row)	100
strawberry, everbearing	1′ (30 cm)	1–1.5′ (0.3–0.4 m)	½ qt/ft of row (1.5 l/m of row)	100

*Depending on training system used.

they are dry. The plants should then be heeled in, or the roots covered with plastic, and set in a cool, shaded (30–60°F; −1 to 15.6°C) area.

When planting, it is generally advisable to plant at the same level at which the plant was growing in the nursery, and this is especially important with strawberries. An exception, however, is currants and gooseberries, which should be planted slightly deeper, as this encourages a strong root system and abundant new shoots from the bases of the branches.

PRUNING AND TRAINING SMALL FRUITS

All the small fruits will produce a better crop if they are pruned and trained. On shrub fruits, pruning may be needed only to remove weak and unproductive wood, whereas on the grape, pruning will be needed to control the productivity and shape of the vine.

Brambles

The pruning of brambles revolves around the knowledge of when the canes produce fruit. On typical brambles, the fruit is borne only on canes produced the previous growing season. These canes, grown from the roots the first year, form flower buds that fall, bloom and fruit the following summer, and then die. The fruit is borne on side shoots that grow from buds on the parent cane.

On everbearing raspberries, two crops are produced per cane. The first crop is borne on the tips of that season's canes in fall, followed the next summer by fruiting further down the cane.

Support Systems for Brambles

Most brambles can grow without support; however trailing blackberries and red raspberries have longer canes and require tying to a wire or stake support system to be manageable and relatively easy to pick. If they are grown as individual plants in a row (a system called the *hill system*), support can be provided by either individually staking the plants or by a trellis in which one or two wires pass through all the plants and are attached to posts at either end of the row. If the plants are planted and allowed to grow together to form a hedge, a wire trellis is the only option.

Pruning and Training Raspberries

The pruning method varies greatly depending upon whether the raspberry is red or black, spring-bearing or everbearing.

Red and Yellow Raspberries. These are pruned exclusively during the dormant period in a four-step pruning/training process (Figure 8-5).

1. Remove all the canes that fruited the previous summer, cutting them off at ground level.

Figure 8-5 Pruning of red and yellow raspberry in four steps, all done during the dormant period. (1) Remove all canes that fruited the previous season; (2) Remove weak, broken, and poorly placed canes; (3) Tie canes to stake or along trellis wire; (4) Cut canes back to 5 feet (1.5 meters).

2. Cut off broken and weak canes and those growing outside of the row or hill.

3. Tie the remaining canes to the wire trellis or a 4- to 5-foot (1.2–1.5-meter) stake.

4. Cut them off at the 5-foot (1.5-meter) height. This will make picking easier the following year.

Everbearing Raspberries (Reds, Yellows, and Purples). Everbearing raspberries fruit first on the tips of the current year's canes in fall and further down those same canes the next spring. They should *not* be pinched and should not be pruned to the ground after fall fruiting.

Instead, only dormant spring pruning to eliminate thin weak canes and tip back the canes that fruited in fall should be done. After the spring fruiting of these canes, they can be removed.

Black and Purple Raspberries. These differ from red raspberries in that, although they do not require support, they must be pruned twice per year: in summer and in winter or whenever they are dormant. Pruning encompasses two steps (Figure 8-6):

1. Remove the tips of new canes in summer.

2. In winter, remove canes that fruited and shorten the side branches of canes produced the previous summer.

The summer pruning is to prevent the canes from growing to excessive lengths that would make picking difficult by creating a thicket. To this end, the tops are cut back 2 to 3 inches (5–8 centimeters) as they reach 14 to 20 inches (35–50 centimeters) for black raspberries

new canes this
season, tipped

fruit-producing cane

cut back side
branches

remove old canes

Figure 8-6 Prunng of black and purple raspberries and erect blackberries. (1) In summer, tip new canes; (2) in winter, cut out old canes and cut back branches on new canes.

unsupported, 24 to 40 inches (60 centimeters –1 meter) for black raspberries staked to trellisses, and 22 to 32 inches (unsupported) to 32 to 52 inches (80–130 centimeters) (supported) for purple raspberries, as they are generally more vigorous. Because the canes grow at different rates, summer tipping will be a one-month process with pinching at weekly intervals.

Winter pruning first removes dead canes that fruited, and of course weak, broken, and poorly placed canes, as well. Then branches that formed on canes tipped the previous season are shortened, and this improves fruit quality. Weak branches should be left 2 to 4 inches (5–10 centimeters) long so that they will produce only a light load of fruit, but strong branches can be left 14 to 18 inches (30–45 centimeters) long.

Erect Blackberries. These are self-supporting and need no stakes or trellisses, but require pruning two times per year: in summer and in winter. Erect black-

berries are pruned in a manner identical to that for black and purple raspberries (Figure 8-6) except that the blackberry canes are summer tipped at 3 to 4 feet (1–1.2 meters) instead of lower, and side branches are cut to 12 to 18 inches (30–45 centimeters) in winter.

Trailing and Semierect Blackberries. These very vigorous plants need support always, but pruning only once per year in the dormant season. Pruning can be accomplished in three steps (Figure 8-7):

1. Cut out the old canes that fruited the previous summer.

2. Select and tie to the support 8 to 10 strong canes and remove the rest.

3. Shorten the tops of these selected canes to 6 to 8 feet (1.8–2.4 meters) and their side branches to 12 to 18 inches (30–45 centimeters).

Figure 8-7 Pruning of trailing blackberries in three steps, all done during the dormant period. (1) Cut out old canes that fruited the previous summer; (2) select and tie up 8–10 strong new canes—remove the rest; (3) shorten the tops and side branches on the tied canes.

Blueberries

The blueberry bears fruit on the growth produced the previous season, and the amount of pruning will vary with the species.

Highbush. Highbush blueberries overbear and, if not pruned annually, will yield small berries. In addition, the energy expended in producing the overabundant crop will inhibit vegetative growth and drastically reduce the next year's yield. Accordingly, highbush blueberries should be pruned "hard" annually in early spring before new growth starts; that is to say, a significant amount of wood should be removed.

Upright-growing cultivars will benefit from thinning out the older center branches (Figure 8-8). This will reduce the berry load, admit more light, and stimulate growth. Drooping lower branches on spreading cultivars should be cut back to avoid breakage during fruiting.

Rabbiteye. Rabbiteye blueberries do not require pruning. The bushes are strong enough to mature large crops of fruit without damaging the plant. An exception to the no-pruning rule should be made for older bushes, which benefit when older stems are thinned out lightly.

No trellising or other training is used on blueberries, which grow naturally into large, attractive shrubs bearing white, bell-shaped flowers in spring.

Currants and Gooseberries

Currants and gooseberries, like brambles, grow into shrubs from canes produced by the roots. The fruit is borne in one-, two-, and three-year-old canes, with older canes becoming progressively less productive.

Pruning both currants and gooseberries should be done once per year before spring growth begins. The aim is to remove wood more than three years old. Excess weak wood should also be removed and the bush shaped into a compact shrub 3 to 5 feet (1–1.5 meters) tall. No training (other than pruning to a shrub form) is practiced.

Grapes

There are many methods of training grape vines. The most commonly used are the following:

Four-Arm Kniffen System. This system uses a two-wire trellis in which the wires are strung between posts in the grape row. It is popular throughout the United States for vigorous native American grapes such as 'Concord.' To train a vine, the strongest new shoot of the vine should be tied to the bottom wire and the rest of the shoots pruned off the first winter after planting (Figure 8-9). The next summer the shoot should reach

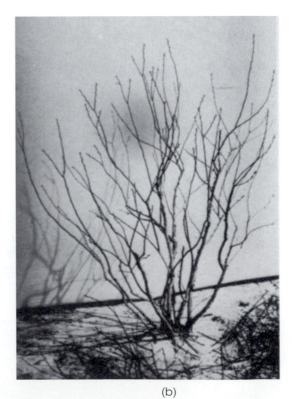

(a)   (b)

Figure 8-8 (a) Before dormant pruning of highbush blueberry, (b) after pruning. *(USDA photo)*

(a)

(b)

(c)

Figure 8-9 Training a grapevine using the four-arm Kniffen system. (a) First winter after planting, (b) the following summer, (c) the following winter after pruning.

Figure 8-10 A grapevine trained to an overhead arbor. *(Photo by the author)*

Annual Pruning

Grape vines bear their fruit only on canes that grew the previous summer. Heavy annual pruning should remove most of the wood growing from the trunk and leave only four shortened canes for fruit bearing, one to grow in each direction along the trellis wires (Figure 8-11).

The canes chosen to remain for fruit bearing should be slightly larger than pencil thickness for best fruit production; they should not necessarily be the largest canes on the vine. Each should be cut to a length that allows for 15 buds. In addition to these four canes, four canes shortened to two to four buds should be left near

the top wire, where it is again tied and tipped (Figure 8-9) to encourage side shoots. The side shoots that will then develop are trained and tied out on the wire, with one cane in each direction along each wire. After this is done, annual pruning as described in the following section will be sufficient.

Arbor Training. Arbor training (Figure 8-10) produces both fruit and shade in the landscape. Because an arbor-trained vine will need to grow to a large size, only a vigorous grape cultivar should be used.

Arbor training a vine should start at planting. The vine should be trained as described under "Four-Arm Kniffen System." During each growing season that follows, the main trunk shoot is continually trained upward. Gradually it will produce side shoots (fruiting canes) to cover the arbor. These fruiting canes should be spaced by pruning, tied to the support, and replaced annually, as described in the following section.

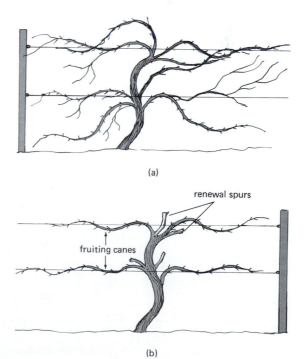

(a)

renewal spurs

fruiting canes

(b)

Figure 8-11 (a) Before annual pruning of a grapevine trained to a four-arm Kniffen system, (b) after pruning.

the two wires. These short canes (called *renewal spurs*) will produce the fruiting canes for the following season.

Renovation Pruning

Grape vines that have been neglected for many years require severe pruning to produce quality fruit again. The plants should be cut back severely before growth starts, removing as much of the accumulated older wood as possible and following the specifications for annual pruning. Any old wood that cannot be removed at this pruning may be taken out with regular pruning the following year.

Strawberries

Three systems of training are used for strawberries: the matted row, spaced runner, and hill systems. Except in the hill system, flowers that form on plants during the year they are planted should be removed instead of being allowed to develop fruit. This will increase the harvest the following year.

Matted Row (Figure 8-12). The matted row system involves transplanting the young strawberries 2 feet (60 centimeters) apart in a row, with the rows 3 to 4 feet (1–1.2 meters) apart. The runners produced in late spring and summer are allowed to root at random,

forming a dense mat about 2 feet (0.6 meter) wide. This training system is the least productive but the easiest.

Hill System (Figure 8-12). The hill system is the best training method for everbearing strawberries and is also useful for spring bearers. The plants are set 1 to 1½ feet (30–45 centimeters) apart in rows 1 to 1½ feet (30–45 centimeters) apart. All runners that appear during the planting year are pinched off, resulting in a strong mother plant that will bear profusely the following season. Yields from strawberries trained to a hill system can sometimes double that from a matted row. The system is used primarily in warm areas because plants are more likely to be damaged by cold.

Renewing the Strawberry Patch

Properly renewed, a strawberry planting can last up to four years, although the yield will not approach that picked the first season.

Renewal of the patch should begin as soon as harvesting is complete. The old foliage is mowed off to 1 inch (3 centimeters) above the crown, and the debris is raked off. A balanced garden fertilizer should be broadcast over the area, all weeds removed, and the rows narrowed to 1 foot (30 centimeters) wide by hoe or rototilling. The patch should be watered thoroughly. New runners for the next year's crop should appear within one month.

Figure 8-12 Strawberry planting systems. (a) hill system, (b) matted row. *(Courtesy of Dr. Danny L. Barney, Sandpoint ID)*

GENERAL MAINTENANCE OF PERENNIAL SMALL FRUITS

Fertilizing

Most small fruits are shallow-rooted and will benefit from an annual spring application of balanced fertilizer. Side-dressing with granular fertilizer no closer than 6 inches (15 centimeters) from the crown of the plant is advisable for brambles, blueberries, grapes, currants, and gooseberries. Strawberries can be broadcast fertilized and should be watered thoroughly afterward to keep fertilizer granules from burning the leaves. Blueberries should be fertilized annually with an acid-forming fertilizer (such as ammonium sulfate) to keep the pH of the soil in an acceptably low range. Rotted manure applied in fall will also help achieve a low pH.

Actual fertilizer rates will vary with the natural soil fertility. General recommendations for plants fertilized with a balanced formula of not more than 12 percent nitrogen are listed in Table 8-3.

Watering

The amount of supplemental watering that small fruits require will depend on the natural rainfall. Provisions for supplemental watering should be made in any climate though, because the shallow-rootedness of these fruits makes them particularly susceptible to drought damage.

 An irrigation specialist could work on small fruits, tree fruits, or vegetables. He or she would advise the farmer on the best type of irrigation, design the system, and supervise the installation. Generally only the western states with prolonged periods without rain use irrigation specialists.

It is possible to overwater small fruits and decrease the fruit quality by making the fruits larger but the flavor less intense. This is particularly a problem in strawberries, which should not be irrigated more often than is necessary to keep the soil moist.

Mulching

Small fruits should be mulched in summer with a plastic or organic mulch for weed control and soil moisture retention. In strawberry plantings trained to the hill system, plastic mulch is often preferred because it warms the soil and can increase and/or speed up fruit production by increasing root growth. Normally the strip of plastic is laid over the prepared soil and the plants are planted in holes cut through it.

The aim of fall mulching with organic matter is to prevent freezing and thawing of the root systems. Fall mulch should be applied late in the year after temperatures have dropped into the 40°F (5°C) range or lower. If the plants are mulched sooner, the mulch may retain sufficient heat near the roots to continue growth, and resulting tender shoots will be killed.

The shrub-type small fruits (brambles, blueberries, currants, and gooseberries) should be mulched deeply around and covering the crowns and lower stems of the plants.

Strawberries are mulched by completely covering them with straw or a similar lightweight material such as wood shavings. The mulch should be raked from the patch promptly when new growth begins in the spring to admit light for photosynthesis.

Grapes should be mulched thickly in a circle extending out about 4 feet (1.2 meters) from the trunk. This will assure that the entire root zone is protected.

Table 8-3 Fertilizer Recommendations for Small Fruits*

Fruit	Amount of fertilizer annually
Brambles	½ cup (120 ml) per plant
Blueberries	½–1 cup (120–240 ml) per plant
Currants	¼ cup (60 ml) per plant
Gooseberries	¼ cup (60 ml) per plant
Grapes	1 cup (240 ml) per plant
Strawberries	1 cup (240 ml) per 20 ft (6 m) of 2-ft (0.6 m)-wide row

*Recommendations are made for mature plants and are based on a balanced (1-1-1 ratio) fertilizer containing not more than 12 percent nitrogen.

SELECTED REFERENCES FOR ADDITIONAL INFORMATION

Anonymous. *Fruits and Berries.* Burpee American Gardening Series. New York: Prentice-Hall, 1995.

Childers, Norman F., and L. Gene Albrigo, eds. *Nutrition of Fruit Crops: Tropical, Sub-Tropical, Temperate: Tree & Small Fruits.* Ann Arbor, MI: Books on Demand, 1966.

Columbia Basin Nursery. *Retail Catalog.* P.O. Box 458, Quincy, WA 98848.

Edinger, Philip. *All About Fruits, Berries, Nuts.* San Ramon, CA: Ortho, 1996.

Elridge, Sherri. *The Berries: Strawberry, Raspberry & Blackberry.* Salisbury Cove, ME: Harvest Hill Press, 1997.

Farmer's Seed and Nursery Company. *Retail Catalog.* Faribault, MN 55021.

Fathauer, Elizabeth. *Small Fruits in the Home Garden.* Carbondale: University of Illinois Office of Agricultural Communications and Education, 1998.

Galletta, Gene J., and David G. Himelrick, eds. *Small Fruit Crop Management.* Englewood Cliffs, NJ: Prentice-Hall, 1994.

Gough, Robert E., and E. Barclay Poling, ed. *Small Fruits for the Home Garden.* Binghamton, NY: Hawthorn Press, 1997.

Hill, Lewis. *Fruits and Berries for the Home Garden.* Williamstown, MA: Storey Communications, 1992.

Holmes, Roger, and Frances Tenebaum, eds. *Taylor's Guide to Fruits & Berries.* New York: Houghton Mifflin Company, 1996.

Kelley Brothers Nurseries. *Retail Catalog.* 940 Maple St., Dannville, NY 14437.

New York State Fruit Testing Association. *Retail Catalog.* Geneva, NY 14456.

Otto, Stella B. *The Backyard Berry Book: A Hands-On Guide to Growing Berries, Brambles & Vine Fruit in the Home Garden.* Maple City, MI: OttoGraphics, 1995.

Otto, Stella B. *The Backyard Orchardist.* Maple City, MI: OttoGraphics, 1995.

R. H. Shumway Company. *Retail Catalog.* Rockford, IL.

Savage Farm Nursery. *Retail Catalog.* P.O. Box 125, McMinnville, TN 37110.

Smith, Miranda. *Backyard Fruits & Berries: Everything You Need to Know about Planting & Growing Fruits & Berries in Your Own Backyard.* New York: Book Sales, Incorporated, 2001.

Storey Publishing Staff. *Taylor's Fruits and Berries.* New York: Random House, Incorporated, 1997.

Striblings Nursery, Inc. *Retail Catalog.* 1620 W. 16th St., Merced, CA 95340.

Swenson, Allan A. *The Gardener's Book of Berries.* Guilford, CT: Globe Pequot Press, 1994.

Waynesboro Nurseries. *Retail Catalog.* Box 987, Waynesboro, VA 22980.

See also Cooperative Extension Service publications from individual states.

World Wide Web Addresses

http://fruitsandnuts.ucdavis.edu/crops.html   University of California. "Fruit and Nut Research and Information Center" 2001.

http://ohioline.osu.edu/b861/   Funt, Richard, C. Michael A. Ellis, and Celeste Welty. "Midwest Small Fruit Pest Management Handbook Bulletin 861" 1997.

http://www.smallfruits.org/   The Southern Region Small Fruit Consortium website. Nov 6, 2001.

http://www1.uwex.edu/ces/pubs/showpubs.cfm?theid=17   University of Wisconsin Cooperative Extension Service. "Publications" Horticulture: Berries. 2001.

http://www.umass.edu/fruitadvisor/   University of Massachusetts Extension. "U Mass Fruit Advisor" 2001.

(Photo by the author)

CHAPTER 9

FLOWER AND HERB GARDENING

Flower gardening is a popular type of home gardening, and many gardeners specialize strictly in growing annual and perennial ornamentals. Closely allied to flower gardening are herb, shade, rock, and container gardening. Each is discussed within this chapter, along with roses and flowers for cutting and drying.

TYPES OF GARDEN FLOWERS

Annuals

Every spring there is a brisk sale of annual seedlings for transplanting into home gardens. Although they live for only one season, during that brief time they more than justify their expense by flowering continuously and profusely with minimal care (Table 9-1, Figures 9-4, 9-5, 9-6, 9-7, 9-8, and 9-9).

 Annual flowering plants (and vegetable transplants as well) are called *bedding plants* in the horticultural industry. A bedding-plant producer usually grows these plants from seed over about a two-month period in advance of the sales season (spring, or spring and fall in mild-winter areas). A bedding-plant producer grows the crop in a greenhouse and may market the plants through a roadside stand or sell to a nursery center, discount outlet, or cooperative.

Commercial breeding of annuals for brighter colors, larger flowers, and different sizes is a continual process. There is an annual to fit almost every garden situation: sun or shade, poor or good soil, 3 inches (8 centimeters) or 3 feet (1 meter) tall. New cultivars are evaluated and introduced yearly. Those judged most valuable are given the All-America Selection (AAS) award for their beauty and adaptability to diverse climates.

Most annuals are purchased as greenhouse-grown transplants in packs of four, six, or eight plants. By buying seedling plants instead of sowing seed, the gardener saves several weeks of growing time and can transplant nearly mature plants into the garden for immediate bloom. A few annuals that either do not transplant well or mature very rapidly are grown from seed. Among these are nasturtium, California poppy, and balsam.

Most annuals are not frost-tolerant. However, a few, called *cool-season annuals,* are tolerant of light frost and grow better in temperatures in the 50–60°F (10–16°C) range. These annuals (Table 9-1) can be planted in spring before the last frost date. In mild-winter areas they are planted in fall for winter and spring bloom.

Perennials

Many garden flowers are herbaceous perennials (Figures 9-10, 9-11, and 9-12) with foliage that dies yearly but with a crown and roots that renew the plant the next season. Only two common perennials, roses and tree peonies, are woody.

A perennial-plant nurseryman generally has an established nursery with propagation facilities and outdoor growing areas for field growing of the plants to salable size. Many perennial-plant nurseries sell by mail order. Others sell in containers to garden centers or discount centers.

Unlike annuals, which flower continually, most perennials have a 2- to 3-week bloom period and flower less profusely. However, they increase in size every year and, in established clumps, create a massive display of flowers and foliage. Perennials are more expensive than annuals. They are sold individually in containers or can be ordered from mail-order nursery catalogs and shipped while dormant. Annuals must be replanted yearly, but perennials are permanent and will continue to flower every year with minimal care. For this reason, they are the mainstay of a flower garden (Table 9-2).

Biennials

Only a few popular flowers are biennials. Foxglove and sweet william are two of the better-known examples. Biennials are sold as one-year-old plants ready for blooming or can be grown from seed.

Cold-Hardy Bulbs

Spring-flowering tulips and summer-flowering lilies are two examples of cold-hardy bulbs. This group also includes plants like crocus that are not bulbs in a botanical sense but function similarly.

Table 9-1 Selected Flowering Annuals for Sunny Locations

Common name	Botanical name	Cool season/ warm season	Remarks
Ageratum	*Ageratum houstonianum*	Warm or cool	Pink, white, or most commonly blue flowers like small powderpuffs; not frost-tolerant
Snapdragon	*Antirrhinum majus*	Warm or cool	Frost-tolerant
Calendula	*Calendula officinalis*	Warm or cool	Yellow and orange shades of flowers of a daisy-like shape
Bachelor button	*Centaurea cyanus*	Warm or cool	Blue, white, pink colors, frost-tolerant
Marguerite daisy	*Chrysanthemum frutescens*	Warm	White or yellow daisies; very fast growing to 3 ft (1 m) and floriferous
Cosmos	*Cosmos bipinnatus*	Warm	Red through white, purple, and orange; many tall and some short cultivars
Cape marigold	*Dimorphotheca sinuata*	Warm	Frost-tolerant; sow in place; daisy-like flowers in orange and yellow
California poppy	*Eschscholzia californica*	Warm or cool	Frost-tolerant; does not transplant well
Gaillardia	*Gaillardia pulchella*	Warm	Orange, yellow, red combinations; good in poor soils
Balsam	*Impatiens balsamina*	Warm	Spike flowers in many colors; fast growing from seed
Sweet pea	*Lathyrus odoratus*	Cool	Pink, white, red, or lavender flowers; vining or bush types; grow from seed
Sweet alyssum	*Lobularia maritima*	Warm or cool	Mat-like habit; white or purple fragrant flowers; frost-tolerant
Stock	*Matthiola incana*	Cool	Pink, white, red, lavender, or yellow spike flowers; very fragrant
Petunia	*Petunia × hybrida*	Warm	All colors, bushy or vining cultivars
Moss rose	*Portulaca grandiflora*	Warm	Succulent and low-water-requiring; many colors of delicate flowers; creeping habit
Salvia	*Salvia splendens*	Warm	Red and, occasionally, pink or purple shades; many heights available
Marigold	*Tagetes erecta* and *patula*	Warm	White, orange, yellow, red, and combinations; many heights available
Black-eyed Susan vine	*Thunbergia alata*	Warm or cool	Reliable vine; flowers orange-yellow with black centers; perennial in zones 9 and 10

Figure 9-1 Shasta daisy, a perennial for full sun. *(Photo courtesy of the Perennial Plant Association, Hiliard, Ohio)*

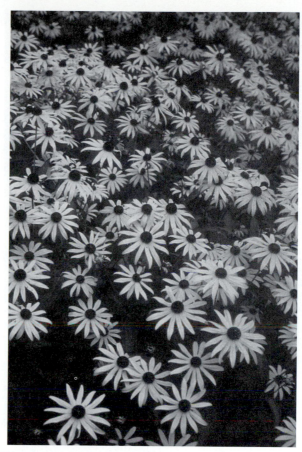

Figure 9-3 Black-eyed Susan, a perennial. *(Photo courtesy of the Perennial Plant Association, Hiliard, OH)*

Hardy bulbs differ from perennials in several ways. Unlike perennials that can be planted from spring until fall, bulbs should be moved only in late summer and fall when dormant. They then generate roots in fall and winter and flower the following year. Also, unlike other perennials, which retain their foliage throughout the growing season, bulb foliage dies in midsummer, leaving only the underground bulb through the latter part of summer.

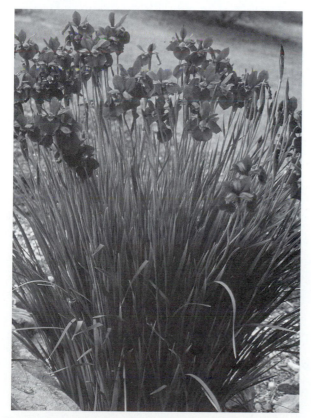

Figure 9-2 Iris, a perennial. *(Photo by the author)*

 Hardy-bulb producers propagate and grow their crop out of doors in the ground. The bulbs are grown to the size at which they will produce flowers, allowed to go dormant, and marketed in the fall by mail or through retail outlets.

The usual source of garden bulbs is a local nursery or mail-order catalog. Although the *major bulbs* like tulip, hyacinth, and daffodil are best known, there are many *minor bulbs* that are easily grown and are a worthwhile addition to any garden. Table 9-3 lists a number of these lesser-known bulbs, as well as major bulbs.

Figure 9-6 Impatiens, a favorite annual for shade. A double-flowered cultivar is shown on the left, and a single on the right. *(Photo courtesy of the National Garden Bureau, Downer's Grove, IL)*

Figure 9-4 Marigolds are among the most popular and dependable annuals. *(Photo Courtesy of National Garden Bureau, Downer's Grove, IL)*

Figure 9-5 A planting of petunias from a pony pack will bloom continuously all summer. *(Photo courtesy of National Garden Bureau, Downer's Grove, IL)*

Figure 9-7 Bachelor buttons can be blue, white, pink, or red. *(Photo courtesy of the National Garden Bureau, Downer's Grove, IL)*

Table 9-2 Selected Flowering Perennials

Name	USDA zones	Height	Bloom period	Remarks
Astilbe, false spirea *Astilbe* spp.	3–9	18–36″ (0.5–1 m)	Mid-to late summer	White, pink, lilac, or red feathery plumes
Aster, michelmas daisy *Aster* spp.	1–10	6″–4′ (15 cm–1.2 m)	Late summer to frost	Many colors
Bellflower *Campanula* spp.	2–10	1–3′ (0.3–1 m)	Late spring to summer	White or blue bell-shaped flowers
Chrysanthemum *Chrysanthemum* × *morifolium*	3–10	12–48″ (0.3–1.2 m)	Late summer to fall	All colors except blue, pinch off flower buds to delay bloom until fall
Feverfew *Chrysanthemum parthenium*	4–10	2–3′ (0.6–1 m)	Early summer to fall	Small, white flowers with yellow centers; produced profusely
Shasta daisy *Chrysanthemum* × *superbum*	2–10	24–36″ (0.6–1 m)	Midsummer to fall	White daisy with yellow center, profuse flowers
Delphinium *Delphinium elatum*	1–10	24–48″ (0.6–1.2 m)	Mid-to late summer	Blue, white, purple spikes
Pinks *Dianthus* spp.	2–10	8–18″ (20–50 cm)	Early to late summer	Pink through white and red carnation-like flowers
Leopard's bane *Doronicum cordatum*	4–10	24″ (60 cm)	Early spring	Yellow daisy-like flowers; foliage dies in summer
Daylily *Hemerocallis* spp.	2–10	15″–4′ (0.4–1.2 m)	Early to late summer, depending on variety	Yellow through wine red flowers; not suitable for cutting, as flowers close at night
Bearded iris *Iris* spp.	5–10	12–36″ (0.3–1 m)	Early summer	Many colors
Siberian iris *Iris* spp.	6–10	24″ (60 cm)	Spring to early summer	Flowers differently shaped than bearded iris
Lythrum *Lythrum salicaria* and *virgatum*	2–10	2–4′ (0.6–1.2 m)	All summer	Pink through red-purple flowers; tolerant of poor growing conditions including wet soil
Bee balm *Monarda didyma*	3–7	36″ (1 m)	Midsummer	Pink, lavender, red, or white flowers; leaves have mint smell
Peony *Paeonia* spp.	4–9	36″ (1 m)	Late spring to early summer	Large globular flowers in white through red
Oriental poppy *Papaver orientale*	2–8	24–36″ (0.6–1 m)	Early summer	Pink through white and red carnation-like flowers
Hardy phlox *Phlox paniculata*	4–10	24–40″ (0.6–1 m)	Midsummer	Large flower heads in many colors
False dragonhead *Phystostegia virginiana*	6–10	36–48″ (1–1.2 m)	Midsummer to fall	Pink or lavender spikes; very vigorous

Table 9-3 Selected Bulbs and Related Plants

Name	Cold hardiness (USDA zones)	Time of bloom	Remarks
Ornamental onion, allium *Allium* spp.	4–10	Late spring	Many species in white, blue, and pink; globe-shaped flower heads
Belladona lily *Amaryllis belladonna*	5–10	Summer	Pink lily-like flowers
Anemone *Anemone × hybrida*	6–9	Midspring	3–15″ (10–40 cm) tall with pastel flowers
Tuberous begonia *Begonia × tuberhybrida*	9–10 perennial, 2–10 summer bulb	All summer	Single and double flower varieties and many colors; good in hanging baskets
Caladium *Caladium × hortulanum*	9–10 perennial, all others summer bulb	Foliage only all summer	Grown for foliage only, which is multicolored
Canna *Canna* hybrids	7–10 perennial, all others summer bulb	All summer	Tall (36–48″) (1–1.2 m) background plants in red, yellow; dwarfs available
Glory-of-the-snow *Chinodoxa luciliae*	3–10	Early spring	Blue, pink, or white flowers; less than 1′ (30 cm) tall
Autumn crocus *Colchicum autumnale*	4–10	Fall	Planted in August for fall flowers; leaves appear the following spring
Crocus *Crocus* spp.	3–10	Early spring	Plants 2–4″ (5–10 cm) tall
Dahlia *Dahlia* hybrids	9–10 perennial, summer bulb other zones	Summer	Easy to grow, many colors and sizes
Winter aconite *Eranthis* spp.	4–9	Early spring	Yellow buttercup-like flowers; plants 2–4″ (5–10 cm) tall
Fritillaria *Fritillaria imperialis*	3–10	Midspring	Unusual and attractive, 2½–4′ (0.7–1.2 m) tall
Gladiola *Gladiolus × hortulanus*	8–10 perennial, summer bulb other zones	Summer	Tall to 48″ (1.2 m), with spike flowers
Amaryllis *Hippeastrum* hybrids	9–10 perennial, houseplants all others	Winter	Lily-like flowers on tall stem
Hyacinth *Hyacinthus orientalis*	4–10	Midspring	Fragrant blooms in many colors, 6–12″ (15–30 cm) tall
Bulbous iris *Iris* spp.	5–10	Spring	Purple, blue, yellow, white
Garden lily *Lilium* spp.	3–10	Early summer	Many species and cultivars 12–36″ (0.3–1 m) tall
Grape hyacinth *Muscari* spp.	2–10	Early spring	Blue, white or pink flowers on 3–4″ (7–10 cm) plant
Daffodil and narcissus *Narcissus* spp.	4–10	Spring	Many species available; 12–18″ (30–45 cm) tall
Oxalis *Oxalis* spp.	6–10, depending on species	Winter in 9–10, spring all others	White, pink, yellow flowers on clover-shaped leaves; plants 2–5″ (5–12 cm) tall
Ranunculus *Ranunculus*	8–10 perennial, all others summer bulb	Midspring	Delicate flowers in pastel colors on 12–18″ (30–45 cm) stems
Scilla, squill *Scilla* spp.	4–10	Spring	Blue and other colors; plants about 6″ (15 cm) tall
Tulip *Tulipa* spp.	3–7	Early to late spring	Multitude of cultivars 3–18″ (10–50 cm) tall
Calla lily *Zantedeschia aethiopica*	8–10 perennial, summer bulb all other zones	9–10 winter, other zones summer	Outstanding foliage and white to yellow flowers; 36–48″ (1–1.2 m) tall

Figure 9-8 Gaillardia, an annual with flowers ranging from yellow through red. *(Photo courtesy of the National Garden Bureau, Downer's Grove, IL)*

Figure 9-11 Garden pinks, or *Dianthus*, an old-fashioned perennial favorite. *(Photo courtesy of All-America Selections, Downer's Grove, IL)*

Figure 9-9 Wax begonia, a short annual for shade. *(Photo courtesy of National Garden Bureau, Downer's Grove, IL)*

Figure 9-12 Daylily, a favorite perennial. *(Photo courtesy of the Perennial Plant Association, Hiliard, OH)*

Summer or Nonhardy Bulbs

Although perennial in mild climates, nonhardy bulbs cannot survive winter temperatures in most parts of the country. Consequently, they are planted in spring for summer flowering and either discarded or stored in fall for replanting the following year. Tuberous begonias (Figure 9-13), dahlias (Figure 9-14), cannas, and gladiolus are included in this group. They should be dug from the garden immediately after the first killing frost, dried in the shade, and stored in dry peat moss in a cool 50–60°F (10–16°C) area. Table 9-3 lists selected summer bulbs.

Figure 9-10 Astilbe, a perennial for shade. *(Photo courtesy of the Perennial Plant Association, Hiliard, OH)*

 Summer-bulb producers also grow their crops outdoors. Unlike hardy-bulbs, in some cases the summer bulbs may be pregrown in containers and marketed at near-blooming stage in the spring.

Figure 9-13 Tuberous begonia, a well-known summer bulb for shade. *(Photo Courtesy of All-America Selections, Downer's Grove, IL)*

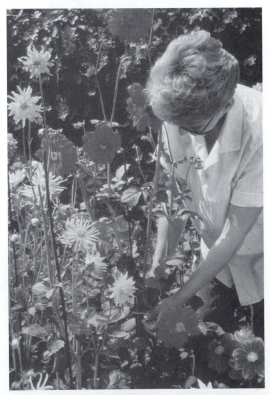

Figure 9-14 Dahlia, a summer bulb in much of the country, but a perennial in USDA zones 9–11. *(Photo courtesy of National Garden Bureau, Downer's Grove, IL)*

PLANNING THE FLOWER GARDEN

Site Selection

Good soil drainage is an important factor in selecting a site for a flower garden. Few plants can be grown in wet soil, and it is particularly unsuitable for bulbs. When drainage is questionable, a raised bed is a solution that will not only aid drainage but will elevate the flowers for better viewing.

The amount of sunlight an area receives will determine which plants can be successfully grown. A full-sun area is preferable for growing the greatest variety of flowers. Many flowers can be grown in partially or fully shaded locations, but they bloom less profusely and produce more foliage. Others weaken from the lack of light and die. Accordingly, if a shady location is chosen for a flower garden, plant selection must be made carefully and limited to species adapted to shade (Table 9-4).

Finally, the site chosen for a flower bed should be readily viewable. The area around a patio or deck is a good choice.

Flower Bed Styles

The Flower Border

A border of flowers with shrubs or a fence used as a background is a popular traditional bed style (Figure 9-15). The border can be as much as 10 feet (3 meters) wide and is designed to be viewed from only one side. Accordingly, tall plants are used in the rear, moderate-sized ones in the center, and short ones along the front.

The Freestanding Bed

A freestanding or island flower bed (Figure 9-16) is well adapted to modern landscape designs and is meant to be viewed from all directions. Tall plants should be placed in the center of the bed and gradually shorter plants stepped down to the edge. If a curving bed style is to be used, a garden hose can be moved in different ways to create a pleasing outline. The edge of the bed can then be marked by spading along the hose.

The Formal Bed

The formal bed or border (Figure 9-17) originated in Europe and was used for flowers and herbs around palaces and large manor houses. It is a tightly restricted, symmetrical style and not commonly used today. Many gardeners prefer formal beds for herb gardening; compact annuals can be easily included also.

Table 9-4 Selected Shade Garden Plants

Common name	Botanical name	Classification	Remarks
Ferns	*Adiantum, Athyrium, Polystichum,* etc.	Perennial	Many cold-hardy species of varying appearance
Ajuga	*Ajuga reptans*	Perennial	Low growing with blue to pink flowers in spring, depending on variety
Columbine	*Aquilegia × hybrida*	Perennial	Unusual-shaped flowers in blues, yellows, pinks, and so on; plants to 30″ (0.7 m) tall
Astilbe	*Astilbe × hybrida*	Perennial	Pink or white plumes atop fernlike foliage; plants up to 3′ (1 m) tall
Tuberous begonia	*Begonia × tuberhybrida*	Summer bulb	Large weeping plants with outstanding flowers in many colors
English daisy	*Bellis perennis*	Perennial	Small plant which blooms in spring to early summer with white to red flowers
Bergenia	*Bergenia × schmidtii*	Cold-tender perennial	Pale pink flowers on an evergreen groundcover plant
Caladium	*Caladium × hortulanum*	Summer bulb	Grown for its foliage; which can be a mixture of red, pink, white, and green
Periwinkle	*Catharanthus roseus*	Annual in all but the mildest climates	Continual blooms in pink to white; grows 8–18″ (20–45 cm) tall, depending on variety
Coleus	*Coleus × hybridus*	Annual	Grown for colorful foliage in red, green, yellow, and all shades between
Lily-of-the-valley	*Convallaria majalis*	Perennial	Fragrant bell-shaped white flowers; spreads to form a dense carpet
Cyclamen	*Cyclamen persicum* and other species	Cold-tender perennial	Clumping plant producing nodding flowers in pink, red, lavender, or white
Bleeding heart	*Dicentra spectabilis*	Perennial	Distinctive red and white flowers on plants 1–3′ (0.3–1 m) tall
Fuchsia	*Fuchsia × hybrida*	Annual in all but the mildest climates	A reliable shade performer with flowers in shades of white through red; cascading and upright forms are sold
Coral bells	*Heuchera sanguinea*	Perennial	Continual blooms in red to white; rise to 2′ (0.6 m)
Hosta or Plantain lily	*Hosta* spp.	A perennial for shade in most climates	Grown for its attractive rosette of green or variegated leaves, and for stalks of lavender to white flowers
Impatiens	*Impatiens walleriana*	Annual in most climates	Very reliable bloomer producing red, pink, white, or orange flowers
Forget-me-not	*Myosotis sylvatica*	Annual	Bright blue flowers on a plant up to 2′ (0.6 m) tall
Primrose	*Primula* spp.	Cold-tender perennial	Spring flowers in many colors in a tight head
Violet	*Viola odorata*	Perennial	Tiny white, yellow, or blue flowers on plants to 6″ (15 cm) tall
Pansy	*Viola × wittrockiana*	Annual	Low-growing favorite; many varieties available
Calla lily	*Zantedeschia aethiopica*	Cold-tender bulb	Large glossy leaves and white funnel-shaped flowers. Pink and yellow spp. also available

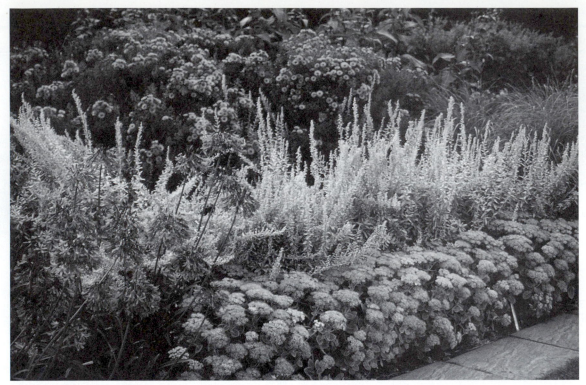

Figure 9-15 A flower border. *(Photo courtesy of the Perennial Plant Association, Hiliard, OH)*

The Cottage Garden

The cottage garden style is an informal massing of flowers of many species and sizes (and sometimes herbs and small vegetables as well) to achieve the effect of profusion of color and variety of form. The effect is a jumble of eye-catching color that arrests the attention of the viewer and causes him or her to wish to stop and look at each of the individual plants in turn. Originally an English style, it is becoming popular in North America because many people now occupy homes such as condominiums that have limited gardening space.

A cottage garden frequently includes other *design elements* such as benches and paths for viewing the flowers. Plants normally chosen are traditional old-fashioned species such as hollyhock, foxglove, and iris. The taller, less compact cultivars are chosen to amplify the informality that is the hallmark of this type of garden.

Figure 9-16 A freestanding flower bed. *(Photo by George Taloumis)*

Figure 9-17 A formal flower border. *(Photo courtesy of National Garden Bureau, Downer's Grove, IL)*

Plant Selection

Most gardeners prefer a mixed garden of herbaceous and woody perennials, bulbs, annuals, and sometimes herbs. A mixture combines the virtues of each: the dependability of perennials, the floriferousness of annuals, and the earliness of the hardy bulbs. The following criteria should be used to select the species that will be planted.

Height. Knowing the ultimate size of the flowers will enable a gardener to select plants of a variety of heights and to locate them in the correct place in the bed. Many annuals are available in several heights, and varieties of the same species can be used as either a foreground or a background plant.

Amount of Sunlight Required. A few versatile flowers grow in any light level, but most are restricted to either shade or sun. Choosing plants that are proven performers under the light available will maximize the beauty of the garden.

Continual Bloom. Planning for continual bloom involves choosing a variety of plants so that flowers will appear in succession all season. Bulbs are the usual choice for early spring and chrysanthemums the choice for fall blossoming. The other perennials generally have 2- to 3-week blooming periods from late spring through summer, and their flowering sequence relative to each other can be found in reference books on flower gardening. Because annuals flower continually, they tie together the bloom periods of perennials.

Disease and Pest Susceptibility. Although flower gardens nearly always require more maintenance than do landscapes of shrubs and groundcovers, there are degrees. For example, by planting carefully selected species it is possible to avoid all need to spray pesticides under normal circumstances. However, the gardener who elects to grow hybrid tea roses should anticipate spraying every 10 days throughout the growing season.

Water Requirement. Although flowers generally require frequent, copious watering to bloom well, people living in areas with water shortages can also landscape with flowers provided they select the right plants. *Xeriscaping,* landscaping with drought-resistant plants including drought-resistant flowers, is practical, environmentally conscious, and beautiful. It is not, as some people suspect, landscaping with cacti and rocks. Instead, floriferous improved succulents such as moss rose, and naturally drought-tolerant common garden flowers such as nasturtium, verbena, lantana, and gaillardia are mixed with newly improved native species to create a garden that few would suspect has been designed to be water conserving.

PREPARING AND PLANTING THE BED

The preparation needed for a flower garden will depend on the present condition of the soil. Recently gardened soils may need little preparation other than the incorporation of fertilizer. Areas that have been uncultivated or in turf will require more work. The area should be spaded to a depth of about 18 inches (0.5 meter), the average root depth for perennials, incorporating a large volume of amendment such as compost or peat moss. About ½ cup (100 milliliters) of balanced fertilizer per 10 square feet (1 square meter) should be spread over the area and mixed into the spaded layer.

If the bed adjoins a lawn area, invasion by grass will be a constant problem. Edging material outlining the perimeter will reduce this problem and give a neater appearance. A row of bricks or ornamental concrete set flush with the ground level or slightly higher is one attractive edging (Figure 9-18). Thin edging boards (bender boards) or plastic edging material is also satisfactory.

 A horticultural writer for a major newspaper, home-oriented magazine, or gardening magazine might have written this paragraph. The writer will follow leads to come up with new and unique stories, arrange for photography, and coordinate with the main editor for publication of the pieces at the appropriate season.

Figure 9-18 Ornamental concrete used as an edging for a flower bed.

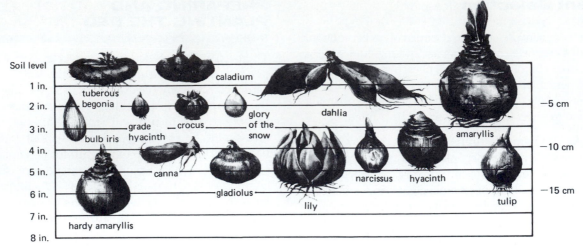

Figure 9-19 Suggested planting depths for common bulbs.

Flowers should be transplanted into the soil at the same level at which they were originally growing, and watered deeply. Bulbs should be planted according to Figure 9-19 or by following the directions on the package. Mark bulb areas with small stakes to avoid disturbing them accidentally when working in the bed.

Annuals grown from seed should be sown according to package directions and thinned to final spacing before they become crowded. Only the largest, most vigorous plants should be kept. The others should be pulled out or transplanted.

 Independent seed producers generally distribute their products through catalogs. Many successful small businesses produce and retail small amounts of hard-to-obtain items such as obscure flowers, antique vegetable varieties, and the like to the public.

FLOWER GARDEN MAINTENANCE

Watering

When natural rainfall is not sufficient, watering will be a frequent part of flower garden maintenance. Watering should soak the root zones of the plant as often as necessary to prevent wilting. Because annuals are relatively shallow-rooted, annual beds will require more frequent but shallower irrigation than do perennial or bulb beds. In mixed beds, watering should be deep enough to soak the most deeply rooted plants.

Mulching

Mulching with compost, grass clippings, or other organic materials will conserve water, reduce weed problems, and give a neat appearance to the bed. Mulch should be applied in spring as soon as annuals have been transplanted and new growth of perennials appears. Over the summer the mulch can be reapplied on top as the old mulch decays underneath.

In cold-winter areas an additional layer of mulch can be laid down over the bed after the cold weather begins. This layer will reduce freezing and thawing of the bed, a common cause of *winterkill,* the death of plants over the winter. The mulch layer should be pulled back from the crowns of perennial plants in spring as new growth starts.

Fertilizing

Fast-growing annuals will benefit from frequent fertilizer applications. Topdressing every four to six weeks starting at transplanting will assure that sufficient nutrients are available for rapid growth.

Frequency of fertilization of perennials will depend on climate and soil type. An annual fertilization in spring will be sufficient for perennial beds in cold-winter climates with loam to clay soils. On sandy soils, which lose nutrients rapidly, and in warm climates where growth continues all year, fertilization should be more frequent. Fertilizing every two months may be necessary on perennial beds, topdressing with a granular fertilizer.

Bulb fertilizing can be done at planting or when the bulb is actively growing. Some gardeners mix about a teaspoon (15 milliliters) of phosphorus-rich granular fertilizer or bone meal into the soil beneath each bulb as they plant it. Others wait until foliage appears and fertilize by topdressing. Either method is satisfactory.

Pruning

Pruning of annuals, perennials, bulbs, and herbs encourages maximum bloom and keeps the garden neat. Pinching or clipping off dead flowers should be a routine, particularly with annuals. This prevents energy from being diverted into seed development and thus increases bloom production.

Dead leaves should be removed from all plants as necessary to improve the appearance of the garden and to eliminate a source of disease. The leaves of hardy bulbs will yellow and die naturally in midsummer, and they can be cut off at ground level at that time. Leaves should never be cut back until they are almost fully yellow, because to do otherwise would remove the source of energy for the next year's blooms.

An alternative to waiting and cutting the leaves when they are yellow is bunching them. In this technique the gardener waits until flowering is finished then collects the bulb leaves together in a bunch while they are still green and folds them over on themselves, securing the bunch with a rubber band. This neatens the garden but still allows the bulbs to continue photosynthesizing to a limited extent. When the leaves turn yellow, they are cut off.

Division

The crowns of herbaceous perennials increase in size every year and, after several years, may become overcrowded. Bloom size and number may decrease due to this overcrowding, and therefore the clump should be divided into several smaller pieces and replanted. Chapter 4 gives instructions for division of perennials and bulbs.

Staking and Supporting

Tall flowers and those plants with profuse or heavy flowers may require staking to keep them from breaking or falling over. Staking will be especially necessary where frequent rains weight the blooms or where wind is common.

One or several stakes may be required to support a plant, depending on its size. Figure 9-20 shows ways of supporting plants by staking. Regardless of which

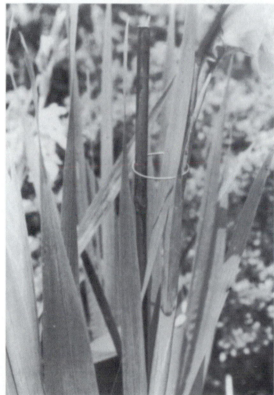

Figure 9-20 Two methods of staking to support a flower. *(Photos by George Taloumis)*

method is used, the stakes should be placed behind or close to the plant so they are hidden by the foliage. The plant should be secured to the stake with twine, strips of cloth, or covered-wire twist tabs.

SPECIALTY GARDENS

Rose Gardens

Roses are the most popular garden flower, so gardens devoted exclusively to rose growing are very common. Many types of roses are grown; the most well-known types are listed:

Types of Roses

Hybrid Teas. Hybrid tea roses (Figure 9-21) produce large blossoms singly on moderate-sized, bushy plants and are favorites for cut-flower use. They are the most prone to disease problems and are not always reliably cold-hardy.

Floribundas. Floribundas bear their blooms in clusters through summer and have blooms generally smaller than those of teas.

Grandiflora. *Grandiflora* (Figure 9-22) is the name given to the roses that are crosses between floribundas and hybrid teas. They produce fewer blooms per cluster than floribundas, but each blossom is nearly as large as a hybrid tea bloom.

 Roses are the most popular flower in North America, and there are horticulturists who specialize in only this flower. Examples of careers focusing on roses include cut-flower production in the greenhouse, field-nursery production of garden plants for sale in spring, and selling of roses in the wholesale cut-flower markets of large cities.

Climbing and Rambling Roses. These have long canes that grow from the base of the plant. They grow very large and are trained along fences and garden arches called *pergolas* (Chapter 10 opening photo).

Tree Roses. Tree roses can be hybrid tea, grandiflora, or floribunda roses. They are grafted onto long-cane hardy species to achieve a tree form. Tree roses are frequently grown in pots for patio decoration in summer. In freezing areas they are then buried, still potted, in a sheltered place for overwintering.

Figure 9-21 A tea rose. *(Photo © Susan McKessar justourpictures.com)*

Miniature Roses. Despite their frail appearance, miniature roses are very cold-hardy. They produce thumbnail-sized blooms singly or in clusters and can be raised indoors in winter on a sunny windowsill.

Rose Culture

Roses are often high-maintenance plants, requiring regular spraying for diseases and insects and annual pruning. Their cultivation is the subject of many books that deal in depth with the subject. The following gives only a brief overview.

Soil. Quick- and thorough-draining soil is essential for roses. Large quantities of amendment should be added if the soil contains a large percentage of clay. Soil should be worked and amended to a depth of 20 to 24 inches (50–60 centimeters), and pH should be slightly acid, in the range of 6.5 to 7.0.

Fertilizing. Granular balance fertilizer should be applied in spring and after each bloom flush.

Watering. Soil should be kept moist to a 2-foot (60-centimeter) depth. Rose foliage should not be wet because this encourages the spread of fungus diseases such as blackspot, rust, and mildew.

Overwintering. Being woody, rose canes live over the winter and renew the foliage in spring. The canes of some types are not reliably cold-hardy though. Bush roses should be mounded with soil to 1 foot (30 centimeters) deep in areas where winter temperatures drop below 10°F (−12°C). Rambler and climber canes may be pinned to the ground and covered with soil in areas where they are not reliably winter-hardy.

Pruning. Pruning should remove only dead wood, weak canes, and crossing branches. Death of the top por-tions of canes during winter is a very common problem and will necessitate shortening the canes to living wood in spring. Unless the rose is very vigorous, pruning should be minimal. The removal of large quantities of wood will slow growth and reduce the number of flow-ers produced.

Figure 9-22 A grandiflora rose. *(Photo courtesy of the Rose Garden, New Zealand Gardens Online. http://gardens.co.nz/RoseGarden/)*

Figure 9-23 A climbing rose. *(Photo by the author)*

Figure 9-24 A rock garden. *(Photo courtesy of Zurn Industries Incorporated)*

Rock Gardens

Rock gardening (Figure 9-24) combines rocks and plants to create a natural-appearing area for the cultivation of species native to rocky or alpine regions. In some cases a gardener can make use of naturally rocky land with some reorganization of the available rocks, but more often large rocks must be brought to the site and arranged artistically to give the appearance of a natural formation. Herbaceous perennials, succulents, bulbs, annuals, and even shrubs and trees (see Table 9-5) are then planted in the small crevices and soil pockets formed between the rocks. Most rock garden plants are small or even minute compared to other garden plants. When shrubs or trees are used, they are generally dwarf cultivars.

Planting of a newly created site should be postponed until after the soil has settled among the imported rocks. Little soil amending beyond the addition of small amounts of organic amendment will be needed, because most rock garden plants are native to soils of poor natural fertility. However, if acid-loving plants are to be grown, pockets between rocks can be amended with peat moss to create the proper soil environment.

Care of a rock garden consists of weeding, dividing large plants, and watering as needed when the soil becomes dry. In general, rock garden plants will need little moisture. Fall care in very cold-winter areas consists of covering the area (with the exception of evergreens) with a mulch layer of hay or other material. The mulch minimizes heaving of the ground and uprooting of the plants.

Container Gardens

Container gardens (Figure 9-25) are a popular way of growing flowers and vegetables without land. Almost any plant can be grown in a pot, including trees, vines, vegetables, flowers, and shrubs.

Because of the limited volume of soil available for root growth in containers, the soil should be of the best possible quality. Artificial media (discussed in Chapter 15) are the best choices, or a mixture of natural soil and a large amount of organic soil amendment (see Chapter 5). Pure garden soil is seldom satisfactory because of its tendency to pack and drain poorly when confined to a pot.

Supplying water to a container garden is the most time-consuming chore. Water is used up very quickly from the soil and must be replenished often. Daily watering will be necessary during hot weather or when clay or small pots are used. However, automatic irrigation systems designed for pots are available (see Chapter 20).

Frequent fertilization is essential for container-grown plants. A soluble fertilizer (see Chapter 15) applied in the irrigation water is one solution. Top dressing with slow-release or granular fertilizer is also satisfactory.

Overwintering perennials, trees, and shrubs in the North in containers may need winter protection. Being aboveground, the roots will be subjected to lower temperatures than roots growing in the earth, and death from cold is likely. If possible, the pots should be planted in the ground in fall and mulched heavily. If this is not practical, they can be overwintered in a garage or placed in a sheltered location next to a building.

Cutting and Dried-Flower Gardens

Although many gardeners harvest flowers directly from their display beds, others prefer cultivating a separate area for flowers to be cut and brought indoors. A cut-flower garden is usually located in the service area of the landscape out of view and includes varied, tall annual and perennial cultivars that lend themselves to arrangement and last well when cut.

 Dried-flower producers field grow their own crops, then dry them at the peak of bloom in drying sheds. The dried product is then marketed to floral shops, gift stores, craft stores, or other outlets. Some growers have operations in which the dried materials are made into arrangements and sold to retail outlets or directly to the public at a roadside stand.

Table 9-5 Selected Rock Garden Plants

Common name	Botanical name	Type of plant	Remarks
Bugleweed	*Ajuga genevensis* and *reptans*	Perennial groundcover	Grows 2–6″ (5–15 cm) high, depending on variety, and thrives in sun or shade; produces blue flowers
Sea pink or thrift	*Armeria maritima*	Evergreen perennial	Small grasslike clumps of deep green leaves produce pink ball-like flower heads
Basket-of-gold	*Aurinia saxatilis*	Perennial groundcover	Low growing and covered with bright yellow flowers
Crocus	*Crocus* spp.	Hardy bulb	Bell-shaped flowers on tiny plants in early spring
Broom	*Cytisus* spp.	Small shrubs, deciduous or evergreen	White to yellow pea-shaped flowers on bright green plants
Garden pink	*Dianthus* spp.	Herbaceous perennial	Flowers pink, white, red, purple on short plants, to 12″ (30 cm) tall
Heaths and heathers	*Erica* spp., *Calluna* spp.	Evergreens	Bell-shaped flowers on plants 1–2′ (30-60 cm) tall; need acid soil
Candytuft	*Iberis sempervirens*	Evergreen groundcover	Deep green leaves and white flower heads
Iris (bulb type)	*Iris reticulata* and other small species	Hardy bulb	Yellow, white, or purple flowers in very early spring on plants 4″ (10 cm) tall
Juniper	*Juniperus* spp.	Needle evergreen	Many types available with blue-green to deep green foliage; choose low or groundcover varieties
Alyssum	*Lobularia maritima*	Annual in all except the mildest climates	White, pink, or purple flowers cover the creeping mat-like plant; reseeds naturally
Grape hyacinth	*Muscari* spp.	Hardy bulb	Purple, blue, or white flowers on tiny plants, up to 4″ (10 cm) tall
Creeping phlox or moss pink	*Phlox subulata*	Herbaceous perennial	White or pink flowers produced profusely over a mat-like plant
Dwarf alberta spruce	*Picea abies* (dwarf cultivars)	Dwarf evergreen tree	Very compact, slow growing, with deep green foliage
Mugo pine	*Pinus mugo* cv. *mugo*	Needle evergreen shrub	Shrub-like pine growing to 18″ (45 cm) high but spreading widely
Lavender cotton	*Santolina chamaecyparissus*	Herbaceous perennial	Finely cut silver-gray foliage is its outstanding trait; yellow to white flowers in summer
Stonecrop	*Sedum* spp.	Perennial succulent	Species vary from 1″ (2 cm) to 2′ (60 cm) high and have bright flowers
Hen-and-chickens	*Sempervivum* spp.	Perennial succulent	Small rosette plants in green or with tinges of purple or red
Thyme	*Thymus* spp.	Herbaceous perennial or annual	Dark green to gray leaves make up a creeping plant
Species tulip	*Tulipa kaufmanniana*	Hardy bulb	Small plants, about 6″ (15 cm) high, with the traditional tulip flowers

Flower preservatives to lengthen the life of cut flowers are an effective and worthwhile investment. However, even more economical is a mixture of $\frac{1}{3}$ tonic water and $\frac{2}{3}$ water. This solution supplies sugar for respiration, a preservative to slow growth of bacteria in the water (bacteria block xylem and cause flower wilting), and acid to lower the pH of the water (a condition that also prolongs flower life).[†]

[†] This formula given to the author in conversation with Dr. Michael Reid, Professor of Horticulture with specialization in post-harvest physiology of flowers at the University of California at Riverside.

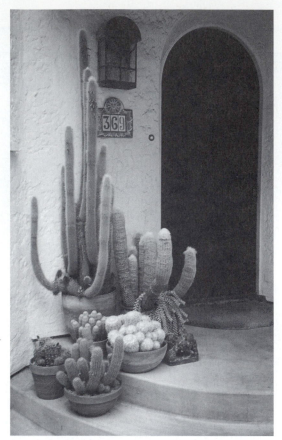

Figure 9-25 Containers of cacti and succulents grace the entrance to a home. *(Photo by the author)*

Figure 9-26 Bells of Ireland, a pale green flower excellent for both cutting and drying. *(Photo courtesy of W. Atlee Burpee and Co., Warminster, PA)*

Flowers in the same area can be grown for drying and use in permanent arrangements (Figure 9-26). These should be harvested at their peak of bloom, bunched loosely with the stems together, and hung upsidedown in a dry area with good air circulation. Hanging out of direct sunlight will help preserve the natural color. Table 9-6 is a list of flowers suitable for drying and cutting.

Herb Gardens

Herb gardens are among the oldest types of gardens and have recently come back into favor. They are both ornamental and utilitarian, supplying fragrant leaves and seeds for decorating, cooking, teas, and perfuming.

 Herb growers can either grow herbs for bedding plants or produce crops to edible maturity in the field. They can market the herbs at farmers' markets, roadside stands, or fresh to large spice producers. Alternately, they can dry the herbs on site and market the dried product to wholesalers or retailers.

 Horticultural plant producers, such as the ones listed in the references at the end of this chapter, publish annual catalogs and employ horticulturists to write plant descriptions, evaluate plant photographs for publication, design the catalog, and supervise its printing.

Table 9-6 Selected Garden Flowers for Cutting and Drying

Common name	Botanical name	Annual/perennial	Remarks
Yarrow	*Achillea filipendulina* and *millefolium*	Perennial	Yellow, white, or pink umbrella-shaped flowers; use fresh or dry
Snapdragon	*Antirrhinum majus*	Annual	Spike flowers in many colors; select tall varieties for fresh use
Aster	*Callistephus chinensis*	Annual	For fresh use, with large heads in pink through red and purple; pick disease-resistant cultivars
Cockscomb	*Celosia cristata cultivars*	Annual	Red or yellow blooms in plume or cockscomb shape; use fresh or dried
Bachelor button	*Centaurea cyanus*	Annual	Blue, pink, or white button-sized flowers for fresh use
Marguerite daisy	*Chrysanthemum frutescens*	Annual in all but the mildest climates	Very fast growing and produces enormous amounts of daisy flowers in white and yellow; for fresh use
Chrysanthemum	*Chrysanthemum × morifolium*	Annual or perennial	Many colors and forms for fresh use
Larkspur	*Consolida orientalis*	Annual	Spike-edged white, pink, or purple flowers for fresh use
Cosmos	*Cosmos bipinnatus*	Annual	Tall-growing plant which may need staking; pink through red colors for fresh use
Dahlia	*Dahlia* hybrid	Summer bulb	Both dwarfs and full-size cultivars are useful; long lasting for fresh use and available in most colors
Gladiola	*Gladiolus* hybrid	Summer bulb	Tall spike flowers in many pastel shades; fresh use only
Globe amaranth	*Gomphrena globosa*	Annual	Cut fresh or for dried use; large cloverlike blossoms in white, pink, red, or purple
Baby's breath	*Gypsophila elegans* and *paniculata*	Annual or perennial	Fresh or dried use, in white or pink; grows into a shrub 3′ (1 m) tall
Strawflower	*Helichrysum bracteatum*	Annual	Use fresh or dried; 2″ (5 cm) blooms in yellow, red, white, purple, and shades between
Helipterum	*Helipterum manglesii*	Annual	Daisy-like flowers in white or pink are used for drying
Statice	*Limonum sinuatum*	Annual except in very mild climates	Papery flower clusters in many colors; useful fresh or for drying
Bells of Ireland	*Molucella laevis*	Annual	Green bell-shaped flowers on a 3′ (1 m) plant; use fresh or dry
Narcissus	*Narcissus* spp.	Hardy bulb	Many types available; useful for fresh cutting in early spring
False dragonhead	*Phystosegia virginiana*	Perennial	White, pink, or purple flowers in spikes; a rampant grower for fresh-cut use
Blue salvia	*Salvia farinacea* and *Salvia patens*	Annual in all but mild climates, but reseeds readily	Deep blue spike flowers for use fresh or dried
Pincushion flower	*Scabiosa fischeri*	Annual	White, pink, or blue flower heads on slender stems; for fresh use
African marigold	*Tagetes erecta*	Annual	Yellow, orange, or white flowers for fresh use; choose tall cultivars; easy to grow
Immortelle	*Xeranthemum annuum*	Annual	Fluffy, papery flowers in pink through red; use fresh or dried
Zinnia	*Zinnia elegans*	Annual	Flower heads 1–4″ in diameter (2–10 cm), depending on cultivar; available in any color; for fresh use

Table 9-7 Herb Garden Plants

Common name	Botanical name	Annual/perennial hardiness	Height	Uses
Garlic	Allium sativum	Tender perennial zones 9–11 (treat as annual)	To 2′ (0.6 m)	Bulbs used for seasoning
Chives	Allium schoenoprasum	Perennial zones 2–11	To 2′ (0.6 m)	Leaves used for seasoning, very ornamental flowers
Dill	Anethum graveolens	Annual all zones	To 4′ (1.2 m)	Leaves used fresh for seasoning and seeds dried for pickling
Tarragon	Arthemisia dracunculus	Perennial	To 2′ (0.6 m)	Leaves used for seasoning
Caraway	Carum carvi	Biennial zones 3–11	To 2′ (0.6 m)	Seeds used for seasoning
Corlander or cilantro	Coriandrum sativum	Annual	To 3′ (1 m)	Leaves used fresh for seasoning and seeds also
Cumin	Cuminum cyminum	Annual	To 6′ (15 cm)	Seeds used as seasoning
Bay	Laurus nobilis	Woody shrub zones 7–11	To 2′ (0.6 m)	Leaves used for seasoning
Lavender	Lavandula angustifolia	Perennial zones 5–11	To 3′ (1 m)	Leaves and buds used for sachets
Mint	Mentha spp.	Perennial zones 3–11	To 2′ (0.6 m)	Leaves used for seasoning and tea
Sweet basil	Ocimum basilicum	Annual all zones	To 2′ (0.6 m)	Leaves used for seasoning
Sweet marjoram	Origanum marjorana	Perennial zones 9–11 (treat as annual in other zones)	To 2′ (0.6 m)	Leaves used for seasoning
Oregano	Origanum vulgare	Perennial zones 3–11	To 2½′ (0.8 m)	Leaves used for seasoning
Parsley	Petroselinum crispum	Biennial treated as annual	To 15″ (0.4 cm)	Leaves used for seasoning and as garnish
Anise	Pimpinella anisum	Annual	To 2′ (0.6 m)	Seeds used for seasoning
Rosemary	Rosemarinus officinalis	Perennial zones 6–11	To 5′ (1.5 m) (prostrate forms common)	Leaves used for seasoning
Sage	Salvia officinalis	Perennial zones 3–11	To 3′ (1 m)	Leaves used for seasoning
Thyme	Thymus vulgaris	Perennial zones 3–11	To 1′ (0.3 m)	Leaves used for seasoning
Ginger root	Zingiber officinale	Perennial zone 10; (annual or houseplant in other zones)	To 3′ (1 m)	Root used for seasoning, especially in Chinese cooking

The growing requirements of herbs are less stringent than those of other garden flowers. Full sun is required for strong growth and full flavor, but herbs will perform well in relatively poor, rocky soils and in drought. A few, including the mints, will even grow in wet soils. Table 9-7 lists the more common herbs, their hardiness, height, and uses.

SELECTED REFERENCES FOR ADDITIONAL INFORMATION

Cairns, Tommy. *All about Roses.* Des Moines, IA: Meredith Books, 1998.

Crockett, James A. *Crockett's Flower Garden.* New York: Little, Brown, 1981.

Edinger, Philip. *Flower Garden Plans.* San Ramon, CA: Ortho Books, 1991.

Hill, Lewis, and Nancy Hill. *Successful Perennial Gardening: A Practical Guide.* Williamstown, Storey Publications, 1988.

Jones, Patrick. *Practical Rock Gardening.* North Pomfret, VT: Trafalger Square, 1993.

Nau, Jim. *Ball Perennial Manual: Propagation and Production.* Batavia, IL: Ball Publishing, 1996.

O'Sullivan, Penelope. *The Easiest Flowers to Grow.* Des Moines, IA: Meredith Books, 2001.

Parks Seed Company. *Annual Retail Catalog.* Greenwood, SC.

Robinson, Peter. *Rock and Water Gardening: A Practical Guide to Construction and Planting.* New York: Anness Publishing, Incorporated, 2001.

Sinnes, A. Cort. *Shade Gardening.* San Ramon, CA: Ortho Books, 1996.

Sinnes, A. Cort, and Larry Hodgson. *All About Perennials.* San Ramon, CA: Ortho Books, 1996.

Strong, Graham and Alan R. Toogood. *The Mix-and-Match Color Guide to Annuals and Perennials: Over a Million Ways to Create Glorious Summer Color.* Alexandria, VA: Time-Life Education, Incorporated, 2001.

Wayside Gardens. *Retail Catalog.* Greenwood, SC.

World Wide Web Addresses

www2.garden.com   A website that includes a garden doctor, garden planner, chat, and shopping.

http://www.parkseed.com/webapp/commerce/command/ExecMacro/psStoreHome.d2w/report?merchant rn=541   Park Seed Company. Park Seed Company website.

http://www.waysidegardens.com/webapp/commerce/command/ExecMacro/psReferAFriend.d2w/Start?merchant rn=542   Wayside Gardens. Wayside Gardens website.

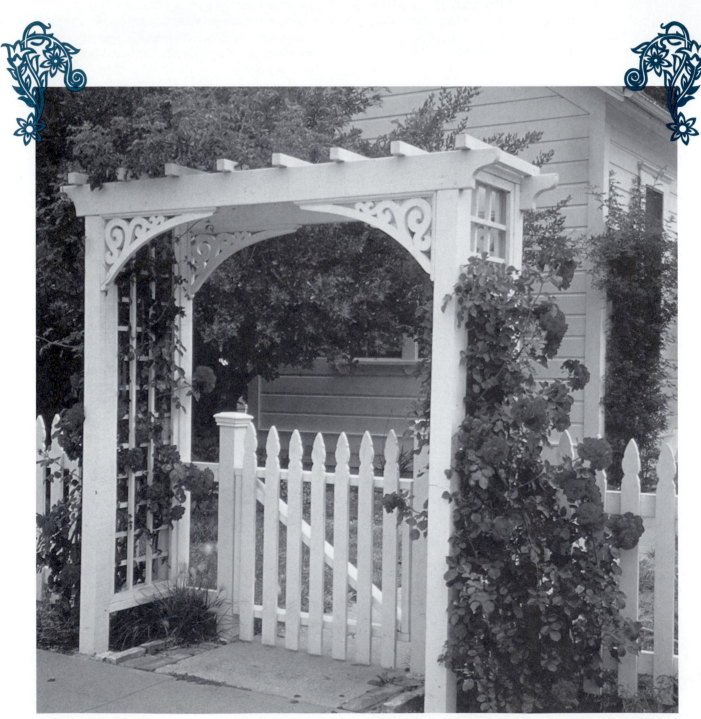

(Photo by the author)

CHAPTER 10

HOME LANDSCAPE PLANNING AND INSTALLATION

Landscaping one's home is not usually thought of as an act of ecological awareness, but it is. Global warming, or the greenhouse effect, has recently come to the foreground of our society's consciousness as an environmental change that could radically affect all our lives. The use of fossil fuels such as gasoline, oil, and coal; accumulation of propellant and refrigerant chlorofluorocarbons; and rapid deforestation, particularly in tropical areas such as the Amazon basin, are all causes of increased temperatures in the world. The warming is significant; the five warmest years of the century were in the 1980s. Although global warming was anticipated, the rate is three times higher than was originally predicted, and there is justification to say that North Americans are largely responsible because we produce nearly one-fourth of the world's carbon dioxide from burning fossil fuels.

All this may sound daunting and not a matter that can be remedied by individual action, but such is not the case. Global Releaf is a public education campaign whose aim is to increase people's awareness of individual actions they can take to help combat this environmental problem. Their suggestions for steps to take start with *planting trees!* Although trees absorb carbon dioxide and thus directly reverse its accumulation in the atmosphere, their main effectiveness is through the shade they give in summer. By shading buildings and paved areas (which generate reflected heat onto buildings) and by their transpiration, trees decrease the need for air conditioning. This indirect cooling reduces carbon dioxide emissions by decreasing electrical use for air conditioning by 10 to 50 percent.

Trees form the backbone of most landscapes, but other landscape plants can also be ecologically beneficial. Groundcovers prevent erosion of topsoil. In winter, evergreens can shield a house from heat-robbing winds and lessen heating bills. Where smog is a problem, plantings of shrubs and trees absorb pollutants.

Finally, trees and shrubs shelter squirrels, birds, and other wildlife, and their berries and nuts serve as food. By adding a source of water such as a pool or fountain, all the requirements of small wildlife can be met and many species will inhabit the yard.

The importance of planning your landscape cannot be overemphasized. As land becomes more expensive and the lots for houses smaller, haphazard planting of shrubs and trees not only creates a cluttered effect but becomes impractical. In the end it is nearly as expensive and labor-consuming as a landscape based on a plan would have been.

A thoughtfully planned landscape offers many rewards. Most obviously it enhances the design and beauty of the home, thus increasing its value. But a landscape has other functions that are not as immediately apparent. For example, it can extend the living area of the home by a patio or deck screened for privacy and protected from wind and sun. An outdoor play area can provide an outlet for overflowing childhood energy and the noise that normally accompanies it.

THE PROFESSIONAL LANDSCAPE

Landscaping a home is often expensive because, unless one is blessed with a home in a naturally beautiful setting with enough land to assure privacy, extensive planting and construction will frequently be necessary.

 Professional landscapers design the planted areas around homes and small commercial buildings. The plans can include paved areas, fencing, walls, plants, and irrigation if needed.

If cost is not the main consideration, a professionally designed and executed landscape is easiest. The plan will be artistic, the shrubs and trees transplanted correctly, and the entire package will come with a guarantee.

But for many homeowners the cost of a professional landscape is prohibitive. For these people, two alternatives should be considered: (1) paying for a professional plan but installing the landscape personally and (2) planning and installing the landscape yourself. Each has advantages and drawbacks.

A professional design may cost many hundred dollars, but it is usually money well spent because most people lack the experience and imagination to design a professional-looking landscape. It is, however, possible for a homeowner to design an attractive landscape, provided considerable time and effort are devoted to the project. The following section gives directions and suggestions for the do-it-yourself landscape designer.

THE HOMEOWNER-DESIGNED LANDSCAPE

Needs Analysis

The first step is to compile a needs analysis, that is, an inventory of what should be accomplished by means of the landscape and what the landscape will include, based on the preferences and life style of the owners.

To facilitate the needs analysis, the property should be divided into three areas (Figure 10-1). The *public area* includes all property viewable from the street: the drive, walk, front door, parking area, and sometimes the sides of the house. The *private living area* includes the patio or deck, play areas for children, bar-

becue and outdoor eating areas, and any game areas. The *service* or *utility area* is the area reserved for necessary but unattractive features such as recreational vehicle storage, dog kennel, and clothesline. A needs analysis for the public area should answer such questions as: "Is privacy needed from the street?" and "Does the entryway need lighting for night entry?" To design the private living area, the lifestyle of the owners should be taken into consideration, asking such questions as: "Does the family entertain frequently outdoors?" and "Is a play area needed for children?"

 A landscape estimator is responsible for determining the cost of executing the design prepared by the landscape designer. He or she must determine costs of paving, plants, lawn installation, and so on, and figure in the profit for the job as well.

Figure 10-2 is a checklist covering most of the options one would consider including in a typical landscape. Desired options should be checked on the list, and the completed form compared to the final design to make sure all requirements have been met.

Site Analysis

Site analysis, the second stage of planning, is an inventory of the property in terms of house architecture, views, soil, land slope, lot size, existing landscape, and the like. It lists both good and bad points of the property, so that they will be dealt with effectively in the design. For example, are there established trees that should be incorporated into the design or an air-conditioning unit that should be screened from view? Are there poorly draining areas where nothing grows or a steep slope where erosion is a problem? Figure 10-3 is a checklist much like the needs analysis sheet.

When site analysis is done, the landscape designer must take into consideration the climate of the area. It is more environmentally sound to work *with* rather than *against* the natural climate when planning a landscape. For example, in dry climates, planting species that require frequent irrigation to thrive will use valuable water resources for ornamental plantings that may be needed elsewhere for food production. The homeowner's water costs could also be high. In such areas (chiefly the West), it is preferable to orient toward

Figure 10-1 The public, private living, and service areas of a typical landscape.

Public area

night lighting
privacy from street
extra parking
improved walkway/entryway
entry court
private living area
irrigation system
shade
lawn
terracing for changes in level

Private living area

patio/deck (specify size and normal use times)
roof for patio/deck
shade
screening from neighbors
irrigation system
night lighting
plants attractive to birds
settings for ornaments (statuary, birdbath, and the like)
ornamental or swimming pool
barbecue grill/fire pit

children's play area (specify included equipment)
game area (specify games and space needed)
view emphasis
wind control
terracing for changes in level
auxiliary patio/deck (specify location)
lawn
night lighting

Service/utility area

recreational vehicle storage
vegetable garden
fruit/nut trees
pet area
tool storage shed
clothesline
trash can storage
irrigation system
firewood storage
cut-flower garden
greenhouse
compost pile

Figure 10-2 Needs analysis checklist.

Overall lot

1. Is the overall soil condition good, average, or poor?

2. Are there any areas with poor drainage where water stands after a rain?

3. Are there steep or gradual slopes to incorporate into the design? rocks or rock outcrops?

4. Are there good or objectional views, and from where can they be seen?

5. Is the property higher than, lower than, or equal in elevation to surrounding properties (for screening purposes)?

6. Could grading improve drainage or add interest to the lot?

The house

1. Is the home one story, two story, or split level?

2. Is the design modern or old? Are there historic considerations in designing the landscape?

3. Could changes in the house (painting, extending windows) improve the effectiveness of the landscape?

4. Where are the windows that will look out on the landscape?

5. Are there unattractive permanent architectural features that should be concealed?

Weather and microclimates

1. In what USDA hardiness zone is the house located?

2. Are there frequent strong winds, and from what direction?

3. Where are the shady and sunny areas of the property?

4. Are there microclimate areas, and where? (See Chapter 1 for a discussion of microclimates.)

5. What is the pattern of sun and shade in the private living area from dawn to dusk?

Existing plantings

1. Are there large existing trees; where are they located?

2. Do any trees need to be removed for shade, disease, or design reasons?

3. Does the existing lawn area need improvement?

4. Is there any native vegetation to be worked into the design?

5. Are there any plants sufficiently desirable to justify being saved (other than trees)?

6. What are the species of any existing plants that will be saved?

Existing construction

1. Are existing fences and plantings sufficient to provide privacy?

2. Should the drive be replaced or rerouted?

3. Is the entryway to the house sufficient, or should it be redesigned on a larger scale or in a more convenient location?

4. Are there utility meters, natural gas tanks, or air conditioners that require screening?

5. Is any existing construction in need of repair or replacement (fences, walks, retainer walls, and the like)?

6. Does the patio or deck area need enlargement?

Figure 10-3 Site analysis checklist.

xeriscaping, the selection of low-water-use plants adapted to the area. In recent years, much breeding has been done of plants native to dry areas, both from North American and from dry areas of South Africa, the Mediterranean coast, Australia, and South America. The new plants that have been developed are drought-tolerant, but very showy and floriferous, and are an environmentally sensitive alternative to landscaping with heavy-water-use plants.

Hydrozoning is another technique for water conservation. In hydrozoning, plants with similar water requirements are grouped together by their need for irrigation. Then the irrigation plan is designed to irrigate some areas more and others less, depending on the plants in that zone. For example, it is possible to design a landscape with a lush entryway of high-water-requiring ferns, flowers, and perhaps a small lawn, and to design outlying areas to incorporate primarily drought-tolerant grasses, succulents, palms, and the like.

 A landscape architect differs from a landscape designer in that he or she must have training in architecture, engineering, and horticulture. Landscape architects are employed to do large-scale designs where there is to be earth grading, large retaining walls, extensive outdoor structures, and other landscape components involving extensive knowledge of drainage and engineering. Landscape architecture is a separate specialization from horticulture in most universities.

The Preliminary Drawing

The first step in designing the landscape is to create a preliminary drawing that shows the basic layout and permanent features of the property and includes the information from the site analysis. A scale drawing of the lot, house, and valued plants should be made on a large sheet of graph paper. This drawing is the skeleton of the design. Sheets of tracing paper can then be laid on top for experimenting with different ideas.

The drawing need not look professional to be serviceable; it is more important that it be accurate and include all the necessary information. Unless there are trees or particularly valued shrubs that will definitely remain, it is best not to include plants in the preliminary drawing. They limit creativity and may influence the designer to see the property only in terms of the present design. However, permanent features such as retainer walls, power lines, good and bad views, and the windows and doors of the house should be included, because modifying them would involve major changes.

Area Layout

After the basic drawing is complete, the parts of the property to be included in the public, private, and service areas should be chosen. In most established yards these areas will already have been defined through use, but unless there are absolutely no alternatives, it is worthwhile to experiment with shifting the locations of these areas. A service area could be moved from the back to the side of the house, or a private living area from the back to an enclosed courtyard in front. Figure 10-4 shows how a property could be designed with the service area in three different locations.

Public Area Design

The public area is the first area a visitor sees. This area conveys an initial impression about the residents of the house. If poorly designed and ill-kept, it will give an impression of sloppiness; if sparsely planted or rigidly pruned, it will convey formality. If walled as a courtyard, it will project a sense of seclusion.

The main functions of the public area landscape are to blend the house with its surroundings and to provide a pleasant and readily accessible entry to the house. To perform these functions, most landscapes include several basic elements: the entryway and walkway, the driveway, and the auxiliary plantings.

Entryway and Walkway Areas

In most homes the builder includes a standard narrow walkway leading from the drive and an undersized concrete slab at the door to serve as a waiting area for guests.

Many homeowners consider these permanent features and are reluctant to use a design that involves removing and replacing them. Although it is true that removing concrete is a major undertaking, if the walk is too narrow and the waiting area at the door uncomfortably tiny, the labor of removing these features and replacing them with a gracious entry is justified. They can also be covered and extended with a new paving material.

At a minimum, any walk should be 4 feet (1.2 meters) wide, the least width that allows two people to walk side by side. Five or six feet (1.5 to 1.8 meters) wide is even better. At this width a person can walk without watching his feet to avoid stepping off the paving. A walk of this width may at first sound excessively wide and unattractive, but the monotony can be broken by small trees framing the walk, groundcover along the edge, or an interesting paving material.

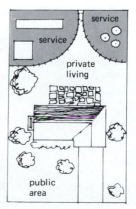

Figure 10-4 Three drawings of the same house, showing how the service area could be positioned in several locations.

Figure 10-5 An entry patio requires little maintenance and is visually pleasing.

Steps that need to be included should be wide and rise slowly, with treads at least 18 inches (45 centimeters) deep. Monotonous flights of evenly spaced steps should be avoided. Instead, several steps followed by a section of walkway and then more steps should be used to complete the level change. Changes in direction to create a curving walk should be considered.

One interesting and little-used public area design is the entry patio. It combines the walk and entryway into one paved area interspersed with beds of landscape plants (Figure 10-5). Such patios require little maintenance and create a feeling of spaciousness.

A variation of the entry patio is the closed entry courtyard (Figure 10-6). A wall or fence is used to delineate an area leading to the front door. The area is then planted with shrubs and small trees or paved and designed to include planting beds. An entry courtyard is useful when the front of the house is not overly attractive, when the house is situated close to the street, or when the home has large front windows that would

Figure 10-6 An entry courtyard in Spanish style.
(Photo by the author)

otherwise have an unattractive street view. An entry court can also double as a private living area.

Before constructing an entry court, it is wise to check the zoning ordinances in the community regarding the construction of walls or fences in the public domain. Occasionally, walls of any type will be prohibited, but more often there are height restrictions.

The Driveway

Unless the driveway is a sweeping circular drive or leads through a large property, it will not contribute to the appearance of the public area and should be as inconspicuous as possible. Unnecessary curves should be eliminated, and the edges should never be outlined with flowers. Such treatment serves only to emphasize its presence.

However the designer should beware of attempting to minimize the driveway by limiting its size. This is not practical as it limits its functionality and the ease with which it can be used. Figure 10-7 gives sample driveway and parking area designs along with the minimum measurements acceptable for each type.

In many homes the driveway doubles as a walkway to the point where the regular walk begins. This is completely acceptable provided the driveway is not blocked by cars; in fact it is preferable to bisecting the yard with a walk to the street. Another effective driveway treatment is to extend the driveway/walkway with a strip of paving (Figure 10-8) merging into a patio attached to the house. Visitors parking in the drive can then step onto a paved area instead of the yard.

The selection of an attractive paving material can do much to improve the appearance of this otherwise mundane landscape element. Blacktop may be cheapest in terms of labor and materials, but if a homeowner can do the installation, more decorative paving materials may be affordable. Table 10-1 lists paving materials.

Auxiliary Plantings

Auxiliary plantings are the third element of the public area design. They encompass the trees, shrubs, and groundcovers that are the living part of the landscape.

Trees form the backbone of the landscape, framing and forming a backdrop for the house. If a house has large, well-placed trees, it will always be inviting, regardless of what else is planted in the yard. The cool, restful feeling of a tree-shaded front yard is unequaled.

If large trees are already present, their location must be worked around when designing the landscape. But in a treeless yard the selection and placement are of utmost importance.

Figure 10-7 Specifications for driveways and small parking areas of various shapes. *(Taken from* Landscape Design, A Practical Approach, *2nd ed. by Leroy Hannebaum, 1990. Reprinted by permission of Prentice Hall, Englewood Cliffs, NJ)*

Figure 10-8 Extending a driveway/walkway with a strip of paving creates a graceful entrance and avoids bisecting the lawn. *(Photo by Andrew Schweitzer)*

Table 10-1 Paving Materials

asphalt or blacktop
brick or brick pavers
concrete (of any color)
exposed aggregate concrete
precast colored concrete slabs
interlocking precast concrete pavers
gravel
railroad ties
stone
tile
combination of brick and concrete

Evergreen trees such as pines are useful because they retain their foliage throughout the year and are a permanent screen from wind, sun, and street traffic. Deciduous trees, on the other hand, offer shade during the summer months but do not screen out welcome winter sunlight.

For easy maintenance and protection from lawnmower nicks, it is helpful to locate trees in beds of shrubs or groundcover or to surround them with mulch. This also eliminates the need for hand trimming grass.

Trees of moderate to large size should not be planted closer than 20 feet (6 meters) from the house, because their spread and height will increase as they

Figure 10-9 A planting that screens the house from the street.

mature. Small trees can be planted closer because their ultimate size is not as great, but they should never be planted under the eaves of the house. In a similar manner, trees should never be planted where they will grow into overhead lines.

Auxiliary plantings of shrubs and groundcovers are used in most landscapes. Beds can be used to screen the house from the street (Figure 10-9), to balance other large plantings in the yard, or as accents. They can be used to incorporate shade trees, because they are not usually close to the house.

Planting beds connected to the house can be either corner or foundation plantings. Corner plantings extend out from the front corners of a house, blending it into the site and eliminating the harsh right angle where the house meets the ground. They are among the most common landscape plantings, often including a small tree, several shrubs, and an attractive groundcover (Figure 10-10).

Foundation plantings were a necessity in the 1930s when homes were sometimes built with about 18 inches (45 centimeters) of unattractive foundation showing above ground level. The foundation had to be covered, and the foundation planting concept was formed. The idea persists today, and most homeowners still plant a line of shrubs along the front of their house. A foundation planting should no longer be considered essential, and planting beds can be designed to cover none, all, or part of the front of the house.

Regardless of whether the bed is freestanding or connects to the house, most beds follow several simple rules. First, the shrubs are used in masses of three or more of the same species and are planted to grow together to form a mass of foliage. Second, the planting bed is always delineated by edging or paving, and all the included area is covered with groundcover or mulch. The clean lines and mulch- or groundcovered area create a finished, professional appearance. Third, sweeping lines are used to form large, curved, or geometric beds. The homeowner is frequently timid when designing planting beds, making them small and close to the house. A better effect can be achieved by large beds with bold shapes that jut out into the yard.

Guidelines for Public Area Design

There are many clues that separate the professionally designed from the obviously homeowner-designed landscape. By using a few techniques from professional landscapers, the finished design will be more pleasing.

 A landscape contractor is a person who coordinates large-scale landscaping projects such as business parks and housing projects. This is a business-oriented occupation, with many people working for the landscape contractor: designers, installers, irrigation specialists, tree movers, and so forth.

First, do not try to save every plant in the yard and all the presently paved areas. If the design was poor initially, adding on will not improve it.

Second, if the lot is flat, consider construction such as terracing, or mounding, or building fences, walls, or

Figure 10-10 A corner planting softens the edge of the house. *(Photo from* Landscape Design, A Practical Approach, *2nd ed. by Leroy Hannebaum, 1990. Reprinted by permission of Prentice Hall, Englewood Cliffs, NJ)*

raised beds. The changes in level that construction creates add interest to the yard even before the plants grow and will eliminate the flat, monotonous feeling.

Third, use only a few species of plants, but many of each of those species. For an average front yard, two to three species of shrubs, one species of shade tree, one or two species of small trees, and one groundcover should be sufficient. Using individual plants of dozens of species creates an unharmonious and unattractive landscape.

Fourth, use groundcovers and mulches. Often the finishing touch can be put onto a landscape by underplanting trees and shrubs with groundcover or mulching with bark or stone (see Chapter 11 for mulch suggestions).

Fifth, avoid most ornaments. Birdbaths, gazing balls, wrought-iron chairs, and the like attract attention to themselves, deemphasizing the house and the remainder of the landscape. Ornaments should be restricted to the private living area or an enclosed courtyard and incorporated within decorative plantings.

Finally, do not be reluctant to use appealing design ideas from other people's landscapes or from books. Many do-it-yourself landscaping books give sample designs that, with slight altering, could be adapted to any house. The references at the end of the chapter include several idea books that are helpful in this regard.

Private Living Area Design

The private living area of your landscape should be just that: PRIVATE. Screening neighbors from both view and hearing is very important if the living area is to be used to full potential. Screening from wind and shading from sun become likewise important to overall comfort and will affect the usefulness of the area. A well-designed private living area can be pleasant for a Saturday morning breakfast, an afternoon nap in the shade, a dinner buffet at 6, and a party until 2. The elements of the private living area vary enormously with the family and climate, but almost everyone wants a deck or patio, screening for privacy, and shade.

Patios and Decks

A patio or deck is an outside room with a floor, walls, and a ceiling. The floor is the paving or decking. The walls can be fences, plants, or walls around it, and the ceiling can be the sky, a canopy of tree branches, a constructed roof, or an umbrella.

Most builders locate the patio area behind the house and adjacent to the living and dining areas. Here it is conveniently accessible from the house, an important consideration. However, other factors such as weather and privacy should also be considered when deciding on the location of this important area.

The orientation of the patio on the north, south, east, or west side of the house affects its livability and determines the amount of construction and shading needed. A patio situated to the east of the house is best in most cases, because it is warmed early in the morning by the sun but shaded in the afternoon by the house. A west patio, by comparison, will remain cool through the morning and require shading in the afternoon for comfort.

South-facing patios receive direct sun all day. In warm climates shade will be needed almost all day. However, in cold climates the continual sunniness warms the patio and extends its usefulness into early spring and late fall. Conversely, a northern patio is shaded through most of the year by the house and remains relatively cool. This orientation could be ideal in the Southwest.

The size of a patio can greatly determine its usefulness. Homes are usually built with 12 × 12-foot (4 × 4-meter) patios, barely adequate for a table and two chairs and room to walk around them. Preferable minimum size is 15 × 15 feet (5 × 5 meters). The extra 3 feet (1 meter) per side will accommodate a table and four chairs or a picnic table easily, with room left for a lounge chair and/or two regular chairs. Although 15 × 15 feet (5 × 5 meters) is suggested as a minimum, building an even larger patio is desirable. A spacious patio enhanced by incorporated planting beds (Figure 10-11) will be well used and will become a focal point for gatherings of family and friends.

While a main patio should be large and spacious, auxiliary patios or decks off a bedroom or bath need not be as large. Often they are designed for viewing from inside the house with only occasional usage. A 6 × 10-foot (2 × 3-meter) area will be adequate to provide room for a small table and two chairs.

Visual and Environmental Screening

Most homes located in an urban or suburban area need screening, not only for privacy but also to keep animals and children in or out and sometimes for wind protec-

Figure 10-11 A raised brick planter such as this one can delineate a patio.

tion. The traditional fence along the property line does make maximum use of the available land but can create a boxed in feeling if not lavishly planted. In addition, local zoning laws may prohibit fences higher than 5 or 6 feet (1.5–1.8 meters) along the property line; if neighboring properties are higher, this may not be sufficient to ensure privacy. Alternatives to property line fencing include berms, raised beds, interior fences, and visual screening by plants.

Berms are mounds or ridges of soil 1 to several feet (0.3–1 meter) high (Figure 10-12). Although some properties have natural berms, usually berms are created by grading. They are useful because they add interest to the landscape with elevation changes and, when planted with shrubs and trees or topped by a fence, can stay within zoning restrictions and still provide privacy.

Raised planters (Figure 10-11) of wood, stone, brick, or other materials holding tall shrubs and groundcover can also screen a patio. Almost any plant that will grow in the ground will grow in a raised planter. Moreover, a raised planter is ideal for growing plants with special soil requirements because the bed can be filled with the soil mixture of the owner's choice. In areas with alkaline soil, for example, a raised planter

Figure 10-12 A raised berm used to screen. (*Photo by Andrew Schweitzer*)

Figure 10-13 An interior fence used to delineate and screen a service area.

can be filled with an acidic mixture of soil and peat moss for growing azaleas and rhododendrons.

Interior fences are constructed within rather than along the boundary of the property. Where existing property line fences are too short, an interior fence can create a small private area and eliminate the necessity of replacing existing fencing. Interior fencing is also useful for screening the service/utility area from the view of the private living area, and it provides an attractive background for plants and garden ornaments (Figure 10-13).

Plants alone can be used to provide screening, but the effect is not immediate as it is with a fence. For the homeowner with patience, plants are among the best visual screening materials.

Plants are also better than most fences for controlling wind. If a fence is used for wind control, the wind frequently goes over the top and comes immediately back down to the ground within 2 to 3 yards (2–3 meters) of the fence. Plants, on the other hand, break up wind into small eddying currents of less velocity.

When planning for wind control, the seasonality of the wind should be considered; in other words, from which direction does it come during each season? Most people would not want to restrict a cool summer breeze in late afternoon, for example, but would like to decrease or eliminate a cold northern wind in fall with the hope of extending the usable season of the patio by a few weeks.

Many species of plants can be used for screening. Pines provide year-round evergreen screening; they can grow up to 100 feet (30 meters) high. A dense hedge of barberry plants will keep animals and trespassers out because of the sharp spines. When choosing plants for screening, ultimate height, denseness, growth rate, and whether the plants are evergreen or deciduous should be considered.

For noise problems, plants are not the best choice. The main factors that influence noise level are distance from the source and the mass (weight) and height of the material used as a noise barrier. In view of these facts, significant noise control will really only be achieved by erecting a wall of brick, stone, or concrete of sufficient height so that the source of the noise is 3 to 6 feet (1–2 meters) below the top. For noise from traffic, this means a wall of 6 feet (2 meters) if the source is primarily automobiles, or even higher if trucks are the problem.

Shade

Afternoon shade is essential to outdoor comfort in most climates. Trees are one way to provide shade, and their location should be chosen carefully to ensure they will cast their shade at the right time of day and over the desired area. The disadvantage of using trees for shade is that the effect is not immediate. The homeowner must wait several years until his or her trees reach an appreciable size.

Roofs offer immediate shade, although their cost is relatively high. They can create a rain-proof area for protecting patio furniture that trees cannot. Aluminum and fiberglass are two popular roofing materials. Wood roofs made of 2 × 6's or lath spaced in varying patterns give a feeling of protection without complete shade, and they are architecturally pleasing in both public and private areas.

An alternative product for use in shading a patio is woven or knitted sunscreen fabric (Figure 10-14). Such fabrics were originally used in nurseries for lowering the light intensity for shade-loving plants. However, now they are available in a variety of colors (formerly they

Figure 10-14 Knitted shade fabric allows light through but filters the sun and makes it less intense. *(Photo courtesy of Weathashade, a Division of the Gale Group, Apopka, FL)*

were always black), which makes them much more suitable in a landscape. A sunscreen fabric has the advantage of cooling a patio while still letting air and a significant amount of light through from above.

Optional Elements of the Private Living Area

For a family with youngsters, a play area is an essential feature of a landscape plan and can be included as part of the private living or utility area. The size of the area depends on the number of children who will use it and the total size of the property. For example, 12 feet × 12 feet (4 × 4 meters) is adequate for one or two children, although more room could be included if the yard was large. Surfacing can be grass, although daily use makes it difficult to maintain in good condition. Useful and safe grass substitutes include smooth pea gravel, bark chips, and sand, all applied thickly to cushion against falls. A paved ring for tricycle riding might also be useful. When locating the play area in the landscape, consider it as a temporary feature used only for a few years when the children are small. Accordingly, it should be placed so it can later be included into the landscape gracefully.

Game areas such as for croquet and badminton require large grassy areas, 24 × 54 feet (7.3 × 16.5 meters) for the latter and 30 × 60 feet (9.1 × 18.3 meters) for the former. This size area may be difficult to include

in a small lot, however. A tetherball pole requires only a 20-foot (6-meter)-diameter circle and could be included in most landscapes. Paving material such as concrete or gravel is recommended for tetherball circle areas.

Emphasis of a view may be part of the landscape plan. With mountain or lake properties, plantings can be situated to frame the view (Figure 10-15), with taller plants to the sides and short ones in front.

The Service/Utility Area

A utility area is the part of a landscape that provides room for necessities such as clotheslines, garbage cans, and pet runs. The design inside the area and the space allocated for each element should be functional, and the area should have convenient access. The narrow strips along the sides of most houses are convenient service areas, because they tend to be difficult to landscape. If a vegetable garden is included, care should be taken that the site chosen will provide maximum sunshine.

A main consideration for the utility area is that it be screened from the private living area. Fences, beds of dense shrubs, and wire supports with climbing vines all perform this function. A baffle effect with overlapping sections of fence can substitute for a gate and give easy access while providing complete visual screening (Figure 10-16).

Figure 10-15 Landscaping to frame a view. Note how attention is focused on the area between the two plantings and how the shapes of the trees mirror the form of the view.

Figure 10-16 Sections of fence used as baffles to hide a service area.

Choosing Plant Species for the Landscape

Selecting plants is the final step in designing a landscape. Establishing the basic design, with patios, walkways, and the like drawn in the plan, is preliminary; the plants to be included at this stage should be represented only by circles or similar symbols and labeled "tree", "shrub", "groundcover", or whatever. The actual choice of plant species is made later to avoid the confusion of coordinating plant species and design elements at the same time. Regardless of whether the plant is a tree, shrub, or vine, certain general guidelines should be taken into account before specifying plants in the design.

General Guidelines

Adaptability to Climate. Winter hardiness is the most limiting climatic factor and the one over which there is least control. Choosing a plant that is not completely cold-hardy condemns the homeowner to extensive pruning of winter-killed wood every spring and may result in the complete death of the plant during a particularly cold year.

Rainfall. Natural rainfall in the eastern and southern parts of the country provides sufficient moisture for most plants. Irrigation is often necessary in the West, where water use for landscaping is fast becoming a luxury. Using drought-resistant plants to eliminate the majority of landscape water use works with, rather than against, the natural climate.

Smog Tolerance. Although smog will rarely kill a plant, some species, especially pines, can be severely damaged by it. Most often leaf margins or areas between the veins brown, leaving the plants burnt-looking and unattractive. Because there is no protection against smog damage, choosing smog-tolerant plants is recommended where there is significant air pollution.

Soil Conditions. pH, the acidity or alkalinity of the soil, can be a serious restriction to plant growth. Plants native to acid soils (azalea and rhododendron, for example) planted in alkaline soils will turn yellow from inability to absorb nutrients at that pH range (see Chapter 13).

These plants should be avoided in favor of more adaptable species or placed in raised planters in which the soil can be modified to the correct pH.

Sunlight. The relative amount of sun and shade the property receives should be considered when selecting plants. For example, the yard of a house built in a wooded area will be largely shady, so shade-tolerant plants should be selected.

Resistance to Pests and Diseases. Some species have a natural susceptibility to insect or disease problems, whereas others remain unaffected. To a large extent, disease and insect susceptibility is determined by the area in which the plant is being grown; species grown out of their adapted habitat are frequently prone to problems. A local nurseryman can offer advice regarding the adaptability of specific ornamentals to your area.

Maintenance. Not all landscape plants require the same amount of maintenance. Selecting the correct species can minimize the amount of time spent on maintenance chores like pruning and raking. Intensive-maintenance plants are those that drop large quantities of dead flowers and fruits or require frequent pruning of dead wood and unruly growth.

Mature Height and Spread of Plants. The consideration of mature height and width is often overlooked when homeowners choose plants. Only when the mature sizes of plants are known can they be intelligently placed in the design to blend without overgrowing their allotted spaces. If plants are placed too close together, frequent pruning will be necessary to keep them within bounds; if they are placed too far apart, they will never grow large enough to create the intended effect.

Growth Rate. Growth rate is a function of both the species and the amount of water and fertilizer received. If the plant has a specific function to fulfill, such as screening, fast-growing species will be desirable. Similarly, fast-maturing trees are frequently desirable because they provide shade quickly. However, ordinarily, moderate- or slow-growing plants are preferred over fast-growing ones because the latter tend to be structurally weaker plants with short life spans.

Personal Preference. Preference for the plant is, of course, important when selecting plant species to be included in the landscape. The attractiveness of the foliage, size of individual leaves (called *fine* or *coarse* texture), and color should be noted, along with outstanding flowers, fruits, or fall color. An assortment of plants with varying foliage textures and colors is preferred to give interest to the landscape throughout the year.

Retail nursery salespeople assist customers in selecting the right plants to suit the growing conditions around their homes. They must be knowledgeable about the plants grown in that area and like to work with people.

Choosing Tree Species

Trees can be divided into three sizes: large (45+ feet; 14+ meters), medium (30–45 feet; 9–14 meters), and small (less than 30 feet; 9 meters). Large and moderate-sized trees are used primarily for shade, although they can be highly ornamental and are then called *specimens*. Trees in the small category can be up to 30 feet (9 meters) tall but are frequently in the 10- to 15-foot (3- to 4.5-meter) range. Small trees yield some shade when mature and are often used by patios and decks. *Patio trees* frequently have ornamental flowers or fruits and are pruned when young to be multitrunked (Figure 10-17).

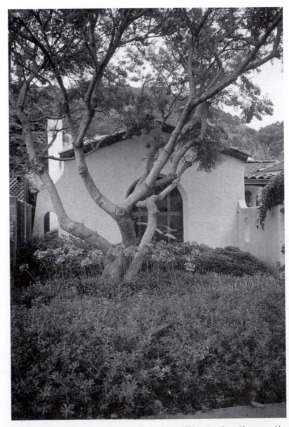

Figure 10-17 A multitrunk tree. *(Photo by the author)*

Shade trees should be selected so that they will complement the scale of the house when fully grown. One-story houses are complemented by trees in the small and medium categories. Owners of two-story, split-level, and larger houses should choose shade trees of moderate or larger size.

Trees vary in the density of the shade they cast. Those with round heads of large leaves cast deep shade, whereas trees with an open branch structure and small leaves cast *filtered shade*. Trees that will provide the proper amount of shade for the climate should be selected; the denser the shade, the more difficult it will be to grow a lawn or landscape plants underneath.

Root systems are another consideration in tree selection. Some trees have *invasive roots* that grow vigorously and can clog a septic tank or sewer pipe. Planting such trees should be avoided within 30 feet (9 meters) of wells or sewer systems.

Other trees produce surface roots radiating from the trunk. The roots can crack driveways or walkways (Figure 10-18). One solution is not to plant such trees next to a paved area; however, this restriction may mean that parking areas and patios are deprived of shade. A better solution is to install a thin sheet of metal or herbicide-treated plastic between the tree and the paved area at planting (Figure 10-19). This will deflect the young growing roots downward and prevent the problem.

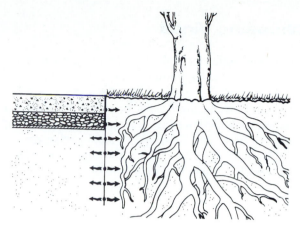

Figure 10-19 Tree roots will not disturb paving if the tree is initially planted with a root-deflecting barrier of a thin, impervious material such as metal or fiberglass. *(Courtesy of Biobarrier®)*

can, however, be pruned to remove their bottom branches and used as small trees.

Along with small trees, shrubs are the main flowering plants in landscapes, and a design should include at least a few flowering species. Evergreen species are essential as well. Their foliage persists throughout the winter, making them a part of the backbone of the landscape.

Choosing Shrub Species

Whereas trees provide height and shade, shrubs provide mass and bulk. They are divided into three sizes: small (less than 3 feet; 1 meter), medium (3 to 6 feet; 1–2 meters), and large (6 to 12 feet; 2–3.6 meters). Small and medium-sized shrubs are most frequently used in home landscapes. Those in the large category can be up to twice as wide as they are tall and are suited to commercial buildings and large properties. They

Selecting Groundcover Species

Groundcover plants are low-growing species that add variation in texture and unify groupings of larger plants. They give a finished appearance to a landscape, prevent weed establishment, and act as a living mulch. Although most groundcovers are less than 6 inches (15 centimeters) tall, they can be considerably larger: up to about 2 feet (0.6 meter). Shorter groundcovers usually are well suited to use on small properties; taller species are better suited to large properties because they are more in scale.

A groundcover should be selected for disease and insect resistance and suitability to climate; it should also blend well visually and in terms of its growing requirements with the other landscape plants. A groundcover should be unobtrusive—visually harmonizing rather than competing with other plants in the landscape. The choice of an evergreen versus a deciduous groundcover is a matter of personal preference. In an area with persistent snow cover, the groundcover will be hidden throughout winter anyway, but in other areas the evergreen foliage will be pleasing in the winter.

Figure 10-18 Tree roots that have cracked a paved area. *(Photo by the authors)*

Choosing Vines

Vines are not often included in landscapes, although they can be very attractive features trailing along the ground, climbing up a wall, or supported by a trellis or arbor. Vines can form the floor, walls, or ceiling of the garden. They can provide screening and shade, supply colorful flowers or fruits, and soften a wall, among other uses.

The methods with which a vine climbs will determine whether it will require a supporting framework. The stems of some vines grow small *holdfasts* that look like small suction cups and anchor them to any flat surface. Others have adventitious roots along the stems. These vines adhere to a wall or climb a fence without additional support.

Other vines climb by twining. The entire vine may grow spirally or may send out tendrils that twine to support the rest of the vine. Vines in this category need a trellis or other means of support.

OBTAINING LANDSCAPE PLANTS

Most people purchase the plants they need for their landscape, but plants can also be propagated at home or transplanted from the wild or from friends' yards.

Home Propagation

Propagating woody plants by cuttings can be difficult for a home gardener. However, cuttings of some shrubs and most herbaceous groundcovers are relatively easy to root. If a source of cuttings can be found, it is practical to propagate the groundcover needed for the landscape at home.

Transplanting from the Wild

Transplanting established trees and shrubs can be successful if the plants are small. Trees should be less than 10 feet (3 meters) tall and preferably growing in full sun, because a move from a shady woodland to a sunny yard can cause serious injury. Cool, rainy weather is best for transplanting (early spring in most cases, winter in warm climates where the ground does not freeze). If the tree is deciduous, it should be moved while leafless. Evergreens should be moved before starting active growth.

The first step is to cut through the roots with a spade in a circle about 15 inches (40 centimeters) from the trunk. The tree can then be uprooted, keeping as much

soil around the roots as possible. The soil ball should then be wrapped in plastic or wet burlap for transport to the new location. Replanting should be done immediately, following the suggestions given for balled and burlapped plants in this chapter. Spraying with an anti-desiccant (available at nurseries) will slow transpiration in evergreens and improve chances of survival.

Purchasing Plants from Local and Mail-Order Nurseries

Both local and mail-order nurseries are good sources of plants, but each has advantages and disadvantages. Local nurseries may not have the selection of mail-order ones, but the buyer can see the plant and be assured of its quality. Often plants from a local nursery have been grown in containers and can be transplanted with very little chance of loss. Mail-order nurseries dig the plants from the field and ship them with the roots wrapped in damp wood shavings. The plants may be subjected to extremes of heat or cold during shipping and are less likely to survive transplanting.

 Owning and operating a small retail nursery is one way a person with horticultural training and experience can have his or her own business. Responsibilities involve customer service, bookkeeping, and plant maintenance. The nursery may propagate and grow all or some of its own plants. A nursery may specialize in just houseplants, just landscape plants, or a combination of both.

 Owners of mail-order nurseries frequently propagate and grow their own hard-to-find varieties or species and market them through catalogs or on the Internet. Mail-order nurseries that specialize in rare or specialty plants can operate with low overhead costs because they can be located in remote areas.

As a compromise, many gardeners patronize local nurseries for more common plants but use mail-order nurseries for hard-to-find species. However, you should not be reluctant to ask for a species or cultivar that a local nursery does not normally carry. The salesperson will usually be willing to order it, particularly if you are willing to buy a sizable number (10 or more) of the species.

The nursery section of a discount store is another source of plants, and bargains are available there with careful shopping. These stores order rather than grow their own plants. You should ask the manager when the nursery shipment is due and inspect the plants a day or two after they arrive. Plants may receive poor care in discount stores, and what were originally healthy plants may deteriorate to poor quality within a couple of weeks after arrival.

NURSERY PRODUCTION*

There are two types of nursery production: *field* and *container*. *Field stock* is either direct-seeded in the ground or transplanted to the field as seedlings, then "lifted" (dug out) without soil as *bare-root stock* for use as "nursery liners" (young ornamental or fruit tree plants to be grown further to saleable size). Field stock is also grown to become "balled and burlapped" (B&B) trees.

Container stock is propagated from seed, rooted cuttings, or field-grown seedlings. Fifty years ago, most ornamental plants were grown in the field, then dug up for sale. Today, 80 percent of ornamental plants are grown in containers. The switch occurred for several reasons:

- Container-grown trees have a greater chance for survival after transplanting than do trees produced in the field.
- Container production does not require good soil.
- Container production takes less acreage.
- Containerized stock lets the grower extend the planting season, because the stock can be moved into plastic greenhouses.
- Land that is unsuited for field production can be used for container production.
- Planting time is not dependent on the weather, because the plants can be moved to protective structures such as plastic greenhouses to provide winter protection (see Chapter 20).
- Some operations (like root pruning, digging, and burlapping) can be eliminated by containerized production.
- Transportation costs are lower because of light-weight media.

The disadvantages are:

- Small containers need frequent watering.
- Nutrients are rapidly depleted from the containers.
- Plants require winter protection if the area is cold because the roots will not survive above ground.
- Plants easily become root-bound.
- Trees can be knocked over by wind.
- Containers are costly.
- Labor costs to pot up plants are high.
- Roots are stressed because of temperature extremes.

Choosing the Site for a Nursery

There are five major considerations in determining where to establish a new nursery: (1) soil, (2) climate, (3) water availability, (4) market, and (5) labor supply.

Soil

Soil productivity is not important when growing only containerized products, but relatively level land with good drainage is still necessary.

Climate

A concern of the beginner is learning the length of time required in that particular climate to produce saleable crops, and how to schedule so that the proper number of each species is available for the first year of sale and each year thereafter.

Water Availability

All nurseries rely on irrigation, and so an adequate, economical source of water is essential.

Market

When deciding whether to start a nursery business, a grower will need to conduct a market analysis to determine what opportunities exist to sell plant materials in the local area. Most new firms begin with only a few acres (hectares) of production and initially market primarily within a 50-mile (80.5 km) radius, unless growing for mail order or on contract. Part of this market analysis includes finding out what crops other nurserymen grow successfully in the region, what crops are most in demand, and what the competition is like.

*Excerpted with permission and revised from Appropriate Technology Transfer for Rural Areas (ATTRA) Fayetteville, AR. Website: http://www.attra.org/attra-pub/nursery.html#genprod

Marketing really begins with a decision on what to produce and at what volume. The primary considerations are as follows:

- Deciding who the customers are
- Determining what species and size plants these customers want
- Keeping up with trends in buyers' preferences
- Knowing what combination of plants maximizes profits.

Ornamentals fall into general categories such as shade trees, conifers, perennials, vines, shrubs, bulbs, and annuals. Although most nurseries grow a range of plants, there appears to be a trend toward specialization. For example, growing only native groundcovers, or only daylilies, are viable niche markets. The production of specialty crops (for example, hardy bamboo, disease-free apple stock, native plants, and the like) and specialization in plants in short supply (for example, native plants, uncommon plants, very large trees) are niche markets that even small growers can serve. A mail-order website can be one way to capitalize on a niche market.

Labor Supply

Nurseries of medium and large size normally use a supply of minimum-wage labor. There must be a pool of minimum-wage workers available in the area to run the nursery. The local employment development department will be able to assist with this information.

Nursery Operations

Irrigation

The two most widely used irrigation systems are *overhead sprinkler* and *drip* (or *trickle*).

Overhead irrigation is designed to cover a large area, and these systems are the least expensive to install. However, this method produces uneven water distribution, which can slow plant growth. Additionally, watering that wets the foliage encourages disease. Third, there is the issue of runoff. A container nursery using overhead irrigation can use from 15,000 to 40,000 gallons of water per acre (60,000 – 160,000 liters per hectare) per day in warm weather.

On the other hand, large containers watered with a drip or trickle system use 60 percent – 70 percent less water than when irrigated by an overhead system. Drip irrigation systems cost more to install than overhead systems, but have superior application uniformity and efficiency. These systems are also less affected by wind (which can blow the sprinkler irrigation water away

from the plants) and crop canopies (which can block the water from reaching the roots). They also have less runoff. Another advantage is that workers can continue working while the plants are being irrigated.

The biggest disadvantage to trickle irrigation, besides the initial cost, is keeping the pipes and emitters clean. Solids or particles in the water can block them. A filtration system is effective, but even so the emitters (see Chapter 7) must be checked frequently.

A third, less-used type of irrigation system is *subirrigation* using capillary sandbeds. In this system, water rises into containerized plants through capillary action. Usually, the sandbed is covered with at least 1 inch (3 centimeters) of fine sand and slopes very slightly from one end to the other. Water is released at the high end and slowly percolates to the low end. These systems cost the most to install, but they have no runoff or leaching.

Sandbeds are normally constructed using wood sidewalks, a plastic bed liner, sand, a small tank, a drainpipe and a float valve. They do not require the use of any electrical parts, and they provide a uniform and consistent supply of water without forming a saturated water table at the base of the soil column in the container. In short, use of sandbeds provides efficient and uniform crop growth with less water, less fertilizer and less pesticide. It also requires less labor, as sprinkler heads, timers, pumps, valves, and water-treatment systems need not be monitored.

The biggest disadvantage of sandbeds is that weeds and the roots of containerized plants grow into them. There is a product designed to alleviate this problem; it is a mat called an Agroliner™ that has been treated with Spin Out™, a product that prevents root growth in general. The mat is placed over the sand and under the containers.

Fertilization

There are four basic ways to fertilize containerized plants: (1) incorporate, (2) topdress, (3) liquid feed, and (4) foliar feed. In a nursery container, fertilizer incorporation in the mix combined with liquid feeding should provide sufficient nutrition.

Adequate levels of nutrients in the container medium must be maintained for optimum growth and maximum profit from woody ornamentals. The levels of soluble nutrients in containers can be significantly reduced after three or four irrigations because of limited container volume and frequent application of water. To overcome this problem, two fertilizing systems are used: slow release and liquid.

Slow-release and controlled-release synthetic fertilizers, such as Nitroform™ and Osmocote™, are common in container production systems. For best results, they should be incorporated into the growing media before planting rather than topdressed. Slow-release fertilizers are often used in combination with liquid fertilization. (See Chapter 20 for discussion of greenhouse production.)

Disease and Insect Control

Integrated pest management (IPM) is a sustainable approach to managing pests by combining biological, cultural, physical, and chemical tools in a way that minimizes economic, health, and environmental risks. An IPM program involves using resistant cultivars, building up populations of beneficial organisms, monitoring numbers of pests and developing treatment thresholds, and using spot treatments of pesticides that are least harmful to beneficial organisms and the environment. It is important to identify pests early, so that appropriate measures can be taken quickly. (See Chapter 13 for a more in-depth discussion of IPM.)

Weed Control

Weed control efforts should focus on two areas: in the pot and under the pot.

When deciding how to treat a weed problem, the grower should ask himself or herself several questions: Are the weeds mostly annuals or perennials? Are they easy to pull? Does the problem exist only in a small group of plants, or is it widespread? Are the weeds monocots (grasses) or dicots (broadleaf)? The answer to this question will determine whether herbicide can be used and which one. What time of year is most effective for controlling these weeds?

Sanitation is the least costly and most effective method of controlling weeds. To prevent weed seeds from blowing into pots, attention to maintaining a vegetation-free zone both on and surrounding the production bed is critical. To keep weeds from growing under the pot, more and more growers place the containers on geotextile weed barriers (often called *fabric weed barrier* or *landscape cloth* (Figure 10-23)).

Modern landscape cloths can last for 10 to 12 years in full sun, so they are durable and long-lasting. They do an excellent job of controlling weeds, yet they are permeable to water from irrigation and rainfall, so that drainage is not a problem. Although the initial cost is high, the expenses can be pro-rated as an annual weed control investment.

Hand weeding is costly, but it may be appropriate in a small nursery setting. Weeds must be removed when they are still small, because large quantities of growing media are lost when big weeds are pulled out of containers.

Herbicides, on the other hand, are widely used in container nursery production. Even though weed-free media is used to establish nursery plants, wind, birds, and surface irrigation water are sources of weed seeds that get deposited onto the pot surface. Broadleaf and grassy weeds thrive in a container nursery because the growing conditions in a media-rich pot are absolutely perfect. Thus, pre- and post-emergent herbicides are commonly used in commercial nursery production to control them.

In 1991, Monrovia Nursery compared hand weeding to spraying herbicides and found that a combination of the two was the least costly method[*]. When it used no preemergent herbicides, it took workers 10 hours of hand weeding per acre (5 hours per hectare), performed 10 times a year. By using a preemergent herbicide once in the spring and once in the fall, the workers were able to perform hand weeding only seven times a year for one hour per acre.

Fabric weed-barrier disks can be used as an alternative to herbicides to control weeds in containers. The fabric disks are pre-slit and fit on top of the pot around the stem. They prevent weeds from growing in the containers by excluding sunlight and inhibiting weed germination. The fabric disks are permeable to air and water but prevent germination of troublesome container nursery weeds like oxalis. An added benefit of the discs is that they reduce evaporation.

Tex-R Geodiscs® are fabric disks that have been treated with Spin Out™ and work the same as the fabric disks mentioned above. They prevent weed growth by excluding light and pruning the roots of weed seeds that are blown onto the fabric. They provide effective weed control for up to three years and can be moved from pot to pot.

PLANTING THE LANDSCAPE

Landscapes are planted in spring in most parts of North America. At this time the plants will be actively growing and will transplant with minimal shock. The ground will be thawed (in the North), and the temperature will be relatively cool, thereby slowing transpiration and preventing wilting. Fall is the second-best time for transplanting, again because of the cool temperatures. This prime planting period extends through winter in

---

[*]Suttle, Walter. 1998. "Weeding out costly controls." *American Nurseryman*. October 15. p. 24–29.

any climate where the ground does not freeze. Summer is the least preferable time for planting a landscape, because sun and hot temperatures make the problems of wilting more acute. However, if container-grown plants are used exclusively, the roots will be practically undisturbed and will grow even in summer.

Transplanting Trees and Shrubs

Shrubs and trees are sold *bare root, balled and burlapped,* or *container grown.* The selling condition determines the price, the chances of surviving transplanting, and the seasons for transplanting.

 A landscape installer plants the landscape but must have knowledge of construction of walkways, walls, fences, and patios because these are all included in a landscape plan. Although some landscape installers have minimal experience and training, a supervising crew chief is generally trained to junior college level in horticulture.

Bare Root

Only deciduous plants are sold bare root, and they are always dug and sold while dormant. Most bare-root plants are sold only in spring; however, occasionally they are available in fall or throughout winter in areas where the ground does not freeze. Bare root is an inexpensive way to buy plants, because they are grown in fields with minimal maintenance and simply dug up when the selling season arrives. Most mail-order nurseries utilize this selling method because it minimizes shipping weight.

 A nursery owner who produces bare-root plants usually is the owner of a fruit tree or landscape tree nursery for deciduous species. Plants are propagated from seed and may also be grafted. The plants are then field grown for several years to saleable size, then dug with equipment and stored in a cooler until sale. Sale can be through mail order, or to garden centers or retail nurseries.

Because bare-root plants are dug from the ground, many roots are lost, so that a bare-root plant is more likely to die during transplanting than a container-grown one. It must be given particularly good care to assure its survival and establishment in the landscape. When se-

short stakes anchor roots

sections of hose-covered wire prevent bark injury

basin for water

topsoil

Figure 10-20 A bare-root tree after transplanting.

lecting a bare-root plant from a nursery, you should pick the sturdiest plant with the largest root system. The roots should be covered with or bagged in wood shavings and be damp to the touch. If they have dried, it is likely that the plant may be dead. You should also scratch a tiny area of bark with your fingernail. If the plant is alive, a green layer will be visible beneath the surface.

A bare-root plant should be planted as soon as possible. If planting cannot take place for several days, the plant should be left in a shady area with the roots covered with moist soil. This *heeling in* prevents drying and the death of the roots.

At planting time the roots should be placed in a bucket of water and the tree taken to the planting site. A hole should be dug large enough to accommodate the roots easily without crowding (Figure 10-20). The topsoil should be piled on a piece of canvas or plastic next to the hole and not mixed with the subsoil. Organic amendment can be mixed with the soil to make a mixture to refill the hole. However, research has shown that in many cases this does not improve growth, and therefore it is probably not necessary unless the soil is of very poor quality.

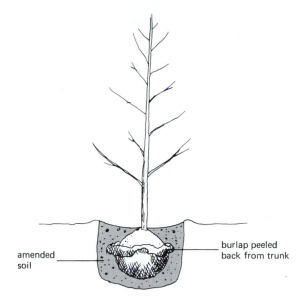

Figure 10-21 A balled and burlapped tree after transplanting.

amended soil

burlap peeled back from trunk

Next, part of the mixture should be shoveled back into the hole so that when the plant is placed in the hole, it will be sitting at or slightly above its original growing level.

Root pruning should take place next, with any badly broken roots clipped off with hand pruners. The remaining roots should be spread outward in the hole and covered with soil to ground level. A small ridge of soil should be mounded around the planting hole as a basin for water. Finally, the plant should be watered in well by filling the basin three or four times.

Balled and Burlapped Plants

Balled and burlapped (Figure 10-21) refers to trees and shrubs grown in a field but dug to keep a ball of soil around the roots. Both evergreens and deciduous plants are balled and burlapped; because the process involves considerable labor, plants sold this way are expensive. Because the digging disturbs only part of the roots, balled and burlapped plants usually transplant successfully provided the ball remains moist and proper planting procedures are used.

Hole and soil preparation is basically the same for balled and burlapped plants as for bare-root plants, with the stipulation that the hole should be twice as wide as the ball and one to one and one-half times as deep. Balled and burlapped plants are heavy, and care must be taken when transplanting to prevent the soil ball from breaking. Once the plant is positioned in the hole, the strings can be cut and removed and the burlap peeled back from the trunk. If plastic burlap has been used, it should be removed com-

pletely. Even ordinary burlap may have been treated to prevent rotting, and it is best that it be removed.

 A nursery man or woman who produces evergreen shrubs or trees will normally field grow them to small, medium, or specimen (large) saleable size and then dig them up by machine before wrapping the root ball or boxing them. The nursery man will then sell to a nursery or garden center. In some cases, a landscape designer may order specimen-size plants of unusual species directly from the grower.

Container-Grown Plants

Many plants are grown and sold in containers. Because the root system develops within the container, the plant is only slightly disturbed at transplanting and can be moved whenever the soil is workable.

As with balled and burlapped plants, a planting hole at least twice as big as the container should be dug. Removal of the root ball from the container without breakage is the only difficult part of transplanting. With a plastic container, the plant can be turned upside down and will slide out easily with little shaking.

Boxed Trees

Occasionally, you will see large trees for sale with their roots in boxes. These plants may have been container grown or field grown. With the latter, the boxing is in lieu of balling and burlapping.

Planting hole preparation should be the same as for balled and burlapped plants. Removal of the box must be done carefully to avoid breaking the soil ball. First, the tree should be turned on its side and the bottom of the box removed. Using a board as an inclined plane, the box should then be slid onto the planting hole. Once in place, the sides of the box can be removed and the hole filled with soil.

Pruning Newly Transplanted Trees and Shrubs

Pruning newly transplanted trees and shrubs encourages them to develop proper form (follow the instructions for pruning in Chapter 11). In the case of bare-root plants, it reduces the amount of top growth to a level that the sparse root system can support. Many times, bare-root plants are sold prepruned and need no further pruning until the following season. However, if not, about one-third of the top should be removed at transplanting by removing the weaker branches and clipping the remaining branches back to one-half to two-thirds their original length. Branches low on the trunk should *not* be removed, as they strengthen it.

Staking and Wrapping Trees

Staking may be needed either to anchor the root system of the tree until it becomes established or to support the trunk in an upright position, usually for a maximum of one year. Unless it can be determined that staking is fulfilling one of these functions, it is not necessary and will cause weakening of the trunk.

For root anchorage of bare-root trees, two short stakes (1–1½ feet; 30–45 centimeters) should be inserted on opposite sides of the trunk (Figure 10-20). The tree should then be secured to each with strips of rubber, cloth, or wire covered with a section of garden hose. Uncovered wire, rope, or twine is not recommended, because friction will cause it to rub off the tender bark.

Trees that require trunk support should be staked as low as will allow them to stand upright under calm conditions. Again, stakes should be positioned on opposite sides of the tree and the trunk tied to the stakes at a single point near the top.

Wrapping (Figure 10-22) involves twining tree-wrapping paper, vinyl, or burlap secured with twine around a tree trunk from the base to the lowest branches. The wrapping protects the trunk from sun damage, a problem on trees grown in rows with their trunks shaded by neighboring trees. Wrapping should be left in place about a year, after which it should be removed.

Transplanting Groundcovers

Groundcovers are herbaceous perennials or woody shrubs. The following recommendations apply to herbaceous groundcovers. Shrub groundcovers should be planted as recommended in the previous section.

Because many herbaceous groundcovers spread by trailing along the surface of the soil, it is important that all soil in a groundcover area be conditioned. The area should be spaded deeply to at least 12 inches (30 centimeters), and amendment should be worked throughout. The groundcover is then transplanted with a trowel throughout the prepared bed and watered thoroughly.

Spacing of herbaceous groundcover plants is determined by economics and how quickly coverage is desired. Spacing plants 2 feet (60 centimeters) apart will cost less than spacing them 8 inches (20 centimeters) apart but will take up to a year longer to cover the area. As a general rule, spacing plants 1 foot (30 centimeters) apart gives coverage at a moderate rate and expense.

When landscaping with groundcover, buying *landscape fabric* to cover the groundcover area is a wise investment (Figure 10-23). It is a water-permeable,

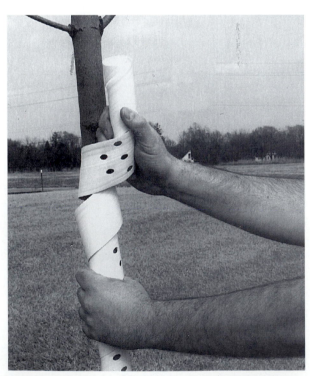

Figure 10-22 Wrapping a tree trunk with vinyl tape. *(Photo courtesy of A.M. Leonard Co., Piqua, OH, (800) 543-8955)*

Figure 10-23 Planting a groundcover through a landscape fabric will make subsequent maintenance much easier. *(Photo courtesy of Jason Eaton, DeWitt Co., Sikeston, MO)*

woven or felted synthetic material that is laid over the area to be planted. Holes are then cut for each ground-cover plant; the plants are inserted through the fabric for planting. The advantages of landscape fabric are permanent, total weed control while still allowing water to penetrate. In this regard, it is different from the black plastic mulch described in Chapter 11.

POSTPLANTING CARE OF THE LANDSCAPE

The most important part of postplanting care of a landscape is regular watering. During the first growing season, plant roots occupy only a small volume of soil and need constant moisture to become established in the new location. Slow soaking with a hose is preferred for shrubs and trees, which should be watered every week to 10 days. Slow watering allows all the moisture to be absorbed into the soil and provides sufficient water to moisten the entire root area.

Groundcover watering is done by hand or sprinkler as frequently as required to keep the soil moist. The generous supply of moisture will enable the groundcover to fill the area within a short period. It is advisable to check the depth of water penetration after watering with a trowel, digging to the depth of the roots and examining the soil moisture. In this way there is no possibility of moistening the surface while the roots remain dry.

Mulching should be a part of postplanting care. It not only keeps in soil moisture but also prevents weed growth, a problem in newly planted groundcover areas. Mulching of landscape plants is discussed in Chapter 11.

SELECTED REFERENCES FOR ADDITIONAL INFORMATION

Botanica Editors. *Botanica's 100 Best Flowering Shrubs for Your Garden.* San Diego, CA: Advantage Publishers Group, 2001.

Chesshire, Charles. *Flower Shrubs: American Horticultural Society.* New York: Dorling Kindersley Publishing, Incorporated, 1999.

Fisher, Kathleen. *Taylor's Guide to Shrubs: How to Select & Grow More Than 500 Ornamental & Useful Shrubs for Privacy, Ground Covers & Foundation Plantings.* New York: Houghton Mifflin Company, 2001.

Hightshoe, Gary L. and Harlen D. Groe. *Native Trees, Shrubs, and Vines for Urban and Rural America.* New York: Wiley, 1987.

Johnsen, Jan and John C. Fech. *All about Trees.* Des Moines, IA: Meredith Books, 2001.

Kellum, Jo and Ortho Books Staff. *Ortho's All about Landscaping Decks, Patios, & Balconies.* Des Moines, IA: Meredith Books, 2001.

Newbury, Tim. *The Ultimate Garden Designer.* New York: Sterling, 1995.

Ortho Books Staff. *Ortho's Guide to Creative Home Landscaping.* Des Moines: Meredith Books, 1991.

Phillips, Roger. *The Random House Book of Shrubs.* New York: Vintage Books, 1989.

Reader's Digest Editors. *A–Z of Evergreen Trees and Shrubs.* New York: Reader's Digest Association, 1998.

Reader's Digest Staff. *Practical Guide to Home Landscaping.* New York: Reader's Digest Association, 1996.

Sinnes, A. *Cort Shade Gardening, Revised.* Des Moines, IA: Meredith Books, 1996.

Smith, Katie Lamar. *Ortho's All about Ground Covers.* Des Moines, IA: Meredith Books, 2001.

Sunset Magazine and Book Editors. *Fences and Gates.* Menlo Park, CA: Sunset-Lane, 1996.

Thomas, R. William. *All about Vines & Climbers.* Des Moines, IA: Meredith Books, 1998.

Zucker, Isabel, and Derek Fell. *Flowering Shrubs.* New York: Barnes and Noble, 1995.

World Wide Web Addresses

http://aggie-horticulture.tamu.edu/extension/xeriscape/xeriscape.html   Welsh, Douglas, William C. Welch, Richard L. Duble. "Landscape Water Conservation... Xeriscape". 2000.

http://www.ag.ohio-state.edu/~ohioline/hyg-fact/1000/1014.html   This is an extension bulletin from Ohio entitled *Preparation and Planting of Landscape Plants.*

ipmwww.ncsu.edu/urban/horticulture/res_landscaping.html   This is an extension bulletin from North Carolina entitled *Residential Landscaping.*

http://www.ext.msstate.edu/pubs/is965.htm   This is an extension publication from Mississippi entitled *Transplanting Trees and Shrubs in the Landscape.*

(Photo courtesy of A. M. Leonard Co., Piqua, Ohio)

CHAPTER 11

LANDSCAPE MAINTENANCE

Regular maintenance by pruning, fertilizing, watering, and weed control is a necessity if home grounds are to remain attractive and plants healthy. But maintenance does not need to be time-consuming. A landscape planned without a lawn and with a goal of infrequent maintenance can be kept in condition with only about two to three hours of labor per month.

 A grounds maintenance person works outdoors. The tasks are varied but include pruning, maintaining any irrigation system, planting seasonal flower displays, cleaning walkways with blowers, spraying for pests and diseases, and fertilizing.

FEATURES OF A LOW-MAINTENANCE LANDSCAPE

Landscapes designed to remain attractive with minimal maintenance include some or all of the following features:

1. Large paved or decked areas that require only occasional sweeping
2. Raised planters that require less bending for maintenance

3. Groundcover areas in lieu of lawns to eliminate mowing
4. Mulches or groundcovers around all trees and shrubs to eliminate hand trimming (which grass requires) and suppress weed growth
5. An automatic irrigation system where natural rainfall is not sufficient
6. A limited number of carefully placed plants rather than large numbers of plants scattered throughout the landscape
7. Plants that do not require frequent pruning and do not drop leaves, fruits, or dead flowers
8. Plants that, when mature, will not require extensive pruning to keep them the proper size
9. Flowering shrubs rather than maintenance-requiring annual and perennial flowers
10. Plants that are naturally disease- and insect-resistant and have been proven to grow well in the locale

TOOLS FOR LANDSCAPE MAINTENANCE

Every gardener needs a selection of tools such as trowels, rakes, and pruners to maintain the landscape. The following sections describe the most frequently used and valuable gardening tools and equipment.

Soil-Working Tools

Trowel. A trowel (Figure 11-1) is used for digging small holes for planting bulbs, setting out transplants, and other similar tasks. When purchasing a trowel, check to make sure the handle attaches firmly to the blade and does not bend.

Round-Point Shovel. A round-point shovel (Figure 11-1) is used for digging larger holes such as for planting trees and shrubs and other general garden digging. The tip is pointed to make it easy to push into the soil.

Hoe. A hoe (Figure 11-1) is used for shallow cultivation of a previously prepared area and will penetrate the soil up to 4 inches (10 centimeters) deep. It is frequently used in vegetable gardening to prepare rows for seeding. Hoes are also used for weeding. For this purpose, they should be used with a scraping motion that severs the weeds at the soil line. If used in a cultivating manner, additional weed seeds will be exposed, germinate, and intensify the weed problem.

Spading Fork. A spading fork (Figure 11-1) is used to move piles of leaves or refuse, turn compost, and break up large clods in the soil. Although a shovel can be used for these purposes, a spading fork is more efficient and will move a greater volume of leaves or compost with each bite. A spading fork is distinguished from a pitchfork in that it has stronger, flat tines.

Watering Equipment

Hose. Many grades of garden hoses are available. Inexpensive plastic ones are stiff and develop permanent kinks that restrict water flow. Reinforced plastic hoses remain pliable, do not kink, and are lightweight. Rubber hoses are the most durable, but they are heavy and expensive.

Pistol Nozzle. A pistol nozzle (Figure 11-2) provides on-off control and can be adjusted to give from a strong forceful stream to a mist. Unfortunately, when the stream is directed, it is too forceful for most watering, and when emitted at a lower volume, it sprays in a circle and cannot be accurately directed. However, some gardeners find the forceful stream useful for washing off insects and the mist ideal for watering seedlings.

Flaring Rose Nozzle (Fan Nozzle). The flaring rose nozzle (Figure 11-2) often does not provide on-off control. However, it applies a large volume of water without force and is good for general overhead watering.

Oscillating Sprinkler. The oscillating sprinkler (Figure 11-3) is the most adaptable to general yard use and can be adjusted to cover any size area by altering

hand
trowel

round-
point shovel

weeder
hoe

spading
fork

Figure 11-1 Most commonly used soil-working tools for the garden.

Figure 11-2 Types of hose nozzles.

Figure 11-3 An oscillating sprinkler.

the water pressure and the oscillation mechanism. Because the head moves back and forth, water is applied over the area at a slow rate, and runoff is less likely than with a non-oscillating type.

Watering Can. Either plastic or galvanized metal watering cans are acceptable. The can should contain a removable sprinkler head so that either a fast volume or a slow sprinkling of water can be applied. Although metal cans last longer, they are heavier and more expensive than plastic cans.

Rakes

Leaf Rake. A leaf rake (Figure 11-4) can be made of lightweight metal or bamboo and has long flexible teeth. It is used for raking leaves, grass, and other lightweight materials.

Garden Rake (Bow Rake). A garden rake (Figure 11-4) is generally metal with short, rigid, widely spaced teeth. It is used for smoothing the soil surface in preparation for seeding a lawn or vegetable garden, removing thatch from the lawn, and other purposes.

Fertilizing Equipment

Hose-End Sprayer. The hose-end sprayer (Figure 11-5) attaches to a hose and automatically mixes any liquid concentrate into the water at dilute strength. It is used for fertilizing lawns, flowers, and other shallow-rooted plants and for applying certain insecticides. However, it is not accurate enough for most pesticide applications.

leaf rake

garden or
bow rake

Figure 11-4 Types of rakes.

Root Feeder. A root feeder works like a hose-end sprayer, proportioning liquid fertilizer into the irrigation water. It is equipped with a long, needlelike tube for injecting the fertilizer into the root zone of shrubs and trees.

Pruning Tools

Hand Pruners. Hand pruners (Figure 11-6) will cut branches up to ¾ inch (20 millimeters) in diameter. They are the most frequently used pruning tool. When using hand pruners, it is important never to twist while cutting

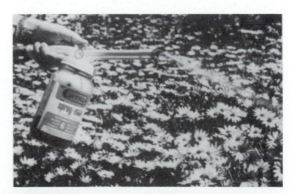

Figure 11-5 Hose-end sprayer. *(Photo courtesy of Vocational Education Productions, San Luis Obispo, CA)*

Figure 11-6 Pruning tools. Hedge shears (left), pruning saw (top), hand pruners (middle), and lopping shears (right).

in an attempt to cut a large branch. This will pull the cutting blades out of alignment and ruin the pruners.

Lopping Shears. Lopping shears (Figure 11-6) will cleanly remove branches up to about 1½ inches (4 centimeters) across in a single cut. Loppers are used on larger shrubs and trees.

 A park or zoo landscape supervisor is responsible for overseeing the turf and all landscaping of the area. Responsibilities include developing a maintenance schedule, supervising maintenance workers, and arranging for tree maintenance by arborists when necessary.

Pruning Saw. A small, curved pruning saw (Figure 11-6) can be used for removing larger branches and is specially designed to fit in tight places. It can substitute for lopping shears but will require more labor. Both lopping shears and pruning saws are available in pole-mounted models for pruning high in trees.

Hedge Shears. Hedge shears (Figure 11-6) are needed only if formal hedges are to be maintained.

Power Equipment for Professional Landscape Maintenance

Professionals use a number of gasoline-powered or electric power tools to efficiently maintain commercial and residential landscapes (see Chapter 12 for more turf maintenance equipment):

Figure 11-7 An electric hedgetrimmer. *(Photo courtesy of John Deere Co., Moline, IL)*

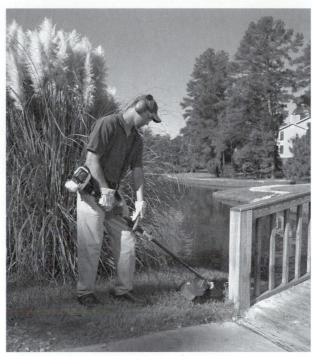

Figure 11-9 A string trimmer. *(Photo courtesy of John Deere Co., Moline, IL)*

Hedge Trimmer. An electric hedge trimmer (Figure 11-7) is essential to the maintenance of long sections of formal hedges.

Power Blower. A power blower (Figure 11-8) is a mainstay for the maintenance of walkways and other paved areas.

String Trimmer. A string trimmer (either gas or electric, Figure 11-9) is used to trim grass from around trees and beside paved areas.

Edger. An edger (Figure 11-10) is used to maintain the clean line between turf and nonturf areas.

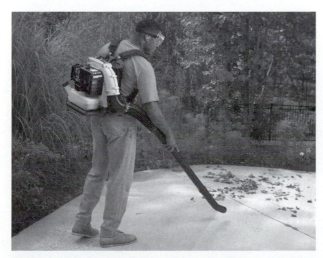

Figure 11-8 A backpack leaf blower. *(Photo courtesy of John Deere Co., Moline, IL)*

Figure 11-10 An edger should always be used with eye protection. *(Photo courtesy of John Deere Co., Moline, IL)*

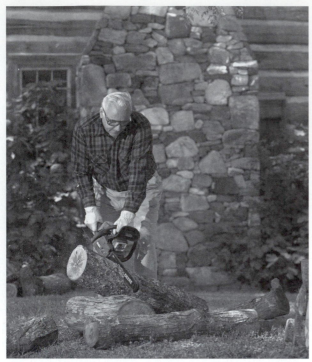

Figure 11-11 A chain saw *(Photo courtesy of John Deere Co., Moline, IL)*

Figure 11-12 A power pruner. *(Photo courtesy of A. M. Leonard Co., Piqua, OH)*

Chain Saw. A chain saw (Figure 11-11) may be necessary to remove large shrubs and trees.

Power Pruner. This gasoline-powered miniaturized chain saw (Figure 11-12) will allow one to reach and remove larger branches without climbing. It is a much safer way to remove high branches than using a chain saw.

Safe Use of Power Landscape-Maintenance Equipment

The first rule of safe operation of the power equipment listed above is to wear approved eye protection such as safety glasses, goggles, or facemask. Objects struck by a lawn edger or string trimmer fly great distances and may cause injury. Survey the area and remove sticks, stones, and the like before starting. Keep people and pets at least 50 feet (15 meters) away. Lawn edgers should not be used in graveled areas.

Hearing protection may also be necessary, especially for equipment with a gas (two-stroke) engine. Safety footwear is also advisable.

String Trimmers

Wear protective clothing when using a string trimmer to protect against flying debris. Long pants, closed-toed shoes, and gloves are best.

Use a string trimmer only for weeds and grass. It is not designed to trim shrubbery, climbing ivy, or other greenery not at ground level. The rule of thumb is not to trim any plant material above waist level.

For two-cycle gas/oil engine string trimmers, inspect for loose fasteners, fuel leaks, or a cracked or chipped string head before each use. Never start or run an engine inside a closed area. This creates fumes that can cause asphyxiation and death.

Follow fueling and premixing instructions carefully. Wipe up spilled fuel quickly. To prevent the possibility of fire, immediately replace the fuel cap. This will not only reduce the chance of fire because of a spark from the engine, it will also minimize pollution.

In some states, string trimmers have spark arrestors to prevent accidental brush fires. If you are in a climate where seasonal fires are common, be aware of this problem.

Blowers

A dust mask may be necessary in addition to eye and hearing protection when using a blower. For safety, point the blower away from people and pets, and reduce blower speed temporarily for passing pedestrians. Before storing the blower, let the engine cool until it is only warm to the touch.

Hedge Trimmers

Always hold the hedge trimmers with both hands.

Do not trim hedges that are higher than your shoulders with a hedge trimmer, and keep both feet on the ground. Do not overreach or cut from a ladder, as you may lose your balance.

Keep your fingers away from the hedge trimmer blades. Never try to clean out trimmings from the blade while the trimmer is running. Don't wear loose clothing or jewelry when using a hedge trimmer. Keep long hair pulled back.

Before removing something caught or jammed in the blades of the trimmer, turn the off-on switch to the "off" position and remove the spark plug wire.

Chain Saw

"Kickback" is the most dangerous hazard when operating a chain saw. Kickback is unexpected forceful pushing backward of the saw while it is running. It is due to the chain hitting an obstacle and stopping abruptly. This can make the user temporarily lose control of the saw. Some chain saws have an antikickback device on the tip of the guidebar.

Never cut with the tip of the guidebar; always keep your body away from the line of cut; always cut with the left arm straight and with a firm grip (thumbs and fingers around the handle). Wear nonslip rubber or leather gloves for better grip.

Bring the saw to full speed before letting the chain touch the wood. This partially prevents "bucking" or kickback. Once the cut is started, keep cutting at a steady speed and pressure. To avoid being pulled off-balance by the saw, do not speed up or slow down the chain while cutting.

When felling a tree, plan a retreat from the path of fall before you start cutting. When cutting a tree limb

that is under tension, be alert for the fact that the limb may spring back when the tension is released.

Keep the saw regularly serviced. Professional service personnel should perform all service work other than maintenance specified in the operator's manual.

Do not cut from a ladder, this is extremely dangerous. Use a pole pruner instead.

PRUNING LANDSCAPE PLANTS

Pruning is probably the most misunderstood maintenance task. Considerable time can be spent on seasonal clipping of trees and shrubs, only to find that in a few months the plants are again in need of pruning. However, this waste of time can be avoided by understanding how to prune to encourage controlled growth, and time invested in pruning can be held to a minimum.

Purposes of Pruning

Size Control

Most people use pruning for size control—to keep plants from growing over windows or a walkway, for example. However, not much pruning should be necessary if the plants are selected with their mature size in mind and positioned with forethought as to obstructions they might later create.

Health Improvement

Plant health is maintained or improved by such pruning practices as removing dead branches and diseased limbs. It also includes pruning trees when young to encourage strong branch structure and thus decrease the chances of branches breaking in storms.

Appearance Improvement

Pruning to remove scraggly branches, clip off dead flowers, and encourage bushy compact growth all improve plant appearance.

Pruning Trees

Young Trees

Pruning trees at transplanting and in the first few years that follow will greatly affect their strength and health in later life. During this early period, the tree develops the main branches that will form the support structure

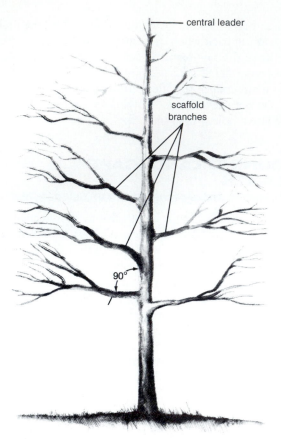

central leader

scaffold
branches

90°

Figure 11-13 A well-shaped tree with a central leader form. Note the even spacing of the scaffold branches down the trunk and the wide branch angles.

A

B

Figure 11-14 Bark ridges on two branches. *(Photo from* Arboriculture: Care of Trees, Shrubs and Vines in the Landscape *by Richard W. Harris. Reprinted by permission of Prentice Hall, Englewood Cliffs, NJ)*

for the foliage. Pruning directs the growth of those main branches in order to create a strong tree.

Removal of Double Leaders. Most trees have a *central leader,* that is, they grow with one main upright shoot (leader) and side branches called *scaffold branches* (Figure 11-13). Occasionally two leaders develop, causing a fork. The angle between the two leaders is a potential weak point of the tree, because as each leader increases in diameter, it exerts pressure against the other at the base. Eventually, conditions conducive to splitting of the trunk develop, and in a wind storm the weight of the leaders can pull the trunk apart. Accordingly, the weaker of the two leaders on a double-leader tree is removed while the tree is young. The remaining one will then develop normally.

Selection of Scaffold Branches. Removal of a double leader is not usually necessary in tree pruning because it will have been done at the nursery, but removal of excess scaffold branches is usually necessary. Ideally a young tree should have only a few branches spaced

widely and evenly along the trunk. Numerous small branches are not desirable because they will enlarge and crowd each other as the tree grows. Scaffold branches should be selected so that they are evenly spaced up the trunk and not growing directly above one another. Branches growing from the trunk at a wide angle approaching 90° (Figure 11-13) are preferred because they are stronger, although many trees have narrow-angled branches when young.

When a double leader or extra branch is removed, the branch should always be cut just beyond the *bark ridge,* also called the *shoulder ring* (Figure 11-14). This ridge is an area of raised bark or a ring that surrounds the point at which the branch grows from the trunk. It contains cells that are a barrier to infection of the cut and proliferate rapidly to close the wound. Therefore even though a small stub may be left in order to leave the bark ridge intact, it is preferable to cut the branch just past the bark ridge.

Topping. Topping is a practice used by some gardeners in the hope of forming a tree that is full-headed and dense. The central leader is removed; the side branches below the cut then grow rapidly. The ultimate effect is a flat-headed tree with an unnatural appearance and weak branch crotches. Topping is *not recommended* except for modified-central-leader-trained fruit trees.

Lion tailing. Similarly, lion tailing is not a recommended tree pruning practice. When a tree is "lion tailed," the leaves and smaller branches in the center are pruned off, leaving the inner parts of the tree stripped bare. This gives the tree an unnatural form with tufts of leaves at the branch tips, like the brush at the tip of a lion's tail. In addition to giving a peculiar appearance, such pruning will weaken the larger branches, stimulate excess new growth in the inner portion of the tree, and allow sunlight into the center, which can sunscald the bark. The tree's balance of foliage to wood is altered when so much foliage is removed, and this imbalance compromises the tree's energy reserve that it needs for future growth.

Limbing Up. Limbing up is the removal of lower branches of a tree that are positioned too low on the trunk and thus create a hazard or inconvenience. Removal is necessary because the height at which a branch originates is the height at which it permanently remains. The trunk does not lengthen and elevate the branch, as is popularly believed. A young tree 4 feet (1.2 meters) tall with branches 2 feet (0.6 meter) off the ground would have large branches still 2 feet (0.6 meter) off the ground 20 years later if left unpruned.

Limbing up should be delayed for three to four years after planting, if possible, because removal of foliage from a young tree will slow its growth and weaken the trunk. Once the tree is established and growing rapidly, one or two branches can be removed each year until the desired trunk height is reached.

Desuckering. Desuckering is the removal of shoots that grow from the base or lower trunk of trees. If the tree is grafted, these branches often grow from the vigorous root-stock. In other cases, buds on the trunk are located far enough from auxin-manufacturing buds that they are not reached by the dormancy hormone. Such branches are called *suckers,* or *watersprouts,* and should be removed as they appear with hand pruners. If left in place, they give the tree an unkempt appearance and use up energy that should be diverted to top growth.

Although most watersprouts grow from the trunk at ground level or slightly above, they are sometimes found on scaffold branches. They are easy to locate because of their vigorous vertical growth and should be removed. If left, they will grow until they cross and rub against the scaffold branches.

Mature Trees

A tree that has been properly pruned when young should need little, if any, pruning when mature. Desuckering may be necessary, along with the removal of branches broken by wind or growing toward the center of the tree (Figure 11-15). Small branches growing close together may rub one another (Figure 11-16); the smaller or more poorly positioned ones should be cut back to their parent branches to avoid injury to the bark and grafting together of the limbs.

Figure 11-15 Branches that begin to grow inward toward the center of the tree should be cut out as shown.

 Pruning of large trees is usually a job for an arborist (see chapter introduction photo). An arborist can be university trained in landscape horticulture with a specialization in arboriculture or can have received on-the-job training with an arborist. Arborists remove trees, climb up to prune unwanted branches, and use power spray equipment to spray for pests and diseases. Telephone and power companies employ arborists to keep the branches of mature trees away from power and telephone lines. Cities also employ arborists to maintain trees beside public roads.

Physiological Responses of Trees to Branch Removal

Wood and bark are made of living, dying, and dead cells with different functions. The living cells in wood and bark are connected in a network called the *symplast* that is found in the cambium between the wood and bark, in an outer bark zone called the phellogen, and in meristems. The symplast tissue allows the tree to defend itself *actively* against microorganisms. When it dies in the natural course of growth of the tree, it becomes *protective wood* such as heartwood that has a *static* protective function. Discolored wood is frequently infected by non-decay

Figure 11-16 The weaker and more poorly positioned of these two rubbing branches should be removed, even though they are both major branches. *(Photo courtesy of Vocational Education Productions, California Polytechnic State University, San Luis Obispo, CA)*

causing fungi. In the early stages, discolored wood is protective wood, but later as more organisms infect it, the wood may lose its protective ability. It can develop soft rot and the secondary layer of wood can become infected.

Previous concepts about tree pruning involved the principle that tree wounds "heal," and that in order to heal most quickly it was necessary to expose as little heartwood (xylem) as possible to invasion via air by microorganisms. Therefore "wound paints" were recommended to seal the wood and theoretically prevented entry of these organisms. However, many of the organisms that are first on a fresh wound are those that "keep away" the more destructive types.

It is now known that a tree reacts to injury, whether natural or man-made through pruning, by strengthening old barriers to decay or forming new barriers to decay. These barriers enable the tree to resist the decay microorganisms that would destroy the xylem and phloem systems for a longer time, but not indefinitely. Nature encourages the survival of both the tree and the microorganism species, and there is an interaction between the tree protecting itself and the microorganisms that survive by decaying the wood of the tree. A tree will survive as long as it can "compartmen-

talize" an infection, that is restrict the spread of the infection to a limited volume within the tree. The microorganisms survive as well, and the lives of both species are extended. The tree wound cannot therefore considered to be "healing" but rather the tree is both compartmentalizing the infection and growing new tissues to counteract the detrimental effects of decay microorganisms. This concept is conceptualized in a model called CODIT. CODIT is an acronym for Compartmentalization of Decay in Trees. Decay in this case being defined as the process where ordered wood becomes disordered and no longer functions.

As a tree branch dies, the trunk wood associated with the dying branch depletes its supply of elements that are essential for life. It can become so depleted of elements that few microorganisms will grow in it. This gives it a protective function. One of the major protection methods used by tree is to move the nitrogen-based substances out to the younger symplast as the wood ages.

As microorganisms invade the trunk via branch openings, the tree may compartmentalize the infecting microorganisms. However, over time the trunk wood walls off more symplast. As a result, space for storage of energy reserve is also walled off. This is a major way root-

Figure 11-17 The three-cut pruning method for removing large branches. This method prevents the branch from tearing down the trunk.

rotting fungi slowly kill a tree: the tree continually loses space for carbohydrate storage, and as energy storage capacity decreases, so does a tree's ability to defend itself.

A major protection boundary that determines the longevity of many trees is the protection zones at the base of branches: the branch collar, bark ridge, or shoulder ridge. As branches die or are removed, the openings in the bark are infection entryways for wood-inhabiting microorganisms. Tree species that have the strongest branch-protection boundaries are those that usually live the longest.

Removal of Large Branches. Occasionally, large branches may need to be removed because of disease, incorrect pruning when young, or some other reason. Such branches are frequently heavy and, while being sawed off, can break and rip down the trunk. To avoid the chance of injury to the tree, the three-cut pruning method (Figure 11-17) is recommended for removing branches

more than 2 inches (5 centimeters) in diameter. The first cut is made under the branch about 1 foot (30 centimeters) from the trunk and goes halfway through the branch. This cut provides insurance against tearing in case the branch should fall. The second cut is made farther out on the branch and cuts completely through and removes the branch. The third cut removes the stub down to the bark ridge, leaving it intact.

If it is necessary to remove a stub that was left because the tree was pruned improperly years ago, the branch collar should be left and only the part of the stub extending beyond it removed (Figure 11-14) even though this may leave a considerable amount of stub. The collar is like the bark ridge in that it prevents entrance of infection.

Removal of Girdling Roots. A girdling root (Figure 11-18) usually results from the incorrect initial planting of a tree. If the roots are not spread out in the hole,

Figure 11-18 A girdling root on a young tree. *(Photo by Rick Smith)*

they can enlarge and press against the trunk at or just below ground level. The resulting pressure halts the flow of water and nutrients within the trunk and can eventually kill that side of the tree.

Signs of damage from a girdling root include flattening of the trunk on one side and, on deciduous species, dropping of the leaves on the affected side earlier in fall. The only cure for a girdling root is to cut it away with a chisel or hatchet.

Pruning Vase-Shaped Trees

Many patio trees, weeping trees, and some larger shade trees like elm do not naturally have a central leader, and it is a mistake to prune them to have one. Such trees may split into two leaders, forming a vase shape, or have all their branches arising at one point on the trunk (Figure 11-19). Pruning for these trees should include only desuckering and removal of crossing and inward-growing branches.

Crown Reduction (Drop-Crotch Pruning)

When a tree is mistakenly planted where it will not have adequate growing room, it may be necessary to lower or drop the crown height. This is the case with trees under power lines especially. When crown reduction is necessary, the tallest terminal (vertical) branches are cut back to a large lateral (sideways-growing) branch. This encourages the tree to grow outward rather than upward. The diameter of the lateral branch to which the terminal branch is removed *must* be at least 1/3 the diameter of the terminal branch being cut back (Figure 11-20).

Pollarding

Pollarding (Figure 11-21) is a pruning technique for deciduous trees and a common pruning practice in certain areas of the country. It restricts the height of the tree and creates a denser head of foliage than the species would normally have. It is an ancient practice with the original purpose of encouraging production of large numbers of whiplike limbs used for animal food, fencing, or firewood. It is currently used only for a specialized decorative effect in a formal setting, and it can only be done on certain species.

Pollarding is usually begun after the tree has attained the maximum height desired. When the tree is dormant, it is cut back severely, and the main branches sawed off deeply into the old wood. When the tree

Figure 11-19 A vase-shaped tree with all main branches arising from one point on the trunk. *(Photo by Rick Smith)*

breaks dormancy, buds will sprout below each cut, producing vigorous watersprouts and creating the dense foliage mass. The next winter and each winter following, the recurring watersprouts are removed. Eventually the ends of the branches become gnarled, giving the plant a distinct appearance.

Although pollarding does give a tree a full, pleasing appearance during the growing season, the tree will be very unattractive when dormant. In addition, pollarding generally requires the yearly services of a tree maintenance person who has the equipment needed to reach high in the tree and remove the watersprouts. Consequently, pollarding cannot be recommended for normal home landscapes.

Pruning Needle Evergreen Trees

Pines are the most familiar needle evergreens, all of which are characterized by a pyramidal, or "Christmas tree," form with a central leader. Evergreens generally do not require pruning except to remove broken or dead branches and occasionally to eliminate a double leader. They may be limbed up above eye level if necessary.

Figure 11-20 Crown reduction. The branches shown in grey will be removed to lateral branches to reduce the tree height.

Figure 11-21 A pollarded tree in winter after pruning. *(Taken from* Arboriculture: Integrated Management of Trees, Shrubs and Vines, 2nd ed. *by Richard W. Harris. Reprinted by permission of Prentice Hall, Englewood Cliffs, NJ)*

Christmas tree growing is a horticultural enterprise that requires a time and labor investment of five years or more before the first profits are made, and therefore would need to be started as a sideline to some other crop or employment. A person interested in growing Christmas trees should seek an area where land can be held in long-term lease or is inexpensive to purchase.

When pruning needle evergreens, remember that, unlike most other trees, they do not branch if cut back to old wood, so pruning into the woody portions has a permanent effect. The only pruning that can be used to limit size or make the foliage thicker is the removal of some of the new growth, or *candles* (Figure 11-22), while the tree is still immature. Breaking out a portion of the center candle and/or side candles will cause the growth to become shorter and fuller.

Pruning Shrubs

To preserve their natural appearance, most shrubs are pruned by either of two pruning methods: *thinning* or *heading back*.

Thinning

Many shrubs are composed of a clump of woody stems or one or more short trunks coming out of the ground. They grow by the increase in height and number of

Figure 11-23 Pruning a shrub by thinning.

these stems or trunks. As the oldest stems age and lengthen, they will have a tendency to shade out the bottom foliage, and the shrub will become bare at the base. Thinning consists of removing the oldest stems at ground level. This shortens the shrub and encourages new shoots to grow and refoliate the bottom. Thinning can be done with hand pruners, if the stems are small, or with loppers, if the shrub is heavily overgrown with large stems (Figure 11-23).

Heading Back

Heading back (Figure 11-24) is done with hand pruners to remove straggly growth or to limit size. It consists of cutting twigs or small branches back to a point where the cut is hidden by the remaining foliage. If possible, the stem should be cut back to just above an outward-pointing bud or branch (Figure 11-25). This will encourage new growth to develop outward and eliminate crossing branches.

Figure 11-22 Pruning a needle evergreen by breaking out the unexpanded candles. *(Photo by Rick Smith)*

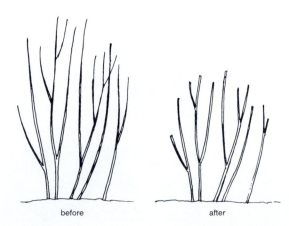

Figure 11-24 Pruning a shrub to increase its density by heading back.

Figure 11-25 Pruning to an outward-pointing bud.

Renewal Pruning

Renewal pruning (Figure 11-26) is used to rejuvenate and shorten overgrown shrubs. It is always done in early spring before new growth starts and consists of cutting back all the branches to stubs 2 to 3 inches (5–8 centimeters) long. This forces new growth from dormant buds on the stubs, which can then be pruned as needed by heading back and thinning to maintain the proper size.

Renewal pruning is a drastic measure and will make the plant unsightly for at least one season. As an alternative, half the branches can be removed the first year and the remainder the second year. In this way the plant will always keep some foliage and have a somewhat better appearance.

Pruning Hedges

Natural hedges are lines of shrubs planted close together to create a thick mass of unpruned foliage. Pruning natural hedges is basically the same as for other shrubs, that is, thinning and heading back once or twice a year.

Formal hedges are pruned to a specific shape (Figure 11-27) and require frequent cutting to maintain a neat appearance; shearing every two weeks during the growing season is typical.

Selection of the correct hedge shape is very important. Too often a hedge is pruned to be narrow at the base and wide at the top. This is incorrect because the wide top will shade the lower portions, causing them to drop their leaves. Pruning the top narrower than, or at the same width as, the bottom (Figure 11-27) will avoid shading of the bottom, and the plants will remain full and leafy to the bases.

Pruning Groundcovers

Most groundcovers do not require regular pruning to improve their appearance or health, although shrub types such as junipers may require heading back to control spread (Figure 11-28). However, occasionally an older bed of groundcover will develop long leafless stems and can be improved by renewal pruning. If the area is small, it can be clipped back severely with hand pruners. For large areas, a lawnmower set to cut as far from the ground as possible (and with a catch bag) can be used during the period of active growth. Hedge

before pruning   after first pruning   after second pruning

Figure 11-26 Renewal pruning.

good shape   acceptable shape   good shape   poor shape

Figure 11-27 Shapes for formal hedges.

shears can be used on woody ground covers. Although the area will be unsightly at first, it will usually refoliate in two to three months.

Timing Pruning

The time of year at which maintenance pruning takes place will not affect the health of a plant and need not be restricted to a particular season. However, early spring is a traditional pruning time for several reasons. First, it is easier to determine which parts of deciduous plants should be removed if the branch structure is not obscured by leaves. Second, by pruning before new growth starts, the growing direction of new shoots can be controlled. By contrast, summer or fall pruning removes poorly placed shoots produced that season.

There are exceptions to the early spring recommendation for pruning. If spring-blooming shrubs are pruned early in the year, some of the flower buds that have already formed will be cut off. Consequently it is best to prune these shrubs immediately after flowering and before they develop buds for the next year's bloom.

Deciduous trees that experience sap flow in early spring are the second exception. Pruning wounds will exude sap liberally in early spring. This loss of fluid will not seriously injure the tree, but it is unsightly and can be avoided by pruning later in the year.

In conclusion, light pruning twice or three times annually is preferable to severe pruning once a year to maintain an attractive landscape.

Special Shrub-Pruning Techniques

Espalier

Espalier (Figure 11-29) is the stylized training of a shrub or tree flat against a wall or trellis, usually in a symmetrical branching pattern. Because not all plants can be adapted to this type of pruning, consult a knowledgeable nurseryman or one of the references listed at the end of the chapter before espaliering a plant.

Espalier pruning must be started when a plant is small. The trunk is tied to the trellis, and the sites where future branches are desired are selected. The buds or branches closest to these points are then tied to the trellis, and all others are removed. This permits only these preselected branches to grow and achieves the desired

(a)   (b)

Figure 11-28 Heading back a groundcover juniper to control its spread (a) before pruning and (b) after pruning. Note how the branches are layered with longer branches on the bottom to prevent shading out and death of the base of the plant.

Figure 11-29 Espalier (left) and topiary (right) pruning styles. *(Photo courtesy of Monrovia Nursery Company, Azusa, CA)*

shape in several years. Annual pruning will still be required to remove stray growth and keep the plant from outgrowing the trellis.

Topiary

Shrubs and small trees trained into pompoms or animal shapes are examples of topiary pruning (Figure 11-29), a novelty pruning style that originated in formal European gardens. Topiary training can be thought of as a combination of a formal hedge and an espalier. Branches are left at selected locations on the trunk and then sheared repeatedly to encourage fullness. As with espalier pruning, a topiary should be started while a plant is young, shaping it gradually to its final form over several years.

Topiary plants in the landscape should be considered specimens; one or two are sufficient to enhance any landscape. They can be used effectively in Oriental or modern landscapes and can be purchased pretrained at garden centers.

WATERING

The goal of watering is to replenish the moisture in the soil surrounding plant roots. A considerable volume of water is needed to accomplish this, and underestimating this volume is a common error. An inexperienced gardener may sprinkle a plant with a hose for a few minutes and consider it watered. In actuality, the sprinkling has only wet the top 2 to 3 inches (5–8 centimeters) of the soil; the majority of the roots remain dry.

Correct watering of shrubs should give a slow flow of water at the base of the plant for 20 to 30 minutes. The water will penetrate evenly throughout the root area and wet the soil deeply. This will encourage the roots to grow downward into the moisture-retaining soil layers instead of along the surface. Trees should be watered deeply over the entire area covered by the branches, because roots will be growing throughout the area. Exception to the rule of deep soaking must be made for established plants accustomed to sprinkler irrigation. The root system of these plants will be shallow, and the plants will need to be maintained with sprinkler irrigation.

Watering Frequency

The frequency with which a plant should be watered depends on its age, its drought tolerance, and the ability of the soil to retain moisture.

Plant Age. More newly transplanted trees and shrubs die from lack of water than from any other cause. When first transplanted, the root system of a plant occupies a limited volume of soil and must extract all its water and nutrients from that soil. Until it is established and is able to absorb moisture through an extensive root system encompassing a considerable volume of soil, watering should be frequent. Watering every seven to ten days for the first year and every two weeks the second year is advisable. After that, irrigation will be needed only during periods of drought lasting more than three or four weeks.

Drought Tolerance. Tolerance to drought varies immensely among plant species. Some are able to extract minute quantities of water present in the soil and remain healthy, whereas others under identical conditions would wilt to the point of death.

Cacti and succulents are the best-known examples of drought-tolerant plants, but they are not the only such plants. Trees with tap-roots and shrubs native to arid climates are similarly drought-tolerant and can be used to create a low-maintenance and ecologically sound landscape where natural rainfall is scarce.

Soil Moisture-Holding Capacity. The proportions of sand and clay in a soil greatly affect its ability to retain moisture and consequently affect the watering frequency. Soils containing a large percentage of clay will be the most water-retaining, with loams intermediate and sandy soils least able to hold water. A sandy soil may need watering twice as frequently as a clay.

If a soil is sandy, its moisture-holding capacity can be greatly improved by adding organic matter, as described under "Water Conservation" and in Chapter 5.

Watering Techniques

Sprinklers

Sprinkler irrigation, whether by permanent or portable sprinkler heads, was the main method of irrigation used for many years in areas lacking regular rainfall. It watered large landscaped areas at one time and, if left on long enough, would eventually saturate the ground to the desired depth. The main disadvantage of sprinklers is that they waste water. They wet all the soil instead of only that which contains plant roots, and also waste water by evaporation from the droplets as they pass through the air.

 In areas of the country where rain does not fall throughout the year, an irrigation system is an integral part of all horticultural growing enterprises. Careers in irrigation include selling irrigation equipment, designing irrigation systems for wholesale nurseries or areas around commercial buildings, and installing irrigation systems.

Today, watering by sprinklers can be recommended only for groundcover, lawns, and densely planted areas such as flower beds. However, sprinkler irrigation should not be discontinued on landscapes established by this watering method, because the plant root systems will have spread throughout the sprinkler pattern area.

Hand Watering

In landscapes where supplemental watering is seldom required, hand watering with a hose is practical. The hose can be adjusted to low pressure and moved from plant to plant every half-hour, or the watering basin made at planting can be filled several times.

Leaching

Leaching is the application of a large volume of water to the soil, much more than would be necessary to wet the root zone, for the purpose of flushing accumulated *minerals* or *salts* from the soil. It is sometimes necessary because an excess concentration of minerals accumulates in the soil, whether from constant application of mineral-rich irrigation water, excessive fertilization, or a naturally rich concentration of minerals in the local soil. Under these conditions plant roots are unable to extract water from the soil even though an adequate amount is available because the natural osmotic process for absorption through the root hairs is disrupted due to excess minerals. To alleviate the problem, it is necessary to dissolve these salts and move them to the deeper soil layers by profuse and prolonged application of water.

One way of leaching is to water a plant deeply, wait a few hours, then repeat the process several more times. During the waiting the minerals dissolve in the water held in the soil, and when more water is applied they are flushed away. Another method is to leave a hose trickling at the base of the plant for half a day or so. This will accomplish the same goal but probably use more water.

The necessity of leaching depends on several factors and is normally most necessary in areas with little natural rainfall, such as the Southwest. In these areas lack of natural leaching by mineral-free rainwater, combined with dry air (which necessitates frequent irrigation with normally saline irrigation water) creates a soil mineral buildup that should be leached every few months.

General symptoms calling for leaching include slowed growth, yellowing of foliage, and sometimes wilting, even when the soil is moist. Some species are more resistant to excessive soil salts than others, and this is another factor to consider in deciding if leaching is necessary. There are attractive plants native to salty soils, and a homeowner should consider planting them if soil salts are a recurring problem.

Drip or Trickle Irrigation

Drip or trickle irrigation (Figure 11-30) is an irrigation technique whose advantages include low cost, ease of installation, and conservation of water. An indirect advantage of drip irrigation is that because it applies water only at one spot (instead of throughout an area, as a sprinkler), the germination and growth of weeds are inhibited.

Figure 11-30 Putting together a trickle irrigation system. Note the small microtubes that will carry water to each plant. *(Photo courtesy of Chapin Watermatics, Watertown, NY)*

A typical drip system would consist of five major parts: a filter, a pressure regulator, ½- to ¾-inch (12- to 20-millimeter)-diameter flexible tubing, and ⅛-inch (3-millimeter)-diameter microtubing and emitters. Water from an outside tap is first filtered to remove particulates and then passed through the regulator to lower the pressure to about 10 percent of normal. Next it passes through the ½-inch (12-millimeter) tubing located near the plants to be irrigated. Short lengths of the microtubing inserted into the main lines carry water to emitters at each plant, with all plants watered at once at a slow, even rate. The "drip" and "trickle" names come from the slow flow of water from each microtube, a total of about 1 gallon (4 liters) per hour.

Trickle irrigation is very easy to install, even for a person with no knowledge of plumbing. Because it operates at low pressure, no glue or threaded connections are needed. The splices between the main lines are held together with self-sealing plastic connectors. Likewise, the microtubes can be inserted into the main line by simply making a hole and pushing in the microtube until it fits snugly. Unlike a sprinkler, no digging is needed to install a drip irrigation system. The main lines are laid on top of the ground and covered with mulch.

The main disadvantage of drip watering is the tendency of the microtubes to clog with mineral deposits from the water, soil, or algae. These problems can usually be remedied by flushing the system periodically at high pressure.

Drip systems are most often used on tree and shrub plantings, because each individual plant can have its own microtube. However, the system can be adapted to water groundcover and grass areas with microsprinkler heads (Figure 11-31) that spray over a limited radius.

Water Conservation

Even where water is abundant, conserving water used for irrigating landscape plants should be considered. There is never enough of this valuable natural resource to justify thoughtless use. The following techniques will reduce the amount of water used in a landscape planting.

Figure 11-31 A microsprinkler head used with a drip irrigation system. *(Photo courtesy of Chapin Watermatics, Watertown, NY)*

Improvement of the Water-Holding Capacity

Though the natural soil in an area may have a poor water-holding capacity, it can be improved with the addition of organic matter. Compost, peat moss, shredded bark, coarse sawdust, and similar materials can be incorporated at planting. Herbaceous groundcover beds can be noticeably improved by mixing in organic matter to a 1-foot (0.3-meter) depth, because the root systems of the plants are naturally shallow and will remain within the improved layer.

Prevention of Evaporation

Much of the water lost from landscape plantings leaves by evaporation from the soil. This evaporation rate can be slowed by the use of a mulch and by watering in the early morning when the temperature is lowest.

Use of Proper Irrigation Techniques

The proper irrigation equipment should be used: trickle for trees and shrubs, plus sprinkler for lawns and groundcover. The system should then be used wisely. Be cautious when using automatic timing devices, because they do not take weather conditions into account and may apply water in a rainstorm. Frequent checks for water runoff will prevent one of the most common means of wasting water.

FERTILIZING LANDSCAPE PLANTS

Many homeowners never fertilize their trees and shrubs, assuming that fertilizers are needed mainly for vegetables and lawns. But just as a fertilizer promotes the strong rapid growth of these plants, it will also improve the growth of landscape plants. A newly installed landscape that is fertilized regularly will grow much faster than an unfertilized one.

Timing of Fertilizer Applications

The time of year when fertilizer should be applied to landscape plants varies with the climate, but generally it should be restricted to just before or while a plant is actively growing.

In most parts of the country, fertilizer is applied annually in spring so the nutrients will be available for the new growth. However, for faster growth, fertilizer could be applied two or three times during the growing season at one- to two-month intervals—for example, on April 1, May 5, and July 1.

In mild climates such as those found in Florida and parts of California, plants grow continuously. Fertilizer will need to be applied every two to three months through the entire year to replenish the nutrients absorbed by the plants.

Rates of Fertilizer Application

Trees

Generally speaking, established trees do not need to be fertilized with phosphorous or potassium unless it has been determined that the soil is naturally deficient in these nutrients. Their main requirement is for nitrogen, which can be applied on a yearly basis by drill hole or surface application (see Chapter 5). Small and newly planted trees are normally fertilized only by surface application.

The yearly amount of fertilizer recommended for trees is about 6 pounds (3 kilograms) of actual nitrogen per 1,000 square feet (100 square meters) of area under the branches. Actually, the amount of fertilizer applied is not critical, provided it falls within a reasonable range. In some cases it will not be practical to compute a fertilizer weight for a small tree with, for example, a 4-foot (1.2-meter) branch span. Estimating is completely satisfactory in most cases. About ½ cup (120 milliliters) of ammonium sulfate or sodium nitrate or ¼ cup (60 ml) of urea or ammonium nitrate is adequate for small trees less than 12 feet (4 meters) high. This allotment can be increased to 1 cup (240 milliliters) for 15-foot (5-meter) trees, 3 cups (700 milliliters) for 20-foot (6-meter) trees, and 4 cups (950 milliliters) annually for trees 30 feet (9 meters) or greater in height.

Shrubs

Shrubs, whether evergreen or deciduous, are fertilized in most circumstances at the rate of 2 to 4 pounds (1–2 kilograms) of nitrogen annually per 100 square feet (10 square meters) of bed area per year, using the surface application method. The fertilizer can be the same as is used for trees and can be split into two applications if desired: one in early spring and the second immediately after flowering. If acid-loving plants are being grown, an acid-forming fertilizer can be used (see Chapter 5).

If fertilizer calculations are not going to be made, a moderate amount of fertilizer can be estimated to equal ½ cup (120 milliliters) of 15-5-5 for shrubs less than 2 × 2 feet (0.6 × 0.6 meter), 1 cup (240 milliliters) for shrubs up to 4 feet (1.2 meters) tall and 3 feet (1 meter) across, and 1½ cups (350 milliliters) for larger shrubs. The fertilizer should be watered in after it is applied.

Groundcovers

Groundcover plants will benefit from yearly fertilization in spring using a surface application. Two to four pounds (1–2 kilograms) per 1,000 square feet (100 square meters) watered thoroughly afterward is advisable, or ¾ cup (200 milliliters) can be estimated to be sufficient for a 10 × 10-foot (3 × 3-meter) area. Either a balanced or a high-nitrogen fertilizer is acceptable.

WEED CONTROL

Weeds growing in a landscape planting can seriously detract from its appearance. They also compete for the available light, water, and nutrition; in an establishing groundcover, weeds will thrive at the expense of the desirable plants. For these reasons weed control is essential, whether chemical or cultural.

 A horticulturist interested in weed control may do lab research in weed-controlling chemicals, sell weed-controlling chemicals to growers and advise them on their use, or be a pesticide applicator. In many areas of the country special certification and training are required to advise about and apply pesticides such as herbicides.

Chemical Weed Control

With few exceptions, chemical weed control is impractical in landscape plantings for the home. Many species are used in a typical landscape, and each will react to weed-controlling chemicals (herbicides) in a different way. An improperly applied weed killer may ruin not only your own plants but also those in neighboring yards.

Exceptions can be made. Herbicides of the *pre-emergence* type kill only germinating seedlings and not established plants. Such chemicals could be applied, for example, to a newly planted groundcover area to alleviate the inevitable sprouting of weed seeds that occurs when the soil is disturbed. *Post-emergence* herbicides may also be used to kill existing weeds prior to planting.

Table 11-1 lists some of the herbicides used in ornamental plantings.

Bioherbicides*. A recently introduced organic weed control is corn gluten meal (CGM), a by-product of corn syrup processing. CGM is a pre-emergent herbicide, applied in early spring. It works best when applied to the top ¼ inch of soil. It has no carryover into the second year of growth, so it must be applied every year. CGM contains 10 percent nitrogen and acts as a slow-release fertilizer for the crop. CGM has been patented and is currently being sold as an herbicide, but using it to treat a large area can be quite expensive. Wheat gluten meal has many of the same effects as CGM, but it has not been patented and so may be more affordable.

More recent research has revealed that corn gluten hydrosylate (CGH), which is made from corn gluten meal, is more effective that corn gluten meal for controlling weeds and may be applied at less than half the rate of CGM for effective weed control. At this writing, CGH is available from only one source, Grain Processing Company in Iowa.

There are some new, environmentally friendly contact herbicides that break down quickly and provide options for weed control in container nurseries around irrigation risers and perimeter areas, as well as for general use in field nursery production. These are made from pelargonic acid, a fatty acid found in plants and animals. They work by rapidly lowering the pH of any plant sprayed, which weakens the cell walls and kills the weed, usually within two hours. These products (Weed Eraser™, Scythe™) are not systemic or selective, so that they kill any plant part sprayed. Several applications may be necessary to kill perennial weeds. Nature's Glory™ Fast Acting Weed and Grass Killer contains acetic acid (vinegar) and lemon juice. It also works very quickly as a contact herbicide and controls such weeds as crabgrass, chickweed, ragweed, plantain, quackgrass, and wild carrot. The commercial formula contains 25 percent acetic acid and is available from Ecoval Company.

Cultural Weed Control

Cultural weed control consists of hand pulling and cultivating with a hoe, using ornamental plants competitively, mulching, and solarization.

Hand pulling and cultivation are effective for controlling annual weeds. But they are only temporarily effective in controlling perennial weeds, because the weeds often break off at the soil line and grow back

*Information extracted from and courtesy of Appropriate Technology Transfer for Rural Areas (ATTRA), Fayetteville, AR. http://www.attra.org/attra-pub/nursery.html#weed.

Table 11-1 Selected Herbicides for Commercial and Homeowner Use in Ornamental Plantings

Chemical	Weeds controlled	Remarks
Clethodim (Envoy®)	Postcontrol of many grasses	Especially useful for control of young annual bluegrass
Corn gluten	Preemergent control of many weeds	For use in nurseries and landscapes; organic product that is partially effective
DCPA (Dacthal®)	Germinating annual grasses and some broadleaf weeds; will not control established weeds	For use in herbaceous plants; most effective in sandy soils
Dichlobenil (Casoron®)	Germinating annual broadleaf and some grassy weeds including quack grass; will not control already established weeds	For use around established deciduous trees and shrubs and around conifers
Dichlofop-butyl (Fusilade®)	Actively growing annual and perennial grasses except annual bluegrass	Does not harm most broadleaf plants except 'Bar-Harbor' juniper. Two applications may be necessary
Dithiopyr (Dimension®)	Preemergent control of grasses and some broadleafs	For use in professionally maintained turf areas—controls crabgrass both before and after
EPTC (Eptam®)	Germinating annual grasses and some broadleaf weeds; will not control established weeds	For weed control in flower plantings; irrigate after application
Ethofumesate (Prograss®)	Pre- and postcontrol of annual bluegrass; controls other turf weeds	For use in turf primarily for annual bluegrass control
Glufosinate (Finale®)	Postcontrol of mainly annuals; suppresses perennials	For use inside greenhouses and in landscapes
Glyphosate (Kleenup®, Roundup®)	Actively growing annuals and perennials	Post-emergence spray or wiping on for weeds around woody ornamentals
Halosulfuron methyl (Manage®)	Postcontrol in turf and landscapes	Excellent control of yellow and purple nutsedge
Isoxaben (Gallery®)	Precontrol of many broadleaf and grassy weeds	For use in landscapes and container and field nurseries
Metolachlor (Pennant®)	Precontrol of many broadleaf and grassy weeds	For use in field and container nurseries
Oryzalin (Surflan®)	Germinating grasses and broadleaves	For use in woody ornamentals and herbaceous plants
Oxydiazon (Ronstar®)	Germinating broadleaf weeds and some grasses	For use in a wide variety of ornamentals and to prevent annual bluegrass in lawns
Oxyfluorfen (Goal®)	Pre- and early postcontrol of many broadleaf and grassy weeds	For use in nurseries and as an ingredient in other granular products in landscapes; especially for control of some glyphosate-resistant weeds such as cheeseweed
Pelargonic acid (Scythe®)	Early postcontrol of most weeds	For use in landscapes, nurseries, and greenhouses
Pendimethalin (Pennant®)	Precontrol of many weeds	For use in landscapes, nurseries
Sethoxydim (Poast®)	Annual and perennial grasses	Post-emergence only
Triclopyr	Annual and perennial broadleafed weeds, especially oxalis	Post-emergence spray
Trifluralin (Treflan®)	Germinating annual grasses and some broadleaf weeds; will not control established weeds	For flower and shrub plantings; irrigate after application

from the roots. Cultivation exposes fresh soil and additional weed seeds to favorable germination conditions and necessitates repeating the cultivation several times each season. Although pulling and cultivation may be the only remedies for large weeds, mulching is more practical and labor-saving over a long period.

Using ornamental plants competitively can also reduce weed problems. By planting ornamentals thickly and using groundcovers on bare soil areas, weeds will not have an opportunity to become established. They will be crowded out by the desirable plants.

Mulching is the third cultural weed control. A layer of mulch excludes light from the soil, checking the growth of both seeds and existing small weeds. A mulch can be any material used to cover the soil surface.

Mulching plants in the landscape serves other purposes besides weed control. It provides a decorative background for plants, conserves soil moisture, and maintains the soil at an even temperature.

Solarization is a useful technique for weed, nematode, and soil pest control. Solarization consists of covering well-prepared soil with a thin layer of clear plastic that is left in place for several weeks during warm weather. In addition to the suppression of pests, solarization may also favor the development of heat-tolerant beneficial organisms. In order for the solarization process to be successful, the soil must be prepared as for a fine seedbed. You must ensure that clods are broken up. The soil is then thoroughly moistened and tightly covered with thin, clear plastic for several weeks. The moist-heating process kills many soil pests and can give excellent weed control, though large-seeded weeds may not be completely controlled. Hot, sunny weather is essential for this process to be effective. After the process is complete, the area can be planted; however, take care not to disturb the soil deeply.

Organic Mulches

The most popular organic (plant-derived) mulches used in landscape plantings are shredded bark and bark chunks (Figures 11-32 and 11-33) and pine straw. They are applied 2 to 3 inches (5–8 centimeters) thick throughout shrub beds and under trees and give these areas a neat, rustic appearance. Because they are organic, they decompose slowly over a number of years and will need to be renewed occasionally by adding a fresh layer to the existing mulch. Organic mulches useful for the landscape are not limited to bark. Rice hulls, cocoa bean hulls, pine needles, and oak leaves are attractive mulches, but are available in only a few areas of the country. Likewise, corncobs, compost, and grass clippings are also effective but are generally considered somewhat unattractive for landscape use. They are discussed in Chapter 6 as vegetable garden mulches.

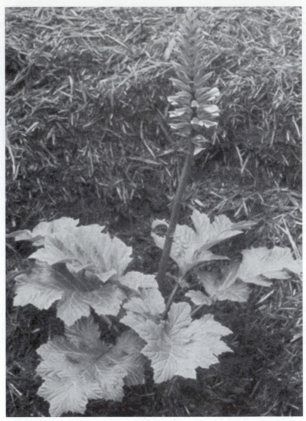

Figure 11-32 Shredded bark mulch. *(Photo courtesy of Barry Glick, Sunshine Farm and Gardens, (www.sunfarm.com) Renick, WV 24966 US)*

Rock Mulches

Rock mulches can be considered permanent because they do not decompose. Included in the inorganic mulches are stone materials such as lava rock, marble chips, smooth stones, and coarse or smooth pebbles (Figures 11-34, 11-35, and 11-36). Their only

Figure 11-33 A chunk bark mulch. *(Photo courtesy of Mountain West Bark Products, Inc. and Development Workshop, Inc. of Idaho Falls, ID)*

Figure 11-34 Lava rock mulch. *(Photo courtesy of Mountain West Bark Products, Inc. and Development Workshop, Inc. of Idaho Falls, ID)*

Figure 11-35 A marble chip mulch. *(Photo courtesy of Mountain West Bark Products, Inc. and Development Workshop, Inc. of Idaho Falls, ID)*

Figure 11-36 A flat stone river rock mulch. *(Photo by the authors)*

disadvantage is that, if replanting of the bed is required, they must be removed and replaced afterward. Organic mulches, on the other hand, could be simply worked into the soil. Because rock mulches gradually sink into the soil, they should be applied over plastic or landscape fabric.

Synthetic Mulches

Plastic and landscape fabric (discussed in Chapter 10) are some of the most effective mulches because they are completely impervious to weed growth. Although unattractive by themselves for use in landscape plantings, they can be covered with a decorative layer of bark or stone. They will also be concealed when used with a groundcover shrub, which will spread and hide the material soon after planting.

Plastic and landscape fabric generally come in rolls of varying widths and are rolled onto the area to be planted with edges overlapping. Plants are then transplanted into slits cut through the material. Finally, the edges are covered with soil to hold them firmly in place in case of wind. For decorative areas such as landscapes, a layer of bark mulch can be applied over the entire area to cover it.

Water will run off plastic during rainfall and be diverted through the holes to the plants. However if plastic wider than 3 feet (1 meter) is used, additional holes should be made for water penetration into the soil.

There is no such problem of water penetration with a landscape fabric, which is an advantage. The fabric is also much less likely to tear, which would permit weed growth through the cut. For these reasons landscape fabrics are becoming more popular than plastic in spite of their higher cost.

SELECTED REFERENCES FOR ADDITIONAL INFORMATION

Haas, Cathy, and Michael MacCaskey. *All About Lawns.* San Ramon, CA: Ortho Books, 1994.

Harris, Richard W. *Arboriculture: Integrated Management of Landscape Trees, Shrubs and Vines.* 2nd ed. Englewood Cliffs, NJ: Prentice-Hall, 1992.

Pirone, P. P. *Tree Maintenance.* 6th ed. New York: Oxford University Press, 1988.

Rice, R. P. *Nursery and Landscape Weed Control Manual.* 3rd ed. Fresno, CA: Thomson Publications, 2001.

World Wide Web Addresses

http://www.alca.org/. Associated Landscape Contractors of American. "ALCA Website." (December 20, 2001)

http://www.anla.org/. American Nursery and Landscape Association. "ANLA Website". (December 20, 2001)

http://www.asca-consultants.org/. American Society for Consulting Arborists: "American Society for Consulting Arborists Website." (December 20, 2001)

http://www2.champaign.isa-arbor.com/ International Society of Arboriculture. "Arboriculture On-Line." (December 16, 2001)

http://www.chesco.com/~treeman/SHIGO/. Shigo, Alex. "Shigo on Tree Systems." (December 20, 2001)

http://edis.ifas.ufl.edu/MENU_WG. University of Florida Extension, Institute of Food and Agricultural Sciences. "Weed Management Guide." (December 20, 2001)

http://www.natlarb.com/. National Arborist Association. "National Arborist Association Website". (December 20, 2001)

http://www.pgms.org/. Professional Grounds Maintenance Society. "PGMS Website." (December 20, 2001)

(Photo courtesy of Toro Company; Irrigation Division; Riverside, California)

CHAPTER 12

LAWN ESTABLISHMENT AND CARE

Lawns have long been a traditional part of the home landscape. Many landscape designers minimize or eliminate them altogether; others assert that a well-kept lawn is the most attractive setting for a home. Although it is true that care of the lawn is one of the most time-consuming maintenance activities, lawns are among the best surfaces for game and play areas.

 Golf course design, a specialization within landscape design, is perfect for those who enjoy golf because, in order to design a course, one should have in-depth knowledge of the game. A designer of golf courses will make extensive use of topography alterations to create the best possible course for players. He or she may design courses around the country and even internationally.

ESTABLISHING A NEW LAWN

For owners of a new house, establishing a lawn is often the first landscape activity. Usually a lawn is seeded or sodded, but in some regions techniques called *plugging* and *stolonizing* are used. Each of these establishment methods is discussed in the following sections.

Establishment by Seeding

Establishing a lawn by seeding has advantages and disadvantages. First, seeding can require several months to make an attractive turf, whereas a sodded lawn is practically instant. More effort is required to establish the seedlings, and there is greater chance of failure. However, seeding is inexpensive, and when there is a large area involved, cost can be important. Some lawn grasses such as hybrid Bermuda grass and zoysia cannot be established from seed, so vegetative means must be used.

 Production of grass seed is a large business. Opportunities for employment in this field might be in producing the seed in the field; overseeing harvest, cleaning, and packaging; or marketing the seed.

Time of Year

The best time of year to seed a new lawn is dependent on the species of grass and the region in which it is being planted (Figure 12-1). In warm areas like the South and Southwest, warm-season turfgrasses (Table 12-1) are popular because they grow best in the temperature

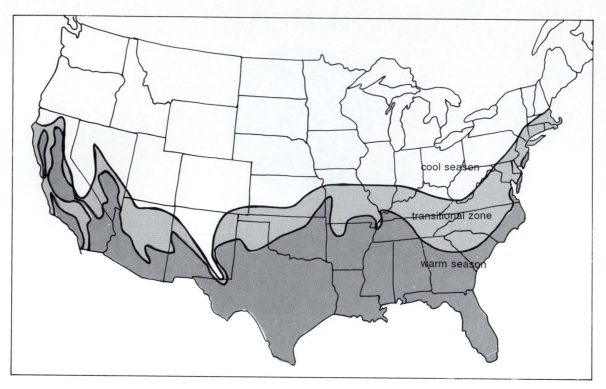

Figure 12-1 Areas for warm- and cool-season grasses.

Table 12-1 Selected Turfgrass Species

Name	Cool/warm season	Characteristics
Tall fescue *Festuca arundinacea*	cool	tough, drought tolerant, coarse textured (new cultivars like 'Bonsai' and 'Twilight' are finer)
Creeping red fescue *Festuca rubra*	cool	fine texture, shade tolerant, good bank cover, drought tolerant
Chewings fescue *Festuca rubra 'commutata'*	cool	same as red fescue but lacks rhizomes
Kentucky bluegrass *Poa pratensis*	cool	medium texture, cold tolerant, thick, attractive turf
Perennial ryegrass *Lolium perenne*	cool	fast establishment but short-lived, use for overseeding
Annual ryegrass *Lolium multiflorum*	cool	same as perennial ryegrass but shorter-lived
Creeping bentgrass *Agrostis palustris*	cool	finest texture, luxuriant turf but requires high maintenance, not usually used for home lawns
Bermudagrass *Cynodon dactylon*	warm	tough, medium to fine texture, hybrids best but must be vegetatively propagated, full sun needed
Buffalograss *Buchloe dactyloides*	warm	drought and cold tolerant, low maintenance
Zoysia *Zoysia japonica*	warm	fine texture, slow to establish, long dormant period, prickly blade, low maintenance
St. Augustinegrass *Stenotaphrum secundatum*	warm	coarse texture, not cold tolerant, thick and tough
Centipedegrass *Eremochloa ophiuroides*	warm	low maintenance, medium texture

range of 80 to 95°F (27–35°C). Spring is best for establishing these grasses, because the hot summer temperatures to follow will supply ideal growing conditions.

 A golf course superintendent is responsible for supervising construction and maintenance of golf course(s), supervising maintenance and repair of construction and maintenance equipment, giving professional advice, opinions, assistance and services to the board of directors or managing body, and participating in all long-range planning meetings.

Cool-season grasses such as Kentucky bluegrass (Table 12-1) are the primary lawn grasses in areas with cold winters. These grasses grow best between 60 and 75°F (15 and 24°C). There, late summer is the preferred seeding period. Sown at this time, the grass will receive cool fall temperatures for establishment, and there will be less competition from annual weeds. Spring planting is also acceptable if fall is not practical.

Site Preparation

Site preparation is the same whether seeding, sodding, plugging, or stolonizing is used, and it is a critical factor in assuring a high-quality lawn. Depending on the condition of the area, some or all of the following operations will be a part of site preparation.

Debris Removal. Debris removal is usually required around a newly constructed house. Pieces of wood, chunks of concrete, and other building materials will doubtless have been left at the construction site. Worse yet, they may have been buried and will later cause sunken areas or interfere with root growth. Debris removal should eliminate any foreign materials in the top 4 inches (10 centimeters) of the soil surface.

Elimination of Existing Vegetation and Preventative Weed Control. The best way to have a weed-free lawn is to prevent initial weed establishment. To accomplish this, the area should be evaluated in terms of the weeds already present and the potential for invasion from outside the lawn area. Noting whether the species are annual or perennial will pay dividends after the grass is planted. Perennial weeds can often be recognized by thick, fleshy roots or creeping underground stems. Generally speaking, you can assume that existing annual weeds will be killed during soil preparation, but that seedlings of the same species will soon reappear. Perennial weeds, on the other hand, will reestablish quickly from roots, tubers, and rhizomes after soil preparation is completed. Spading these weeds under or rototilling will spread them, resulting in a severe problem later. It is therefore important that perennial weeds be controlled prior to grading.

Perennial weeds may be controlled in several ways. If a small number of perennials are present or if time is short, they can be removed by digging. All portions of roots and stems must be removed, because even a small section will sprout a new plant. If the infestation is too widespread for hand digging, herbicides can be utilized. In selecting an herbicide, avoid those that kill only the tops, because the roots will survive and resprout. An ideal herbicide translocates through and kills the entire plant without leaving a harmful residue in the soil. The herbicide glyphosate known commercially as Kleenup®, Roundup®, or several other names will kill both annual and perennial weeds and can be recommended for this purpose.

Annual weeds are much more abundant than perennials, but they are much easier to control. One way to decrease the annual weed population in an area to be seeded is to allow the weeds to germinate prior to seeding, kill them, and then sow the grass. Using this weed-control procedure, called the *stale seedbed technique*, it is necessary to prepare the seedbed in the normal fashion until it is ready for seeding. At this point, irrigation is begun and continued for 7 to 10 days. If the weather is favorable, the weeds will sprout and can then be easily destroyed with a *contact herbicide* (one that kills the plant when sprayed on the leaves and stem). Be sure the herbicide selected does not leave a residue in the soil that will harm the grass. Care should be taken not to disturb the soil afterward, because this will cause new seeds to sprout. If the soil has become hard, a very light ¼-inch (6-millimeter)-deep raking may be used to break the surface crust. The grass can then be seeded, covered with a mulch, and irrigated. Resulting turf will generally contain 50 percent or less of the usual annual weed population. With some turf species, a *preemergence herbicide* (one that controls weeds as they germinate) can also be used at planting.

Some weeds will always be present after the grass has emerged, despite all precautions. These can be controlled by pulling or, after the lawn has been mowed once, by the application of a *postemergence herbicide* (one that controls a weed that has already germinated). Several postemergence herbicides will kill broadleaf (nongrass) weeds in turf; however, controlling emerged grassy weeds must still be done by hand.

Recommendations for specific herbicides are given in Table 12-2 or can be obtained from a local nurseryman, county Cooperative Extension Agent, or Farm Advisor. Like all pesticides, herbicides are dangerous to humans,

Table 12-2 Lawn Herbicides

Name	Time of application	Weeds controlled	Remarks
For newly seeded lawns			
Bromoxynil	After grass has emerged and weeds are in the 2- to 4-leaf stage	Many broadleaf weeds	Larger weeds require higher rates
Glyphosate (Kleenup®, Roundup®)	Anytime prior to planting	All	Useful for killing all existing weeds
Siduron	Immediately following seeding	Emerging annual grasses	Water immediately after treatment
For established lawns			
Benefin	Spring, before weeds germinate	Crabgrass and other annual grasses	Does not leach readily; do not use on bentgrass
Bensulide	Spring, before weeds germinate	Crabgrass and other annual grasses, a few broadleaves	
Clopyralid (Lontrel®)	Postemergence	Many broadleafs	May be used on dichondra lawns; use eye protection
Corn gluten	Preemergence	Broadleafs and grasses	Organic material suppresses many weeds
Dacthal	Spring, before weeds germinate	Crabgrass and other annual grasses	Less effective on heavy soils
Dicamba	Postemergence	Many broadleafs	May leach into root zone of desirable trees and shrubs
Dichlorprop	Postemergence	Many broadleafs	Often combined with other postemergence materials
Dithiopyr	Pre- and postemergence (crabgrass)	Grasses and broadleafs, especially crabgrass	For professionally maintained turf areas
DSMA	Postemergence	Young crabgrass, chickweed, oxalis, dallisgrass, sandbur	Do not use on St. Augustine-Bahiagrass, or centipedegrass; apply 1–3 times 5–7 days apart
Ethofumesate	Pre- and early postemergence	Many grasses and broadleafs	Particularly useful for pre- and postemergence annual bluegrass control
Fenoxaprop-ethyl	Preemergence	Many grasses	Good crabgrass control
Glufosinate (Finale®)	When weeds are actively growing	All	Nonselective
Glyphosate (Kleenup®, Roundup®)	When weeds are actively growing	All	Must be applied *only* to the weed (spot treatment) with a paintbrush
Halosulfuron-methyl	Postemergence	Nutsedges and some broadleaf weeds	Excellent sedge control
MCPA, MCPP	Postemergence	Many broadleaf weeds	Often tank mixed with other broadleaf herbicides
MSMA, DSMA	Postemergence	Crabgrass, chickweed, dallisgrass, and sandbur	Apply 1 to 3 times at one-week intervals, suppresses kikuyugrass when combined with triclopyr
Metsulfuron	Postemergence	Many broadleafs	
Oxadiazon	Preemergence	Many broadleafs and grasses	For use in bermudagrass, St. Augustinegrass, and zoysiagrass by commercial applications only
Pendimethalin	Preemergence	Many broadleafs and grasses	
Prodiamine	Preemergence	Many broadleafs and grasses	
Triclopyr	When weeds are actively growing	Annual and perennial broadleaf weeds	Use on cool-season grasses only
2,4-D amine	When weeds are actively growing	Most annual and perennial broadleaves	Do not apply when windy; do not use on St. Augustinegrass

pets, the environment, and desirable plants if used improperly. Read the label thoroughly and follow all directions precisely. Particular care should be taken to prevent herbicides from blowing to areas adjacent to the lawn (such as shrub beds), where damage to plants may occur.

Grading. Grading is generally done by the construction crew so that the ground slopes evenly away from the building at not less than a 1 percent grade. However, in some cases this has not been done or has been done improperly, or grade changes are called for in the landscape plan. In these cases care should be taken to remove the existing topsoil, grade the subsoil layer, and then replace the topsoil. When the topsoil is missing, it should be replaced with purchased topsoil of good quality applied to a depth of 4 inches (10 centimeters).

Soil Amending. Depending on the condition of the topsoil, soil amending can be needed to alter pH, improve fertility, or increase or lessen water-holding capacity with organic matter.

Ideally, a soil test should be made, and the proper pH- and fertility-correcting chemicals added as described in Chapter 5. However, it is usually sufficient to assume the pH is in an acceptable range and give a general fertilizer application of 2 pounds (1 kilogram) actual nitrogen per 1,000 square feet (100 square meters) each of nitrogen, phosphorus, and potassium. The fertilizer should be worked throughout the topsoil layer.

Amending the soil with organic matter need not follow any formula. Any of the amendments listed in Chapter 5 can be used, remembering the basic rule that a large amount of amendment is usually needed to cause any appreciable change in the soil.

Surface Preparation. The surface on which seed is to be sown should be free of clods and as smooth as possible. Clods can usually be broken up by rototilling or, in a small area, by hand. A light raking will remove remaining clods and provide the final surface.

Finally, the area should be rolled with a lawn roller to compact the soil surface. Lawn rollers can usually be rented and are filled with water to provide the necessary weight.

Seeding and Mulching

Seed Selection. Selecting the proper variety or mixture of grass for the area is important to growing an attractive lawn and should be done carefully. Selected grass species are listed in Table 12-1; because of the wide variety of grasses grown throughout the country, a discussion of the merits and drawbacks of all is not possible. The state Cooperative Extension Service or Farm Advisor's Office will be able to assist in the choice.

Buying seed is not simply a matter of finding the greatest amount of seed for the money. Information on both *germination rate* (percentage of the seeds that will germinate under ideal conditions) and *purity* (percentage of the seeds of the specific variety or mixture being purchased) should be noted. The label will give the germination test date, which should be within the current year. Otherwise the seed may be more than a year old and have a decreased germination rate.

The aim in buying seed is to obtain the highest germination rate and purity at the best price. In no circumstances should purity or germination rate be sacrificed in favor of low price. Problems of weeds and poor germination will far outstrip any small savings in seed cost.

Other factors to be considered include the amount of sunlight the area receives and the soil condition. Some seed mixtures are specially formulated for shade, poor soils, wet soils, and so on. Choosing a suitable seed mix can avoid problems later.

Seeding Techniques. When seeding any area greater than 30 by 30 feet (9 × 9 meters), a mechanical seeder (Figure 12-2) should be used to assure even coverage. However, for small areas hand seeding is satisfactory. No matter which method is chosen, the seed label should be carefully followed as to the quantity of seed planted in an area. Because grass seed is lightweight and easily blown by the wind, seed should be sprinkled on a windless day or during early morning or late afternoon, when the wind is likely to be at a minimum.

When seeding (or fertilizing with granular fertilizer), a pattern such as the one shown in Figure 12-3 will assure even coverage. To avoid missing strips between passes, the wheel of the spreader should be run just inside the previous wheel track. A steady walking pace is necessary to prevent double dosing and insufficient coverage problems.

Mulching. Mulches are applied to newly seeded areas to hold in moisture and to control erosion and washing away of the seed during rainfall or irrigation. Unlike mulches used for weed control, mulch over seed must be applied in a thin layer so that the germinated seedlings can grow up through it. Coarse sawdust, shredded bark, peat moss, wood shavings, and straw are several mulches for this purpose; they should be applied so that the ground remains visible through the mulch (Figure 12-4).

Postseeding Care

Irrigation. Frequent irrigations to keep the soil surface moist are the key to obtaining good germination. Drying out even once during the germination and initial establishment stages can kill the majority of the seedlings and necessitate reseeding. Watering may be needed as

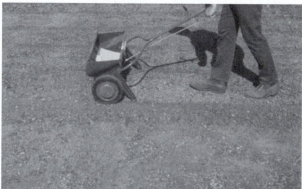

Figure 12-2 Two types of mechanical seeders for distributing seed or fertilizer: hand spreader (top), and push type (bottom). *(Photos by [top] Rick Smith and [bottom] Vocational Education Productions, California Polytechnic State University, San Luis Obispo, CA)*

often as three or four times a day to keep the soil surface moist. As the roots penetrate down into the ground, less frequent but longer waterings should be made; watering every two to three days will be sufficient.

Fertilizing. A light topdressing of about ½ pound of nitrogen per 1,000 square feet (0.25 kg/100 square meters) of lawn is recommended as the first fertilizing of turfgrass seedlings. The fertilizer should be applied only when the majority of the seedlings are 1½ to 2 inches (4–5 centimeters) high and should be immediately watered in.

Mowing. The first mowing should take place as soon as the seedlings exceed the normal cutting height. Afterward, mowing should be done regularly.

Figure 12-3 A push-type mechanical seeder being run in a pattern to assure even coverage.

overlap spreader wheel with previous pass

Figure 12-4 Straw mulch applied to a newly seeded lawn. *(Photo by George Taloumis)*

Sodding

Sod consists of long strips of turf cut from the ground with the roots intact (Figure 12-5). It is heavy because of the weight of attached soil and is delivered on pallets to the site where it will be used. A new method of sod production is now being used by some growers in which the sod is grown in an organic medium without soil. This produces a lightweight, good-quality sod that is easier to lay at a cost of only a few cents more per square foot. Sodding can be done any time during the growing season, and it creates a high-quality lawn immediately.

 A sod producer grows acres of high-quality sod of various grass varieties from seed. Production involves watering, mowing, disease and pest control, and harvesting. The sod is then marketed to landscape installers.

Buying Sod

Sod is expensive and should be selected carefully. An ideal sod is thick and uniform, free of weeds and diseased areas, and has only a small layer of thatch. You should always ask to see a sample of the sod you are purchasing. Specific arrangements for its delivery can then be made to allow adequate time for preparing the site in advance.

Site Preparation

The same methods of site preparation that precede seeding should also precede sodding. Negligence in site preparation may initially be covered up by a healthy sod, but the sod will not remain healthy if the soil into which it sends new roots has not been thoroughly prepared.

Laying Sod

Sod must be laid as soon as possible after it is harvested. If it is stacked or left in rolls for more than a day or two, it will develop heat and disease problems and deteriorate quickly. If the sod must be left stacked, the

Figure 12-5 Laying sod. *(Photo courtesy of Vocational Educational Productions, California Polytechnic State University, San Luis Obispo, CA)*

outside edges should be lightly watered; the sod should not be heavily watered.

When laying sod, the strips should be staggered (Figure 12-5) so that the ends do not form a line that will be noticeable later. Avoid stretching the sod, because it will shrink later and leave bare strips between the pieces. Cut the sod with a sharp knife to fit tightly around sprinklers and other obstructions and tightly against lawn edges. Gaps left between the sod strips will fill with weeds and become a maintenance problem.

After the sod is laid, the area should be rolled to ensure contact between the sod and the prepared soil beneath. Roll with a water-filled lawn roller in the opposite direction the sod was laid.

Care after Transplanting

Immediately after being laid, sod should be watered thoroughly enough to wet both the sod itself and the first several inches (7–10 centimeters) of soil beneath. During the first two weeks, it should be lightly irrigated once each day during the warmest period to minimize transpiration during reestablishment of the roots. Avoid walking, playing, or other heavy traffic for a couple of weeks.

After the sod has rooted it will benefit from occasional light application of fertilizer. Mowing should begin when the grass has grown sufficiently to require it. The mowing height should be set relatively high until the sod is well established.

Lawn Establishment by Stolonizing and Plugging

Stolonizing and plugging are used to establish new lawns of vigorous grasses such as Bermuda and creeping bentgrass. Stolonizing consists of spreading small stem sections or sprigs of the turf variety over a prepared area and then rolling or lightly covering with soil. The stolons root quickly and form a turf in several months.

In plugging, small 2- to 4-inch (5- to 10-centimeter)-diameter pieces of sod are transplanted into a prepared area 6 to 14 inches (15–35 centimeters) apart (Figure 12-6). The closer the plugs are spaced, the sooner the area will fill in. Warm-season grasses such as Bermudagrass, zoysiagrass, St. Augustinegrass, and buffalograss are the species most commonly established by plugging.

LAWN RENOVATION

Renovation is used when a lawn is in poor condition but has not deteriorated to the point where it should be completely reestablished. The poor condition may be

Figure 12-6 A plug of turfgrass ready to be planted in a lawn. *(USDA photo)*

due to insufficient soil fertility, weed invasion, thatch accumulation, soil compaction, or any combination of these and other causes. Although renovation can be started at any time during the growing season, late summer is best for cool-season grasses and late spring for warm-season grasses.

Determining the cause of the poor growth is the first step in renovation. If there is excessive weed growth, footpaths, or pale color, neglect may be suspected. However, if only isolated areas are dying, other causes should be suspected. Excess shade, poor drainage, diseases, and insects may be problems. Sending a soil sample for analysis will indicate nutritional or pH problems.

The next step should be weed control by hand or with the chemicals recommended in Table 12-2. Excess thatch should be removed, if necessary, using a machine rented for this purpose (see Dethatching).

Aeration of the soil can be done with a rented machine or a hand aerator. Finally, bare spots should be reseeded, followed by fertilizing as indicated by the soil test and a deep, thorough irrigation.

Given minimum care afterward, the turf should remain in good condition. If additional problems develop, or if renovation does not rejuvenate the lawn, consult a specialist.

TURF GRASS MAINTENANCE

Mowing

Mowing is the most frequent maintenance chore in caring for a lawn, and it must be done correctly for the turf to look its best.

Greenskeeping for golf courses combines an interest in sports with an interest in horticulture. There are currently many job opportunities in this field for trained personnel. Horticulture education with coursework in turfgrass is required to be a greenskeeper.

Table 12-3 Preferred Cutting Heights of 19 Turfgrasses as Determined by the Resulting Turfgrass Quality and Vigor[*]

Relative cutting height	Cutting height (in.)	Turfgrass species
Very close	0.2–0.5	Creeping bentgrass
		Velvet bentgrass
Close	0.5–1	Colonial bentgrass
		Annual bentgrass
		Bermudagrass
		Zoysia grass
Medium	1–2	Buffalograss
		Red fescue
		Centipedegrass
		Carpetgrass
		Kentucky bluegrass
		Perennial ryegrass
		Meadow fescue
High	1.5–3	Bahiagrass
		Tall fescue
		St. Augustinegrass
		Fairway wheatgrass
Very high	3–4	Canada bluegrass
		Smooth brome

[*]Extracted from James B. Beard, *Turfgrass: Science and Culture*, Prentice-Hall, Englewood Cliffs, NJ, 1973.

Mowing Frequency

Mowing is essentially a pruning operation done on grass. It removes a portion of the photosynthesizing part of the plant, which in turn lessens the supply of carbohydrates going to the roots for growth and respiration. Thus, roots are as affected by mowing as the blades of the grass.

Mowing should be as frequent as necessary to remove only 30 percent of the blade length at any one time. Cutting that removes more than 30 percent of the blade is detrimental to the roots because of the abrupt drop in carbohydrate flow.

Mowing Height

The height at which grass should be maintained varies with the species and may range from ½ to 4 inches (1.3–10 centimeters). Table 12-3 lists suggested mowing heights for grasses grown throughout the United States.

Mowing at a height less than suitable for a particular grass is called *scalping* and leaves the turf an unattractive brown stubble. Occasional scalping will not permanently harm grass. But if repeated on a susceptible area (such as a berm top), it will eventually kill the turf.

Returning Clippings

Many people use grass catchers to remove clippings during mowing. Although this does give a neater surface, it is not essential to turf maintenance. Returned clippings supply additional nutrients when they decay and reduce fertilization requirements. Clippings need only be removed if they are overly heavy and long and would inhibit passage of light to the grass underneath.

Mower Operation

Mowing only when grass is dry is suggested for two reasons: first, because the mower is less likely to clog with soggy clippings and the clippings will distribute evenly rather than in clumps; second, because grass blades are more susceptible to disease invasion through the cut surface when wet.

The importance of sharpening mower blades so they cut grass cleanly is often overlooked. A clean cut minimizes the damaged area and reduces tip browning of the blades. It also decreases moisture loss and eliminates bruised blades, which are a site of disease entry.

Power and Tractor Mower Safety

Basic familiarity with the controls of the power or tractor mower and its capabilities is essential to safety. Therefore, reading the operator's manual is the first step in safety. Wear proper protective boots, a long-sleeved shirt and pants, and eye protection. Check to see that the area to be mowed is free of sticks, stones, holes, stumps, irrigation valves, and debris. Mow only in daylight.

Never use fingers to remove grass or debris from lawn mower blades or the discharge chute. Stop the engine (disconnect the spark plug wire for safety) and use a stick to remove debris.

When using a grass catcher, stop the engine and wait until the cutting blade comes to a complete stop before removing or attaching the bag.

Self-Propelled and Push Power Mowers

When working with a self-propelled or push power mower, avoid situations that require you to pull the lawn mower backward, as this is often the cause of foot injury. Wear rubber-soled, closed-toe shoes (preferably steel-toed shoes).

Never make a wheel-height adjustment on the mower while the engine is running.

If your mower is self-propelled, be sure the drive clutch is disengaged when starting the mower. Never stand in front of a self-propelled mower.

Tractor Mowers

- When working with a tractor mower: Be sure the tractor has a rollover protective structure (ROPS) and a seat belt. If not, have it fitted with one.

- Check that the power take-off (PTO), mower input driveline, drive belts, chains, and gears are all properly guarded. Repair or replace any of these items, if necessary.

- Adjust the seat, fasten the seat belt, set the parking brake, place the shift lever in neutral or park, and disengage the power take-off (PTO) before starting the engine.

- Check the discharge chute where clippings are expelled to see that it is present and pointed downward. Fill the fuel tank outdoors with the engine stopped and cool. Do not smoke while filling the tank.

- Be sure your hands and shoes are clean and dry before mounting the tractor to prevent slipping. Use the steps and handbars.

- Raise the mower blades to the high position and use low rpm before engaging power to the blades.

- Use a ground speed appropriate to the length and density of material being cut—normally between 2–5 mph (3–8 km/hr).

- Mow up and down slopes only with rear-mount, pull-type (Figure 12–7), and wing-type mowers. Mow across slopes with side-mount, offset, and sicklebar mowers.

- When stopping, disengage the PTO, place the shift in neutral or park, set the parking brake, turn the engine off, and wait for all movement to stop before getting off.

Irrigation

As for irrigation of any plant, watering of grass should be done according to need rather than on a fixed schedule that does not take weather conditions into account.

Figure 12-7 A pull type mower for large areas such as golf courses. *(Photo courtesy of Teagle Machinery Ltd. Blackwater, Truro, Cornwall, England.)*

The aim is to apply water just before wilting, wet the top 6 to 8 inches (15–20 centimeters) of soil where the roots grow, and wait until wilting approaches before watering again.

Determining exactly when wilting is about to occur is difficult. One technique called *footprinting* is advocated by Dr. James Beard of Michigan State University. Walk across the grass, then observe how quickly the turf stands upright again. Turf with adequate water will recover rapidly, whereas turf approaching wilting will recover very slowly.

Where rainfall is insufficient, a permanent sprinkler system can be installed before surface preparation and seeding. Pop-up sprinkler heads, which are below cutting level normally but are elevated above the turf by water pressure during operation, are the most popular. They can be used in conjunction with plastic pipe to create a durable inexpensive system easily installed by a homeowner. To minimize water loss by evaporation, sprinklers should be used late in the afternoon or early in the morning. The former is preferable when disease problems are common, because the foliage will not remain wet and susceptible to infection through the night.

Subsurface irrigation is a relatively new technique in use primarily on golf courses. Underground pipes spaced closely throughout the turf area ooze water at a slow rate. The water is moved by capillary action to the grass roots. The main advantages of a subirrigation system are the elimination of evaporation and runoff. Disadvantages include the possibility of clogging and buildup of salts. In addition, not all soils have sufficient capillary action to utilize the system.

Fertilization

Choosing the Fertilizer

Because the goal in fertilizing a lawn is the production of vegetative growth, the fertilizer chosen should be high in nitrogen. Fertilizers with two to four times as much nitrogen as either phosphorus or potassium are suggested. In addition, it is best to buy fertilizers with part of the nitrogen in a form that is immediately available and another part in a slow-release form.

 Lawn maintenance services maintain turf areas in an attractive healthy condition for homeowners or commercial businesses. A lawn maintenance specialist makes weekly visits to each site for which he or she is responsible to mow and monitor overall plant health, apply fertilizers and pesticides, dethatch, aerate, and water as necessary.

Combinations of fertilizer plus herbicide, fungicide, or insecticide are widely sold for use on lawns. Their use eliminates separate applications of each chemical, but extra care must be taken to apply them at the proper rate.

Timing and Rate of Fertilizer Applications

The frequency at which fertilizers should be applied to a lawn depends on whether the goal is to establish a young lawn or to maintain an older one, on the soil type, and on the weather conditions. The following recommendations are based on a loam soil with irrigation or precipitation every one to two weeks. In lighter soils or with more frequent irrigation, fertilizer should be applied more often.

Young Lawns. Establishing lawns requires frequent fertilizer for rapid growth. As a general guide, a ½ to 1 pound per 1,000 square feet (0.25–0.5 kg/100 square meters) application of immediately available nitrogen should be applied when grass reaches 2 inches (5 centimeters). This should be followed by applications at the same rate every three to four weeks during the first growing season. In cold-winter climates applications should be stopped about one month before the onset of winter dormancy.

Established Lawns. Established lawns require less frequent fertilization to maintain an attractive appearance.

Two fertilizer applications per year, one in early spring and one in fall, are normally sufficient. Two pounds (1 kilogram/100 square meters) of nitrogen per 1,000 square feet is recommended at each application.

In areas with mild winter temperatures, grass frequently does not have a dormant period and will grow throughout the year. Accordingly, an additional fertilizer application will be needed in midwinter to supply nutrients for healthy growth.

Fertilizer Application Methods

Fertilizer is normally applied to lawns in granular form using a broadcast and water-in method. A fertilizer spreader is best to assure even coverage over large lawns. Hand broadcasting or applying liquid fertilizer with a hose-end applicator is practical for small areas. The latter method is used by some commercial lawn care companies and is valuable for very sandy soils with a poor nutrient-holding capacity. A diluted solution can be mixed and applied in place of plain water at recommended intervals.

Dethatching

Thatch is a layer of stems and decaying organic matter that accumulates between the turf surface and the soil (Figure 12-8). The accumulation occurs when the rate at which the turf sloughs off dead parts is greater than the rate at which they decompose.

Figure 12-8 A thatch layer in turfgrass. *(Photo courtesy of The Lawn Institute)*

Figure 12-9 Dethatching with a vertical mowing machine. *(Photo courtesy of the Cooperative Extension Service of the University of California)*

Although a thin layer of thatch is not harmful, any layer greater than ½ inch (1–2 centimeters) can cause problems. The heavily thatched area may dry out fast, diseases and insects may invade, and there may be decreased tolerance to heat, cold, and drought. Vigorous grass varieties are the most susceptible to thatch accumulation because of their rapid growth rate.

Excess thatch is detected by examining a cross section of turf. It can then be removed mechanically by a *vertical mower* (Figure 12-9). This machine cuts and pulls the turf up in narrow strips. The material is then raked off and the lawn watered and fertilized. Timing of the mowing should allow at least 30 days of good growing conditions afterward for recovery of the turf. Early spring is suitable for most grasses.

For small lawn areas, renting a vertical mowing machine may be impractical. Hard raking with a stiff metal rake will loosen and remove the thatch.

Preventing future buildup of thatch involves altering the care of the turf to balance the accumulation and decomposition rates. If irrigation or fertilization has been excessive, it should be reduced to slow the growth of the turf. Liming may be needed, because the pH of thatch is generally too low for the optimum decay rate. Finally, topdressing the turf with soil will hold in the moisture necessary for decomposition and provide a source of microorganisms for the breakdown process.

Aeration

Aeration is needed when a lawn has been compacted by foot or vehicular traffic. Compaction is particularly prevalent in clay soils. Under compacted conditions, the movement of oxygen and other gases into and out of the soil is restricted. Water will soak in very slowly or run off. Turf roots will eventually begin to die.

Aeration removes small cores or slices of grass and soil from the turf area, penetrating to a depth of up to 4 inches (10 centimeters) from the soil surface. The remaining soil then can expand into the open area and relieve the compaction problem. Engine-driven aerators can be rented if a large area is to be treated. For spot treating or small lawns, a home aeration tool, which is used like a shovel, is more practical.

Weed Control

Weeds in a lawn are undesirable because they detract from its even appearance. Their varying textures and colors make them stand out, and their tendency to grow faster than the surrounding turf increases the mowing frequency. Seasonal weed species (crabgrass, for example) crowd out desirable turf and then die, leaving bare spots in the lawn. When weeds flower and set seed, they are even more conspicuous in a lawn and cause additional problems. Seed heads of common Bermudagrass, for example, generally will not be cut by reel-type mowers but must be removed by hand. Others have seeds that are spiny (bur clover and puncture vine) and will injure bare feet.

The objective of lawn weed control should be to keep weed populations within a tolerable range, not necessarily to have a flawless lawn. A few weeds seldom detract from the lawn enough to require control measures. In fact, many homeowners prefer a lawn containing a combination of plants including clover and ground ivy.

A healthy lawn will generally not have severe weed problems. However, because of disease, insects, or poor cultural practices, weeds sometimes become established in large numbers. Thus, the first line of defense against weeds is to keep the grass healthy. Correct mowing height and frequency, fertilization, and insect and disease control will prevent weed seedlings from becoming established and even enable the grass to outcompete established weeds. For the few remaining, hand pulling is a sound control method. Pulling after a rain or irrigation will help remove plants with taproots.

Occasionally, weeds such as crabgrass become a severe problem in even the healthiest of lawns. Pre-emergence or postemergence herbicides can then be utilized. Crabgrass is most often controlled with a granular herbicide (often combined with a fertilizer) applied preventatively in spring before the crabgrass germinates. Weeds escaping preemergence treatment can be hand pulled later or sprayed with a selective postemergence herbicide that kills only the weeds without harming the grass. Several applications may be necessary for complete control. Table 12-2 lists herbicides used for weed control in lawns.

LAWN SUBSTITUTES

Dichondra

A number of plant species will tolerate some foot traffic and can be planted as grass substitutes. Dichondra is the best known of these, although its use is limited to the warmest climates of the country.

Dichondra (Figure 12-10) is an extremely attractive grass substitute with a flat, matlike habit of growth. It is restricted to areas where the winter temperature does not drop below 25°F (−4°C) and is tolerant of the very warm summer temperatures such as those found in the Southwest.

Figure 12-10 Dichondra, a popular lawn plant in California and Florida. *(Photo by Rick Smith)*

A dichondra lawn must be established in spring or summer for soil temperatures to be sufficient for seed germination and rapid growth. Soil preparation, sowing, and postseeding care are essentially the same as for grass. Mulches of coarse sawdust, shredded peat moss, or other relatively fine material are preferred over straw because the creeping growth habit of the dichondra will not readily obscure a straw mulch.

Plug planting can be used to establish a dichondra lawn and is useful when only a small area is to be planted. In large areas the expense is prohibitive. As with plug planting of grasses, the area should be thoroughly prepared beforehand to facilitate rapid spreading. Flats of dichondra can be cut into plugs 2 inches (5 centimeters) square or larger and transplanted 6 to 12 inches (15–30 centimeters) apart.

The dichondra lawn requires regular watering and frequent fertilization to remain attractive. From this standpoint, it is not a low-maintenance plant. Fertilizing every two to three weeks is best, using a high-nitrogen fertilizer at 1 pound per 1,000 square feet (0.5 kg/100 square meters) of area.

Mowing frequency depends on growing conditions. Shade and abundant moisture increase dichondra lushness and height, and mowing may be required once per month. In full sun, growth will stay low, and mowing will never be required.

Primary weeds that invade dichondra are oxalis, chickweed, and miscellaneous lawn grasses such as Bermudagrass and crabgrass.

Other Lawn Substitutes

Dichondra is the only widely used lawn substitute for large areas. However, a number of uncommon plants will tolerate limited walking such as a front yard might receive. Most will be hard to find, and all are established by plugging. It is advisable to try a small area before using the species to cover an extensive lawn area.

Table 12-4 lists a number of lawn-substitute plants, along with their cold hardiness and ideal growing conditions. Many of the plants mentioned here are suited to growing in shade, where grass is particularly difficult to maintain.

Table 12-4 Lawn-Substitute Plants

Common name	Botanical name	Cold hardiness	Remarks
Beach strawberry	*Fragaria chiloensis*	−20 to −10°F (−29 to −23°C)	Native plant good for sandy soils and spreading by runners; white flowers and small edible fruits in spring
Chamomile	*Chamaemelum nobile*	−20 to −10°F (−29 to −23°C)	Often sold under the name *Anthemis nobilis* and used commonly as a lawn plant in Europe; withstands drought and emits a pleasant odor when stepped on; fernlike; bright green leaves and good traffic tolerance; daisylike white flowers
Colchis ivy	*Hedera colchica*	−10 to −5°F (−23 to −21°C)	Relatively rare ivy with heart-shaped leaves which tightly overlap over the ground; shade-tolerant
Corsican sandwort	*Arenaria balerica*	−5 to 5°F (−21 to −15°C)	Needlelike leaves; grows up to 3″ (7 cm) high; best in shade with moist soil; produces white flowers in May
Creeping speedwell	*Veronica repens*	−10 to −5°F (−23 to −21°C)	Grows to 4″ (10 cm) tall, with a mosslike appearance and shiny foliage; bluish flowers in spring
Creeping thyme	*Thymus serphyllum*	−35 to −20°F (−37 to −29°C)	Very flat growing (less than 1″; 3 cm); evergreen; will withstand poor soil, shade, drought; small purple flowers in summer; many species of thyme are sold under this name, including *Thymus lanuginosus* and *Thymus lanicaulis,* and since these grow taller, they are less tolerant of foot traffic
Dwarf lily turf	*Ophiopogon japonicus*	5–10°F (−15 to −12°C)	Dark evergreen foliage is grasslike, to 6″ (15 cm) tall; covers an area rapidly, grows in shade or sun, and produces lilac to white flowers in summer; good in sandy soil
Garden lippia	*Phyla nodiflora*	0°F (−18°C)	Sometimes sold under the name *Lippia repens* or *Lippia canescens*; has a matlike growth habit with gray-green leaves and is good in heat and sun; mow occasionally to maintain an even surface; very drought- and traffic-tolerant
Mazus	*Mazus reptans*	−35 to −20°F (−37 to −29°C)	Deciduous plant, although leaves hang on long in fall; plant grows 1″ (3 cm) tall with 1″ (3 cm) leaves; purple flowers in May; prefers moist soil and will grow in partial shade
Mock strawberry	*Duchesnea indica*	−10 to −5°F (−23 to −21°C)	2″ (5 cm) high with yellow flowers in spring; strawberrylike leaves
Moss sandwort	*Arenaria verna*	−35 to −20°F (−37 to −29°C)	Bright green needlelike foliage; mat-forming, and develops lumps unless thinned occasionally
Mountain sandwort	*Arenaria montana*	−20 to −10°F (−29 to −23°C)	Grows 1″ (3 cm) high with grasslike leaves; tolerant of poor soil and flowers profusely in spring with white blooms
Pearlwort	*Sagina subulata*	−20 to −10°F (−29 to −23°C)	Mat-forming with needlelike leaves, similar to *Arenaria;* grows up to 4″ (10 cm) tall and is good in shady areas
Pennyroyal	*Mentha pulegium*	−5°F (−21°C)	Very flat-growing, and has a strong mint scent when stepped on
Rusty cinquefoil	*Potentilla cinerea*	−35 to −20°F (−37 to −29°C)	Strawberrylike leaves and yellow flowers on a plant 2 to 4″ (5–10 cm) high
Self-heal	*Prunella vulgaris*	−35 to −20°F (−37 to −29°C)	Native to temperate Northern Hemisphere; leaves are semievergreen, lying flat on the ground; purple flower spikes in summer; best in damp soil and spreads rapidly

SELECTED REFERENCES FOR ADDITIONAL INFORMATION

Ali, A. D., and C. L. Elmore. *Turfgrass Pests*. University of California Cooperative Extension Publication 4053. 1989.

Brandenburg, Rick L. *Handbook of Turfgrass Insect Pests: A Guide to Successful Insect Pest Management*. Lanham, MD: Entomological Society of America, 1995.

Brede, Doug. *Turfgrass Maintenance Reduction Handbook: Sports, Lawns and Golf*. Chelsea, MI: Sleeping Bear Press, 2000.

Crotta, Carol A. *Lawns: Quick and Easy Grasses and Groundcovers*. Minnetonka, MN: Creative Publishing, 2000.

Daniels, Steve. *The Wild Lawn Handbook*. Indianapolis, IN: Hungry Minds, 1997.

Daniels, Stevie. *Easy Lawns: Low-Maintenance Native Grasses for Gardeners Everywhere*. New York: Brooklyn Botanic Garden, 1999.

Fermanian, Thomas W., and Malcolm C. Shurtleff. *Controlling Turfgrass Pests*. Paramus, NJ: Prentice Hall, 1996.

Meredith Books. *All About Lawns*. Des Moines, IA: Meredith, 1998.

Organic Gardening Editors. *Lawns*. Emmaus, PA: Rodale Press, 2001.

Rice, R. P. *Nursery and Landscape Weed Control Manual*. 3rd ed. Fresno, CA: Thomson Publications, 2001.

Schultz, Warren. *A Man's Turf: The Perfect Lawn*. New York: Crown Publishing Group, 2000.

Sprague, Howard B. *Turf Management Handbook*. Danville, IL: Interstate Publishers, 1996.

Thomas, Taylor. *Weekend Guide to Lawns*. New York: Houghton Mifflin, 1997.

Turgeon, A. J. ed. *Turfgrass Management*. 3rd ed. Madison, WI: American Society of Agronomy, 1994.

Turgeon, Alfred J. *Turf Weeds and Their Control*. Madison, WI: American Society of Agronomy, 1994.

University of California Cooperative Extension. *Guide to Turfgrass Pest Control*. Leaflet 2209. 1988.

Vengris, Jonas, and W. Torello. *Lawns*. 3rd ed. Fresno, CA: Thomson Publications, 1982.

Waddington, D. V., R. N. Carrow, and R. C. Shearman. *Turfgrass*. No. 32 in the series *Agronomy*. Madison, WI: American Society of Agronomy, 1992.

White, Bud. *Turf Manager's Handbook for Course Construction, Renovation and Grow-In*. Chelsea, MI: Sleeping Bear Press, 2000.

Winward, Lane L. *The Healthy Lawn*. Guilford, CT: Globe Pequot, 2001.

Zeman, Anne M. *Installing and Renewing Lawns*. Chattanooga, TN: N K Lawn & Garden Co., 1993.

World Wide Web Addresses

http://www.pueblo.gsa.gov/press/nfcpubs/lawn.txt This is a Department of Agriculture publication entitled *Healthy Lawn, Healthy Environment*.

http://www.gcsaa.org This is from the Golf Course Superintendents Association of America.

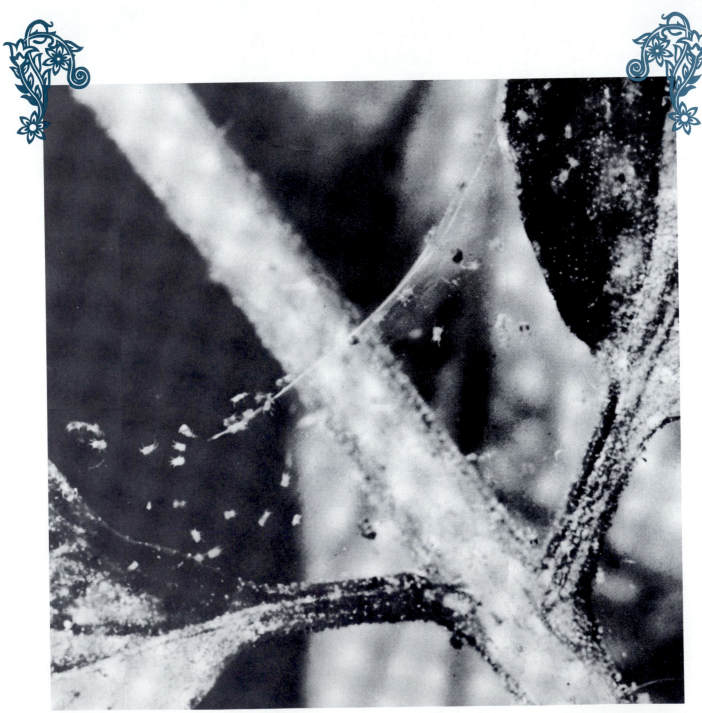

(Photo courtesy of USDA)

CHAPTER 13

DIAGNOSING AND TREATING OUTDOOR PLANT DISORDERS

The diagnosing of some plant disorders is easy; for example, wilting usually indicates lack of water, and caterpillars found chewing through a leaf are proof of an insect infestation. But as often as not, a problem is hard to diagnose, and a systematic approach to solving the problem should be used.

IDENTIFYING THE AFFECTED PLANT

The first step in diagnosing a plant problem is identifying the plant that is affected. A common name is sufficient for fruits, nuts, or vegetables, because those names are standard throughout the country. However, with ornamentals, the botanical name is often essential.

Identifying the plant will narrow the list of possible causes considerably, because species vary in their susceptibility to different maladies. A species often will be prey to the same problem repeatedly, and a knowledgeable nurseryman can often diagnose a problem given only the plant name and the symptoms.

DETERMINING THE CAUSE OF THE PROBLEM

Plant problems are almost always attributable to either a parasitic organism or an environmental condition. Parasitic organisms include insects, rodents, and microorganisms such as bacteria and fungi. Environmental conditions include heat, drought, cold, and the like. Each of these causes of plant injury will be discussed, along with its typical symptoms, in the following sections.

 A person interested in problem solving might be interested in becoming a diagnostician of plant problems or a pest control advisor (PCA). Universities and large and small businesses offer diagnostic services for different crops such as flowers, vegetables, or turf when the cause of a decline in plant health is not easily determinable. Lab work with diseased plants will be involved, as well as occasional on-site visits to the plants affected.

Insects Damaging to Plants

Insects pass through several stages in their maturation process, and in one or more than one of these stages, they may damage plants. Butterflies are the best-known example of this phenomenon called *metamorphosis*. They begin life as eggs laid on a host plant. The eggs hatch into a larval or caterpillar form, and during this stage the insect feeds on and causes damage to plants. After reaching full size, the caterpillar makes a cocoon (pupates), later emerging as the adult butterfly, a nectar-feeding insect form. However, although the butterfly itself is harmless, it lays the eggs for the next generation of caterpillars.

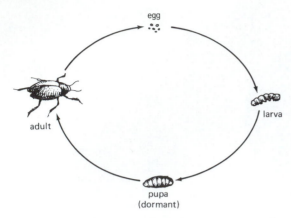

Figure 13-1 The typical maturation sequence of a metamorphosing insect.

This egg-larva-pupa-adult sequence (Figure 13-1) is common to many insects, with the larva stage often referred to as a "worm," "caterpillar," or "grub." The adult will vary, with adult beetles, flies, or moths emerging as commonly as butterflies.

Damage to plants by insects and related pests is almost always from feeding of two types: chewing or sucking. Chewing pests include many larvae (such as caterpillars and grubs) as well as grasshoppers, boring insects, snails, and beetles. These pests eat plant parts rapidly, becoming larger and more damaging as they mature. Sucking pests, on the other hand, do not chew holes. They feed on the sap either by inserting their mouthparts into the phloem or by rasping a small area and feeding on sap oozing from the wound. Sucking pests include insects such as aphids and scales, and mites. Their feeding damage is harder to diagnose.

Damage by chewing pests is relatively easy to identify. Irregular holes (Figure 13-2) appear on leaves, or a stem may be severed completely. The insects may still be present and eating or may have left droppings on the plant.

If chewing pests are suspected but not present, an effort should be made to determine what insect is causing the damage, for this affects the control to be applied. The plant should first be inspected under the leaves and near buds and feeding sites. If no pests are present, they may be in the soil at the base of the plant. Many pests hide there during the day, coming out to feed at night. For the same reason, checking the plants after dark will often reveal the insect.

Sucking pests present a different detection problem. They are usually small (some microscopic), and their damage varies. Typical symptoms include puckering and bleaching of the leaves (Figure 13-3). A hand lens is helpful in detecting these pests.

Recognizing damage by pests feeding on roots is not as easy as recognizing aboveground infestations. Grubs and other larvae are most often the culprits, chewing the roots of a plant until the root system is insufficient to sus-

Figure 13-2 Damage due to a chewing insect, in this case a caterpillar. Similar damage is caused by slugs and snails. *(USDA photo)*

Figure 13-3 The leaf on the bottom is puckered from a sucking insect feeding on it. The top leaf is normal. *(USDA photo)*

tain the foliage. The resulting symptoms are the same as for any condition harming the roots: stunting, yellowing, wilting, and eventual death of the top portions. When digging up the plant, signs of insect damage (and often the insects themselves) will be apparent.

The following sections include most of the chewing and sucking pests found in home gardens.

 Entomologists are scientists who specialize in insects. A horticultural entomologist would diagnose, recommend treatments, and perhaps research pests of horticultural crops.

Chewing Pests

Caterpillars. Caterpillars are chewing insects that consume amounts of leaves equaling several times their body weight each day, and sometimes even change color depending on what they have eaten!

Grubs and Borers. Grubs and borers are usually the larvae of beetles. They live underground eating plant roots (grubs) or burrow into stems and fruits (borers). Some, such as the Japanese beetles, are destructive in both larval and adult stages.

Leaf Miners. Leaf miners are miniscule wormlike insects that tunnel between the upper and lower leaf surfaces. They are most commonly the larvae of flies. Their feeding develops scribblelike white patterns on affected leaves.

Grasshoppers. Grasshoppers are chewing, defoliating insects throughout their life cycle. They have no distinctly different larval, pupal, and adult stages; young grasshoppers resemble the adults, but lack wings.

Beetles. Beetles damage large numbers of plants by chewing and are particularly troublesome in late summer.

Snails and Slugs. These pests feed at night or during wet weather, chewing holes in leaves, flowers, and fruits. Because they are not insects, they cannot be controlled with most insecticides.

Earwigs. These are nocturnal feeders. They are brown, up to an inch long, and are easily identified by the large pinchers on their abdomen.

Sucking Pests

Aphids. Aphids are small, sucking insects generally found clustered on the growing tips of plants. Their feeding causes disfigurement of the leaves and buds (Figure 13-4) and produces honeydew. They may also transmit viruses.

Whiteflies. Whiteflies (Figure 13-5) are sucking insects related to aphids. All whiteflies inhabit the undersides of leaves. The adults are white and covered by wax. They lay their eggs on the undersides of leaves, and the eggs hatch into crawlers that move around, find a feeding site, insert their mouthparts, lose their legs, and then resemble small, nearly clear scales. Eventually they molt and

Figure 13-4 Typical symptoms of aphid damage to a delphinium (note distorted new growth). *(USDA photo)*

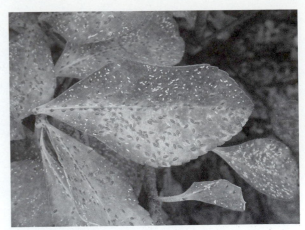

Figure 13-6 Scale insects *(Photo courtesy of Takumasa Kondo)*

(Figure 13-6) look much like raised spots, and mealybugs (see Figure 18-2) like dots of cotton. Their unique appearance stems from the waxy secretion that oozes from the back of the insect after it has settled and begun feeding. Once protected in this manner, the pests are unaffected by most insecticide sprays. Most also produce honeydew.

Mites. Spider mites, which are not insects but are more closely related to spiders, cluster on the underside of plant leaves sucking juices and often give a pinprick pattern to the top surfaces. In heavy infestations, some species spin fine webs. Because they are nearly microscopic, a magnifying glass and strong light are helpful in diagnosing a mite problem. Mites on fuchsias, cyclamen, African violets and other plants can cause deformation as in Figure 13-7.

the adult winged whiteflies emerge to complete the life cycle. Whiteflies are damaging because they reproduce rapidly, forming large populations that suck large quantities of plant sap. Additionally, they transmit many viral diseases and produce sticky honeydew that attracts ants and upon which a black fungus called *sooty mold* grows. Sooty mold is unattractive and prevents light from entering leaves, thus decreasing photosynthesis.

Scales and Mealybugs. These sucking insects spend their lives in colonies attached to leaves or stems. Scales

Figure 13-5 A closeup of whiteflies, which are often a pest on fuchsias.

Figure 13-7 Spider mites and their damage. *(USDA photo)*

Figure 13-8 Cuban laurel thrips on an ornamental fig (*Ficus* sp). *(Photo courtesy of the Cooperative Extension Service of the University of California)*

Figure 13-9 Root-knot nematode damage to a tomato. *(USDA photo)*

Thrips. Thrips are sucking pests only slightly larger than mites. They attack leaves or flowers and cause damage in the form of streaks, small bleached spots, or distorted leaves (Figure 13-8). They also transmit viruses including *Impatiens* necrotic sport virus.

Gall-Forming Insects

Both mites and aphids, as well as other insects and diseases, can cause the abnormal swellings known as "galls." Although unsightly, most are not seriously deleterious to plant health.

Nematodes (Eelworms)

Nematodes are microscopic or nearly microscopic worms. Most are soil dwelling, burrowing into roots and decreasing their growth. This results in yellowing and general lack of vigor. Root nematodes can sometimes be seen with a magnifying glass, but one of the most common types, the root-knot nematode, is identified by the characteristic swellings formed on roots by its feeding (Figure 13-9).

Damage from foliar nematodes is less common. But some species emerge from the ground and crawl up to the leaves on a film of water such as from dew or rain. They then enter the leaves through the stomata and begin feeding. Typical symptoms include death of irregular sections of leaves and spotting.

Microorganisms Damaging to Plants

Several types of microorganisms damage plants, causing diseases that kill the foliage or rot the roots. These microorganisms can be divided into four types—fungi, bacteria, viruses, and viruslike organisms. Each is discussed in the following sections.

 A plant inspector is responsible for determining that plants are disease and insect free before they are permitted to enter certain countries or states. Plant inspectors also test shipments of horticultural crops arriving from outside the United States and Canada for pesticide residues.

Fungi

Fungi are single- or multicelled organisms that lack chlorophyll. Therefore they rely on other sources for carbohydrates and parasitize green plants. A typical fungus has an extended threadlike body of mycellia that increases in size by cell division. At maturity, a reproductive system is formed that produces dustlike spores that spread the fungus.

Fungi are responsible for more plant diseases than any other type of microorganism, but fungus diseases are also the most easily treated. The diseases vary widely, but most can be classified into one of the types listed in Table 13-1.

Bacteria

Bacteria are single-celled, lack chlorophyll, and frequently grow in colonies. They are the cause of relatively few plant diseases, but diseases caused by them are difficult to treat. Leaf spots, rots, cankers, blights, and wilts may all be caused by bacteria. In addition, bacteria can cause galls. They are often spread by splashing water, contaminated garden tools, or by insect "vectors" such as sharpshooters. A vector is an organism that transmits disease.

Viruses

Viruses (Figure 13-14) are among the smallest microorganisms, and only a few cause plant diseases of importance to the home gardener (Table 13-2).

They may be spread among plants by insects. The insects feed on a virus-infected plant, ingest the virus, move to an uninfected plant, and spread the infection. Aphids, whiteflies, mites, leafhoppers, and nematodes can spread plant viruses. Some viruses are also spread by rubbing against infected plants and then touching healthy plants, or by pruning infected plants and then pruning noninfected plants with the same unsterilized shears. Vegetative methods of propagation such as grafting, budding, and cuttings also commonly spread viruses. Some viruses can even be transmitted by seed.

Table 13-1 Common Fungus Diseases of Cultivated Plants

Disease	Symptoms and parasitizing pattern
Mildew (Figure 13-10)	The fungus grows on the surfaces of plant leaves and stems. The threadlike body can be easily seen with the naked eye as a powdery white covering. Infected leaves eventually wither and drop, and the whole plant dies.
Wilt	Fungi invade and multiply in the phloem and xylem of the plants, clogging them and restricting translocation. The entire plant or only a section will wilt and not recover.
Rot	Roots, underground storage organs, and the lower portions of stems are attacked by these soil-inhabiting fungi. Cells in infected areas die and decay, disrupting absorption and translocation and causing the death of the plant.
Leaf spot (Figure 13-11)	Blotches or spots appear on the foliage and sometimes stems or flowers, spreading rapidly to cover the plant. The spots are usually very distinctive in appearance, having, for example, a dark center surrounded by rings of different-colored tissue. The spots gradually reduce the photosynthetic area of the leaves and cause the death of the plant.
Rust (Figure 13-12)	The fungi that cause rusts are distinct by virtue of requiring two different species of host plants to complete the life cycle. Often one host will be a cultivated plant and the other a native plant or weed. Common symptoms of rusts are reddish-brown spots on the foliage and stems. On some species, the fruiting body that bears the spores is a 1-inch ($$$2.5 centimeters) mass of orange gelatinous horns with a bizzare appearance.
Canker (Figure 13-13)	Relatively uncommon disease affecting woody plants. Fungi cause a sunken bark area that girdles the branch or trunk and eventually kills that portion of the plant above the infected area.
Smut	Uncommon disease most often encountered on corn. Invaded portions swell grotesquely, eventually bursting open and dispersing a multitude of dusty black spores.
Witches'-broom	Mainly a disease of woody plants and not frequently encountered. Fungi infect the entire plant, but symptoms appear on small branches as unrestrained twiggy growth resembling a bird nest. The disease is sometimes fatal.
Blight	Usually affects woody plants and causes quick wilting and death of young and growing tissues such as flowers and twigs. Can spread downward to kill more mature portions of the plant.

Figure 13-10 Mildew on zinnia leaves. *(Photo courtesy of Dr. R. E. Partyka)*

Figure 13-11 Leaf spot disease on strawberry leaves. *(USDA photo)*

Figure 13-12 Rust Disease of roses, shown here on the underside of the leaf. *(Photo courtesy of Baldo Villegas)*

Figure 13-14 Watermelon mosaic virus on summer squash. *(Photo courtesy of Dr. O. W. Barnett, Clemson University)*

Figure 13-13 A bark canker along the upper part of the stem. Normal, healthy bark is at the bottom. *(Photo courtesy of Dr. R E. Partyka, Columbus, OH)*

Table 13-2 Typical Virus- and Viruslike-Organism-Caused Plant Diseases

Disease	Causal organism	Symptoms
Stunt	Virus or mycoplasma	Slowed or stopped growth under favorable growing conditions and with no other apparent cause
Curly top	Virus	Causes twisting and distortion of foliage; symptoms are similar to weed-killer injury
Mosaic	Virus	Mottled light and dark areas on foliage or fruit
Yellows	Mycoplasma	Yellowing of foliage, stunting, and witches'-broom shoot growth under good growing conditions and due to no other apparent cause

Virus symptoms vary but include stunting, uneven coloration of leaves and flowers, dead spots, and yellowing. The virus may eventually be fatal to the plant, although death may be quite slow. To date, viral diseases are largely and for practical purposes untreatable, so that infected plants should be removed to prevent the spread of the virus, and insects that spread viruses should be controlled.

A plant pathologist specializes in plant diseases. He or she diagnoses, recommends pesticides, and/or could do research on diseases of one type of horticultural plants such as ornamentals, vegetables, or fruit trees.

Viruslike Organisms

Viruslike organisms are very similar to viruses. Of the many types that are known, only a few cause plant diseases. Symptoms are similar to those of virus disease and are difficult to diagnose. They are spread in the same manner as viruses but, unlike viruses, some can be treated using pesticides.

Figure 13-15 Gopher damage. *(Photo courtesy of Vocational Education Productions, California Polytechnic State University, San Luis Obispo, CA)*

Large Animal Pests Damaging to Plants

A number of rodents and similar animals cause damage in the garden. The more common ones and their symptoms are discussed in the following sections.

Rabbits

Rabbits are troublesome in many areas of the country. They nibble vegetable gardens and the tender spring growth of many ornamentals including peony and dahlia.

Mice

Mice are a problem chiefly in the winter when food is scarce, burrowing under mulch or snow layers to feed on the bark of fruit trees. The feeding can girdle the tree and cause its death.

Pocket Gophers

These burrowing rodents create a nuisance by digging holes and tunnels throughout the garden (Figure 13-15). They also feed on roots and can kill a large plant in a day or two. Pocket gopher damage can be easily diagnosed; a plant will wilt suddenly and, when lifted, will come out of the ground easily and have no roots.

Moles

Moles also burrow, but feed mainly on insects they find in the soil. The damage results from their burrows, which leave a raised pattern throughout an area with mounded exit holes. Moles are particularly damaging to lawns because they burrow through grass roots, causing the death of the turf above.

Deer

Deer are an increasing problem in rural and even semi-urban areas. They eat twigs from shrubs or trees and will also attack vegetable and flower plantings. They usually feed late at night.

Birds

Birds are a problem where fruit plantings are grown. They will eat cherries, raspberries, and other fruit crops.

Dogs

Dog urine is toxic to plant foliage, causing browning or yellowing. Both male and female dogs frequently damage the lower leaves of shrubs or leave circular spots on turf.

Environmental and Other Nonparasitic Causes of Plant Injury

Frost Damage

Plants differ in their susceptibility to frost according to age and species: young tissues are more tender than mature ones; woody plants are more frost-resistant than herbaceous ones. Accordingly, a light frost may strike sporadically, damaging one plant and not the next. A key symptom is sudden overnight wilting, particularly on the uppermost leaves and buds.

Winterkill

Winterkill is the partial or total death of normally hardy perennial plants due to cold temperatures. In minor cases a few branches will die back, and the plant will resprout from buds on the lower stems or crown. With complete winterkill, both tops and roots die. This occurs during very cold winters when temperatures fall below the critical temperature for root survival.

Winterkill manifested as "burning" of evergreens is a common occurrence in cold climates. Leaf tissues become brown, and because evergreens do not refoliate yearly, the appearance of the plants can be permanently marred. Burning is usually due to lack of soil moisture during the winter combined with continuing water need and drying of the needles by winter winds. Mulching and watering during the winter when the soil is not frozen are therefore recommended.

Drought

Wilting is the main symptom of drought, and watering revives a plant in a drought condition. Drought-struck plants will usually revive completely unless their leaves are already leathery or crispy. Even beyond that stage, many species will regenerate from the roots after complete death of the aboveground parts.

Poor Drainage and Flooding

Poor drainage, in which roots remain in water-saturated soil, will kill most plants. Those native to bog and other lowland areas are exceptions. Death is due to lack of oxygen. The length of time a plant can remain in standing water without injury varies, but most will tolerate several days to a week of occasional flooding.

Lack of Soil Nutrients

Lack of adequate soil nutrients will stunt and bleach plants but will not usually kill them very quickly. Lack of a required element will cause different symptoms on dif-

Table 13-3 Deficiency Symptoms for Commonly Lacking Nutrients

Element	Symptoms
Nitrogen	General leaf yellowing, showing up first on older leaves; deciduous trees defoliate earlier in fall
Phosphorus	General yellowing of foliage, with young expanding leaves and shoots frequently showing purple discoloration
Potassium	Older leaves die at the margins, and there is reduced shoot growth and leaf size; the marginal burn may be preceded by yellowing and is most severe late in the growing season
Iron	Yellowing of areas between the veins on young leaves
Magnesium	Retarded growth followed by yellowing of areas between veins on old leaves; yellowing spreads later to young leaves

ferent species, and this makes nutrient deficiencies difficult to diagnose. However, probably 80 percent of all deficiencies are due to a lack of nitrogen, phosphorus, potassium, magnesium, or iron. Descriptions of deficiency symptoms for these elements are given in Table 13-3.

 A nutritional specialist for horticultural plants works in a diagnostic lab to determine the cause and remedy for plants suspected of having nutrient disorders. This career involves mainly lab work with some visits to growing sites.

Unavailability of Soil Nutrients Due to pH

Even if an essential element is present, a plant may be unable to absorb it because of an excessively high or low soil pH. Iron and manganese are most commonly affected by pH. Symptoms will be the same as for normal deficiency (Table 13-3).

Excess Soil Salts

A nutrient excess is frequently called a *salt buildup*. Nitrogen can be the primary damaging element if the toxic buildup stems from overfertilization, but compounds present in normal irrigation water and alkaline soils can be as damaging.

Symptoms of excess salts are wilting and browning of leaf margins. This is due to death of the roots from dehydration by salts, and the effect is often indistinguishable from that of drought. The cure is by leaching, as described in Chapter 11.

Excess Heat (Heat Scorch)

Excess heat damage (see photo, Figure 1-4) frequently occurs on plants growing in paved areas or on south-facing walls; it is due to the reflection of sun-generated heat from these surfaces. The leaf margins and areas between the veins brown, and succulent new growth may die. Tolerance to excess or reflected heat varies among species. (See Chapter 1 for further information.)

Soil Compaction

Constant foot or equipment traffic causes soil compaction around plants. The air spaces normally present in the soil are compressed, and water absorption and air penetration are inhibited. Plants in compacted soil grow slowly, and portions may die back. In severe cases, the entire plant may die.

Improperly Applied Chemicals

Pesticide sprays can injure plants if applied improperly, in excess concentration, or to plants for which they are not intended. The lawn weed killer 2,4-D is among the most frequently damaging. Spray drift can carry the chemical to neighboring areas, and curling and distorting of plant foliage (Figure 13-16) will result. Similar damage can appear when one sprayer is used for both weed killer and insect or disease sprays. Only a small amount of weed killer residue is needed to cause injury.

Figure 13-16 2,4-D herbicide damage on squash leaves. *(Photo courtesy of Marge Coon)*

Another herbicide that commonly injures plants is glyphosate, which is sold as Roundup® or Kleenup®. Roundup® is absorbed by leaves and then translocates in the plant to growing points. Typical injury is yellowing of growing points followed by browning and death. When Roundup® is used, extreme care is necessary to prevent its blowing to desirable nearby plants. Also, when spraying the base of plants, the chemical can be absorbed by suckers and by thin green bark of young trees and shrubs. If you accidentally spray Roundup® on a desirable plant, wash it off immediately or prune off the sprayed part.

Chlorosis (persistent yellowing without leaf death) of plant foliage is a symptom of weed killer injury by chemicals such as Simazine. This weed killer injury can be distinguished from nutrient deficiency by the suddenness of the symptoms: weed killer injury will appear in two to three weeks, whereas nutrient deficiency will progress over several months. There are no practical remedies for weed killer injury, and the chances of its killing the plant depend on how much chemical was absorbed. Often the foliage will be marred for one or two seasons, but the plant will grow out of the damage and regain its former appearance.

Insect and disease sprays applied in excess concentrations or on plants for which they are not approved show different symptoms. Sudden browning of foliage to which the chemical was applied is the most common and may be followed by *defoliation* (leaf drop). A single defoliation from chemicals, disease, or insects is seldom fatal.

Grade Changes

Raising the soil level over the roots of plants more than a few inches (5 centimeters) can be fatal because water and oxygen flow to the roots will be restricted. For this reason, tree wells are necessary when grade changes are planned. They should be made as large as practical, extending to the branch tips if possible. Supplemental drain tiles may need to be installed (Figure 13-17).

Air Pollution

Air pollution damage is a problem only in a few areas of the country. It affects only susceptible plants, such as conebearing evergreens. The damage takes many forms, but usually involves some browning and death of the foliage. Pollution damage is difficult to diagnose and should only be suspected if the area frequently accumulates smog.

Allelopathy

Allelopathy is the injury of one plant by another via the excretion of toxic substances. Both black walnut and eucalyptus trees are allelopathic to some other species.

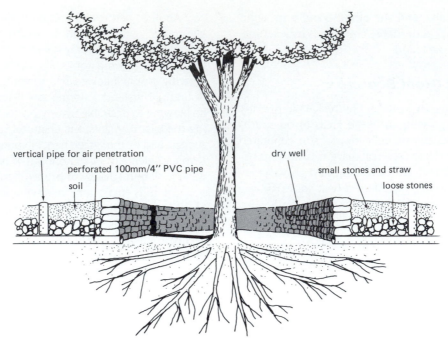

Figure 13-17 A tree well with drainage tiles.

vertical pipe for air penetration

perforated 100mm/4″ PVC pipe

soil

dry well

small stones and straw

loose stones

TREATING PLANT DISORDERS

The steps in restoring a sick plant to health relate directly to the cause of the disorder. For environmental problems, there are two choices: correct the condition harming the plant, or find a plant that will tolerate the situation. Excess salts, drought, soil compaction, and pH problems, for example, can all be corrected. Air pollution, winterkill, and reflected heat problems, on the other hand, cannot be eliminated, but plant selection can be changed.

Animal damage to plants can be dealt with by either preventing the damage or removing or killing the animal. Fences will keep out rabbits, dogs, and deer, although an 8-foot (2.4-meter) fence will be needed to prevent intrusions by deer. Blood meal may also work as a deer and rabbit repellent or nighttime motion sensors may be triggered to turn on sprinklers. Netting or noise-making devices are most successful with bird problems. Girdling of trees by mice and rabbits can be avoided by pulling away mulch layers and enclosing the trunk with a tall collar of tin or hardware cloth (Figure 13-18).

Damage by moles, gophers, and ground squirrels is harder to prevent. Poison baits are the best recourse against pocket gophers and ground squirrels, but moles are less likely to accept baits. Trapping with specially designed mole traps is frequently successful, but to be effective the trap must be placed in a main bur-

Figure 13-18 A hardware cloth collar to protect the base of a tree from rodent damage. *(Photo by George Taloumis)*

row used repeatedly by the animal. When moles infest a lawn, it is often because the lawn contains grubs, a favored food. Applying an insecticide drench will destroy the food source, and the moles will retreat.

Treating Microorganism-Caused Diseases

Plant disease control is both cultural and chemical, and an understanding of both is essential to restoring and maintaining plant health.

Cultural Disease Control

Cultural control involves using plant care techniques that minimize disease infection and spread.

Watering. Water is a key factor in plant disease attack and spread. Many fungi spores lie dormant until in the presence of water, germinating only then to invade the host plant. Accordingly, keeping water off plants susceptible to foliage diseases lessens the chances of infection. Excessive soil water similarly favors the growth of disease organisms causing root rot, and overwatering will trigger that disease.

Splashing water is often a factor in foliage disease spread, carrying the microorganisms from infected to uninfected plants. Careful watering, with the water applied only to the soil, can decrease disease problems on susceptible species. The timing of watering can also reduce or enhance disease spread. The foliage of plants watered late in the day may stay wet overnight and be susceptible to disease. Plants watered from morning to midday dry quickly and are less vulnerable.

Sanitation. Sanitation involves removing all disease-harboring materials to reduce the population of microorganisms. Many disease organisms have a seasonal cycle with dormant and active periods. The overwintering site for many foliage disease organisms is the fallen leaves that were infected during the summer. Accordingly, leaf and twig litter under disease-infested plants should be cleared away promptly in fall to prevent it from being a source of infection next season. These infested plant parts should be burned or hauled away, not composted.

Weed removal is another sanitation measure. One disease will frequently attack several related plants. If one of those plants is a weed, it can be a source of infection for cultivated plants. As another example, rust fungi require two plants to complete their life cycle, one of which is often a weed. Without a weed host nearby, the rust will not be able to attack the cultivated species.

Pruning. Pruning can be a valuable control technique for localized diseases such as gall, wilt, and canker. The infected portion of the plant should be pruned out and burned as soon as it is detected, cutting back several inches (5–10 centimeters) into healthy tissue. Between cuts and after pruning, the pruners should be dipped in alcohol to avoid spreading the microorganisms by the blades.

Roguing. Roguing involves pulling out diseased plants to stop the spread to others. It is valuable for annual plants or when only a few plants are infected. In addition, it is the only control for virus diseases and the most practical for virus-like organisms and most bacteria.

Crop Rotation. Crop rotation is usually associated with vegetable plants, but it can apply to flowers as well. It involves planting a different species in an area each year instead of the same one repeatedly. Rotation is a valuable control for soilborne diseases, most of which are hard to control by chemical means. By using crop rotation, the parasitic microorganisms growing in conjunction with a species will be depleted yearly because they are not able to prey on the new crop.

Resistant Cultivars and Species. Planting species and cultivars resistant to disease is one of the soundest disease-control techniques. Many vegetables and flowers have resistance to diseases imparted by breeding, and the small extra amount these cost is well spent. Landscape plants are similarly bred for disease resistance, although some have natural disease and insect resistance. Planting these ornamentals in lieu of disease-susceptible ones can greatly lessen maintenance.

Plant Selection and Culture. An important way to minimize disease is to select plants that are not seriously affected by diseases and that are well adapted to the environment in which they are to grow. For example, a rose planted in a sunny, open environment may have minimal problems with powdery mildew or rust; but when planted in a shady environment, it may need to be sprayed constantly to control these diseases. Additionally, proper plant care can help to minimize disease problems. For example, pruning, can increase air movement within the plant canopy, minimizing diseases that require high relative humidities to thrive.

Chemical Disease Control

Chemicals are very helpful with some disease problems and nearly useless with others. Leaf spots, rust, mildew, smut, and blight diseases can usually be chemically controlled, but wilts, rots, cankers, witches' brooms, and all virus diseases are only marginally controllable or uncontrollable by chemicals. Fortunately, most commonly encountered diseases fall in the controllable group.

A pest control advisor for horticultural crops advises growers of the best chemicals and practices currently available to control their pest and disease problems. Pest control advisors frequently work for large chemical companies and maintain a list of clients whom they visit regularly.

Fungicides. Fungicides are the most common disease-controlling chemicals, and most can be used on a variety of plants for numerous diseases. The container label and Cooperative Extension Service publications will give information regarding which diseases the product controls.

Although most fungicides are used as foliage sprays, a few combat soil-borne diseases and are applied to the root zones of plants. These are called *fungicidal soil drenches*. Still fewer are applied by one of the methods just described but are absorbed into the plants instead of simply forming a protective covering. These are called *systemics*.

Many fungicides are only *protectants* that stop the germination and growth of fungal spores but do not actually kill the fungus inside the plant. A protectant is most useful in the early stages of a fungus infection before the pathogen becomes established enough to do significant damage to the plant. For this reason, monitoring the health of plants is important to catch disease infections before they become epidemics. In addition, protectants must be reapplied frequently in order to continue to give protection to a plant. New growth that occurs after spraying, for example, will not be protected and an additional spraying will be needed.

A few fungicides are classified as *curatives* (chemotheraputants). Curatives may kill existing fungi completely as well as preventing germination and growth of spores.

Bactericides. Bactericides are generally antibiotics such as streptomycin and tetracycline. They are less common than fungicides and are used only on a few bacteria- and mycoplasma-caused diseases including blights. Copper sprays may also control bacteria.

Seed and Bulb Treatments. Often purchased seeds are oddly colored (usually pink or blue) because of treatment with a fungicide before packaging. The treatment will have been applied by dusting or quickly dipping the seeds (followed by immediate drying) in liquid fungicide. The fungicide gives protection against damping off.

If damping off has been a frequent problem, seeds can be home-treated before planting by shaking them in a bag with a teaspoon of powder fungicide recommended against damping off before planting. If bulbs are bought and appear to have mold growing on them

(normally penicillium), they can be treated by dipping them in fungicide mixed to proper application strength before planting. The dip will slow or stop the development of the organism in the soil.

Inoculants. There are a number of fungicidal inoculants available that work on the principal of colonizing the roots or the soil around them and either producing exudates that retard the growth of plant pathogenic fungi or compete with harmful fungi, thereby diminishing their chance of infecting plants. These products either are applied to soil at the time of potting or can be "drenched" onto the plants later by watering with a solution containing the organisms. When combined with good cultural practices, these inoculants may reduce root-disease problems. Two of these beneficial fungi are in the genus *Trichoderma* or *Gliocladium,* they are available under various trade names.

Also effective is the application of certain composts, called *disease-suppressive composts*, that naturally contain organisms that suppress root diseases. As a general rule, soil that supports a healthy microorganism ecosystem will be less supportive to pathogens than are soils that are low in organic matter, high in salts, or polluted through excess applications of fertilizers or other chemicals.

Treating Nematode and Insect Problems

Cultural Controls

The cultural controls for combating insect problems are less successful than those for combating diseases but can sometimes be effective. One strategy is to remove the insect hiding places. Accordingly, weed control and removing litter at the base of a plant are two controls. Pruning off of an insect-infested portion of a plant is also used.

Physically removing large insects such as worms by hand picking during the day or the night, depending on when they are active, can be effective. A strong stream of water can be used to wash off aphids.

Supplying plants with adequate nutrition and other requirements for healthy growth is another cultural control technique. Healthy plants are less susceptible to both insect and disease problems than those under stress.

Plants that lack adequate water, for example, are prone to mite problems. On the other hand, plants that are overfertilized, particularly with nitrogen, produce abundant new growth that is very attractive to aphids.

Another strategy for cultural control is to plan a diverse garden with many different types of plants in close proximity to each other. This accomplishes two things. First, by not having a large number of plants of one species, if a pest attacks one species, it will be a minor outbreak simply because the food source is limited. Second, the diversity of plants will produce a diverse population of beneficial insects as there will be diverse food sources. Remember, many beneficial organisms exist in a garden, including insects, mites, birds, and bats that eat insects, soil-inhabiting fungi that may infect insects and nematodes in the soil, and others. The garden must be looked upon as an ecosystem that must be maintained in a healthy condition by ensuring that the soil is rich in organic matter and that there are habitats for a diverse population of organisms.

 A commercial pesticide applicator is a person trained and experienced in the application of pesticides. Only persons licensed through examination as pesticide applicators are allowed by law to apply pesticides commercially.

Chemical Controls

Many gardeners prefer to use chemicals over other methods of pest control because they are perceived as easy and effective. Nevertheless, chemicals should be used as a last resort in the garden because most are highly disruptive to the garden ecosystem. Once one starts to use chemicals widely in the garden, one often begins a cycle of having to use chemicals over and over because beneficials are killed and natural controls no longer keep the majority of pests under control. Chemicals should only be used when outbreaks are severe, and then a chemical as specific as possible to the pest should be chosen, and spot treatments should be used. Chemicals that break down quickly in the environment are also better because even if they kill beneficials, new beneficials can move in from outside without being killed by active residues on the plants.

Chemicals used to control insect pests are most often applied to plants as sprays, soil drenches, or granules and protect the plant by killing the insects. However, baits, which attract insects and kill them, and repellents are two other chemical controls.

Sprays. Surface sprays either kill on contact or are absorbed by insects as they walk across a sprayed leaf. These sprays are effective on most pests but are not useful for those with protective coverings such as mature scales and mealybugs.

Stomach poisons applied as foliage sprays are useful on chewing insects, which ingest the poison as they feed on the plant. They are less effective on sucking insects, which feed on the untreated sap.

Dormant oils are the third type of spray and are used on dormant deciduous plants. Applied in early spring, they are the primary means of combating scales and mealybugs. The oil coats trunks, branches, and twigs and excludes air, suffocating insects and their eggs. Later in the season, light oils safe for foliage can be used. Sulfur should not be used within 30 days of using an oil spray.

Insect growth regulators (IGRs) are a fourth type of spray. IGRs work by interrupting insect development. IGRs may or may not be toxic to adult insects, so results from their use are often slow but may be long-lasting, because reproduction is stopped. One common IGR is *azadirachtin,* which is the water-soluble extract from a tropical tree called the neem tree (*Azadirachta indica*). When applied to immature insects, azadirachtin interrupts their molting (shedding of the exterior shell to allow growth) and the insects slowly die. It also has repellant properties. Though IGRs can be used alone, when scouting has revealed that there are low populations of adults but a burgeoning population of immature stages, they can be combined with a conventional insecticide to kill adults. The combination will give rapid, long-lasting control with minimal chances for resistance development.

Soil Drenches. Soil drenches combat underground pests. They are contact or systemic and are usually the same chemicals as surface sprays but in weaker concentrations.

Systemics. Systemic insecticides are absorbed into a plant and kill only those pests that feed on the plant sap. They can be applied as foliage sprays or soil drenches (penetrating by root uptake) and, occasionally, by injection or implantation into the trunks of trees (see Chapter 18). Insecticides have both contact and systemic action when applied as a foliar spray. Soil-applied systemics are generally less disruptive to beneficials than are sprays.

Baits. Baits are the most commonly used control material for ants, snails, slugs, sowbugs, earwigs, centipedes, and sometimes rodents. They act as stomach poisons, and careful attention should always be paid to placing them where other animals will not eat them.

Repellents. Few repellents are available commercially except for rodent repellents. However, homemade solutions of varying effectiveness can be made by following the instructions in organic gardening books. Neem, pepper wax, and pyrethrum products repel some insects.

Choosing Pesticides

The criteria for selecting a chemical should be that it controls the pest satisfactorily and is as nontoxic as possible both to people and to beneficial organisms.

The best sources for information on the toxicity and uses of pesticides are government publications (Cooperative Extension, Farm Advisor, United States Department of Agriculture) and the label of the material.

Table 13-4 lists the LD_{50} of many common insecticides, fungicides, and bactericides. LD_{50} is an abbreviation for the phrase "lethal dose killing 50 percent of a population." It indicates how much of a pesticide is required to kill 50 percent of a population of laboratory animals and is expressed in milligrams per kilogram of total body weight of the animal. Therefore, the larger the number, the more of the material was needed to kill the animals and the *less* toxic the material. For comparison purposes, the LD_{50} values of salt, caffeine, and aspirin are included in the table, but this comparison in no way implies that pesticides less toxic than these materials should be used injudiciously.

Another criteria in choosing pesticides that is very important in commercial use but less so when pesticides are used occasionally in home gardens is that pesticides with different modes of action should be rotated to avoid resistance. The "mode of action" is the way in which pesticides kill the pest. For example, pesticides in the organophosphate class (i.e., malathion and acephate) kill by decreasing the concentration of an enzyme called *cholinesterase,* thus interfering with the transmission of nerve impulses. When these chemicals are used repeatedly, insects and mite populations can become resistant to them, and they may cease to become effective. By rotating between insecticides with different modes of action, resistance development is slowed and better pest management results are obtained.

Pesticides and Human Health

There is probably no other topic in horticulture debated as heavily as whether pesticides are harmful to human health. The debate has been characterized by emotion and bad science on both sides. Because the majority of the population does not understand scientific principles, their ability to understand and participate in the debate is limited. Proponents of pesticide use argue that, without pesticides, food could not be produced as cheaply and of such high quality. Therefore, they attest, banning pesticide use would have a negative health effect on the population as people would eat less of the

nutritious fruits and vegetables currently available. Opponents of pesticide use argue that organically produced food tastes better and is safer and healthier than pesticide-treated food. What is the correct position? As in most debates, it is probably somewhere in the middle. Here are some facts that should influence your decision about how you produce horticultural crops and whether you will choose to consume organically grown or conventionally grown fruits and vegetables.

The production and use of pesticides is extensively regulated. Before pesticides can be used, the producers must demonstrate within reasonable parameters that they are safe. This is done through extensive tests over a period of 7 to 10 years, during which laboratory animals are exposed to the pesticides in their food and in their environment. Over many generations, these animals are inspected for cancer, birth defeats, and other problems resulting from exposure. Researchers establish a *no-effect level* below which the pesticide does not cause any danger to the test animals. Then a *tolerance*, or level acceptable in human food, is set at least 100 times below the no-effect level. Food, including that imported from other countries, is regularly inspected for pesticides not *labeled* (legally allowed to be used on the crop) for use in the commodity or for *residue* (the amount of pesticide in the commodity) levels that exceed the tolerance. When these incidences are discovered, the food is discarded and the producer may be fined. It is believed that, as a result of this process of regulation and inspection, the risks associated with pesticide use are understood and dangers mitigated through required safety practices.

Still, it is never possible to know all the risks, and occasionally a new risk associated with a particular pesticide or pesticide combination is discovered. When this occurs, the pesticide is further regulated, or it may be removed from the market completely. There is also no doubt that the use of pesticides can damage the environment and cause risks to workers who come into frequent contact with them. For this reason, extensive safety regulations are in place to minimize these risks.

Some of the risks that may be associated with pesticide exposure are cancers, birth defects, neurological disorders, and endocrine system disorders. For all of these risks, there are studies that show that there may be some association with pesticide exposure in some people. Especially vulnerable are children, developing fetuses, and those who have high rates of exposure. Disorders of the endocrine system (the system of hormone-producing glands in the body that regulates many processes) are a relatively recently discovered risk of pesticide exposure for which screening did not occur previously. In 1996, the Food Quality Protection

Table 13-4 Oral LD_{50} Values of Selected Garden Chemicals

Chemical and associated trade names	Use	$LD_{50}{}^a$
Acephate (Orthene®)	insecticide	700
Anilizine (DYRENE®)	protective fungicide	2710
Aspirin	—	750
Azadirachtin (BioNEEM®)	insecticide (growth regulator)	5000
Bacillus popillae milky spore disease	insecticide for Japanese beetle larvae	5000
Bacillus thuringiensis BIOTROL, THURICIDE®, DIPEL	insecticide	no toxicity at tested doses
Bayleton	curative fungicide	363
Bendiocarb	insecticide	40
Benomyl (BENLATE)	curative fungicide	9590
Caffeine	—	200
Captan (ORTHOCIDE®)	fungicide	9000–15,000
Carbaryl (SEVIN®)	insecticide	500–850
Chlorothalonil (DACONIL®, BRAVO®)	protective fungicide	10,000
Chlorpyrifos (LORSBAN®, DURSBAN®)	insecticide	97–276
Copper	protective fungicide, bactericide	low toxicity
DCPA (DACTHAL®)	herbicide	3000
Diazinon	insecticide	300–400
Dimethoate (CYGON®)	insecticide	215
Disulfoton (DI-SYSTON®)	insecticide	2–12
Dursban	insecticide	96
Endosulfan (Thiodan®)	insecticide	23
EPTC (EPTAM®)	herbicide	1600
Gasoline	—	150
Hexakis (Vendex®)	miticide	2630
Kelthane	miticide	100–1000
light and heavy oils	insecticide	15,000
Malathion	insecticide	1000–1375
Mancozeb (Dithane®)	protective fungicide	4500
Metaldehyde	insecticide, snail, slug killer	600–1000
Metasystox R	insecticide	65
Methoxychlor	insecticide	6000
Oftanol	insecticide for soil insects	20
oryzalin (SURFLAN®)	herbicide	10,000
oxadiazon (RONSTAR®)	herbicide	8000
PCNB (TERRACHLOR®)	soil fungicide	1650–2000
Pyrethrum	insecticide	820–1870
Resmethrin	insecticide	4240
Rabon	insecticide	4000–5000
Rotenone	insecticide	50–75
Ryania	insecticide	750–1200
Salt	—	3320
Simizine (PRINCEP®)	herbicide	5000
Soap (Safer Soap®)	insecticide	very low toxicity
Steinernema carpocapsae beneficial nematodes (Exhibit®)	insecticide	none
Streptomycin	bactericide	9000
Sulfur	protective fungicide	17,000
Tetramethrin	insecticide	4640
Thiophanate-methyl	curative fungicide	15,000
Triforine (Funginex®)	curative fungicide	16,000

[a]Class 1: highly toxic (1–50); class 2: moderately toxic (50–500); class 3: low order of toxicity (500–5,000); class 4: very low toxicity (5,000+).

Act was passed. This act requires that all pesticides be screened for endocrine system disruption, and the law added additional evaluation procedures to further reduce dangers that may be associated with pesticides. Because there may be dangers associated with contact with pesticides and other chemicals, it is prudent to minimize exposure. To do this, one must understand the sources of pesticide exposure.

Sources of Pesticide Exposure

Pesticides can enter the body through a variety of routes, including through the skin or mucous membranes (nose or mouth), inhalation, or ingestion along with food or water. For people who apply pesticides, exposure is most likely to occur through the skin or through inhalation. For the average person who does not apply pesticides, ingestion in food and water is most likely. Washing food before eating will decrease exposure. For those applying pesticides, the use of safe application techniques and wearing protective clothing is necessary. For people working in pesticide-treated areas, simple safety practices including keeping out of treated areas until after the *restricted entry interval* (that period after application when residues are high and workers are legally prohibited from entering), washing hands after touching treated plants and surfaces, and wearing clean clothes when starting work each day will minimize exposure.

Protective Clothing for Pesticide Application

Pesticides enter the body during application most frequently through skin. Occasionally, they enter through the eyes, nose, or mouth. Wearing basic protective clothing reduces the risk of pesticide poisoning. Specialized clothing may be needed when handling liquid concentrates and highly toxic materials, or when clothing is likely to become wet from spray drift. The pesticide label gives recommendations for the type of protective clothing that should be used.

The risk of pesticide poisoning can occur at any time, including during mixing, loading, application, and equipment cleaning and repair. However, applicators face the greatest risk of exposure while mixing and loading. During mixing and loading, the applicator works with full-strength, undiluted chemicals. Pesticides in the diluted, ready-for-application concentration are normally less toxic.

An often-overlooked hazard occurs when people or animals are allowed to enter a treated area during the prescribed "no-entry period". Protective clothing and equipment should be worn when entering a pesticide-treated area before the no-entry period has passed.

The *signal* words on the label—such as "Caution," "Warning," or "Danger"—can help determine the type of protective clothing to wear. Many pesticide labels give additional useful information. For example, dusts, wettable powders, and broken particles from granules are inhaled easily. Labels for many of these products carry statements like "harmful if inhaled" or "fatal if inhaled." When applying such products, wear a respirator.

Labels also describe the formulation, which may affect the readiness with which the pesticide is absorbed. For example, oil-based liquids (emulsifiable concentrates) absorb easily through skin, so extreme care needs to be taken to protect the skin.

The minimum protection needed when working with pesticides is long sleeves, long pants, shoes and socks, rubber gloves, and splash-proof eye protection, regardless of the toxicity of the pesticide. Specialized clothing for moderately or highly toxic pesticides includes rubber boots, face protection, and a respirator. The Environmental Protection Agency's (EPA's) recommendations include wearing a double layer of clothing. This can be accomplished by wearing coveralls over long pants and a long-sleeved shirt, and rubber boots over shoes and socks.

Washable Clothing

Wear heavyweight, tightly woven, washable work clothing with long pants and sleeves, or wear coveralls. *Do not* wear leather belts or shoes, as these absorb pesticide and it cannot be washed.

Gloves

Wear unlined, liquidproof, chemical-resistant, elbow-length gloves. Gloves are mandatory when working with highly toxic pesticides. It is recommended that only unlined rubber or neoprene (nitrile, and the like) gloves must be used. Gloves should be thoroughly washed inside and outside after each use. The glove cuffs should be inside the shirtsleeves for most work. This will keep runoff pesticide from getting into the gloves. However, when working overhead, put the cuffs of gloves outside the sleeves. Check rubber-type gloves for leaks each time they are washed. Do this by filling the gloves with water, folding the cuff over, and squeezing to put pressure on the water in the glove. If there are holes, water will leak out. Discard gloves with leaks. *Never use cloth or leather gloves when working with pesticides.*

Boots

Choose calf-high, unlined, neoprene or rubber boots. Boots should be worn over work shoes or in place of work shoes when mixing or applying pesticides. Pull the legs of trousers over the tops of the boots to help prevent spilled pesticide from getting inside the boots. Wash boots with soap and water after each use. *Never wear cloth or leather boots.*

Apron

Wear a chemical-resistant apron over cloth coveralls when mixing, loading, or handling undiluted pesticides.

Respirator

Respirators (not to be confused with dust masks) prevent inhalation of toxic fumes and mists, and include special chemical cartridges that absorb pesticides during breathing. They should be used when mixing or applying pesticides if the label specifies the need to do so. Choose the correct cartridge for the type of pesticide being used. The manufacturer or supplier can provide guidance on selecting the correct cartridges. Replace the cartridges when the odor of the pesticide becomes noticeable or when breathing becomes difficult. The life of the cartridges varies with the concentration of pesticide in the air around the respirator, breathing rate, temperature, humidity, and type of cartridge.

Respirators *will not* provide adequate protection for a person with a beard. Choose the model and make of respirator that provides a good seal between the face piece and the face to prevent pesticides from leaking into the respirator and being breathed.

Always wash the face piece and straps in soap and water after each use. After drying, place the respirator and cartridges in a clean plastic bag until they are needed again. Select respirators having the approval of the National Institute for Occupational Safety and Health (NIOSH).

Face Protection

Wear an eye shield or a face shield and goggles if the pesticide may splash. It is also necessary to wear splashproof goggles when working with pesticides. Not only can the pesticide be absorbed through the eyes, but the acidity of a pesticide can cause permanent eye injuries. When pouring or mixing concentrates, it is preferable to use a full-face shield to protect the face from splashes. Always wash the goggles and face shield with soap and water after use.

Headgear

A waterproof hat should be worn when mixing or applying pesticides, because pesticides can be readily absorbed through the scalp. The hat should have a brim to keep drift or splashes off ears and neck. Plastic safety hats are ideal for use with pesticides and should be washed in soap and water after each use. Cloth hats absorb pesticides and contaminate the wearer. *Do not wear cloth hats.*

Daily washing of clothing worn by the applicator reduces the risk of pesticide buildup on the clothing. Never wash clothing exposed to pesticides with other household laundry. Clothing saturated with pesticides should be disposed of properly.

Pesticide Safety

Safe pesticide use requires applicators who are familiar with the routes of human exposure and the ways that pesticides can damage the environment. The first key to this is to read and follow the label directions exactly (Figure 13-19). After reading the label, the applicator should plan the application carefully, choosing the proper equipment for the job and ensuring that it is in good operating condition. He or she should then follow the suggestions below to safely apply the pesticide.

Procedures for Mixing and Loading. Before mixing, check protective clothing for breaks and cracks, wear pants outside of boots and shirt sleeves over gloves to decrease the chance of spillage on the inside of the boots or gloves. Mix the pesticide outdoors. If it is necessary to mix it indoors, do so in a well-ventilated inside area. Wear protective glasses. Measure and mix below eye level to decrease the possibility of splashing into the eyes. Empty the container into the spray tank and hold the container in a vertical position for 30 seconds to allow most of the pesticide to drain into the tank. Always make a slurry with a wettable powder before adding it to the spray tank. Rinse the container three times with water equal to 20 to 25 percent of its volume, and add all rinse water to the spray tank. After this triple rinsing, empty the pesticide containers, and store them in a labeled, covered storage container inside a locked storage area until they can be disposed of according to law.

Personal Contact. Remove contaminated clothing immediately. Wash contaminated skin thoroughly with soap and water immediately. If a pesticide gets into the eyes, flush them with clean water for at least 15 minutes; then take the affected person to the doctor. When going to the doctor, bring the pesticide container with you, but carry it outside the passenger compartment (that is, in the trunk or the back of the pickup truck).

PRODUCT NAME

DIRECTIONS FOR USE

This section tells you what the pesticide product controls, where, how and when to use the product. Often, the product manufacturer has included a booklet with the container. Some manufacturers also provide a toll-free number for consumers to use to obtain additional information about their products.

FIRST AID INSTRUCTIONS

This section tells you what to do first if someone accidentally swallows or breathes the pesticide, or gets it on their skin or in their eyes. Labels may also contain a section labeled "Note to Physicians" which provides doctors with specific medical information.

MANUFACTURER'S ADDRESS

The name and address, and sometimes phone number, of the manufacturer, or distributor, of the product.

ESTABLISHMENT NUMBER (EPA Est. No.)

The number which identifies the particular facility where the final phase of production of the pesticide product took place.

PRECAUTIONARY STATEMENTS, HAZARDS TO HUMANS & DOMESTIC ANIMALS

This section describes potential hazards to people or pets and actions you can take to reduce those hazards, for example, wearing gloves. These statements may also provide extra information on how to protect your children or pets.

STORAGE AND DISPOSAL

This section tells you how to best store the product and what to do with the unused portion of the product and the empty container.

The label elements shown in the figure include:

KEEP OUT OF THE REACH OF CHILDREN
DANGER

FIRST AID

(STATEMENT OF PRACTICAL TREATMENT)

IF SWALLOWED _____
IF INHALED _____
IF IN EYES _____
IF ON SKIN _____

ACTIVE INGREDIENTS: _____ %
OTHER (INERT) INGREDIENTS: _____ %
TOTAL: **100%**

THIS PRODUCT CONTAINS XX LBS. OF XXXX PER GALLON

WARRANTY STATEMENT
_____
_____

MANUFACTURER'S ADDRESS
_____
_____

NET WT. NET CONTENTS STATEMENT: _____

EPA Registration No. / EPA Reg. No.: _____

EPA Establishment No. / EPA Est. No.: _____

DIRECTIONS FOR USE

It is a violation of federal law to use this product in a manner inconsistent with its labeling.

PRECAUTIONARY STATEMENTS
HAZARD TO HUMANS
(AND DOMESTIC ANIMALS)
DANGER

ENVIRONMENTAL HAZARDS

PHYSICAL OR CHEMICAL HAZARDS

STORAGE AND DISPOSAL

STORAGE _____

DISPOSAL _____

Figure 13-19 Sample pesticide label. *(Courtesy of United States Environmental Protection Agency, Office of Pesticide Programs. Web Site: "Read the Label First" at* http://www.epa.gov/pesticides/label/*) (continues)*

PRODUCT NAME

DIRECTIONS FOR USE

It is a violation of federal law to use this product in a manner inconsistent with its labeling.

_____
_____
_____

PRECAUTIONARY STATEMENTS
HAZARD TO HUMANS
(AND DOMESTIC ANIMALS)
DANGER
_____
_____

ENVIRONMENTAL HAZARDS
_____
_____

PHYSICAL OR CHEMICAL HAZARDS
_____
_____

STORAGE AND DISPOSAL

STORAGE _____
_____

DISPOSAL _____
_____

KEEP OUT OF THE
REACH OF CHILDREN
DANGER

FIRST AID

(STATEMENT OF PRACTICAL TREATMENT)

IF SWALLOWED _____
IF INHALED _____
IF IN EYES _____
IF ON SKIN _____

ACTIVE INGREDIENTS: _____ %
OTHER (INERT) INGREDIENTS: _____ %
TOTAL: **100%**

WARRANTY STATEMENT
_____
_____

**THIS PRODUCT
CONTAINS XX LBS.
OF XXXX PER GALLON**

MANUFACTURER'S
ADDRESS
_____

NET WT. NET CONTENTS STATEMENT: _____

EPA Registration No. / EPA Reg. No.: _____

EPA Establishment No. / EPA Est. No.: _____

SIGNAL WORDS

On the label you will find one of the following "signal words": "Caution," "Warning," "Danger," or "Danger-Poison." The signal word indicates the pesticide's potential hazard level to humans, with "Caution" being the least harmful, and "Danger-Poison" the most harmful.

ACTIVE INGREDIENTS

Identifies the "active ingredient(s)" in the product. The active ingredient controls the pests listed on the label.

WARRANTY STATEMENT

Statement intended to limit a company's liability, or to act as a disclaimer, or as a warranty for the product.

NET WEIGHT/
NET CONTENTS STATEMENT

Identifies how much pesticide product is in the container.

EPA REGISTRATION NO.

All pesticide products sold in the United states must be registered with the EPA. The registration number shows the product has been reviewed by the EPA and that the EPA has determined the product can be used with minimal risk if you follow the directions on the label properly. The number is not an endorsement or guarantee of product effectiveness.

Figure 13-19 (continued)

Discard or launder separate from the family wash all pesticide-contaminated clothing. Always wear clean clothing when beginning a pesticide application job. Never eat, drink, or smoke during pesticide application. Wash thoroughly before eating, drinking, or smoking after a pesticide application. After each period of pesticide application, shower thoroughly and put on clean clothes.

Avoiding Environmental Contamination. Apply pesticides only as directed, and observe all caution and warning statements. Do not apply pesticides when it is windy, so that drift can be avoided. Apply with the proper equipment that is well maintained in order to avoid accidents such as broken hoses. Apply with nozzles that are appropriate to the type of application, and apply the pesticide at low pressure to keep droplets large so that they will drift less. Be aware of the fate of the pesticide after application. Is the pesticide likely to move to undesired locations after application through runoff or erosion? In the long run, the most important safety rule is that the user should concentrate on the job at hand and understand the product he or she is using. Accidents most commonly occur when the applicator is distracted and fails to observe the recommended safety practices.

Pesticide Transport and Storage

The person transporting a pesticide is legally responsible for its safe transport. For this reason, it is best to have a dealer deliver pesticides directly to an agricultural storage building and thereby reduce this liability.

If pesticide must be transported and used at another site, be sure the container is in an open area of the truck, never in the cab. Keep absorbent material such as cat-box litter in the truck in case of a spill. Use the litter to absorb the spill, and then handle the contaminated litter as if it were the chemical itself. In the event of a large spill, build a dike of sand or dirt to contain the spill and call the fire department to alert a professional chemical-spill cleanup team. Never hose down a spill, as this spreads the contamination.

Storage of pesticides in a separate structure *only* for pesticides is the safest practice. The building should be locked and must give protection against temperature extremes, humidity, and sun. It should be dark, cool, dry, well-ventilated, and insulated against freezing. The building must conform to state and local fire codes for storing flammable or combustible materials.

FEDERAL PESTICIDE LAWS

Application of any material classified as a pesticide is governed strictly at both the federal level (through the EPA) and at the state level.

The EPA has extensive and detailed registration procedures for pesticide chemicals. It also has strict regulations that deal with mixing pesticides, record keeping, transportation, storage and disposal procedures, interval of time between pesticide application and harvest or sale of a horticultural product, and other aspects of pesticide application. These regulations are designed to prevent accidental or inadvertent injury to people or to the environment through improperly applied pesticides. These standards are under constant review, as new pesticides are developed and existing ones are assessed for both their effectiveness (through the development of disease or insect resistance) and environmental impact. Listed in mainly historical order, the major federal legislation dealing with pesticides is as follows:

The Federal Insecticide, Fungicide, and Rodenticide Act (FIFRA) of 1947

The Federal Insecticide, Fungicide, and Rodenticide Act (FIFRA) of 1947 formed the core legislation regarding pesticide use until 1972, when the Federal Environmental Pesticide Control Act (FEPCA) was passed. The latter shifted the emphasis of FIFRA from safeguarding the consumer against fraudulent pesticide products to a role of protecting both public health and the environment.

The Federal Environmental Pesticide Control Act (FEPCA) of 1972

Two years before FEPCA was passed, Congress established the EPA and gave it the authority to regulate all pesticides. Previous to this, pesticides were under the jurisdiction of the U.S. Department of Agriculture. No pesticide may be distributed or used in the United States unless it has been reviewed by the EPA and licensed, registered, and given a product registration number. To date, about 50,000 pesticides have been registered by the EPA.

Old pesticides in use before the founding of the EPA and the establishment of current scientific standards have been evaluated by the "no unreasonable adverse effects" guidelines applied to new pesticides.

Other authority given to the EPA is the right to do the following:

- Approve the information to be put onto the product label

Figure 13-20 Protective gear is always used when mixing concentrated pesticides *(Photo by the USDA)*

- Classify pesticides as either for "general use" by the public or "restricted-use" (Restricted-use pesticides are only made available to licensed pest control applicators.)

- Decide on the penalties for improper use of pesticides and illegally storing or disposing of pesticides. Generally, these penalties are fines of $1,000 to $25,000 and/or prison terms of up to one year, depending on whether the violation was unintentional or intentional.

- Establish maximum tolerances for pesticide residues on raw fruits and vegetables or in processed food under the Food, Drug, and Cosmetic Act of 1938. The EPA is also allowed to monitor food crops for pesticide residues and enforces tolerance levels.

- Make regulations (interpretations of the law) regarding worker protection from pesticides, including mandating the use of protective equipment such as gloves, hats, respirators, or chemical-resistant suits. (Figure 13-20)

- Deny registration or gradually remove a pesticide from the marketplace to encourage the development of alternative chemicals or technologies

- Suspend the use of a pesticide on a regular or an emergency basis if the agency believes the pesticide poses an imminent hazard. Suspension halts the use of a pesticide until a decision on withdrawing its registration can be made.

Occupational Safety and Health Act (OSHA) of 1970

The Occupational Safety and Health Act (OSHA) of 1970 is a major piece of legislation (still in effect) that is administered by the Department of Labor. It requires any employer with 11 or more employees to keep records of all work-related deaths, injuries, and illness, and to make periodic reports. Minor injuries needing only first aid treatment need not be recorded. Records must be made if the injury involved medical treatment, loss of consciousness, restriction of work or motion, or transfer to another job.

The act also requires investigation of employee complaints that may be related to pesticide use, reentry into a pesticide-treated area before it is permissible to do so, or accidental pesticide exposure.

Hazard Communication Standard (HCS) Law

The Hazard Communication Standard (HCS) Law is a part of OSHA. It sets out the following requirements:

- Employers must read the standard and understand the provisions and responsibilities of an employer.

- A list of the hazardous chemicals (pesticides are classified as hazardous chemicals) in the workplace must be made.

- Employers must obtain material safety data sheets (MSDSs) for all hazardous substances on their list. These information sheets specify all risks associated with each pesticide.

- All containers must be labeled.

- A written communication program must be developed and implemented.

- Employee training must be conducted based upon the chemical list, MSDSs, and labeling information.

- Employers must create a hazard communication file and make it available to any employee upon request.

The Endangered Species Act (ESA) of 1973

Within the United States, about 275 animals and 190 plants have been listed as endangered or threatened with extinction. These species are covered by the Endangered Species Act (ESA) of 1973.

The EPA is responsible for ensuring that registered pesticide use is unlikely to further jeopardize endangered species. To accomplish this, the EPA estimates the maximum environmental concentration of each pesticide. If this estimated concentration might affect an endangered species, data about the pesticides is given to the Fish and Wildlife Service, which determines if the pesticide uses are likely to jeopardize the endangered species. That agency can recommend alternatives and/or restrict the use of the pesticide within the habitat of the affected species.

The EPA responds to the Fish and Wildlife Service determinations by changing the pesticide label. The new label may list restrictions, or it may direct pesticide applicators to read an Endangered Species Bulletin with directions for the use of the pesticide where endangered species could be affected.

The Transportation Safety Act of 1974 (Department of Transportation)

The Transportation Safety Act of 1974 (Department of Transportation, or DOT) covers any safety aspect of transporting hazardous materials, including the packing, repacking, handling, describing, labeling, marking, placarding, and routing of such materials. Many states have adopted these federal regulations and are enforcing them. Many pesticides are not defined by the DOT as hazardous, but some are.

Resource Conservation and Recovery Act (RCRA) of 1976

The Resource Conservation and Recovery Act of 1976 specifies that any farmer that accumulates 2.2 pounds (1 kilogram) per month or more of waste containing a regulated *acute hazardous chemical* or 2,200 pounds (1000 kilograms) of waste containing a regulated *hazardous chemical* must register as a generator of hazardous waste, obtain and ID number from EPA, and follow certain disposal requirements. The list of chemicals specified as acute hazardous chemicals or hazardous chemicals is available by calling the EPA RCRA Hotline at 1-800-424-9346.

Small farmers who properly dispose of pesticide wastes, excess pesticides, and triple-rinsed empty containers on their own property are in general exempt from the requirements of this law, although their state requirements are often more strict than national ones and would apply. They must, however, dispose of their triple-rinsed* pesticide containers in *EPA-approved* sanitary landfills.

Superfund Amendments and Reauthorization Act of 1986 (SARA Title III)

The Superfund Amendments and Reauthorization Act of 1986 (SARA Title III) is designed to inform communities regarding hazardous chemicals located in the vicinity so that a community emergency response plan would be in place in the event of an accident. This law

---

*Triple rinsing is an established technique that specifies that when a container is filled with water and emptied three times, the concentration of pesticide remaining as contamination on the container is deemed to be within acceptable levels to avoid unnecessary environmental pollution.

affects pesticide producers, distributors, retailers, and some pesticide applicators.

Among the aspects of this legislation that deal with pesticides is the requirement for emergency-release reporting to describe the safety measures that will be followed when an accidental release (such as a spill) of an appreciable quantity of a hazardous pesticide specified in the act occurs. In the event of a spill, the spill must be reported to the State Emergency Response Commission (SERC) and the Local Emergency Planning Committee (LEPC). A report must also be made to the National Response Center (1-800-424-8802).

The 1992 Worker Protection Standard (WPS)

The 1992 Worker Protection Standard (WPS) is a set of regulations announced by the EPA to protect workers in greenhouses, farms, vineyards, and so forth from accidental pesticide poisoning. It specifies the safety measures that must be taken by pesticide handlers who mix, load, apply, clean, or repair equipment; act as flaggers, and the like; as well as by regular workers. For example, workers who are not wearing protective garments are not permitted to enter an area where pesticides are being applied. When this is the case, the area must be either posted or an announcement made that the area is off limits. Likewise, they are to be protected from accidental contact with pesticides through drift.

WPS also requires that decontamination supplies and emergency assistance be available whenever and wherever pesticides are being applied. Workers must by law be informed about pesticide hazards through safety training, safety posters, access to labeling information, and access to specific information about the treated areas.

The Food Quality Protection Act (FPQA) of 1996

The Food Quality Protection Act of 1996, passed unanimously by Congress, requires all pesticides to be safe for infants and children. The law further stipulates that combined exposures to pesticides be considered when setting safety standards. The law requires the EPA to act to protect infant and child health, even in the absence of total scientific certainly regarding the toxicity or exposure effect of pesticides for the fetus, infant, or young child. This is a dramatic reversal of previous statutory requirements where the EPA had no mandate and arguably could not act to protect the public health, even child health, in the absence of complete data on the risk from a pesticide. Now the law is clear. In the absence of complete and reliable data on pre- and postnatal toxicity and exposure to a pesticide, the EPA must err on the side of child safety and apply an additional tenfold margin of safety to food tolerances for the pesticide. A clause also directs the EPA to test chemicals to determine whether they cause birth defects or reproductive problems. The new law allows on the market only substances that cause no more than one additional cancer case for every million people who are exposed to them over the course of a lifetime. At least 30 chemicals in use at the time the law was introduced did not meet the one-in-a-million standard and are being phased out, and others do not pass requirements that chemicals be safe for children. Under the law, states are barred from passing their own, more stringent food-safety standards.

Pesticide Application

The goal of pesticide application is to evenly disperse the pesticide on the target in the proper concentration. In order to accomplish this, there is a wide variety of equipment available, each designed for a particular type of application. Generally, pesticides that are to be applied to the foliage (such as insecticides and fungicides) are most effective when small droplets are evenly applied to the tops and undersides of the leaves. This requires sprayers that are designed to atomize the droplets and convey these small droplets to the target. On the other hand, pesticides that are to be applied to the soil (such as herbicides, soil fungicides, and soil insecticides) usually do not require the fine atomization and can be applied with equipment that applies relatively large droplets and may even be applied in the irrigation system. In greenhouses and nurseries, soil-active fungicides and insecticides are often applied through the watering system in what is called a *drench*. Because large droplets are least likely to drift off target, the use of larger droplets will result in increased safety for the applicator and the environment. Table 13-5 summarizes some common types of sprayers used in horticulture.

Table 13-5 Some Common Types of Pesticide Applicators Used in Horticulture

Sprayer type	Comments
High-pressure hydraulic sprayer	For use in full-coverage dilute sprays of nurseries, landscapes, orchards, vegetables, and the like. Applies a high volume with possible drift problems.
Electrostatic sprayer	For use in full-coverage sprays of nurseries, landscapes, orchards, and vegetables. Applies a low volume utilizing the charging of spray droplets to assist with coverage and to reduce drift. Pesticide quantities per area are usually reduced because of decreased runoff and wastage. Compressed air assists atomization and carries droplets to the target.
Fogger	Aerosol-sized droplets are produced by a variety of techniques, including heat and high pressure. For use in greenhouses. Excellent coverage and decreased pesticide usage is possible, depending on the type of equipment.
Backpack and tank sprayer	For use in small-scale applications commercially and residentially. Low pressures result in relatively large droplets that do not achieve good coverage and cause runoff and increased pesticide use.
Controlled droplet sprayer	Centrifugal force creates uniform droplets suitable for all types of applications, depending on the nozzle selected. Decreased drift and decreased pesticide use are possible. Nozzles can be boom mounted or used singly.
Air blast sprayer	Spray solution is injected into a high-speed turbulent blast of air that carries the pesticide to the target and assists in coverage. Drift is problematic, but coverage is good.
Hydraulic boom sprayer	High-volume nozzles are mounted on a spray boom. Characteristics vary with nozzle type. Drift may be a problem. For use in both full-coverage and broadcast applications, depending on nozzle type and arrangement
Airplane or helicopter sprayer	Sprays are atomized and injected into downdraft from airplane or helicopter, which creates turbulence and carries spray into the canopy. Drift can be problematic. Useful when soil is too muddy for conventional ground equipment
Granular spreader	Broadcast or gravity distribution of pesticides formulated for application to the soil as dry granules

Sprayer Calibration

Depending on the goal of a pesticide application, the pesticide label may specify two ways of mixing and applying a pesticide. The first, known as a *broadcast application*, requires that a given amount of pesticide be evenly distributed over a given area. Typically, the label directions will say to spray 1 quart per acre, or 1 liter per hectare, for example (an amount of pesticide/unit area). There may be a further suggestion as to a range of amounts of water to use along with the pesticide or, in the case of many preemergent herbicides, the dilution may not matter. The second type of application is one where the label says to mix a given concentration of pesticide and to spray to obtain good coverage of the plant. Typically, the directions will say to apply 1 quart per 100 gallons of water or 1 liter per 100 liters and to spray to runoff. This type of application is known as a *full-coverage spray*. A full-coverage spray does not require that the sprayer be calibrated. However, the pesticide should be mixed to the correct concentration and then skillfully applied so that all parts of the plants are treated.

Generally, broadcast applications require that the sprayer be calibrated. *Calibration* means that the amount of spray delivered per unit area is determined under the conditions of application. Many factors can affect delivery rate, but the most important are nozzle size, type, and condition; application speed; sprayer pressure; and the viscosity of the spray solution.

Calibrate a Sprayer. There are a number of calibration techniques available; however, the most accurate technique is the following:

1. Using a tape, accurately measure off an area of at least 100 square feet (10 square meters) for hand sprayers and one equaling several swaths across the field for tractor-mounted sprayers. The larger the area, the more accurate the calibration will be. The area should be the same type of terrain as that actually to be sprayed, as sprayers will move at different speeds in different terrains.

2. Place some water in the sprayer and test it to be sure that it is functioning properly. Set pressure and adjust nozzles on sprayers that have adjustable nozzles.

3. Place a known quantity of water into the sprayer, making sure that the amount is sufficient to spray the calibration area.

4. Spray the calibration area in exactly the same manner as you would if you had been applying a pesticide.

5. Empty the remaining water in the sprayer into a calibrated container, and record the amount remaining.

6. Subtract the remainder from the amount originally placed in the container. The difference is the amount used for the area sprayed. *Example*: You placed 2 liters (a half gallon) in a sprayer and, after spraying 100 square feet (10 square meters), you had 1 liter (1 quart) remaining. Thus, your sprayer is calibrated to deliver 1 liter (1 quart) per 100 square feet (10 square meters).

7. Based on the calibration above, you would deliver the following:

$$\frac{1 \text{ liter}}{100 \text{ square feet}} = \frac{x \text{ liters}}{43,560 \text{ square feet (1 acre)}}$$

cross multiplying: $100x = 43560$

solve for x: $x = 435.6$ liters/acre

Therefore, if a recommendation was for 5 pounds (2.25 kg) per acre (0.4 hectares) of a preemergence pesticide, you would put 5 pounds of the pesticide in your sprayer and then add enough water to give you a total of 435.6 liters (113.25 gallons). If you then applied the pesticide in exactly the same way as when you calibrated, the sprayer should spray exactly 1 acre before running out.

If you are calibrating a hand sprayer, it is advisable to check your calibration by respraying the calibration area. This should be repeated until you can come close to regulating your speed so as to use the same amount of liquid each time. When you begin actually spraying the pesticide, it is advisable to mentally divide the area into sections and then to spray each section and check the amount of pesticide used. If this is done, the operator can adjust his or her speed accordingly so that accuracy is increased.

Granular Applicator Calibration. Granular applicators have to be calibrated just as sprayers; however, their calibration may be more difficult.

Two general techniques are available for spreader calibration. The first technique requires that a paved area be carefully swept and measured. The granules are applied to the measured area and then carefully swept up and weighed. The weight of granules actually applied is then compared with the desired amount. The aperture of the spreader is then opened or closed accordingly, and the operation is repeated until the desired amount is delivered. This is a time-consuming process, so the spreader setting, once determined, should be noted. Unlike sprayers, which must be recalibrated periodically due to nozzle wear, there is less wear on a granular applicator. Thus, frequent recalibration is usually not necessary. Another, similar way of calibrating a spreader is to weigh the pesticide that is placed into the spreader, apply it to a known area, and then weigh that remaining. After comparing the amount used to the amount recommended, the aperture is opened or closed and the process repeated. This technique is faster than the previous one, but results in wastage of the pesticide.

When whirlybird-type granular applicators are calibrated, a bag can be placed over the opening to catch the pesticide granules, and then the known area can be "treated" as if the pesticide were actually being applied. At the end, the amount in the bag can be weighed and compared with the desired amount. As in the previous methods, the aperture can then be adjusted and the process repeated.

Another method is to place a tray of known area—say 1 square foot—on the ground, apply the pesticide over the area (including that of the tray), and then weigh the amount caught in the tray. If the tray has, for example, caught 10 grams (0.4 ounces) of pesticide and is 1 square foot in size (0.3 square meters), the rate applied is 10 grams per square foot. The rate per acre or hectare can then be calibrated using the technique in step 7 above and the aperture opening adjusted as in previous techniques.

ORGANIC GARDENING

Organic gardening is a popular way of gardening that does not permit the use of any synthetically manufactured granular fertilizers, insecticides, disease sprays, and the like. Organic gardeners contend that such materials work against rather than with nature, upsetting the natural balance of organisms in the environment and creating more problems than they solve. In lieu of conventional manufactured chemicals, they advocate mainly cultural solutions to plant disorders and, when necessary, pesticides, repellents, and baits that occur naturally in the environment. They contend that by virtue of being naturally occurring, chemicals do less harm to the environment and at the same time pose little threat to humans.

The cultural controls discussed for diseases and insects are essentially the same as those used by organic gardeners. In addition, the following pest control methods are advocated by organic gardeners, although the effectiveness of several has not been proven experimentally.

Companion and Repellent Planting

In companion planting, relatively insect-resistant plants are grown next to susceptible ones to repel potential pests from the susceptible plant. Marigold roots, for example, exude a substance that repels nematodes,

and onions and garlic are thought to ward off a number of pests. Garlic has been used successfully on some bacterial and fungal diseases. Companion plants may also harbor beneficial organisms.

Trap Crops

Trap planting is believed to work in the opposite manner as repellent planting. Insects often have definite plant preferences. In theory, planting a more attractive plant nearby will lure potential pests away from the principal crop. Dill supposedly can attract hornworms away from tomatoes, and Japanese beetles are reportedly lured by zinnias.

The effectiveness of trap crops is questionable. It is possible that the presence of two attractive crops compounds rather than alleviates a pest problem.

Traps

Traps can attract insects by providing food or shelter. Molasses, beer, and rotting fruit are reputed feeding lures that attract insects and then drown or stick them in the trap as they feed. Because many insects spend the day under plant refuse, a substitute hiding area can be made by laying a board on the ground. Slugs, sowbugs, and so on will congregate there and can then be destroyed. More sophisticated light and electrocuting traps can also be purchased. Traps can also be used to detect the presence of small numbers of insects so that controls can be used at the most effective time.

Two types of traps are particularly effective. A pheromone trap uses a small quantity of an insect sex attractant (*pheromone*) inside a trap usually coated with a sticky material. The insects are attracted and caught in the trap. Another type of trap, which is effective for whiteflies, fungus gnats, shoreflies, and thrips, is yellow sticky cards or yellow sticky tape. Because these insects are attracted to the color yellow, they fly into the trap and are caught on the sticky material. In the vegetable garden, yellow sticky traps are especially effective when used in combination with reflective mulches such as those with a foil surface. Because aphids, whiteflies, and thrips navigate with ultraviolet light, reflected light disorients them and increases the numbers caught on sticky tape.

Barriers

Barriers prevent insects from reaching a plant. Fly paper, for example, can be placed around a tree trunk to stop insects crawling up from the ground. Wood ashes or diatomaceous earth scattered around the base of a plant will repel snails and slugs. Vegetable garden transplants can be protected from cutworms with paper collars held together with paper clips. Lightweight fabric row covers can also be used to exclude pests from vegetables.

Biological Control

Biological control uses beneficial organisms to control harmful insects. A number of insects use other insects as food or for part of their reproductive cycle. These beneficial insects can reduce the population of pests to a level at which their damage does not justify additional control. Among these are ladybugs, praying mantises, fireflies, lacewings, spiders, certain mites, and many wasps. See Table 13-6 for info on useful beneficial insects.

Beneficial insects (Figure 13-21a,b) can be encouraged in the garden by minimizing the use of broad-spectrum chemicals that kill both pests and *beneficials* (insects that either do no harm or are actually beneficial in a garden). Buying and releasing predators can be effective, although they often don't stay in the garden permanently. Ladybugs, for example, are migratory and probably won't remain in the garden for more than a day or two. During that time they can, however, clean up a population of aphids nearly completely.

Bacteria, fungi, nematodes, and viruses infect and kill insects, but research in this area has been slow, and only a few are available for home use. Ultimately, these may prove to be one of the most valuable forms of insect control.

One disease of insects is *Bacillus thuringiensis (BT)*, a bacterium that kills the larvae of moths and butterflies. Cabbageworms, cutworms, corn borers, and tent caterpillars are among the insects controlled. The bacteria are applied as a spray as soon as worms are detected. After ingesting the bacteria, the worms stop feeding and are dead two to four days later. The dead insects then serve as sources of infection for future worms. *Bacillus thuringiensis* is completely harmless to plants and to animals other than the larvae.

A number of different strains of BT affect different insects. Because they are quite specific, the correct strain must be purchased for the problem at hand. One strain, for example, is very effective for the control of the Colorado potato beetle. Through genetic engineering, the gene for production of this strain has even been incorporated into commercial cultivars of potato causing the potatoes to produce BT themselves, which kills the beetles that feed on them. Another useful strain of BT kills mosquito larvae. Homeowners can purchase this in the form of donut-shaped cakes that are placed into the water and quickly kill mosquitoes present without harming fish or other organisms. The key to the successful use of BT preparations is to use them only when the pest is actively feeding, because they work in the insect's gut and are killed quickly by sunlight.

A third biological control is the beneficial nematode species *Steinernema carpocapsae*. This nematode is

Table 13-6 Common Commercially Available Biological Control Agents

Pest controlled	Biocontrol agent	Comments
Aphids	*Aphelinus semiflavus* (parasitic wasp), *Aphelinus ervi* (parasitic wasp), convergent ladybeetle, lacewings	Parasitic wasps are species specific, so that identification of aphid species is necessary. Ladybeetles are migratory and may have to be reintroduced frequently.
Soil-inhabiting insects (grubs, fungus gnats, cutworms, Japanese beetle larvae)	*Steinernema glaseri, Steinernema carpocapsae*	Beneficial nematodes that are drenched into the soil in infested areas
Caterpillars	*Trichogramma* spp.	Parasitic wasp lays eggs on caterpillars.
Mealybugs	Mealybug destroyer, Lacewing	Mealybug destroyer is a black ladybeetle whose larvae closely resemble mealybugs.
Greenhouse whitefly	*Encarsia formosa* (parasitic wasp)	
Silverleaf whitefly	*Encarsia* spp. *Eretmocerus eremicus* (parasitic wasps)	
Spider mites	*Phytoseiulus persimilis, Neoseiulus* spp. (predatory mites)	Release a few predator mites into each colony.
Western flower thrips	*Neoseiulus cucumeris* (predatory mite)	Mites are often shipped in slow-release packets that are hung on plants and release predators for a six-week period.

(a)

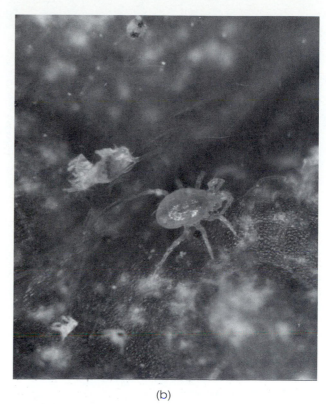

(b)

Figure 13-21 (a) An aphid parasitoid (*Aphidius erui*) emerges from a parasitized aphid. (b) The predatory mite phytoseisulus persimilus, an effective control for the two-spotted spider mite in both outdoor and greenhouse situations. *(Photos by Susan Hofer)*

active in the soil against the larvae of borers, weevils, chafers, and against cutworms, fungus gnats, and some parasitic nematodes. However, it is completely harmless to noninsect species. The nematode enters the body of the host insect and multiplies in the digestive track, killing the larvae. Use of this product will decrease damage from moles and skunks by killing the soil insects upon which they feed.

Biorational Control

Biorational pesticides are chemically similar to naturally occurring material and are minimally disruptive to the environment. Biorational controls can thus be useful tools in pest management. Examples of biorationals are pyrethrin, neem products, fatty acids, and oils. Because discovering new biorationals is an active area of research, new products are expected. The most useful current products are listed below.

Pyrethrin is an insecticide derived from the dried flowers of the pyrethrum daisy that is produced in the upland tropics. The insecticide kills most insects and acts as a repellent but breaks down within a day when exposed

to sunlight. Though it will kill beneficials, they rapidly recolonize treated plants because there is no toxic residue. Some people experience allergic reactions, principally coughing, when they come in contact with pyrethrin.

Neem products are derived from the neem tree, which is native to India. Both oil and water-soluble products are available. Azadirachtin, the water soluble product, acts as an insect growth regulator, preventing molting, and is active on most soft-bodied insects such as aphids, whiteflies, fungus gnats, and caterpillars. It also has repellent effects. The oil product has insecticidal properties and also retards the growth of fungi. Its activity is primarily through smothering of insects, and it can be used to control a wide variety of insects.

Fatty acids or soaps, as they are commonly known, can be insecticidal or herbicidal depending on the particular product. They work primarily by dissolving waxes that coat both plants and insects, causing them to dry out. They do kill beneficials that are directly sprayed but do not kill those that move in after treatment.

Oils work by smothering insects. There are many types and all indiscriminately kill all insects that are contacted. However, like soaps, they have little effect on insects that move onto a plant after the spray has dried.

Oils are commonly classified as either summer (light) oils or winter (heavy) oils. The lighter the oil is, the less *phytotoxic* (damaging to plants) it is likely to be. Winter oils are generally reserved for spraying dormant plants that do not have leaves present, whereas summer oils can be used on most plants when they are actively growing. Because of the potential for plant damage, oils should not be used when weather is hot and humid, within 30 days of application of any sulfur-based product, or too frequently.

Homemade Sprays

Recipes using common household substances for insect control are also available. Buttermilk, soap and water, baking soda, turpentine on rags, and various oils are suggested by organic gardeners as insecticides and repellents.

Botanical repellents can be made at home from a number of common plants. Their effectiveness has never been thoroughly investigated. Onions, garlic, hot peppers, tomato leaves, nettles, and horseradish are all reputed to have excellent qualities when mixed in water and applied as a spray.

INTEGRATED PEST MANAGEMENT

Integrated pest management is an intermediate course between spraying indiscriminately with insecticides and rejecting them completely, as organic gardeners do. It combines many controls—cultural, biological, and chemical—to prevent and treat plant problems. Emphasis is placed on learning as much as possible about the pest or disease problem—the life cycle, host plants, most favorable environment, and so on—and using the information to formulate a control plan. Integrated pest management also involves establishing an amount of damage to plants that can be tolerated and accepted, as opposed to assuming that plants must be completely pest and disease free. In an integrated program, insecticides are used only when cultural and biological controls are ineffective, and they are used when they will be maximally effective.

When an insecticide is necessary, it is chosen carefully with regard to effectiveness, toxicity to beneficial insects, persistence in the environment, and danger to humans. It is unnecessary, for example, to spray a long-lasting, toxic insecticide for pest control when a less toxic and less persistent one will work as well. Similarly, one should not use a multipurpose spray that will kill both harmful and beneficial insects when a systemic drench would kill only insects damaging the plant.

Although it would be difficult for a gardener to be familiar with the life cycles of all the insects and diseases of plants, he or she can use integrated pest management to a limited extent by following these steps.

1. *Determine the cause of the problem.* Decide whether it is due to an insect, disease, rodent, nutritional deficiency, or other source by analyzing symptoms, inspecting for pests, and using a reference book for additional information where necessary.

2. *Determine whether the harm being done is sufficient to justify control.* Set a level of tolerable damage. Mildew attacking annuals in late fall should, for example, be more tolerable than mildew on plants in spring, because the plants will soon be frosted anyway. Treatment might not be justified at all. In the same way, if lettuce has already been severely chewed by insects, spraying may be superfluous because the lettuce will not recover.

3. *Use cultural and biological controls, if practical and available, to treat the problem.* Hand picking a few cabbageworms or trapping slugs in beer is more ecologically sound than spraying and can be nearly as effective. Similar cases can be made for using *Bacillus thuringiensis* on caterpillars and for hosing off aphids and mites with water.

4. *Apply a pesticide as a last resort or if it is the only available control.* Choose the chemical carefully so that it is as nontoxic, nonpersistent, and specific to the pest as possible. For example, miticides should be used for mites, not an all-purpose combination. The chemical should be applied to the damaged plant. Spraying unaffected neighboring plants is not preventative, it is overkill and unnecessarily pollutes the environment and kills beneficials.

SELECTED REFERENCES FOR ADDITIONAL INFORMATION

Agrios, G. N. *Plant Pathology.* 3rd ed. New York: Morgan Kaufmann, 1997.

Bohmont, Bert L. *The Standard Pesticide Users Guide.* Upper Saddle River, NJ: Prentice Hall, 1999.

Driestadt, Steve H. *Integrated Pest Management for Floriculture and Nurseries.* University of California Cooperative Extension Publication #3402. 2001.

Dreistadt, Steve H. *Pests of Landscape Trees and Shrubs.* University of California Cooperative Extension Publication #3359, 1994.

Ellis, B., and F. Bradley. *The Organic Gardener's Handbook of Natural Insect and Disease Control.* Emmaus, PA: Rodale, 1996.

Ellis, Barbara, D. *Organic Pest and Disease Control.* New York: Houghton Mifflin, 1997.

Fermanian, T. W. and Malcomb C. Shurtleff. *Controlling Turfgrass Pests.* Upper Saddle River, NJ: Prentice Hall, 1997.

Flint, Mary Louise and Steve H. Driestadt. *Natural Enemies Handbook.* University of California Cooperative Extension Publication #3386. 1998.

Gardens Alive. *Retail Catalog.* 5100 Schenley Place, Lawrenceburg, IN, 47025.

Johnson, Warren T., and Howard Lyon. *Insects That Feed on Trees and Shrubs.* Ithaca, NY: Cornell University Press. 1991.

Lloyd, John. *Plant Health Care for Woody Ornamentals.* Urbana: University of Illinois, 1997.

Moorman, Gary W. *Diseases of Ornamental Plants.* Available from American Nurserymen, Chicago, IL, 1996.

Olkowski, W., Sheila Daar, and Helga Olkowski. *Common Sense Pest Control.* Newton, CT: Taunton Press, 1991.

Pierce, Pamela K. Environmentally Friendly Gardening *Controlling Vegetable Pests.* Des Moines, IA: Meredith, 1996.

Pirone, Pascal P. *Diseases and Pests of Ornamental Plants.* 5th ed. New York: Wiley, 1978.

Sinclair, Wayne A., Howard H. Lyon, and Warren T. Johnson. *Diseases of Trees and Shrubs.* Ithaca, New York: Cornell University, 1987.

Smith, Michael D., ed. *The Ortho Problem Solver.* Des Moines, IA: Meredith Books, 1994.

University of Arizona College of Agriculture. *Homeowner's Guide to Outdoor Pesticide Safety.* Tucson: University of Arizona College of Agriculture.

Ware, George W. *Complete Guide to Pest Control: With and Without Chemicals.* 2nd ed. Fresno, CA: Thompson Publications, 1996.

Ware, George W. *The Pesticide Book.* Fresno, CA: Thompson Publications, 1999.

Westcott, Cynthia, revised by R. Kenneth Horst. *Westcott's Plant Disease Handbook.* 6th ed. Norwell, MA: Kluwer Academic Publishers, 2001.

World Wide Web Addresses

www.ipm.ucdavis.edu   A site on integrated pest management from the University of California

www.nysaes.cornell.edu/ent/biocontrol   A site on biological control of insects from Cornell University

www.cdpr.ca.gov/docs/dprdocs/goodbuy/organism.html   A website with sources of biological control agents

http://ipmwww.ncsu.edu/biocontrol/biocontrol/biocontrol.html   A website on biological control from North Carolina State University

http://www.cdc.gov/niosh/nasd/menus/topchem5.html   This is from the National Ag Safety Database. "Documents by Topic: Chemicals/Pesticide: Laws" NASD Website, 2001. (May 31, 2001)

http://www.epa.gov/pesticides/label/   This is from the United States Environmental Protection Agency, Office of Pesticide Programs. "Read the Label First" EPA Website 2001. (August 5, 2001)

http://www.urbanext.uiuc.edu/ecosystems/teacheguide3.html.   University of Illinois Extension. "Integrated Pest Management Teacher's Guide" Website. (October 1, 2001)

PART THREE

GROWING PLANTS INDOORS

INTRODUCTION

No other branch of horticulture has undergone the rise in popularity that indoor plant growing has during the past three decades. Almost every home now displays at least a few indoor plants, whether they are hanging baskets, large freestanding specimens, or windowsill pots. Indoor plant enthusiasm remains constant because many people enjoy being surrounded by plants in their homes, at work, where they shop, or where they dine. Lush green foliage creates a cool, tranquil feeling and mixes with any decor from modern to traditional. Indoor plants link the outdoor world of nature and the indoor domain of people in a mutual living space.

Many commercial buildings, more now than ever before, are designed with atriums, skylights, and artificial lighting to accommodate the light requirements of plants. Commercial *interiorscaping* or interior landscaping is a full-fledged multibillion dollar industry with a technological base.

Significant scientific advancements have been made during these decades to make houseplant growing more successful for the average person. Specifically, new cultivars have been selected or generated from tissue culture mutations. These new plants are much more durable and better able to survive—even thrive—in less light and less humidity than the original cultivated species are. They have also been selected for pest resistance, although this has proven to be a more elusive quality than environmental adaptability. Also, new species never commercially produced or even cultivated until relatively recently have been introduced.

In response to consumer preferences, much research has focused on developing new and more dependable flowering plants, ones with longer lasting flowers in a wider range of colors. A florist pot plant (see Chapter 14) that would have been expected to last two weeks indoors several decades ago now will last months in good condition. People can throw out their Christmas poinsettias on the 4th of July!

Lastly, plants have been selected for compactness, which has also been induced by growth retardants. Compactness improves *keeping quality,* the ability of a plant to maintain an attractive appearance over time in unfavorable growing conditions.

PLANTS AS CLEANERS OF INDOOR AIR POLLUTANTS

Beyond aesthetic considerations, research over the past two decades has shown that plants improve indoor air quality. During the energy crisis of the 1970s, building designs that restricted outdoor air exchange were the norm to avoid energy use necessary by constant heating and cooling of outside air. Self-contained ventilation systems that recirculated the air were standard, and most large commercial buildings today are still designed this way. In such buildings, pollutants accumulate to levels that make the air inside these buildings even more polluted in some ways that the outside air of a large industrial city.

Although recirculating air does conserve energy, it has led to a collection of illnesses referred to as "sick building syndrome." Sick building syndrome includes a variety of physical symptoms such as respiratory infections and difficulties, headaches, skin irritations, eye irritation, nervous system disorders, and even psychological disturbances.

The most common chemicals identified as indoor pollutants are carbon monoxide, nitrous oxide, benzene, formaldehyde, and trichloroethylene. There are also problems with acetone, ammonia, toluene, and xylene. Carbon dioxide and nitrous oxide are by-products of smoking, and so is benzene. Benzene is additionally a solvent in ink, paint, dye, plastic, and synthetic fabrics. Formaldehyde is used in the manufacture of particle board, carpet, fire retardants, insulation, some cleaning agents, and paper bags. Trichloroethylene is found in printing ink, varnish, and some adhesives.

Better ventilation is, of course, the obvious solution, but this is not advisable in cities where exterior pollution levels are high. Plus, there remains the problem of the many buildings built with recirculated air systems that are still in use or being built today. For these situations, plants are effective in *scrubbing* the air.

NASA research aimed at scrubbing recirculated air in space vehicles has been the main source of information on the air-cleaning abilities of various plants. NASA researcher B. C. Wolverton tested 50 flowering and foliage houseplants for their ability to remove toxic gases. The plants were put in sealed chambers, and the parts per million of specified gases measured both before and after a 24-hour period. The results were impressive. Almost every plant tested reduced the level of pollutants, but there was variation in which plant was most effective on which pollutant.

Spider plants removed 96 percent of carbon monoxide and 99 percent of nitrous oxide. Formaldehyde was removed well by about half the species. Sanseverias, scheffleras, dracaenas, and a few others were excellent at removing benzene. The study found that, in general, plants that can live in low light and have large leaves remove pollutants best. Table III-1 lists some of the plants tested and found to be effective pollutant removers.

Soil was also found to play a role in removing pollutants, and microorganisms in the soil played a particularly important part. A root-soil complex alone was capable of purifying air to some extent. But when potted plants were combined with activated carbon filters and a fan, the combination was more effective than plants alone. The carbon absorbed pollutants, and the roots grew into the carbon and degraded the chemicals there.

An interesting fact discovered in the course of the research was that plants and soils continuously exposed to air containing toxic chemicals actually improve over time in their ability to clean air. *Aglaonema* 'Silver Queen' initially removed 47.6 percent of benzene in a given time, but after six weeks of intermittent exposure was able to remove 85.8 percent in the same amount of time. This phenomenon was attributed to the well-known genetic adaptability of soil microorganisms to improve their detoxifying potential with exposure to toxic agents.

THE INDOOR ENVIRONMENT

A plant must adapt to many environmental conditions if it is to live indoors. These are light, temperature, humidity, water quality, containerization, and air circulation.

Table III-1 Plants Found Effective in Removing Indoor Air Pollutants

Botanical name	Common name
Aechmea fasciata	vase plant
Aglaonema modestum[1]	Chinese evergreen
Aloe vera[1]	aloe
Chamaedorea erumpens	bamboo palm
Chlorophytum elatum[1]	spider plant
Chrysalidocarpus lutescens	areca palm
Chrysanthemum morifolium[1]	chrysanthemum
Dendrobium sp.	dendrobium orchid
Dieffenbachia maculata	dumbcane
Dracaena[1]	
deremensis 'Janet Craig'	'Janet Craig' dracaena
deremensis 'Warnecki'	'Warnecki' dracaena
marginata	red-margin dracaena
massangeana	corn plant
Epipremnum aureum[1]	pothos
Euphorbia pulcherrima	poinsettia
Ficus benjamina	weeping fig
Gerbera jamsonii[1]	gerbera daisy
Guzmania 'Cherry'	guzmania bromeliad
Hedera helix	English ivy
Neoregelia carolinea 'Tricolor'	tricolor bromeliad
Nephrolepis exaltata	Boston fern
Peperomia obtusifolia[1]	peperomia
Phalenopsis sp.	butterfly orchid
Philodendron	
domesticum	elephant ear
scandens	philodendron
selloum	lacy tree philodendron
Rhapis excelsa	lady palm
Rhododendron sp.	azalea
Sanseveria trifasciata 'Laurentii'	snake plant
Schefflera arboricola[1]	miniature schefflera
Spathiphyllum 'Mauna Loa'	peace lily
Syngonium podophyllum	arrowhead

[1]Species found to be particularly effective at removing pollutants.

In the final analysis, plants will always be at a disadvantage indoors because human comfort will always be the primary consideration. Temperature will be controlled to the liking of people, as well as light intensity, air circulation, and other factors. The conditions that control the health of plants sharing that living space are considered secondarily.

Light

Probably the most important factor is reduced light. Light intensities comfortable for people to live and work in are insufficient for many species of plants. If plants do not receive sufficient light to manufacture the carbohydrate required for respiration, they will eventually die.

Temperature

Outdoors there is usually a 5 to 15°F (3–8°C) difference between day and night temperatures, with the temperature dropping slowly after sunset. These cooler night temperatures reduce the rate of respiration and prevent the rapid depletion of stored carbohydrate reserves generated during the daylight hours and necessary for growth and maintenance. However, plants grown indoors are often not exposed to cool night temperatures. As a result, their carbohydrate reserves are used more quickly, and the plants have little or none to use for new growth.

Insufficient Humidity

Insufficient humidity is a third factor contributing to the less than ideal conditions indoor plants experience. Home heating creates dry air conditions indoors in most parts of the country. This causes tropical plants to lose water more rapidly by transpiration than they would in their native environment. The plants absorb water to replenish that which has been lost, but often the water is transpired more rapidly than it can be replaced. The leaf tips then turn brown, and the plants develop other symptoms indicative of insufficient humidity.

Water Quality

The quality of the water used on indoor plants is also unlike that which they would receive outdoors. Under natural conditions rainwater supplies moisture. It has relatively few impurities and, generally, a neutral pH (neither acid nor alkaline reaction). But tap water is used indoors. Depending on its source and prior treatment, it may contain numerous chemicals such as chlorine, sodium, or fluoride. In addition, it may be acidic or basic, which will over time affect the availability of nutrients from the growing medium.

Containerization

Plant roots restricted to a pot live under unnatural growing conditions. The nutrients and water available to the plant are limited by the small amount of medium in the pot and must be more carefully regulated than for plants outdoors.

Some physical characteristics of the potting medium also change by being restricted. Soil that is loose and drains quickly outdoors may pack down and drain slowly in a pot. Consequently, garden soil is not often suitable for houseplant use unless it is mixed with other materials (Chapter 15).

Air Circulation

In the outdoor environment, air movement is fairly constant because of changing temperatures and natural winds. This circulation moves oxygen and carbon dioxide over the leaves of the plants, aiding in photosynthesis and respiration and drying foliage after rainfall. But indoor air can be relatively stagnant during the winter months when windows remain closed. Any fresh air is likely to come through the opening of outside doors and may be 40°F (22°C) colder than room temperature. Accordingly, plants in the direct path of doors often suffer undiagnosed illnesses from rapid air temperature changes.

Conclusion

With so many modified environmental factors to cope with, it would seem that growing plants indoors would be difficult. But the size and superb condition of many tropical plants growing in homes disprove this. Their owners obviously understand the requirements of plants and meet those requirements. Their skill can be acquired by anyone through reading, observing, and, most important, gaining experience in growing many plants.

SELECTED REFERENCES FOR ADDITIONAL INFORMATION

Blessington, Thomas M., and Pamela C. Collins. *Foliage Plants—Prolonging Quality.* Batavia, IL: Ball Publishing, 1993.

Brown, Ross. "Are Plants the 'Magic Bullet' to Cure Sick Buildings?" In *Interior Landscape Industry,* March 1989, pp. 50–55.

Center for Indoor Air Research. *Final Report: The Role of Foliage Plants in Indoor Air Quality.* Linthicum, MD: Center for Indoor Air Research, 1997.

DeBraak, LaRonna. *A Breath of Fresh Air: A Practical Guide for Filtering Out Indoor Air Pollution Utilizing House Plants.* Denver: LaRonna DeBraak, 1991.

Graf, Alfred Byrd. *Hortica: Color Cyclopedia of Garden Flora and Exotic Plants.* 4th ed. New York: Macmillan, 1992.

Marchant, Brent. "Interim NASA Report Reveals Impact of Roots in Cleaning Air." In *Interior Landscape Industry,* October 1989, pp. 12–14.

Wolverton, B. C. *How to Grow Fresh Air: 50 Houseplants to Purify Your Home or Office.* New York: Penguin, 1997.

Wolverton, B. C., Anne Johnson, and Keith Bounds. "Final NASA/ALCA Research Results." In *Interior Landscape Industry,* January 1990, pp. 12–16.

Wolverton, B. C., and Douglas L. Willard. "NASA/ALCA Test Update." In *Interior Landscape Industry,* June 1989, pp. 8–12.

World Wide Web Addresses

http://www.ext.vt.edu/departments/envirohort/articles/misc/plntclar.html   This article by Diane Relf is entitled "Plants Actually Clean the Air."

http://www.plants4cleanair.org/   This is a home page for a group of organizations that sponsor research on indoor air pollution.

http://www.plants4cleanair.org   (undated) (November 2, 2001) Plants for Clean Air Council. "For Purifying the Air you Breathe, Plants are a Natural!"

http://ohioline.osu.edu/hyg-fact/1000/1227.html.   (November 2, 2001) Martin, Jane C. "Maintenance of Interior Foliage Plants." Bulletin HYG-1227-96.

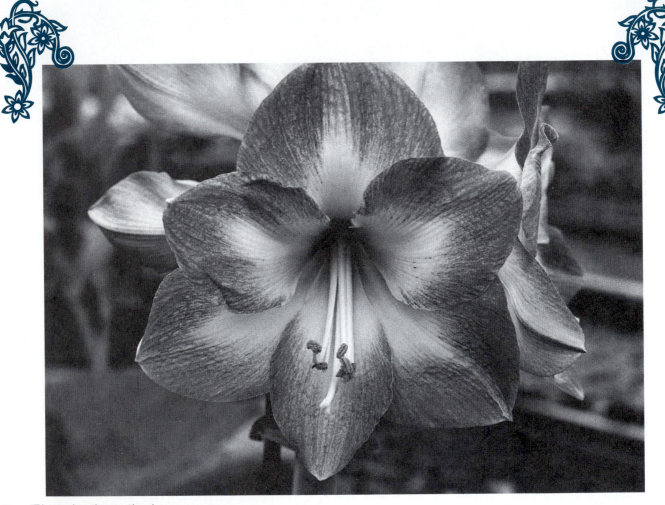

(Photo by the author)

CHAPTER 14

INDOOR PLANT MAINTENANCE

PURCHASING INDOOR PLANTS

With plants, as with any other purchase, it pays to shop around. There are bargains and lemons among indoor plants, and the way a plant looks in the store is not necessarily the way it will look after two or three months under home growing conditions.

> A foliage plant producer would probably be located in a warm winter area such as Florida and California. The plants are normally propagated on-site and greenhouse-grown in containers before being shipped throughout the country for sale

One of the biggest disappointments can occur with purchases of flowering plants. Such plants are usually grown in greenhouses under bright light intensities that trigger flower induction. Under indoor growing conditions the light intensity is often not strong enough to cause new buds to set, so that after its present buds are exhausted, the plant is not likely to bloom again. Table 14-1 lists some reliable houseplants that bloom indoors without the use of supplemental lights.

However, even though most flowering plants will not rebloom in the house, it is still possible to have color in indoor plants by either purchasing ones with *variegated* foliage (striped or marked with yellow, white, pink, or red) or houseplants currently in bloom

that will continue to develop the current crop of buds even though the light level is insufficient to set new ones. These can bloom for several months to the better part of a year and are well worth the investment (see section on florist plants).

When you are considering buying a houseplant, take the time to examine the plant carefully. A well-grown houseplant will be compact and have richly colored foliage, strong new growth, and a full crop of leaves. The stem should be strong enough to easily support the weight of the top and have no weak or discolored areas, especially at the base. An insect-infested plant is never a good buy, so check the growing tips, leaf undersides, and potting medium for hiding pests or signs of their damage.

The price charged for the plant will vary according to where it is purchased, species, cultivar, and, sometimes, proximity to the large indoor plant-growing areas of Florida and California. Mass merchandising outlets such as supermarkets and discount stores normally have the lowest prices, florists the highest, and local plant shops and greenhouses are in between. However, mass merchandising outlets usually have the least selection. Also, if the plants have been displayed in the store for long, they may have been improperly cared for and be in poor condition.

The price charged for a plant is often a reflection of the ease with which it is propagated and the time required to grow it to selling size. Wandering Jew and coleus, for example, propagate very easily from cuttings and grow rapidly. They are among the least expensive indoor plants. Palms, on the other hand, are

Table 14-1 Selected Houseplants That Flower Without Artificial Lighting

Common name	Botanical name
Bromeliad	*Aechmea, Billbergia, Nidularium,* and so on
Begonia	*Begonia* spp.
Spider plant	*Chlorophytum* spp.
Coleus	*Coleus × hybridus*
Crocus*	*Crocus* spp.
Orchid	*Cypripedium, Cattleya,* and so on
Crown of thorns	*Euphorbia milii* cultivars
Rose-plaid cactus	*Gymnocalycium* spp.
Amaryllis	*Hippeastrum* hybrids
Hyacinth*	*Hyacinthus orientalis* cultivars
Pincushion cactus	*Mammillaria* spp.
Grape hyacinth*	*Muscari* spp.
Daffodil*	*Narcissus* spp.
Geranium	*Pelargonium* spp.
Swedish ivy	*Plectranthus* spp.
Primroses	*Primula obconica* and *P. malacoides*
Crown cactus	*Rebutia* spp.
Easter cactus	*Rhipsalidopsis* spp.
African violet	*Saintpaulia ionantha* cultivars
Strawberry begonia	*Saxifraga stolonifera*
Christmas cactus	*Schlumbergera bridgesii* cultivars
Thanksgiving cactus	*Schlumbergera truncata* cultivars
Christmas cherry	*Solanum pseudocapsicum*
Spathe flower or Peace lily	*Spathiphyllum* spp.
Cape primrose	*Streptocarpus saxorum* cultivars
Tulip*	*Tulipa* spp.

*These are spring-flowering bulbs that can be *forced* in winter on a windowsill, then planted outside in spring.

grown from seed, and kentia palm seed takes an average of 8 to 12 months just to germinate.

Before transporting a newly purchased plant, take the time to wrap it in paper (folding the leaves upward) or enclose it in a paper bag before taking it outdoors. It will be less likely to break, and in below-freezing weather the paper will act as insulation against the sudden temperature drop.

Then find out the plant's growing requirements for light, watering, humidity, potting medium, and the like. Check a reliable reference, and then try to match the ideal growing conditions as best you can in your home.

ACCLIMATIZATION AND CONDITIONING

Plant *shock* is a common occurrence in newly acquired houseplants. It usually results from moving plants directly from the production area to home growing conditions without a conditioning period. The sudden change to lower light and humidity disturbs the metabolism of the plant and temporarily halts growth. Leaf drop may occur, and some plants may even die.

Properly grown houseplants are *conditioned* or *acclimated* prior to sale. The light and humidity are gradually lowered to the levels commonly found in indoor conditions, and watering and fertilization are less frequent. Plants that have not been acclimated often betray the fact by their appearance: they are bright green and extremely lush appearing, with much new growth and dense succulent or soft foliage. Basically, they are extremely healthy and vigorous plants and would continue to grow well in a greenhouse. But they are not able to adjust well to indoor conditions.

It is possible to acclimate a tropical plant that was not conditioned before sale. First, give the plant the maximum amount of light and humidity available and keep the soil moist. This will go as far as possible toward approximating the former growing environment and minimize shock. Then gradually move it further from the light and decrease the humidity and water to the recommended range over several months. This long period of adjustment will allow the plant to acclimate to an indoor growing environment, but even then some foliage drop should be expected. After acclimatization, the plant should grow slowly but satisfactorily provided the correct light level and other requirements are met.

MAINTENANCE ACTIVITIES

Routine maintenance is an essential part of houseplant care if the original health and beauty of plants are to be preserved. Two major aspects of maintenance, watering and fertilizing, are significant enough to require discussion in individual chapters, but the less commonly practiced maintenance activities discussed here are no less important.

 Foliage plant maintenance personnel normally have a number of client businesses (malls, offices, or even homes) that they visit throughout the week. At each site the plants are watered, pruned, cleaned, and checked for disease and insect infestation. Plants in poor condition are replaced.

Inspection

Regular inspection for pests and diseases is a part of maintenance that is commonly overlooked, to the decline and ruin of many beautiful plants. Its purpose is much the same as annual checkups for people: to detect and cure minor health problems before they become bigger and more difficult to control. Inspection of houseplants should take place every one to two weeks and should consist of a general search of the plant for any signs of poor health. New growth, in particular, should be carefully scrutinized under bright light for signs of pests such as spider mites (see Chapter 18); a small magnifying glass is helpful for this purpose. Likewise, more mature leaves should be examined both above and underneath, where insects frequently hide. The leaf axils are another favored spot of concealment for pests.

At the same time, the general condition of the plant should be evaluated. If young growth is spindly or many lower leaves are beginning to yellow, the growing conditions should be altered to remedy the problem before the plant deteriorates any further.

The soil surface and roots should be checked for signs of soil-dwelling insects. Possibly toxic salt accumulations (with their characteristic white coating) may be noticeable.

Plant inspection may appear to be time-consuming, but it actually will take only a few minutes. The time will be well spent if an insect infestation can be detected before the pests spread to other plants.

Cleaning

Cleaning is the second aspect of maintenance. Outdoors, rainfall regularly washes the leaves of plants. It removes the dust and dirt that block out light and decrease photosynthesis. But indoors most water is applied directly to the soil, and so the leaves are never cleansed. Periodic washing or dusting is a substitute for natural cleansing by rainfall and should be performed whenever the leaves show an appreciable amount of dust. In addition, it will help remove insects and their eggs, lessening pest problems.

Several methods may be used to clean the leaves, and the one chosen should be determined by personal preference and the size of the plant. Small plants are perhaps most easily cleaned by the spray nozzle found on most kitchen sinks. Excess water will conveniently drain in the sink, and leaves should be sprayed thoroughly on both the tops and the undersides.

A shower is very useful for large-specimen plants. Place the plant in the tub, and adjust the water to approximately room temperature. Then leave the plant under a low-pressure spray for about 5 minutes, turning the plant occasionally to make sure all the leaf surfaces are adequately washed. If there is a particularly stubborn residue on the plant, the leaves can be sponge cleaned with a drop or two of detergent in a bowl of water, but the plant should be thoroughly rinsed afterward.

Many indoor gardeners find that carrying large plants to the shower every month is inconvenient, if not completely impossible, so they use a damp cloth or sponge for cleaning instead. This method is satisfactory provided the leaves are smooth and not easily damaged. It is especially recommended for areas in which the water has an appreciable mineral content. This "hard" water will leave unsightly residue on sprayed or showered plants, so wiping should be substituted.

Dusting is another way of cleaning large-leaved plants. Only a very soft duster, such as a feather or lamb's wool duster, should be used.

Leaf Shines

Many commercial "plant shines" or "leaf cleansers" are now available in aerosol sprays, fiber-tipped applicators, and foil-encased towelettes. All claim to improve the appearance of foliage and enhance the looks of plants. Most garden books, though, contend that they are bad for the health of the plants because they cause "pore" clogging. Neither is completely correct.

Most plant shines and leaf cleaners will not harm houseplants. However, a few will, especially on young foliage, and this is most often attributable to an aerosol propellant. The pore-clogging claim is not valid: first, because the "pores" (technically, stomata; see Chapter 2) are not like human pores and do not respond in the same manner to clogging. In fact, commercial nurseries often use chemical sprays called *antidesiccants* or *antitranspirants* during transplanting. These antidesiccants are solely for the purpose of sealing the stomata and preventing moisture loss. Second, most stomata are found on the undersurfaces of leaves, where leaf shine is not applied. Consequently, the majority will not be affected.

Nor are leaf shines or cleaners necessarily the best way to cleanse plant leaves. Their chemical makeup varies greatly among the different brands. Some leave an artificial-looking gloss and an oily residue that attracts dust, whereas others leave no residue and only a slight shine. People who prefer to use a leaf shine or cleaner should compare several brands before deciding.

Homemade leaf shines such as milk or salad oil were commonly used in the past. Salad oil cannot be recommended because of the resulting greasy film. Milk does not have this objectionable characteristic and is safe and inexpensive.

The hair-covered leaves of tropicals such as purple velvet plant (*Gynura aurantiaca*) and panda plant (*Kalanchoe tomentosa*) can be either cleaned with a spray of water or dusted with a soft paintbrush. Leaf shines should not be used, and neither should hand wiping, which can flatten or break off the hairs. When rinsing the hairy-leaved members of the African violet family, always use water at room temperature. Gesneriads are extremely sensitive to temperature changes. Water more than 10°F (5°C) cooler than room air temperature will cause yellow splotching on the leaves wherever it touches. The discoloration that results is caused from the destruction of leaf chlorophyll and is permanent (see Figure 17-9).

Whereas some indoor tropicals have hair-covered leaves, others possess a natural filmlike coating that should not be rubbed or covered over with leaf shine. Bromeliads are one example. The foliage of many bromeliad species is covered with flaky silverish scales (Figure 14-1). The scales are decorative, but their func-

Figure 14-1 White bands of water-absorbing scales on *Aechmea* bromeliad. *(Photo by Rick Smith)*

tion is highly practical. They are cells for absorbing water directly through the leaves.

Cacti and succulents frequently have a waxy cuticle that rubs off easily with the fingers. The coating serves as an insulation layer against moisture loss and is an adaptation factor that helps them survive in arid environments. As a general rule, leaf polishes or cleansers should not be used on cacti and succulents.

Turning

Normally plants indoors receive their light from a nearby window. The leaves facing the light receive more solar energy, causing the plants to develop more in the direction of the light. To prevent plants from becoming lopsided, it is advisable to turn them a quarter turn every few weeks so that a different section of foliage faces the light.

Pruning

Pruning is commonly regarded as an outdoor plant maintenance chore. Consequently, indoor plant pruning is often neglected, resulting in many poor-looking houseplants.

Pruning has several functions. First, it causes branching and new growth, producing a fuller, more attractive plant. Second, it removes dead, insect-infested, or diseased plant parts to halt the spread of the organisms.

Pinching

The most commonly practiced form of pruning for indoor plants is a technique called *pinching* (Figure 14-2).

(a)　　　　　　　　　　　　　　　　(b)

Figure 14-2 (a) Pinching a coleus. (b) One month later after branching has occurred.

Pinching consists of breaking off up to about 3 inches (8 centimeters) of the growing tips.

The meristems of plants are known to produce a number of chemicals, including a plant hormone called *auxin* (see Chapter 3). This hormone constantly moves down the stem of the plant from its source of manufacture. It inhibits growth of buds lower on the stem and prevents branching. This hormonal effect on growth is termed *apical dominance*, indicating that the top bud (the apical meristem) "dominates" the growth of the buds below it. Buds directly below the growing point are kept fully dormant, because the auxin is in strongest concentration there. As auxin continues moving down the stem, the concentration decreases. The lowest buds on the stem receive auxin in least concentration and sometimes overcome dormancy and grow into shoots.

Auxin manufacturing rates and movement vary among species, accounting for why some plants branch naturally when others do not. Pinching removes the source of auxin manufacture on plants that will not branch of their own accord.

When a growing tip is removed, it is nearly always the buds nearest the tip that begin growing (Figure 14-2). Although sometimes only one will grow, usually two to three will surge into growth at the same time, producing several new leafy shoots. Because they are also growing tips, they soon begin auxin manufacture, keeping the buds below dormant, and the cycle continues.

Pinching must be done regularly on some species to promote constant branching and to develop a full plant. As new shoots lengthen, the tips are removed continually to make the plant full and dense.

Leafy, vining houseplants most often require regular pinching to prevent a "stringy" appearance. Ideally they should be pinched frequently after potting to de-

Table 14-2 Most Common Houseplants Requiring Frequent Pinching

Common name	Botanical name
Coleus	*Coleus* × *hybridus*
Goldfish vine	*Columnea* spp.
Tahitian bridal veil	*Gibasis geniculata*
Purple passion vine	*Gynura aurantiaca* 'Purple Passion'
Pocketbook plant	*Hypocyrta* spp.
Polka-dot plant	*Hypoestes phyllostachya*
Bloodleaf	*Iresine* spp.
Aluminum plant	*Pilea cardierei*
Swedish ivy	*Plectranthus* spp.
Christmas cherry	*Solanum pseudocapsicum*
Wandering Jew	*Tradescantia* or *Zebrina* spp.

velop a large number of cascading stems. The most common vining and upright-growing houseplants requiring frequent pinching are listed in Table 14-2.

Renewal Pruning

Renewal pruning is used to rejuvenate a plant that has become unattractive and bare-stemmed through drought, insufficient light, or some other cause. It involves removing all or nearly all of the foliage, leaving only 1- to 2-inch (3- to 5-centimeter) stems (Figure 14-3). In several weeks new growth will begin from the leafless stubs. With regular pinching *and* improved growing conditions, the plant will become attractive again in a few months.

Renewal pruning is a drastic measure and should be employed only if the plant is basically healthy, with a strong root system and stored carbohydrate reserves.

Figure 14-3 Stages in renewal pruning.

For plants in a severely weakened condition, a different treatment should be used. First, the source of the problem should be corrected. This will encourage healthy growth and will often be sufficient to break the dormancy of lower buds. After the plant is partially refoliated, the remaining unattractive portions can be cut back. By allowing some foliage to remain, further carbohydrate depletion will be prevented.

Pruning by Thinning

Not all indoor plants are pruned by pinching. Many houseplants such as spider plants grow in a form called a *rosette* (see Chapter 2) and produce new foliage from a short section of stem at soil level called the *crown*. Others, such as most ferns, send up new leaves from an underground *rhizome* system (underground stems) and botanically speaking have no "stems" aboveground at all, only leaf petioles. Each new shoot that develops is a single leaf, and the dormant buds are found at the crown at or just below soil level. Pinching a leaf back will never cause branching. Instead, it should be cut off at the base to promote new growth from the crown.

In pruning these types of houseplants, usually only a few leaves are removed at a time as needed to improve the appearance of the plant. However, in cases in which the entire plant is unsightly or heavily infested by insects, renewal pruning may be advisable. As with renewal pruning of houseplants having stems, all the foliage is cut off just above ground level, and the plant is allowed to refoliate under improved growing conditions and care.

Plants That Seldom Require Pruning

A moderate percentage of houseplants seldom require any pruning except removal of dead leaves. Pinching or renewal pruning of these plants will often destroy their form and create an unnatural appearance. Selected houseplants that require minimal or no pruning maintenance are listed in Table 14-3.

Pruning Flowering Houseplants

Flowering houseplants follow the basic rules for pruning that apply to other houseplants. For many, however, the flowering period can be lengthened by removing the flower heads and seed pods as soon as the blooms fade. If the seed pods are not removed, energy will be channeled into seed production rather than into the formation of additional flower buds.

Exceptions to this practice are plants in which the fruit is an ornamental feature, such as Christmas cherry. Blooms should also be left if one is trying to

Table 14-3 Selected Plants Requiring Minimal or No Pruning

Common name	Botanical name
African violet	*Saintpaulia*
Bromeliad	*Aechmea, Neoregelia, Vriesia* etc.
Cactus	*Cereus, Gymnocalycium, Mammillaria* etc.
Chinese evergreen	*Aglaonema commutatum* cultivars
Dracaena	*Dracaena*
Dumbcane	*Dieffenbachia* spp.
Norfolk Island pine	*Araucaria heterophylla*
Orchid	*Cattleya, Paphiopedilum*
Palm	*Chamadorea, Howiea, Phoenix* etc.
Rosette-type succulents	*Agave, Aloe, Haworthia* etc.
Snake plant	*Sansevieria* spp.
Spathe flower	*Spathiphyllum*
Umbrella plant	*Brassaia*

raise seeds to grow additional plants. Table 14-4 lists plants from which blooms should not be removed.

Pruning for Disease and Insect Control

Pruning for the control of diseases and insects is not a common practice because in most cases the entire plant is infected before the problem is discovered. However, these problems often do start on one or two leaves or a shoot tip. Quick removal of the unhealthy parts can be preferable to a series of chemical sprays. Pruning is particularly valuable in dealing with houseplant foliage diseases because chemical cures are rarely available. When pruning for this purpose, all diseased portions should be cut back to healthy areas.

The clipping of browned leaf tips is commonly used to improve the appearance of houseplants such as dracaenas and palms. Although temporarily effective, the leaf tips usually brown again within weeks.

Transplanting and Repotting

Transplanting (or *repotting*, as it is sometimes called) is a routine part of indoor plant maintenance. It is performed repeatedly from the seedling or cutting stage through maturity.

Primarily transplanting is done to provide crowded roots with additional potting medium from which to absorb water and nutrients. Less often, transplanting is

Table 14-4 Plants from Which Faded Flowers and Fruits Should Not Be Removed

Common name	Botanical name	Remarks
Bromeliad	*Aechmea, Nidularium,* and other genera	Colorful fruit often follows blossoming
Chinese evergreen	*Aglaonema commutatum* cultivars	Flowers followed by red berries; seldom flowers indoors
Asparagus fern	*Asparagus* spp.	Ornamental red berries follow bloom
Spider plant	*Chlorophytum* spp.	New plants form on flower stalks
Wax plant	*Hoya carnosa* cultivars	Next season's blooms from old flower stubs
Cactus	*Mammillaria, Rebutia,* and other genera	Red or orange fruits frequently follow blossoming
Christmas cherry	*Solanum pseudocapsicum*	Red fruits form from white blooms

used as a general tonic or cure-all for an undiagnosed plant ailment suspected of being caused by the roots or the potting medium.

Most people repot more often than necessary, switching to bigger pots when frequent fertilization would have been sufficient. Although unnecessary, repotting is seldom harmful, and it involves a considerable amount of extra work.

Timing of Repotting

Several factors must be evaluated to determine if a plant should be moved to a larger container. First, does the plant look top-heavy in its pot and tip over easily? A plant with a large foliage mass will grow best if given an adequate volume of potting medium to support that top growth. Keeping it in an undersized pot will stunt its growth. Exceptions to this are bromeliads and orchids. These can have large aboveground portions easily supported by a very small root mass.

Next, does the plant require frequent watering to keep the soil moist, and does it wilt within a few days of watering? Roots that are overcrowded extract available water and nutrients very quickly, necessitating more frequent watering and fertilization.

Third, is the health of the plant declining with no change in care or growing location? Some houseplant media pack and crust on the surface after a period of months or years. Repotting in fresh media may be advisable.

In extreme cases of underpotting, roots will be found growing on the surface of the medium. Roots may also grow vigorously at the base of the pot, pushing the plant out of the container. This phenomenon is occasionally encountered in dracaenas and spider plants.

Some indoor gardeners use the presence of roots growing from the drainage hole or circling the root ball as an indicator for repotting. This method is not always reliable. Roots grow most vigorously in areas that hold both moisture and air. The soil near drainage holes and around the inside surface of unglazed clay pots provides these conditions most ideally. Accordingly, roots will grow there in abundance, while the bulk of the

potting medium is undisturbed. However, if large numbers of roots have grown in circles, repotting is justified because the bulk of the root system is not in contact with the potting medium.

The season chosen for repotting indoor plants is not important. Most indoor plants go through seasonal dormancy and growth spurts as do outdoor plants, and transplanting during either of these periods is acceptable. Most transplanting is performed in spring or summer, because the increased vigor due to increased light and humidity makes plants outgrow their pots quickly during these periods.

Transplanting of blooming plants should be restricted to a period when they are not flowering, because it is common for a plant to drop buds and flowers prematurely when the roots are disturbed. Repotting blooming plants immediately after flowering is a reliable practice. It provides room for new root growth and, consequently, better foliage growth before the next flowering period.

Unpotting

The amount of difficulty in removing a plant from its pot depends on the container and plant involved. Plants in 8-inch (20-centimeter) diameter or smaller-sized containers can usually be unpotted by a technique known as *knocking out*. To do this, the medium should be moist (but not freshly watered) to keep it from falling away from the roots. Supporting the top of the soil with one hand and the plant stem between the fingers, the pot is turned upside down, and the edge lightly tapped on a counter or table (Figure 14-4). The soil and roots should slide out intact. If they do not, harder tapping, shaking, or running a knife around the rootball may be necessary. Only in very difficult cases should a plant be loosened from its pot by pulling on the stems. Root or stem damage is likely to result.

Unpotting a plant that has been growing in a narrow-mouthed pot is always difficult. A portion of the roots will nearly always be lost when the root ball is pulled through the narrow opening. As with knocking out, the potting medium should be moist and the

Figure 14-4 Knocking a spider plant out of its pot. *(Photo by Rick Smith)*

merged in water and a stream of water directed at the mouth to loosen the medium and flood it from the roots. Eventually enough will flood away to allow the roots through the opening with minimal damage. Because unpotting by flooding causes less root damage, it is valuable for delicate or highly prized specimens.

Unpotting a large plant is often a two-person job. Because it is too large to be held upside down and knocked out, the usual procedure is to lay the plant on its side and ease the roots out horizontally. Foliage breakage can be minimized by folding the leaves upward and wrapping the top of the plant snugly with newspaper tied in place with string.

A good time to inspect the roots is after removing a plant from its container. Healthy young roots are firm and usually white. (A few are other colors, such as the orange of *Dracaena* roots.) But they generally should not be brown or mushy. Such roots are dead and decaying and should be trimmed off before the plant is repotted. Soil insects may also be detected in a repotting operation.

Roots that have circled the root ball should have special treatment at repotting. If repotted in that condition, they may continue to wrap around the soil ball, never branching into the fresh soil and its available nutrients. Wrapped roots may be loosened by pulling them away with the fingers (Figure 14-5) or by cutting several places around the root ball with a knife. This, like pinching, causes branching and the development of new *feeder roots*.

plant dug and eased through the narrow opening. Laying the pot on its side is often an aid in easing the roots out. The medium will fall to the bottom side, and a tool can be inserted in the space above to pull the roots free.

Flooding is another way to unpot a plant from a narrow-mouthed container. The pot should be sub-

Selection of the New Pot

When selecting a new pot for transplanting, both size and drainage should be considered. The new pot should preferably be only 1 to 2 inches (3–5 centime-

Figure 14-5 Loosening the roots of a pot-bound plant. *(Photo by Kirk Zirion)*

ters) deeper and wider than the previous pot. This additional area will hold an adequate volume of fresh potting medium for new root growth. If a larger pot is chosen, it is likely that a large portion of the new potting medium will stay waterlogged because the roots have not grown through it.

Drainage is a characteristic of the potting medium, but where drainage water is deposited is determined by the pot. A container with a drainage hole or holes allows excess water to pass out of the pot. In a pot without holes, water filters through the potting medium and excess eventually collects in the bottom. This excess water keeps the medium at the base of the pot saturated, excluding oxygen necessary for root respiration. For this reason, pots with holes are preferable unless watering is carefully regulated.

A water reservoir can be created in holeless pots to keep drainage water away from the roots. Stones or gravel at least a half-inch (1 centimeter) in diameter should be poured in a 1-inch (3-centimeter) or deeper layer in the bottom of the pot. The layer is then covered with nylon net, a sheet of sphagnum moss, or screen to keep out potting medium and leave air spaces open for water. In repotting large specimens, consideration should be given to the weight that gravel adds. In such cases styrofoam "peanuts" packing material can be substituted.

It is a common misunderstanding that gravel in the bottom of any pot, having holes or not, will improve the drainage. The gravel layer does *not* improve the drainage. Actually, it worsens it by decreasing the depth of soil in the pot, because drainage is a function of both the depth of the pot and of the components that make up the medium. (Medium components' effects on drainage are discussed in Chapter 15, and the physics of water drainage based on pot height are discussed in Chapter 17.) A gravel layer only provides a reservoir in pots from which the water has no way of escaping. Gravel is unnecessary for pots with holes. The only improvement these pots need is a single piece of broken clay pot (round side up) or screen over the drain hole to keep in the potting medium.

Repotting

Repotting is a simple process. The container is filled with potting medium to a level that will permit the plant to be growing as it was previously.

A plant should never be set deeper than it was originally growing with the idea of improving its appearance. This exposes the stems to soil fungi and bacteria that can invade the tissues and kill the plant.

After firming the medium lightly, the plant should be watered to settle out air spaces and lessen shock.

Figure 14-6 Double potting.

Watering may cause the medium to settle somewhat, and it can then be refilled around the top. Enough room should be left to make watering convenient. A space of ½ to 1 inch (2–3 centimeters) is sufficient.

Double Potting

Double potting (Figure 14-6) involves inserting a clay or plastic pot with drainage holes inside a holeless decorative pot. The spaces between the two pots are filled with sphagnum moss, gravel, or other material. Potting by this method provides the advantages of a free-draining pot with the beauty of a decorative one.

Posttransplanting Care

Sometimes plants *shock* after transplanting. They wilt and may remain wilted for days or weeks. To lessen shock, keep newly transplanted plants out of direct sunlight for a few days. Delicate plants can also be enclosed in a plastic bag for a week or two following transplanting. This increases the relative humidity while the roots are reestablishing and regaining their normal absorption potential.

In the rare cases where bagging is ineffective and the plant remains wilted, pruning may be necessary. One-quarter of the foliage should be cut off and the plant should be observed for several days. If it remains wilted, more foliage will need to be removed.

PLANTS WITH SPECIAL REQUIREMENTS

Bonsai

Bonsai is the art of growing and training trees to be miniature versions of the form they achieve in nature (Figure 14-7). It originated before the eleventh century in China, and some bonsai are hundreds of years old. However, a bonsai need not be old to have the appearance of age. Instead, aging is accomplished by pruning, training, and restricting the soil volume and fertilizer available to the roots.

Bonsai are not difficult to create and care for, but they do require a larger investment of time than do many other houseplants. The first step in creating a bonsai is to select the plant. Visit a local nursery and choose an easy-to-grow plant such as pyracantha, pine, or juniper in a 1-gallon can. A large, bushy plant is not necessarily the best choice. Instead, purchase a plant with an irregular form, less foliage, and even some dead wood, because this will help create an aged appearance.

The next step is to select the front or primary viewing angle of the bonsai, the angle from which it is most attractively viewed. If the trunk has a natural curve, the viewing angle should emphasize it. Next, the two lowest major branches are located and all branches growing below them removed flush with the trunk. The ultimate height of the bonsai is then decided, the top removed to that point, and the two lowest branches shortened to be one-half the length of the trunk.

Next, the upper branches are chosen. They should alternate up the trunk and not necessarily be the strongest branches. Smaller, thinner branches may appear more natural. The goal is to space the branches evenly both up and around the trunk and to trim them gradually shorter as they progress upward to mimic branch growth in nature.

Wiring can be done at the same time as pruning. Copper wire of a 10 to 20 gauge is wrapped in spiral fashion around the trunk and branches, which can then be bent to the desired shape. The wiring can be removed in several months, and the tree will retain the new form. Bottommost branches are preferably wired to point either horizontally or downward, and branches progressing up the trunk point sequentially at slightly more vertical angles (Figure 14-7).

The next step is to remove the bonsai from its pot and examine the roots. If there are predominately large (anchorage) roots and few fibrous (absorptive) roots, the root ball will need to be pruned to fit a training pot half the depth of the present container and grow in this pot for several months to develop fibrous roots. If not, and it is immediately planted in a very shallow bonsai pot, the shock of root removal may be too great and might kill the plant.

The final container will need to have its drainage holes covered with screen wired into place, and if it is suspected that the bonsai will need to be wired

(a) the plant in its original state

(b) after pruning and training the top

(c) after potting

Figure 14-7 Steps in bonsai making. *(Photo from* Bonsai: The Art and Technique *by D. Young, 1985. Reprinted by permission of Prentice Hall, Englewood Cliffs, NJ)*

through the roots to be stable in its pot, an additional long strand of wire can be secured through the drainage hole onto a short peg with the ends of the wire left hanging over the edge of the pot. The root ball should then be flattened, and long, thick roots pruned off to make a root ball that will fit the new pot with 1 inch (2–3 centimeters) of clearance on the sides and the bottom. Normally, a bonsai is potted slightly off center, and this must be taken into account when shaping the root ball.

After filling the pot to 1-inch depth (2–3 centimeters) with potting medium (a sterilized mix of equal parts soil, sphagnum peat moss, and coarse sand), the bonsai can be placed in the pot and checked for height. At this point the trunk base should be uncovered to show the point of origin of the first major root. This should be positioned to be slightly above the top of the container with the bulk of the root ball sloping downward to the lip. At this point wiring the tree to the pot for stability can be done if necessary by crossing the wires over the root ball and twisting them taut to form a knot near the trunk. The wire and knot should be covered with moss, which is laid over the entire root area as a final touch.

Watering of bonsai will depend on the species. Because of the limited soil volume, bonsai dry out relatively quickly. They must be watered frequently to prevent death from drought but should not be kept constantly moist because this would encourage undesired, vigorous growth. Watering every day during warm weather is not unusual.

Fertilizer for bonsai should be in liquid or water-dissolving form (see Chapter 15) applied from spring through fall. A dilute solution applied once per month is usually adequate, because the objective is to maintain the plant at its present size, not to encourage growth.

Bromeliads

Members of the family Bromeliaceae (Figure 14-8) are termed *bromeliads*. In nature they are tree-dwelling (*epiphytic*) plants, but they are not parasites as they use the trees only for anchorage. As houseplants they are dependable and easy to care for. All are valued for their foliage, and many have long-lasting flowers and colorful fruits.

Bromeliads survive under a wide range of light levels from bright to dim, but the foliage will show the best coloration under the proper light level for the species. As a general rule, plants in the genera *Ananas, Dyckia, Orthophytum,* and *Tillandsia* are bright-light-requiring. *Aechmea, Neoregelia,* and *Billbergia* are

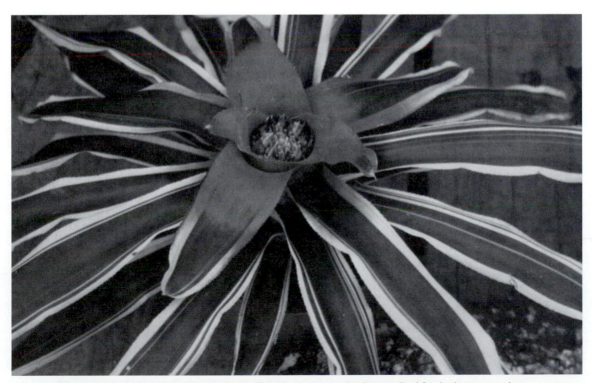

Figure 14-8 *Neoregelis carolinea* 'Tricolor,' a popular bromeliad for indoor growing. The leaves are banded green and white. At flowering, the center cup becomes red. *(Photo courtesy of Marge Coon, Vista, CA)*

moderate- to bright-light-requiring, and *Cryptanthus, Guzmania, Nidularium,* and *Vriesea* are moderate- to dim-light-requiring.

The water requirements of bromeliads differ from those of other houseplants. Potted types use soil moisture but also frequently have a center cup, formed by the leaves, that should be kept half (not completely) full of water. Both potted and plaque-mounted species absorb water efficiently through their foliage, and misting is beneficial. Because misting is the only water source for plaque-mounted bromeliads, daily spraying is advisable when there is little humidity in the air.

Bromeliads should be fertilized with a water-soluble fertilizer mixed at the strength recommended for foliar application. The fertilizer should be applied to the medium of potted specimens and also to the cup (Figure 14-8). Misting of both potted and mounted species with the solution is recommended.

Potted bromeliads generally have small root systems and require repotting infrequently. Use the Cornell epiphytic mix discussed in Chapter 15 or another, similar lightweight medium. Table 14-5 is a list of bromeliads particularly suitable for growing as houseplants.

Table 14-5 Selected Bromeliads for Indoor Growing

Common name	Botanical name
Burgundy vase plant	*Aechmea* 'Burgundy'
Vase plant	*Aechmea maculata*
Urn plant	*Aechmea fasciata*
No common name	*Aechmea gamosepala*
No common name	*Aechmea maculata*
Patriotic plant	*Aechmea mertensii*
Pineapple	*Ananas comosus*
Queen's tears	*Billbergia nutans*
Muriel Waterman	*Billbergia* 'Muriel Waterman'
Earth star	*Cryptanthus* cultivars
Cryptbergia	*Cryptbergia rubra*
Orange star	*Guzmania lingulata*
Flasklike neoregelia	*Neoregelia ampullacea*
Blushing bromeliad	*Neoregelia carolinae*
Fireball	*Neoregelia* 'Fireball'
Black Amazon bird nest	*Nidularium innocentii* 'Innocentii'
No common name	*Nidularium regelioides*
Pink quill	*Tillandsia cyanea*
Blushing bride	*Tillandsia ionantha*
No common name	*Tillandsia xerographica*
Flaming sword	*Vriesea splendens*
Sword plant	*Vriesea* hybrids

Cacti and Succulents

Confusion persists about the difference between cacti and succulents. Basically, a cactus is a member of the *Cactaceae* family, and a succulent is any plant that is fleshy and contains a large quantity of water in its leaves or stems. Almost all cacti are succulents; however, obviously, not all succulents are cacti. Many other genera contain succulent plants.

The presence of thorns or spines does not necessarily classify a plant as a cactus. Many succulents have spine-like organs, and only careful examination and a fairly detailed knowledge of plant anatomy will enable one to classify a plant as a member of the *Cactaceae* with certainty.

Cacti (Figure 14-9) are classified as either *xeriphytic* (adapted to dry conditions) or *epiphytic* (tree dwelling in tropical forests). The former includes all desert cacti and all commonly grown succulents. The epiphytic cacti include Christmas cactus, Easter cactus, orchid cactus, and related species. The care for the two types differs greatly. Xeriphytic cacti require bright light; a coarse, fast-draining potting medium; and infrequent watering for most of the year. The epiphytes grow best in moderate light and a richly organic and constantly moist potting medium.

Improper watering is a major cause of death in cacti and succulents. Xeriphytic species should be watered infrequently from fall through spring, their dormant period. The medium should be allowed to dry out thoroughly between waterings, and if possible, they should be kept at a cool 45 to 65°F (7–18°C) tem-

Figure 14-9 A group of succulents. *(Photo by Rick Smith)*

Figure 14-10 A folded newspaper aids in transplanting a cactus. *(Photo by Rick Smith)*

perature. As soon as new growth begins, the watering frequency should be increased so that the medium is rewet as soon as it approaches dryness.

Because of their sensitivity to overwatering, xeriphytic cacti should be transplanted only when actively growing in spring or summer. They should be repotted in a dampened media in lieu of postpotting watering.

Transplanting spiny cacti and succulents can be difficult; however, they require repotting only every two to four years. The process can be made easier by handling the plants with a band of folded newspaper, as shown in Figure 14-10.

Table 14-6 lists cacti and succulents suitable for growing as houseplants. When purchasing xeriphytic cacti, beware of large cacti at bargain prices. They are often dug from the desert and their roots severed. Such plants are difficult to reroot and usually die over several months. They may also be endangered protected species.

Ferns

Most ferns (Figure 14-11) are moderate- to low-light-requiring houseplants. During winter, if light intensities are poor, they can be placed directly in an east, north, or west window. But in summer, they should be either moved back from east and west windows or given protection by a sheer curtain.

With the exception of a few genera such as *Pellaea* and *Cyrtomium,* ferns will not survive complete drying of the potting medium. The medium should contain some moisture at all times. A quick-draining but

Table 14-6 Selected Cacti and Succulents for Indoor Growing

Common name	Botanical name
Cacti	
Sea urchin cactus	*Astrophytum asterias*
Bishop's cap	*Astrophytum ornatum*
Silver torch	*Chamaecereus silvestri*
Peanut cactus	*Cleistocactus strausii*
Golden barrel cactus	*Enchinocactus grusonii*
Rainbow cactus	*Echinocereus pectinatus* var. *neomexicanus*
Orchid cactus	*Epiphyllum* hybrids
Fire barrel	*Ferocactus acanthodes*
Rose-plaid cactus	*Gymnocalycium mihanovichii* cv. *friedrichii*
Agave cactus	*Leuchtenbergia principis*
Pincushion cactus	*Mammillaria* spp.
Ball cactus	*Notocactus* spp.
Bunny ears	*Opuntia microdasys*
Crown cactus	*Rebutia* spp.
Easter cactus and others	*Rhipsalidopsis* hybrids
Christmas cactus and others	*Schlumbergera* hybrids
Succulents other than cacti	
Agave	*Agave angustifolia*
Aloe	*Aloe* spp.
Ponytail	*Beaucarnea recurvata*
Jade plant	*Crassula argentea* cultivars
Watch chain	*Crassula lycopodioides*
String of buttons	*Crassula perforata*
Hen and chickens	*Echeveria* hybrids, *Sempervivum* spp.
Crown of thorns, pencil plant, and others	*Euphorbia* spp.
Ox tongue	*Gasteria* spp.
Wart plant	*Haworthia* spp.
Wax plant	*Hoya carnosa* cultivars
Panda plant, air plant, and others	*Kalanchoe* spp.
Mother-in-law's tongue/ Snake plant	*Sansevieria trifasciata* cultivars
Burro tail and others	*Sedum* spp
Wax ivy, string of beads, and others	*Senecio* spp.

moisture-retentive potting medium is best for root growth. Most prepackaged houseplant media are suitable. Humidity level is more critical for ferns than other houseplants, and a 30–60 percent relative humidity range is necessary for the healthy growth of most species. Table 14-7 lists selected ferns recommended for indoor growing.

Figure 14-11 A cultivar of Boston fern. *(Photo by Andrew Schweitzer)*

Table 14-7 Selected Ferns for Indoor Growing

Common name	Botanical name
Maidenhair fern	*Adiantum raddianum*
Maidenhair fern	*Adiantum tenerum*
Bird's nest fern	*Asplenium nidus*
Mother fern	*Asplenium bulbiferum*
Holly fern	*Cyrtomium falcatum* cultivars
Boston fern	*Nephrolepis exaltata* cultivars
Boston fern	*Nephrolepis duffii*
Button fern	*Pellaea rotundifolia*
Green cliff brake	*Pellaea viridis*
Hart's tongue	*Phyllitis scolopendrium*
Common staghorn fern	*Platycerium bifurcatum*
Rabbit's foot fern	*Polypodium aureum* cv. *areolatum*
Rabbit's foot fern	*Polypodium aureum* 'Undulatum'
Climbing bird's nest fern	*Polypodium punctatum*
Holly fern	*Polystichum tsus-simense*
Cretan bracken	*Pteris cretica* cultivars
Australian bracken	*Pteris tremula*
Ladder bracken	*Pteris vittata*
Tongue fern	*Pyrrosia lingua*
Leatherleaf fern	*Rumohra adiantiformis*

Florist Plants

The species traditionally called *florists plants* (Figure 14-12) such as chrysanthemum, poinsettia, and forced bulbs are handsome, long-lasting additions to the indoor environment. Recent improvements in keeping quality have greatly extended the anticipated indoor life expectancies of many species. In addition, the range of species used for indoor color and their flower colors have also increased dramatically in recent years. In previous years selection was limited to perhaps a dozen major species and a plant was expected to last only a few weeks; now 50 or more popular species are available and most will last from one to several months indoors with minimal care.

 A florist plant producer owns or works as a grower in a greenhouse specializing in the high-volume florist plants associated with holidays and floral gifts: poinsettias, Easter lilies, pot chrysanthemums, and the like. He or she must monitor the operation of the greenhouse, schedule the production, monitor plant health, and arrange for shipping of the crop to market

The life of the flowers can be extended considerably with correct care in the home. The plant should be placed in bright light but preferably out of direct sunlight and drafts. The potting medium should be kept moist.

Gesneriads

The family Gesneriaceae includes many outstanding houseplants such as episcias, African violets, and gloxinias. Many popular flowering houseplants are found in this family (Table 14-8).

Most gesneriads are of tropical origin and should be grown at temperatures from 55 to 80°F (13–27°C). Nighttime temperatures in the lower half of this range and daytime temperatures in the upper half will provide the best conditions for growth. Light should be moderate to bright to achieve flowering, but long periods of strong sunlight in summer should be avoided. Less intense winter sunlight is acceptable. Gesneriads are also known to grow well with artificial lighting as the only light source.

The potting medium of gesneriads should remain slightly moist at all times. All water applied should be at room temperature, because cold water can cause permanent leaf spotting.

The humidity range most acceptable for gesneriad growing is 50 to 80 percent. They can be successfully grown in less humidity provided other growing conditions are met satisfactorily. Because of the humidity desirable for raising gesneriads, they are particularly suited to terrarium growing.

Figure 14-12 A selection of florist flowering plants, clockwise from top: Kalanchoe, gerbera daisy, exacum, cyclamen, and chrysanthemum. *(Photo by Andrew Schweitzer)*

Table 14-8 Selected Gesneriads for Indoor Growing

Common name	Botanical name
Columnea	*Columnea* 'Chanticleer'
	Columnea 'Early Bird'
	Columnea 'Mary Ann'
Flame violet	*Episcia* 'Moss Agate'
	Episcia 'Acajou'
	Episcia 'Cygnet'
	Episcia 'Jinny Elbert'
Nautilocalyx	*Nautilocalyx forgettii*
Candy corn plant	*Nematanthus wettsteinii*
African violet	*Saintpaulia ionantha* cultivars
Slipper plant	*Sinningia pusilla* cultivars
Cape primrose	*Streptocarpus saxorum*
	Streptocarpus 'Wiesmoor hybrids'
	Streptocarpus 'Constant Nymphs'
	Streptocarpus 'Diana'
	Streptocarpus 'Fiona'
	Streptocarpus 'Karen'
	Streptocarpus 'Marie'
	Streptocarpus 'Louise'
	Streptocarpus 'Paula'
	Streptocarpus 'Tina'
	Streptocarpus 'Margaret'

Hanging Plants

The care of hanging plants differs from the care of those set on tables, windowsills, or the floor, for several reasons. First, the air temperature is usually warmer around hang-ing plants. Heat will rise to the top of a room, and consequently the air surrounding a plant hung at eye level may be 5°F (3°C) warmer than the air at floor level. There is also greater air movement around a suspended plant because there is no supporting surface to block air flow.

Because of these environmental differences, hanging plants lose moisture from both the medium and the leaves at an accelerated rate. More frequent watering will thus be required. Because many hanging plants are drought-sensitive ferns, it is important to check these plants frequently for moisture. Table 14-9 lists a number of excellent houseplants for hanging baskets.

Orchids

Orchids at one time had a reputation for being hard to grow, rare, and definitely not within the range of expertise of normal people for growing as flowering houseplants. However, many are not hard to grow, and many are actually much tougher than many other houseplants.

In the past few years, the hybridization of orchids among genera has resulted in an enormous increase in previously unheard of new genera created by plant breeders. Up to 80 percent of the orchids sold retail today are a result of these crosses. Some of the most common crosses are between cattleyas so famous for their large showy flowers (Figure 14-13), and other genera or between easy-to-grow *Odontoglossum* or *Oncidium* and other genera. The resulting genera would potentially bear long names to incorporate the names of

Table 14-9 Selected Plants Suitable for Hanging Baskets

Common name	Botanical name
Lipstick vine	*Aeschynanthus* spp.
Rattail cactus	*Aporocactus flagelliformis*
Asparagus fern	*Asparagus* spp.
Begonia	*Begonia* spp.
Rosary vine	*Ceropegia woodii*
Spider plant	*Chlorophytum* spp.
Coleus	*Coleus* × *hybridus*
Cissus vine	*Cissus* spp.
Goldfish vine	*Columnea* spp.
Episcia	*Episcia* spp.
Creeping fig	*Ficus pumila* cultivars
Bridal veil	*Gibasis geniculata*
Purple passion vine	*Gynura aurantiaca* 'Purple Passion'
English ivy	*Hedera helix* cultivars
Wax plant	*Hoya carnosa* cultivars
Pocketbook plant	*Nematanthus* spp.
Fern	*Nephrolepis, Adiantum, Davallia,* and other genera
Arrowhead	*Syngonium podophyllum* cultivars
Trailing peperomia	*Peperomia* spp.
Heart-leaf philodendron	*Philodendron scandens oxycardium*
Swedish ivy	*Plectranthus australis* and *oertendahlii*
Easter cacti	*Rhipsalidopsis gaertneri* and *rosea*
Strawberry begonia	*Saxifraga stolonifera*
Devil's ivy	*Epipremnum aureum* cultivars
Trailing sedum	*Sedum morganianum* × *rubrotinctum* and other species
German and wax ivies	*Senecio macroglossus* cultivars
Wandering Jew	*Tradescantia* and *Zebrina* spp.
Christmas cactus	*Zygocactus truncatus*

sometimes up to three parents, such as *Cochlioda* × *Miltonia* × *Odontoglossum*. To avoid this problem, new genera names have been assigned to these intergeneric hybrids. For example. *Vuylstekeara* is the new genus for the three-parent hybrid named above.

In other cases, the names of the genera are abbreviated by first letter, with the understanding that those familiar with developments in orchid breeding will recognize the parents. For example, BLC is used to abbreviate a *Brassavola* × *Laelia* × *Cattleya* cross, and LC a *Laelia* × *Cattleya* cross.

Although many orchids are epiphytes, a number are terrestrials. Selected species of both types grow well as houseplants (Table 14-10). The anatomy and growth pattern of many orchids are unusual. Monopodial orchids (Figure 14-14) have a rosette form, or leaves only on opposite sides of the stem, giving them a two-dimensional appearance. *Sympodial orchids* (Figure 14-14) have storage organs called *pseudobulbs,* with each pseudobulb bearing one leaf. These orchids grow by extension of rhizomes and generally gain one new pseudobulb and one new leaf per year.

Orchids (Figure 14-13) fall into the following three growing-temperature categories:

Cool	(45–58°F)	(7–14°C)
Intermediate	(55–70°F)	(13–21°C)
Warm	(62–80°F)	(17–27°C)

However, as an all-round temperature range, 56 to 62°F (13–17°C) night temperature and 62 to 80°F (17–27°C) day temperature are acceptable for most species.

Epiphytic orchids are potted in osmunda fiber, bark, or similar lightweight potting medium. Watering should remoisten the potting medium just before it completely dries. Because water drains through these

Figure 14-13 Lady Slipper orchid *Paphiopedilum* sp.

Table 14-10 Selected Orchids for Indoor Growing

Botanical name	Common name
Brassavola nodosa	lady-of-the-night
Brassia spp.	spider orchid
Calanthe vestita cultivars	none
Encyclia citrina	tulip cattleya
Cattleya gaskelliana	summer cattleya
Cattleya labiata cultivars	autumn cattleya
Colmanara hybrids	none
Dendrobium phalenopsis	none
Encyclia radiata	none
Epidendrum spp.	spice orchid
Lycaste aromatica	none
Rossioglossum grande	tiger orchid
Oncidium ornithorhynchum	maidenhair orchid
Odontocidium hybrids	none
Phalaenopsis equestris	moth orchid
Paphiopedilum spp.	lady slipper
Vanda spp.	vanda orchid
Vuylstekeara hybrids	none

Source: Appreciation is expressed to David Goldeen of Orchids of Los Osos, Los Osos, CA who assisted in writing this section on orchids.

loose media freely, care must be taken to ensure that the entire medium is wet. Watering thoroughly with a sink sprayer will accomplish this, as will submersion watering (Chapter 17).

Terrestrial orchids may be grown in an all-purpose potting medium suitable for foliage plants. They should be watered whenever the medium becomes nearly dry.

When repotting epiphytic orchids, a moist medium is more convenient. The bark or osmunda should be soaked in water overnight prior to use in repotting. The orchid should be removed from its previous pot and the old medium cleaned from its roots. Next, any dead roots should be cut away and the plant repotted in the fresh medium. Finally, staking should be done, if necessary. The orchid should not be watered until the medium dries.

The light requirements of orchids can be met by placing the plants in a south-facing window during the fall and winter months, and in an east or west window[1] during the spring and summer when light intensities become brighter. Orchids receiving insufficient light will be surprisingly attractive, with sleek, dark green leaves, but they are unlikely to flower.

[1]Avoid west-facing windows in the sunny West unless a net curtain is used.

leaf

root

Dendrobium

(a)

flower bud

leaf

pseudobulb

Cattleya

(b)

Figure 14-14 The monopodial (a) and sympodial (b) growth forms of orchid.

Orchids grow best in a humid atmosphere, which is the most difficult growing condition to meet indoors. Misting will improve the humidity for a few minutes; it is more important as a source of moisture that is absorbed by the foliage. Other methods for increasing household humidity are discussed in Chapter 17.

Fertilizer should be applied to orchids every other week while they are in active growth. The fertilizer should be of a liquid or dissolvable-powder type and applied to the medium, leaves, and pseudobulbs at dilute strength.

 A self-employed horticultural writer produces articles for gardening magazines, books, video scripts, and similar materials. This profession requires broad knowledge of horticulture, access to large numbers of horticultural references and the Internet, and the ability to communicate clearly in the writing tone preferred by the publication. A "self-starting" personality and an ability to work at home are also necessary.

SELECTED REFERENCES FOR ADDITIONAL INFORMATION

Bond, Rick. *All About Growing Orchids.* San Ramon, CA: Ortho Books, 1988.

Catteral, Eric. *Begonias, The Complete Guide.* London: Crowood, 1996.

Chicago Botanic Garden staff. *Indoor Gardening.* Chicago, IL: Chicago Botanic Garden, 1995.

Graf, A. B. Hortica: *Color Cyclopedia of Garden Flora and Exotic Plants Indoors.* Atlantic City, NJ: Roehrs, 1992.

Griffith, Lynn P., Jr. *Tropical Foliage Plants: A Grower's Guide.* Batavia: Ball Publishing, 1997.

Hodgson, Larry and National Gardening Association Staff. *Houseplants for Dummies.* NY: Hungry Minds, 1998.

Hodgson, Larry and Susan M. Lammers. *All About Houseplants.* San Ramon, CA: Ortho Books, 1996.

Koreshoff, Deborah R. *Bonsai: Its Art, Science, History and Philosophy.* Beaverton, OR: Timber Press, 1997.

Watson, Janet. *Growing Orchids.* West Palm Beach, FL: American Orchid Society, 1997.

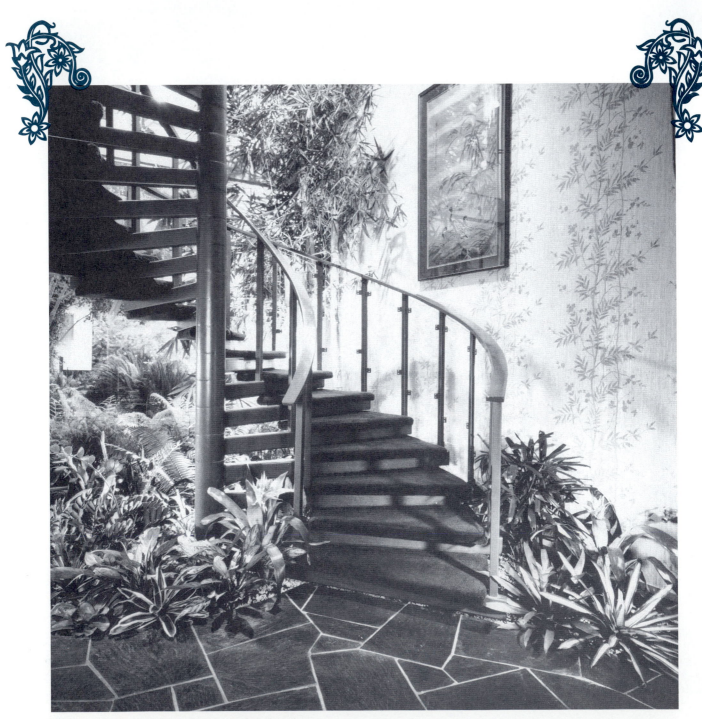

(Photo courtesy of British Indoor Gardens)

CHAPTER 15

POTTING MEDIA AND FERTILIZERS

INDOOR PLANT POTTING MEDIA

The word *medium* (plural, *media*) is not familiar to most people when used in relation to indoor plant growing. A medium is any material used for rooting or potting plants. It is soil-like in that it performs the same functions as soil does for plants outdoors: supporting the roots and acting as a reservoir for moisture, nutrients, and air. But a medium can be a mixture of many ingredients, some manufactured and some naturally occurring. It need not be a natural material like soil. In fact most commercially available media used for houseplants contain no soil at all.

Potting media have several advantages over most pure soils for growing houseplants. First, they are made up of materials that create large pore spaces, and they are therefore less likely to pack around the roots of plants. Therefore, danger of root death due to air exclusion is largely eliminated. Second, they have a loose structure that drains well. Water passes through quickly (the definition of "good" drainage), yet some moisture is retained. Third, they are usually free of disease organisms, weed seeds, and insects that could damage indoor plants. Finally, they are lighter in weight, which makes moving large plants easier and decreases shipping costs.

Overall, potting media are superior to natural soils for growing plants in pots and are the best choice for rooting and growing indoor plants.

Potting-Media Components

Many brands of premixed potting media are available. They are sold under the name "houseplant soil" or sometimes "cacti soil" or "African violet soil," although most contain no soil at all. They are completely satisfactory for indoor plant growing, although they can be relatively expensive when used in quantity. A few, particularly the types sold for cacti, could be improved by the addition of more coarse sand. The sand adds weight to the medium and lessens the problem of a top-heavy plant that tips easily.

Premixed potting media do not have one set formula or list of ingredients. They vary widely among manufacturers, depending on the available materials.

Any potting medium will generally consist of two to three types of components. The first is *organic, plant-derived material*. Its water-absorbing and -holding capacity will be great, and it will frequently have a spongy structure resistant to packing. In addition, its nutrient-holding ability or *cation-exchange capacity* will be substantial, enabling it to function as a reservoir for nutrients.

The second component is called the *coarse aggregate*. The aggregates improve the drainage of the medium because the angular shape of each particle wedges them apart, creating large or *macropore spaces* (Figure 15-1). These pore spaces facilitate the free flow of air and water through the medium and increase the drainage rate. In general, coarse aggregates have a hard structure and poor water-holding and cation-exchange capacities.

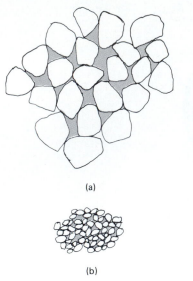

(a)

(b)

Figure 15-1 Macro- (a) and micro- (b) pore spaces.

A third optional component is soil. It is used to hold and supply nutrients and acts as a reservoir of water in addition to the organic material. Soil particles are usually smaller than the coarse aggregate or organic material particles. Consequently, they fit close to each other and other materials, forming *micropore spaces* (Figure 15-1) that restrict the flow of air and water. However, these micropores are efficient at retaining water because of a phenomenon known as *capillary action*. The concepts of pore space, capillary action, and cation exchange are discussed in more detail in Chapter 5.

The following are the most common ingredients used in making prepackaged mixes (Table 15-1).

Organic, Plant-Derived Materials

Peat Moss. Sold under the names "Canadian peat moss," "sphagnum peat moss," and just peat moss, this bog plant is harvested, dried, and used in the United States and

Table 15-1 Common Ingredients of Houseplant Media

Ingredient	Type	Remarks
Compost	Organic, plant-derived	A combination of decayed leaves, twigs, grass, and so on made at home; must be sterilized before use
Activated charcoal	Organic, plant-derived from partially burned wood, but functions as a coarse aggregate	Absorbs odors, chemicals from the soil for several months before becoming saturated; can be reactivated by heating in an oven at 350°F (180°C) for about 30 minutes; black
Ground bark	Organic, plant-derived material	By-product of lumber mills like coarse sawdust; holds water, nutrients
Bark chunks	Organic, plant-derived material	By-product of lumber mills; holds moisture, nutrients; used for orchids
Sphagnum peat moss	Organic, plant-derived material	Holds water and nutrients well, acid pH, breaks down very slowly; water repellent when dry
Leaf mold	Organic, plant-derived material from partially decomposed leaves	Flaky texture; decomposes in the medium, releasing nutrients; must be sterilized before use
Humus or muck soil	Organic, plant-derived material	Dark brown and highly decomposed with a powdery feel; can be difficult to rewet if allowed to dry out; also called *Michigan peat*
Sand	Coarse aggregate	Inert (not chemically reactive); use only coarse grades and not ocean sand
Aquarium gravel	Coarse aggregate	Heavy, inert, and very coarse; excellent for use with cacti and succulents; various colors
Perlite	Coarse aggregate	Lightweight and coarse; contains fluorides that are toxic to some plants; white in color. Perlite dust can be a lung irritant and therefore should be moistened before mixing.
Vermiculite	Coarse aggregate; also a water- and nutrient-holding component	Lightweight, holds water and nutrients; useful as a coarse aggregate, but it packs down over time; beige and shiny
Scoria or lava rock	Coarse aggregate	Found in natural deposits near volcanos; relatively lightweight with a pleasing dark color

Figure 15-2 Milled (left) and unmilled (right) sphagnum peat moss. *(Photo by Rick Smith)*

Canada in enormous quantities each year. In its raw state, it can be bought in mats or bags, is light brown, and has a stringy texture (Figure 15-2). It is also commonly available shredded or "milled," with fibers broken down so that they mix easily with other ingredients.

Peat moss is the main ingredient in many prepackaged houseplant mixes, often making up 50 percent of the volume of the medium. It is even used alone for some purposes such as air layering and rooting cuttings.

When dry, peat moss is a very fluffy, lightweight material. When wet, it is capable of holding a large volume of water without packing down, therefore providing plant roots with the ideal combination of moisture and air.

Peat moss is usually very acidic, with a pH range of 4.0 to 5.0. This range is too low for most houseplants, so pH-increasing materials are normally added when peat moss is part of the potting medium. One of the few disadvantages of peat moss is that if it is ever allowed to dry out thoroughly, it will be very difficult to rewet. Water will run over the fibers rather than being absorbed. The moss or medium must be gradually rewet by submersion or bottom watering or by the use of a chemical "wetting agent" or surfactant. For home use a drop of detergent in a gallon of water will serve this purpose for occasional use. Hot water is also effective.

Leaf Mold. Leaf mold (Figure 15-3) is a second possible component of prepackaged houseplant media. Like peat moss, it is derived from plants and is made up of partially decayed leaves and twigs. It is dark brown with a flaky texture. Leaf mold is capable of holding water and nutrients, although not as much as peat moss. Leaf

mold may be used in premixed potting media in addition to or as a complete substitute for peat moss.

Leaf mold is a natural product in the decomposition of leaves. It is seldom sold as a pure product but can be made at home by composting fallen leaves (Chapter 5). It can also be gathered from the ground in deciduous woodlands.

Humus Peat or Muck. Humus peat or muck is actually a soil type. It is found in low-lying areas and is composed largely of organic material in the form of highly decayed vegetable matter such as leaves, stems, and roots. It is black and fluffy when dry and has a fine, powdery feel. It is commonly available under the name of "garden peat" or "Michigan peat," although it is not

Figure 15-3 Leaf mold. *(USDA photo)*

peat at all. Muck continues decomposing rapidly when used in a potting medium. Its fineness may create aeration problems.

Do not confuse humus peat with sphagnum peat. The latter is a lighter-brown fibrous material and is the more expensive of the two.

Bark. Shredded or ground bark is occasionally used as an organic potting-medium component. Fir, redwood, pine, hemlock, and oak are all satisfactory sources of bark. Bark of incense cedar and walnut should be avoided because it may contain toxic substances.

Compared to peat moss, bark has less water-holding capacity and a higher pH. However, pH-altering chemicals are still needed to raise the pH to an acceptable level. Nitrogen fertilizer is generally included in bark-based potting media to compensate for nutrients temporarily lost as the bark decomposes.

One type of bark derived from pines has been shown to be mildly antibiotic, suppressing the growth of fungi in the potting medium. However, this property does not render it significantly superior to bark from other species.

Hydrating Gel Granules. Hydrating gel granules, synthetic acrylic copolymer granules or spikes, are an occasional component of potting media and are sold for use in sandy soils as well (Figure 15-4). They absorb more than 100 times their weight in water, swelling to become ¼ inch (0.5 centimeter) diameter pieces of nondecomposing gel. Gel granules are most useful as "insurance" in potting media for ferns and similar moisture-loving species that suffer serious damage or die if they dry out. Their value is in their extreme moisture-holding capacity.

The other use for gel granules is as a component of a medium to lessen watering frequency. To this end they are employed in potting mixes for use in commercial interiorscapes to decrease maintenance costs. They might also be useful for home use for people who have large numbers of plants or who travel frequently for extended periods. Research has shown that time between irrigation can be extended slightly with their use.

Coarse Aggregates

Sand. Sand is a naturally occurring material used occasionally in premixed potting media, although heavy shipping weight discourages frequent use. It is primarily quartz that is not chemically reactive in the medium. Not all grades of sand are satisfactory for use in potting media. In most cases only coarser grades known as "builder's sand" or "mason sand" are used in potting media (see exception under University of California "C" Mix).

(a)

(b)

Figure 15-4 (a) Hydrating gel granules. This brand is sold premixed with a microencapsulated fertilizer (see fertilizer section). *(Photo courtesy of Gromax, Salinas, CA)* (b) The root zone of a plant with hydrated gel granules, showing the roots penetrating the granules. *(Photo courtesy of JRM Chemical Co., Inc., Cleveland, OH)*

Perlite. Perlite is a coarse aggregate (Figure 15-5) commonly used in commercial media because of its extremely light weight. It is made by heating volcanic rock until it expands and then grinding the resulting material into irregular chunks ¹⁄₁₆ to ⅛, inch (1.5–3.0 millimeters) in diameter. Like sand, perlite does not

Figure 15-5 Perlite. *(Photo by Kirk Zirion)*

hold appreciable amounts of water or nutrients and does not react in the potting medium. In appearance and size it is very similar to white aquarium gravel.

One of the disadvantages of perlite is its tendency to float to the top of a potting medium when a plant is watered. Another is that, because of its light weight, it may not provide enough base weight for larger plants, and they may tend to fall over when the medium is dry. Cacti potted in a largely perlite medium may tend to fall out of their pots, because large-bodied cacti frequently have sparse root systems that provide little anchorage.

Perlite has also been proved to contain levels of fluoride that are toxic to some houseplants. As a precaution, it should be rinsed in water three times before it is used. Perlite-containing potting media should be avoided for members of the Liliaceae family because they are particularly susceptible to fluoride injury.

Vermiculite. Vermiculite (Figure 15-6) is made by a heat expansion process similarly to perlite, but a mica-like material is used in lieu of volcanic rock. The resulting material is shiny, silvery-brown in color, lightweight, and with a spongy feel. The pH is neutral to slightly alkaline, in the range of 7.0 to 7.5. Vermiculite naturally contains appreciable amounts of potassium and magnesium, two nutrients essential to plant growth.

Vermiculite is unusual among the coarse aggregates in that it is an active part of a medium. It has a good capacity for holding water and nutrients, although the size of the individual particles still facilitates good drainage.

Several size grades of vermiculite are manufactured, although not all of them may be available in gar-

Figure 15-7 Horticultural charcoal. *(Photo by Kirk Zirion)*

den centers. The fine grade is useful for seed germination, and the medium for rooting cuttings or adding to media. The coarse grade is occasionally used for packing plants for shipping. The main disadvantage of vermiculite is its tendency to compress and lose its drainage-enhancing quality with time or when it is watered frequently. Nonetheless, vermiculite is still a popular material for potting media.

Charcoal. Charcoal (Figure 15-7) in small chunk form (such as used for aquariums) is sometimes used as a potting media additive in place of other aggregates. It can be beneficial in some instances in that it absorbs and holds odors, toxic pesticides, and other undesirable chemicals in the short run, but its use is not practical because of its relatively expensive cost and short-term effectiveness.

Scoria or Lava Rock. Scoria (Figure 15-8) is a natural volcanic rock ground to a size of ⅟₁₆ to ⅛, inch (1.5–3.0 millimeters). It is gray or black and intermediate in weight between perlite and sand. Where readily available, it is an inexpensive and satisfactory coarse aggregate.

Soils

Loamy Soil. Loamy soil is a well-balanced combination of the basic soil particles—sand, silt, and clay—in conjunction with a moderate amount of humus or

Figure 15-6 Vermiculite, an occasional component of potting media. *(Photo courtesy of Vocational Education Productions, California Polytechnic State University, San Luis Obispo, CA)*

Figure 15-8 Scoria, a coarse aggregate used in potting media. *(Photo by Kirk Zirion)*

other organic matter. It is the rich topsoil found in many successful gardens. Loamy soil holds water well, does not pack or crust when dry, and can be easily crumbled in the hands when it is moderately damp, a characteristic called *friability*. It is an excellent potting medium additive even in relatively large quantities but should not be used alone.

Clay Soil. Clay soils are composed largely of tiny plate-like particles that pack together closely, forming many micropore spaces. Clay soils are very water- and nutrient-retentive but poorly aerated. They are seldom added to premixed potting media but can be used in re-stricted amounts when balanced carefully with organic and coarse aggregate components.

Sandy Soil. If the sand it contains is coarse enough, a sandy soil may substitute in part for the coarse aggre-gate portion of a potting medium. However, fine sandy soils are less widely recommended because the parti-cles create relatively small pore spaces.

Silty Soil. Silty soil has a floury feel when dry and has poor aeration like clay. It has neither the nutrient-holding capacity of clay nor the coarse aggregate func-tion of sandy soil. It should not be used in media.

Homemade Potting Media

Many people make their own potting media for various reasons: first, buying premixed potting media in bags becomes expensive if there are more than a few plants in the household. Second, premixed media are often formulated with more thought to eye appeal and ship-ping costs than plant growth. They can be heavily com-posed of dark, rich-looking humus but lacking essen-tial aggregates like sand.

There is no one formula or list of ingredients that must be adhered to when making a potting media at home. The mix composition is largely determined by the availability of different ingredients in the area and their relative cost. The organic part of the mix need not necessarily be made up of sphagnum moss, for exam-ple. Leaf mold gathered from the floor of a woods, compost, or bagged humus can be substituted in whole or in part. Likewise, leftover aquarium gravel, crushed charcoal, cinders, or sand from a gravel pit or lake can be used in place of more expensive perlite and vermic-ulite. Ocean sand contains salt even after thorough washing and should not be used.

In addition to the bulk ingredients used in home-made media, other ingredients will sometimes be needed

to alter pH or improve fertility. Although in quantity they constitute a very minor component, they are essential parts of a good potting medium.

pH Alteration

Many potting media formulas call for additions of small amounts of powdered dolomitic limestone to a mix with a peat moss, leaf mold, or bark base. The lime-stone counteracts the acidity of the organic component to give the media the slightly acid pH at which most tropicals grow best. It also contains quantities of cal-cium and magnesium needed for growth.

Fertility Improvement

Many fertilizer materials, both natural and synthetically made, are recommended for incorporating into home-made potting media. The natural materials such as bone meal and dried manure decompose slowly and release their nutrients slowly into the potting medium. Plants are supplied constantly with fertilizer over several months. Other dry chemical fertilizers such as ammo-nium nitrate, super phosphate, or a typical complete fer-tilizer of nitrogen, phosphorus, and potassium are less expensive but effective for a shorter period of time.

Incorporated fertilizers are largely optional except when bark is used as a potting medium component (see previous section on bark). Occasional fertilization should supply all the nutrients required by plants.

Potting Media Formulas

The following media formulas are useful for the major-ity of houseplants and show the variation that is ac-ceptable when mixing a growing medium. They can be taken either as bases from which to experiment or as standard formulas to be followed exactly. For the expe-rienced gardener, the feel of the medium may indicate its suitability, with the desired product being loose, with a rich supply of organic matter to produce a spongy feel.

Soilless or Artificial Media

Mix 1. An adaptation of the foliage plant mix de-veloped by Cornell University:

4 qt (4 liters) shredded sphagnum moss
2 qt (2 liters) vermiculite
2 qt (2 liters) perlite
1 tbsp (15 ml) ground dolomitic lime
1 tsp (5 ml) 10-10-10 granular fertilizer

Mix 2. University of California Mix "C," using fine sand:

1 part (by volume) fine sand

1 part (by volume) peat moss

Mix 3. Bark-based mix:

3 parts milled pine bark

1 part sphagnum peat moss

1 part sand

For mixes 2 and 3, add for each cubic foot (35 liters) of mix:

6 oz (200 g) dolomitic lime

2 oz (65 g) hydrated lime

¼ tsp (1–2 ml) fritted trace elements

Soil-Containing Media

Mix 1. For use with heavy soils that are clay-based:

1 part (by volume) soil

2 parts (by volume) organic matter

2 parts (by volume) coarse aggregate

Plus for each bushel (35 liters), add 6 to 8 oz (200–250 g) ground or dolomitic limestone

Mix 2. For use with medium-weight soils such as loams:

1 part (by volume) soil

1 part (by volume) organic matter

1 part (by volume) coarse aggregate

Plus for each bushel (35 liters), add 6 to 8 oz (200–250 g) ground or dolomitic limestone

Mix 3. For use with light, sandy soils:

1 part (by volume) soil

1 part (by volume) organic matter

Plus for each bushel (35 liters), add 6 to 8 oz (200–250 g) ground or dolomitic limestone.

Mix 4. For stretching commercially prepared mixes:

2 parts (by volume) premixed potting medium

1 part (by volume) loamy garden soil of good quality

1 part (by volume) coarse aggregate

Specialized Potting Media. The formulas just listed are sufficient for most houseplants. However, a few plants require special media, whereas others will do better with slightly different media.

Very organic medium—for ferns and African violets:

3 parts shredded sphagnum moss

1 part sand

Xeriphytic cacti and succulent medium—fast draining and coarse:

1 part coarse sand or aquarium gravel

1 part topsoil

1 part peat moss

1 part premixed commercial medium

1 part coarse sand

Acidic medium—for plants requiring a low pH (5.0–5.5) (see Table 15-2):

1. 3 parts sphagnum peat moss

   1 part vermiculite

   1 part sand

2. Pure sphagnum peat moss

Rooting media—fast draining with a moderate moisture-holding capacity (not for xeriphytic cacti):

1. Pure vermiculite (medium grade).

2. 1 part perlite

   1 part shredded sphagnum moss

Table 15-2 Acid-Loving Houseplants That Commonly Show pH-Caused Chlorosis

Common name	Botanical name
Azalea	*Rhododendron* spp.
Camellia	*Camellia japonica*
Citrus	*Citrus* spp.
Climbing fern	*Lygodium palmatum*
Flame of the woods	*Ixora coccinea*
Gardenia	*Gardenia jasminoides*
Heath, heather	*Erica* spp.
Hydrangea (blue)	*Hydrangea macrophylla*
Sasanqua	*Camellia sasanqua*
Sweet potato	*Ipomaea batatas*
Venus fly trap	*Dionaea muscipula*

Seedling media—fine texture and a moderate moisture-holding capacity:

1. 1 part vermiculite (fine grade)
   1 part finely milled sphagnum moss
2. 1 part sand
   1 part premixed commercial medium
   1 part finely milled sphagnum moss

Epiphytic orchid media—extremely fast draining and well aerated:

1. Pure bark chunks ½ inch (1–2 centimeters) in diameter
2. Pure osmunda fiber (Figure 15-9)

Epiphytic bromeliad media—fast draining:

1. 1 part sphagnum peat moss
   1 part bark
2. University of California "C" Mix

Mixing Potting Media

For the most thorough mixing, the ingredients for a houseplant medium should be damp but not wet. Mixing ingredients that are completely dry will cause dust to fly as well as the finer particles to settle to the bottom of the medium.

Dampening sphagnum peat moss can be difficult when it is fully dry. It can be allowed to soak in a bucket overnight and then squeezed out to the correct dampness the next day. As an alternative, hot water will wet it quickly.

Sterilizing Potting Media

Sterilizing houseplant media is done to prevent the introduction of unwanted disease organisms, insects, and weed seeds. It is particularly important for seedling and rooting media, where soil-borne diseases are common.

Not all media require sterilization to be considered safe to use. Vermiculite and perlite are disease-free because they are heat-treated in the manufacturing process. However, all other ingredients (including

Figure 15-9 Osmunda fiber derived from the osmunda fern. The pole at the top is used to support climbing vines. Loose fiber at the bottom is used for potting orchids and other epiphytes. *(Photo by Rick Smith)*

peat moss, leaf mold, soil, and sand) require sterilization before use.

Heat sterilization is the most common soil sterilization method. The damp mix is placed in a large baking pan, covered with foil or a lid, and baked at about 180°F (85°C) for 30 minutes or more until heated to the center. It is then allowed to cool and sealed in a plastic bag to prevent recontamination.

Soil-containing and soil-less media give off an odor during baking that some people find objectionable. The unpleasantness can be avoided if the medium is baked in a sealed roasting bag instead of a pan.

Any fertilizer that is to be included in the potting medium should be added after it is sterilized. Some fertilizers are affected by hot temperatures and will break down and release all their nutrients in the baking process.

INDOOR PLANT FERTILIZERS

The growing medium in which a plant lives is a reservoir for the nutrients it needs for growth. As these nutrients are absorbed or leached from the medium, they must be replaced with fertilizers.

Major Nutrients for Houseplant Growth

Many nutrients in varying amounts are needed by plants, but the ones used in the greatest amounts are nitrogen, phosphorus, and potassium, and to a lesser extent calcium, magnesium, and sulfur. They are termed *plant macronutrients*.

Nitrogen. Nitrogen (abbreviated *N*) is the nutrient used in greatest quantity by houseplants. Plants absorb nitrogen constantly and use it for manufacturing many chemical compounds within the plant, including chlorophyll and amino acids. Without a constant supply of nitrogen, leaves are unable to synthesize chlorophyll and become pale yellow.

Phosphorus. Phosphorus (abbreviated *P*) is the second essential nutrient for plant growth. It plays a part in many plant activities including energy transfer. Phosphorus is particularly essential for proper flower and fruit formation.

Potassium. Potassium (abbreviated *K*) is the third element nutrient for plant growth. It is involved in carbohydrate translocation and protein synthesis. Potassium also enhances coloration, strengthens stems, and aids in strong root development.

Calcium. Calcium is needed for cell division and expansion. An insufficient supply of this element also slows root growth and can cause death of the meristem.

Magnesium. Magnesium is a constituent of chlorophyll and, like potassium, is necessary for protein synthesis.

Sulfur. Sulfur is a component of many amino acids and two of the B vitamins found in plants.

Of the six elements considered macronutrients, nitrogen, phosphorus, and potassium are most generally included in houseplant fertilizers. Each element is included as part of a chemical salt that can be easily shipped and stored. Nitrogen, for example, can be included as calcium nitrate ($Ca(NO_3)_2$) or ammonium nitrate (NH_4NO_3). Phosphorus can be formulated as phosphoric acid (P_2O_5) or monocalcium phosphate (CaP_2O_5). Potassium can be included as potassium nitrate (KNO_3), potassium chloride (KCl), or potassium sulfate (K_2SO_4).

The *salts* of N, P, and K that are used to make houseplant fertilizers dissolve readily in water, and in fact some are packaged predissolved as concentrated liquids. The dissolving ability makes it possible for them to be carried through the medium in water to reach the roots of plants. This ability is particularly valuable for phosphorus, which will not move readily in a medium if applied to the surface.

Interpreting Fertilizer Labels

Because of the importance of the three macronutrients for healthy plant growth, they are contained in almost all houseplant fertilizers. By law, any product sold as a fertilizer or "plant food" must list the percentage of these elements it contains in bold type and separated by dashes. These numbers are known as the *fertilizer analysis* and are of more help in determining the value of the fertilizer than most claims made by the manufacturer.

For example, a fertilizer with an analysis of 5-10-5 contains 5 percent of its weight in nitrogen, 10 percent in a standardized phosphorus-containing compound, and 5 percent in a standardized potassium-containing compound, for a total of 20 percent nutrients by weight. The 5-10-5 formula is a relatively common analysis for houseplant fertilizers, but they range from 1.6 to 60 percent total nutrient content, depending on the manufacturer. Table 15-3 shows the analyses of a number of houseplant fertilizers that were being sold in a garden center in 2002.

When purchasing a fertilizer, you do not necessarily get what you pay for. Price appears to be governed more by packaging and volume than by the total

Table 15-3 Examples of Houseplant Fertilizers

Safer Fish Emulsion All-Purpose Fertilizer	5-1-1	Liquid concentrate	2.25 lb (1.02 kg)	$4.49
Alaska Fish Fertilizer	5-1-1	Liquid concentrate	1 lb 2 oz.	$4.39
Schultz African Violet Plus (includes micronutrients)	8-14-9	Liquid concentrate	5.5 oz (156 g)	$3.09
Cactus Juice (includes calcium)	1-7-6	Liquid concentrate	9.6 oz (272 g)	$4.99
Green Light Super Bloom	12-55-6	Dissolving powder	1.5 lb (681 g)	$6.89
Schultz Expert Gardener Bloom Plus	10-60-10	Dissolvable powder	1.5 lb (680 g)	$4.49
Miracle-Gro Water Soluble Plant Food	15-30-15	Dissolvable powder	3 lb (1.36 kg)	$8.39
Miracle-Gro Plant Food Spikes	6-12-6	Dissolving spikes	1.1 oz (31 g)	$1.79
Osmocote Slow Release Plant Food (Indoor/Outdoor)	19-6-12	Encapsulated granules	1.25 lb (567 g)	$5.99
Sta-Green 9 month Indoor/Outdoor Plant Food	18-6-12	Encapsulated granules	3 lb (1.36 kg)	$9.99
Growmore Seaweed Extract	0.1-0.1-1.5	Liquid concentrate	16 fl oz	$4.99
Osmocote's Outdoor and Indoor Plant Food	18-6-12	Encapsulated granules	16 oz (454 g)	$5.29
Schultz Plant Food plus micronutrients	10-15-10	Liquid concentrate	12 oz	$4.99
Schultz All Purpose Plant Food Plus (includes trace elements)	20-30-20	Dissolving powder	10 oz	$2.79
Grow More Premium Orchid Food	20-20-20	Dissolving powder	1.25 lb (567 g)	$4.99
Grow More Premium Orchid Food Bloom Formula	6-30-30	Dissolving powder	1.25 lb	$4.99
Oxygen Plus	1-2-1	Liquid concentrate	16 fl oz	$5.79

amount of nutrients the fertilizer contains. A fertilizer with relatively few nutrients may be nearly the same price as a nutrient-rich one. When selecting a fertilizer, look for the largest numbers in the analysis at the least cost to get the best buy.

The comparison of three fertilizers in Table 15-4 illustrates this point. Brand A is the best buy, brand C the second, and brand B the least economical. Comparison shopping obviously can save considerable money.

Although it is not crucial, a *balanced* fertilizer with equal portions of N, P, and K compounds is widely used by commercial houseplant growers with excellent results. A 5-5-5 fertilizer would be an example of one balanced formula. Barring this, a formula should be chosen that has a nitrogen content equal to or greater than its phosphorus and potassium portions, such as 18-6-12.

Fertilizer Forms

Houseplant fertilizers are available in many forms, from tablets to granules and ready-to-use liquids. Each has advantages with regard to price and ease of application, but all are equally acceptable fertilization methods.

Dissolving Powder Concentrates

Powder fertilizers designed to be mixed with water are one fertilizer form. Economically they are usually the best buy because they are the most concentrated. The nutrients dissolve and are applied to plants in place of plain water, are carried down to the root zone in the water, and are immediately available for uptake by plant roots. A disadvantage with older brands was a

Table 15-4 Cost Comparison of Three Houseplant Fertilizers

	Analysis	Nutrients	Weight	Total nutrition	Package price	Price per oz. of nutrient
Brand A	23-19-17	59%	8 oz.	4.72 oz.	$2.50	$ .52
Brand B	5-2-9	16%	2.5 oz.	0.4 oz.	$3.99	$9.98
Brand C	1-2-1	4%	8.5 oz.	0.34 oz.	$2.49	$7.32

Figure 15-10 Houseplant fertilizer in tablet and spike form. (*Photo by Rick Smith*)

Figure 15-11 An encapsulated, slow-release fertilizer for houseplants. (*Photo by Rick Smith*)

poor dissolving ability because of filler materials that settled out, but this is uncommon now. Some brands contain colored dyes that can stain skin or clothes when they are used in concentrated form.

Tablets and Spikes

Dry fertilizers can be purchased as tablets designed to be laid on the medium or as small spikes to be pushed down into it (Figure 15-10). As a plant is watered, these solids break apart, dissolving their nutrients into the water as it flows down to the roots of the plant.

 Manufacturers of horticultural products such as these fertilizers employ horticulturists to design new products, test them, develop packaging, write advertising, and market the products.

Solid fertilizers avoid the bother of mixing and are usually moderate in price.

Liquid Concentrates

Fertilizers in liquid concentrate form are actually dissolvable powders to which water has been added. They are further diluted before use and applied in place of plain water at watering time.

Liquid concentrates mix in water instantly but are more expensive than almost all other fertilizer forms based on the amount of nutrients.

Ready-to-Apply Liquids

Ready-to-apply liquids are poured directly on a plant in place of water and, accordingly, always have a very low nutrient concentration. They avoid mixing incon-

venience but are the worst buy because of the substantial percentage of water they contain.

Encapsulated Slow Release

Encapsulated fertilizer (Figure 15-11) can be worked into the medium surface or mixed into the medium at planting with equal effectiveness. The material consists of pinhead-sized beads containing concentrated liquid or solid fertilizer. Covering each bead is a water-permeable plastic coating. The slow-release feature is made possible by moisture present in the medium seeping in and out of the beads. It carries in it dissolved nutrients that are moved to the roots of the plant at each watering. In the end only the plastic coating is left. The covering does not break down readily and may last for several years in the medium, but it is not harmful.

Slow-release encapsulated fertilizers require the least work and mess of any fertilizer type. The beads do not stain, and one application is effective for three to nine months. The expense is moderate when compared to that of other forms.

Granular Slow Release

Granular slow-release fertilizers look very much like garden fertilizers. They do not possess the plastic coating of encapsulated slow-release types; instead, they release their nutrients slowly because they are a blend of several chemical compounds. One compound releases nutrients immediately, whereas the other must be chemically broken down before it can be absorbed by the plant. Thus, a plant receives nutrition over a long span of time from one application.

Granular slow-release fertilizers are moderately expensive. They have no particular advantage over other fertilizer types applied at proper rates and intervals.

Fertilizer Frequency and Rate

Because of the wide variety of concentrations of fertilizers available, it is impossible to prescribe a standard dilution rate or application frequency. The container mixing directions are the main source of mixing information.

However, many houseplant fertilizers advise weekly or monthly applications of fertilizer on a year-round basis, and it is seriously questionable whether this is advisable. In some cases adherence to container directions could result in overfertilization because the frequency recommendation is based on a rapidly growing plant.

The fertilizer needs of plants growing indoors vary greatly depending on the species and its growing conditions. A naturally slow-growing species like yew pine (*Podocarpus* spp.) would require much less fertilizer than a fast-growing Wandering Jew (*Zebrina* spp.), for example. Fertilizing both at the same rate is therefore not desirable.

Plant nutrient requirements are also greatly affected by the light intensity they receive. In the north, winter light intensities are very poor, and indoor plants experience a slowdown in growth. At this time fertilizer should be applied sparingly or not at all. However, increased light intensity in the spring causes an increased growth rate, boosting water and fertilizer requirements. Fertilizer frequency may need to be increased to keep pace with the rapidly growing plant and to replace those nutrients leached by frequent watering.

As a general rule, fertilization should be done only when a plant is actively growing. The dilution rate should be equal to or weaker than what the package directions stipulate. Application should be infrequent, with the fertilizer applied only during periods of growth and at intervals equal to or longer than those recommended. Few houseplants are harmed seriously by underfertilization, but all are susceptible to damage from overfertilization.

General Guidelines for Fertilizing Houseplants

One of the most important facts to remember when fertilizing a plant is not to increase the fertilization rate or frequency unless you are sure that the plant is suffering from nutrient deficiency. Even if nutrient deficiency is suspected, fertilizer should never be mixed stronger than package directions indicate. Instead, increase the fertilizer frequency slightly, and wait for signs of improvement.

A second precaution is never to apply fertilizer to a dry medium, with the exception of extremely weak ones designed to be used in place of water. A concentrated fertilizer solution can draw water from the roots, and fertilizer "burn" will result.

Third, fertilizing should not be considered a cure for plants that are declining from an unknown cause. It can do actual harm by creating stress on the plant because of excess fertilizer buildup around the roots. Instead, stop fertilizing entirely, resuming it only when the problem is corrected and the plant is growing well again.

Special Fertilizers

A number of special fertilizers for houseplants are sold today. Some are promoted as being strictly for African violets, some are derived from fish, and others claim that one drop in water satisfies all plant requirements.

Organic Fertilizers

Organic fertilizers are extracted from plant or animal sources instead of being obtained by mining or chemical reactions. Fish emulsions, seaweed extracts, and manure formulations are several types of organic fertilizers.

Organic fertilizers contain the same nitrogen, phosphorus, and potassium as other fertilizers, but only in naturally occurring chemical forms. In the process of nutrient uptake, plants absorb fertilizer only as simple inorganic molecules. It is therefore necessary that microorganisms break down organic fertilizer components before absorption can take place. The organisms work slowly, so that organic fertilizers are available over a long period of time.

By analysis, organic fertilizers usually contain a lower percentage of nitrogen, phosphorus, and potassium than inorganic ones, but they sell for about the same price per package. They also contain small quantities of "micronutrients," and these may have a beneficial effect on houseplants. Although organic fertilizers are no less beneficial for plants than inorganic types, they have no particular proven advantages over them. It should be noted that the microorganisms needed to break down organic fertilizers into usable chemical forms may be scarce in sterilized media. This can cause nutrients to become available extremely slowly and create nutrient deficiency problems in spite of regular fertilization.

African Violet and Flowering Plant Fertilizers

Fertilizers sold as specially formulated for blooming plants contain the same N, P, and K compounds as all-purpose houseplant fertilizers but usually in a different ratio. Less nitrogen, a lot of phosphorus, and a moderate to generous quantity of potassium typify a flowering-plant formulation.

Although a deficiency of phosphorus can be a cause of poor flowering, using a phosphorus-rich fertilizer will not guarantee a blooming plant. Other requirements such as light intensity or duration, maturity of the plant, and temperature must be satisfied. Nor can a phosphorus-rich fertilizer guarantee a longer bloom life or be an insurance against bud and flower drop, for other factors are also involved in these processes.

In price, flowering plant fertilizers are about the same as all-purpose fertilizers. There is no research to support claims that these are better than general-purpose ones for flowering houseplants. An evenly balanced (N, P, and K equal) fertilizer will generally supply all the nutrients needed for flowering.

Micronutrient Fertilizers

Whereas macronutrients are used in moderate amounts by plants, the micronutrients are needed only in minute quantities. The micronutrients include such elements as iron, copper, zinc, manganese, boron, and molybdenum. Micronutrient fertilizers can be purchased either with micronutrients alone or as a part of a regular all-purpose fertilizer in dissolvable or encapsulated form.

Potting media usually contain enough micronutrients to supply a plant indefinitely because of the included soil and the decomposition of organic materials. But occasionally plants will develop nutrient deficiency symptoms that are not corrected by all-purpose fertilizers, and a micronutrient deficiency is the likely cause. In these cases a micronutrient fertilizer is advisable, and it can be applied either to the roots or as a spray on the affected foliage.

Many times, micronutrient fertilizers will state that their nutrients are in *chelated* form. Chelated nutrients are used when the pH of the potting medium is causing micronutrients to be "tied up." Even though the micronutrients are present, at an abnormally high or low pH they are held to the medium and are unavailable to a plant. However, chelated nutrients are less affected by pH and will remain available even at unfavorable pHs.

Acid-Forming Fertilizers

Acid-forming fertilizers react in the medium to make it more acidic or lower the pH. They are also sold under the name of "azalea fertilizer," because azaleas are common plants that require a very acidic potting medium. But other plants listed in Table 15-2 are also "acid-loving" and will show deficiency symptoms for iron or manganese if the medium pH rises out of the acid range. For these plants, acid-forming fertilizers are useful to keep the pH in the 5.0–5.5 range.

Foliar Fertilizers

Some fertilizers are sold exclusively as foliar fertilizers, and others give instructions for mixing a solution for use as a foliar fertilizer. These fertilizers usually contain N, P, and K plus micronutrients or just the micronutrients alone.

Fertilizers applied to the foliage are used primarily for two purposes: to apply micronutrients and to fertilize epiphytic plants adapted to receiving nourishment chiefly in that manner. Foliar fertilizers are very weak in concentration and are applied with a mister, spraying until the solution runs off the foliage. Repeat applications are usually necessary before enough nutrient is absorbed to eliminate deficiency symptoms.

The ability of a plant to absorb nutrients foliarly is dependent on its species and the thickness of the waxy cuticle that covers the leaf surface. It is also dependent on environmental conditions when the fertilizer is sprayed. Best absorption takes place in bright light and warm temperatures, but not in direct sunlight, as this dries the fertilizer on the leaves before it has adequate time to be absorbed. High humidity is also an aid to absorption because it increases fertilizer drying time.

When applying a foliar fertilizer, bear in mind that young leaves are more effective at absorbing nutrition foliarly than older ones, and the lower surfaces are more effective than the upper ones. Additionally, a surfactant should be used (if one is not included in the fertilizer) to aid in spreading the liquid over the leaf surfaces to maximize contact and absorption. A drop of detergent in a quart (liter) of mixed foliar fertilizer liquid would be sufficient. Chelated fertilizer formulations are recommended because they increase mobility of the nutrients within the plant.

Oxygen-Generating Fertilizer

Plant roots need oxygen for respiration. Lack of oxygen around the roots causes a deleterious effect because it stops root growth. This in turn restricts growth of aboveground portions of the plant because of decreased root absorption. Normally, adequate oxygen reaches the roots of plants from the soil surface, being conveyed downward through the macropore spaces of the medium. However, in some cases the oxygen content of a medium may be inadequate. This can happen when a plant sets in its drainage water continually, has been growing for too long in the same medium that, over time, has decomposed and compacted, or is potted in a medium that was formulated initially without sufficient pore space.

For these situations in which insufficient oxygen in the root zone is the limiting factor to healthy growth, there is a brand of fertilizer that will help alleviate the problem. Plant Research Laboratories has a houseplant fertilizer called Oxygen Plus® that, through a chemical reaction with the urea peroxide it contains, releases oxygen around plant roots. The actual release of oxygen into the growing medium and subsequent improvement in growth (and sometimes flowering) of a number of species[1] *growing under water-logged conditions* has been proven experimentally at both Plant Research Laboratories of Irvine, California, and the Department of Soil and Environmental Sciences of the University of California at Riverside.[2] Although not all species tested showed improvement, many did, and the product can therefore be used with some measure of confidence to aid in solving plant growth problems associated with water-logged media.

Vitamins and One-Drop-Does-All Chemicals

Vitamins and one-drop-does-all chemicals cannot properly be termed fertilizers because they do not contain the elements that have been proven essential for plant growth. Instead, they contain vitamins and plant hormones (see Chapter 3). Although these chemicals are necessary for healthy plants, they are manufactured in the plant itself and generally do not need to be additionally supplied. Vitamins and plant hormones should not be considered as fertilizer replacements or substitutes.

---

[1]Species on which Oxygen Plus® proved effective were *Syngonium, Chamaedorea elegans, Chlorophytum comosum, Dieffenbachia* 'Exotica,' *Dizygotheca elegantissima, Chrysanthemum × hortorum* cv. Bright Golden Ann, *Monstera deliciosa,* and African violet.
[2]Plant Research Laboratories research conducted by P. D. Gaede and M. P. Boghosian. University of California research conducted by E. M. Herr and W. M. Jarrell.

MEDIUM- AND FERTILIZER-RELATED INDOOR PLANT PROBLEMS

Medium Problems

Any problem present in the potting medium will affect the roots of the plant, and this in turn will produce symptoms on the top growth. Careful observation of the medium for any problems that might be developing is essential when growing plants.

Compaction

Compaction is the excessive packing or settling of potting medium. This packing can exclude air from the roots and results in rotting of roots and decline in plant vigor.

Compaction is a characteristic of a medium or a result of decomposition of the medium over time, and it is common in soil-containing media. The symptoms are yellowing foliage, weak new growth, and a decrease in growth rate. One cure is to remove the plant from its pot, gently break or soak off the majority of the medium from the roots, and repot in fresh medium. To make the new medium less prone to compaction, add a greater portion of peat moss and coarse aggregates and less or no soil. Switching to Oxygen Plus® houseplant fertilizer (see previous section) can also help with the problem.

Poor Drainage

Poor drainage (assuming the pot has a drainage hole) is again a symptom of an incorrectly formulated medium. Media that are prone to compaction are also prone to poor drainage. If a layer of gravel or other coarse material was put in the bottom of the pot at transplanting, this could also be the problem. (see Chapter 14.)

A plant growing in a medium with poor drainage will exhibit the same symptoms as one growing with compaction, and for the same reason: air is excluded from the roots, causing them to decay and die. The cure is also the same: remove the old medium from the roots and repot.

Water Repellency

Water repellency is not common, but it can be a problem in media composed of high percentages of sphagnum moss. Although the moss is capable of holding up to 20 times its weight in water, when completely dried it will shrink and become water-repellent. Water applied to the medium will run through or down the sides of the shrunken soil mass, with little being absorbed. Rewetting can be achieved by submersion or bottom watering with warm water, and the medium should not be allowed to become as dry between waterings again. When the problem is recurring, apply a wetting agent (surfactant) mixed in the regular water.

Unsuitably High or Low pH

Although most media start at a slightly acid pH suitable for most plants, changes in pH can occur over time because of the pH of water used on plants. In such cases, the pH of the irrigation water can be raised or lowered by adding small amounts of common household chemicals: vinegar to lower pH and pure ammonia to raise it. To use these chemicals safely, buy a small, inexpensive pH test kit such as is sold for use on a pool. It should be used initially to test the natural pH of the water and at intervals while adding drops to a gallon or several liters of water to check the progress in raising or lowering the pH to the desired level. The amount of chemical necessary to change the pH can then be recorded and used as a reference when preparing more irrigation water.

If the pH is too high, the level should be adjusted to between 5 and 6 for a month or so and then raised to 6 or 7 to maintain the medium at the acidity level best for most tropicals. If the water is too acidic (a rare situation), ammonia should be used to alter it to 7 or 8, and then after a few months the level should be dropped back to 6 or 7 for maintenance. Acid-loving plants can be maintained on irrigation water in the 5 to 6 range. When using ammonia, be aware that ammonia contains nitrogen, which will probably not then need to be applied in the form of fertilizer.

Fertilizer Problems

Both over- and underfertilization can be injurious to the health of houseplants. The conditions are relatively easy to correct provided the problem can be identified.

Overfertilization (Toxicity)

Overfertilization results from an excess concentration around roots of compounds called *fertilizer salts*. This does not permit entry of water into the roots and creates, in effect, a drought condition. Wilting, browned leaf tips, and stunted growth are symptoms of both maladies. But unlike a drought-stricken plant, an overfertilized plant will not immediately recover when watered, because the cause is still present. The only cure

Figure 15-12 Typical nitrogen-deficiency symptoms on two Wandering Jews. The shoot on the left is bleached looking, whereas the one on the right has a healthy, dark green color. *(Photo by Rick Smith)*

is to remove the excess salts as quickly as possible. This is done by leaching. Large amounts of water should be poured through the medium to flush the fertilizer salts out. (See Chapter 17 for a more complete description of the leaching process.) To prevent a salt accumulation in the soil, always water with enough water so that about 10 percent drains through the bottom of the pot, and discard the drainage water.

Nitrogen Deficiency

Lack of nitrogen shows up first as yellowing of the older leaves of a plant, but it can eventually cause both older and younger leaves to discolor (Figure 15-12). Yellowing is also symptomatic of low light and overwatering, but in these instances the leaves soon drop, whereas in the case of nitrogen deficiency they do not. Additional deficiency symptoms are small new leaves and thin stems.

Phosphorus Deficiency

The first symptoms of phosphorus deficiency are a stunting of growth and the production of undersized new leaves.

Potassium Deficiency

Potassium deficiency appears on older leaves first, frequently as a browning of the leaf margins and tips. The areas between the veins may also be affected.

Iron Deficiency

A lack of iron is the most common micronutrient deficiency found in houseplants. It occurs frequently in acid-loving houseplants growing in media of an unsuitably

Figure 15-13 Iron deficiency in a coffee plant (*Coffea arabica*). *(Photo by Rick Smith)*

high pH. The first symptom is yellowing between the veins of the young leaves (Figure 15-13). The yellowing will eventually include the entire leaf, and the symptoms are known as *iron chlorosis*.

Iron deficiency can be temporarily corrected by foliar sprays of micronutrients and permanently corrected by changing the pH of the irrigation water (see previous section).

Fertilizer Salt Deposits

Buildups of whitish fertilizer deposits on the medium surface or the outside of clay pots (Figure 15-14) are a common occurrence. They are formed when fertilizer

Figure 15-14 Fertilizer salt deposits on the outside of a clay pot. *(Photo by Rick Smith)*

salts dissolved in water move to the outside of the soil ball in response to evaporation of water. The water leaves as vapor, but the salts remain.

Salts on the pot or medium surface do not automatically indicate that there is excess fertilizer in the medium; however, a leaching of the medium is a worthwhile precaution.

SELECTED REFERENCES FOR ADDITIONAL INFORMATION

Banks, David P. *Growing Orchids*. New York: Simon and Schuster Trade, 2001.

Bilton, Ray and Mike Tibbs. *Orchids*. New York: Sterling Publishing, 1998.

Bunt, A. C. *Media Mixes for Container Grown Plants*. Fresno, CA:Thomson Learn, 1988.

Douthitt, Jack. *Bonsai*. New York: Rizzoli International, 2001.

Dreilinger, Sigmund. *Indoor Bonsai*. New York: Brooklyn Botanic Garden, 1991.

Joiner, Jasper. *Foliage Plant Production*. Englewood Cliffs, NJ: Prentice-Hall, 1981.

Mastalerz, J. W. *The Greenhouse Environment: The Effect of Environmental Factors on Flower Crops,* 2nd ed. New York: Wiley, 1977.

Michigan State University Extension. "Watering—The Key to Houseplant Success." *Home Horticulture,* bulletin 03900091 January 1, 1996.

Reed, David William, ed. *A Grower's Guide to Water, Media, and Nutrition for Greenhouse Crops*. Batavia, IL: Ball Publishing, 1996.

World Wide Web Addresses

http://landscaping.about.com/cs/interiors/ Mary Ann Lynch, (webmaster) From "Interior Landscaping." NewHomeOnline.com. 2001. This is a link to sites of interior landscaping.

http://www.aboutflowers.com/# This is a site of the Society of American Florists. Aboutflowers.com. 2001. This site has miscellaneous monthly articles on flowers and indoor plants.

http://www.attra.org/attra-pub/potmix.html Appropriate Technology Transfer for Rural Areas (ATTRA), Organic Potting Mixes, gives recipes for potting media.

http://www.ces.ncsu.edu/depts/hort/consumer/quickref/ houseplants/watering.html This is by Erv Evans, North Carolina State University Cooperative Extension Horticultural Science. "Watering Houseplants." September 2000.

http://www.colostate.edu/Depts/CoopExt/4DMG/Plants/ pottmix.htm   Carl Wilson's, article "Consumers Beware Potting Mix Choices." August 08, 2001.

http://www.ext.nodak.edu/extpubs/plantsci/landscap/ h1123w.htm by Ronald C. Smith, "Interior Plantscaping with Large Plants." North Dakota State University NDSU Extension Service. March 1997.

http://www.ianr.unl.edu/pubs/horticulture/g1022.htm This is by Donald H. Steinegger, Frederick P. Baxendale, and John E. Watkins. "Guide to Growing Houseplants." Nebguide. February 1991.

http://www.plantscaping.net/toc.html   This is from the Great Rocky Mountain Foliage Co. "Interior Plantscaping." July 23, 2001.

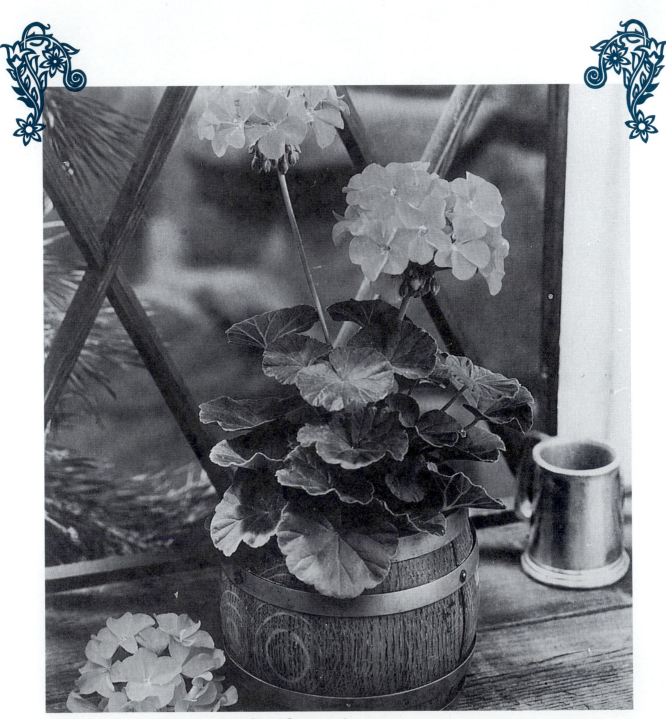

(Photo courtesy of Pan American Seed Company)

CHAPTER 16

LIGHT AND INDOOR PLANT GROWTH

Outdoors, sunlight is not frequently considered when planning gardens. It is simply assumed that unless an area is tree-shaded, sufficient light will be available to grow whatever is planted. When plants are placed indoors, ample light for growth is no longer automatically available. In separating themselves from nature with walls and ceilings, people restrict the available light and must properly manage and supplement the remaining light to grow plants successfully.

EFFECTS OF LIGHT ON PLANT GROWTH AND DEVELOPMENT

Light is essential for photosynthesis, and without it, chlorophyll, carbohydrates, hormones, and many other plant-manufactured chemicals could not be made. Nor would energy be available for respiration and growth.

A logical conclusion would be that plants cannot grow without light. But this is not strictly true. Plants grown in insufficient light do not stop growing, but they do not produce healthy growth. Instead, they develop abnormally long stems with relatively few leaves (Figure 16-1) and develop a stringy or lanky look. During this process, called *etiolation,* the plant respires its stored carbohydrates. When most of the carbohydrate has been respired, it will die.

Phototropism and Similar Responses

Phototropism is the leaning of a plant toward light (Figure 16-2). It is similar to etiolation in some ways, except that only the side of the stem away from the light gains auxin and lengthens. Because the sides of the stem are growing at different rates, the stem curves in the direction of light. Thus, the "seeking" of light by plants is actually a hormonal response.

In other plants, differences in light level change the leaf form. A weeping fig (*Ficus benjamina*) grown in bright light will have large numbers of thick, leathery leaves. When moved to a shadier location, it will drop many of these and refoliate with "shade leaves." The shade leaves will be larger, thinner, and fewer in number than the previous "sun leaves."

Cutleaf philodendron (*Monstera deliciosa*) shows a different reaction to varying light levels (Figure 16-3). The number of splits that new leaves contain decreases at lower light levels. A plant growing in poor light may have no splits at all, whereas one growing in moderate light may have several in each leaf.

Light and Flowering

In addition to affecting photosynthesis and hormone manufacture, light also plays a key role in the flowering of many plants.

Figure 16-1 Etiolation of a plant due to insufficient light. Both plants have the same number of leaves and were propagated at the same time, but the plant on the right received less light. *(Photo by Kirk Zirion)*

Figure 16-2 Phototropism caused by light reaching the plant from only the left side. *(Photo by Rick Smith)*

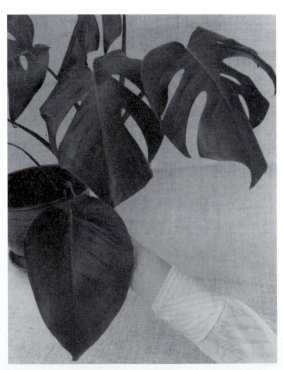

Figure 16-3 Splitting of philodendron (*Monstera deliciosa*) leaves is a function of light intensity. The unsplit leaf developed in dim light, the split leaves in brighter light. *(Photo by Rick Smith)*

Intensity

Insufficient light intensity can inhibit the flowering of many plants grown indoors. Relatively bright light intensities are necessary to trigger flowering in most cultivated plants, and those grown indoors are no exception. Unfortunately, bright light intensity is not often available indoors.

Photoperiod

Correct photoperiod combined with sufficient light intensity causes or hastens flowering in some indoor plants. The word *photoperiod* (discussed in Chapter 3) refers to the day/night ratio in each 24-hour period and can affect not only flowering but also tuber and bulb formation.

Plants can be classified with regard to photoperiod as *short day*, *long day*, or *day neutral*. Many indoor plants are from tropical regions near the equator where day length varies little throughout the year. Accordingly, most are day neutral and flower because of combinations of age, light intensity, and temperature. But a few indoor plants such as chrysanthemum and poinsettia require specific photoperiods to bloom (see Table 16-1 for additional listings).

The number of weeks a photoperiodic plant must be kept under correct day and night lengths to trigger flowering varies among species. If the light/dark cycle is not maintained long enough, the plant may revert to its vegetative state, ceasing bud development in the middle of the flowering process. If kept too long under light/dark control, no harm will usually be done. However, a short-day plant on a long-night schedule too long will lose valuable photosynthetic time and may lose vigor. As a general rule, a photoperiodic schedule should be maintained until the flower buds are completely developed.

Techniques for Controlling Photoperiod. Most houseplants in which photoperiod controls flowering are short-day plants. They require daily periods of uninterrupted darkness before flowering will take place.

Table 16-1 Photoperiodic Houseplants

Common name	Botanical name	Photoperiodic classification	Remarks
Gardenia	*Gardenia jasminoides*	Short day	Causes flowering
Strawberry begonia	*Saxifraga stolonifera*	Long day	Runners and new plantlets form
Spider plant	*Chlorophytum comosum*	Long day	Earlier production of runners and plantlets
Variegated spider plant	*Chlorophytum comosum* 'Vittatum'	Short day	Earlier production of runners and plantlets
Hardy begonia	*Begonia grandis*	Short day	Forms aerial stem tubers
Rex begonia	*Begonia rex*	Long day	Increases leaf area and stem elongation only
Wax begonia	*Begonia semperflorens*	Long day	Flowers under long days at warm temperatures but day neutral at cool temperatures
	Begonia socatrana	Long day/ short day	Flowers under long days; forms aerial stem tubers under short days
Tuberous begonia	*Begonia × tuberhybrida*	Long day/ short day	Forms blooms under long days and underground stem tubers under short days
Pocketbook plant	*Calceolaria crenatiflora*	Long day	Accelerates flowering but is not essential
Cattleya orchid	*Cattleya* spp.	Short day	Causes flowering
Chrysanthemum	*Chrysanthemum × morifolium*	Short day	Flowering may also be hastened by cool temperatures
Poinsettia	*Euphorbia pulcherrima*	Short day	Causes flowering
Fuchsia	*Fuchsia × hybrid*	Long day	Causes flowering in some cultivars
Kalanchoe	*Kalanchoe blossfeldiana*	Short day	Causes flowering and increased leaf succulence
Evening primrose	*Oenothera* spp.	Long day	Flowering accelerated by cool temperatures also
Shamrock	*Oxalis* spp.	Short day	Forms underground tubers
Sedum	*Sedum spectabile*	Long day	Causes flowering
Cineraria	*Senecio × hybridus*	Short day	Short days hasten flowering but are not essential
Christmas cactus	*Schlumbergera bridgesii*	Day neutral or Short day	Day neutral at night temperatures between 50–55°F (10–13°C). At night temperatures of between 55–60°F (13–16°C), it requires 13 hours or more of darkness to flower. Seldom flowers if night temperatures are greater than 70°F (21°C)

Plant species vary in their sensitivity to exposure to light that accidentally "breaks" the necessary dark period. Poinsettias are very sensitive, and bloom will be delayed if the dark period is ever broken by light or if the darkness is not absolute. Most plants, however, are not so sensitive.

One common home procedure for supplying a short-day plant with a long dark period is placing the plant in a closet each day at 5 P.M. and taking it out the next morning at 8 A.M. The schedule can be varied to accommodate personal schedules provided at least 14½ continuous hours per day are spent in darkness. The darkness should be checked by standing inside with the door closed. If light can be seen around the door, the leak should be covered or another closet should be used.

Flowering long-day plants is less exacting. The plant can simply be left in a room that is used in the evenings. The artificial lights will supplement the natural day length with enough light to total 14 or more hours, and blooms will be initiated.

Light and Plant Health

The minimum amount of light a plant must receive to remain in a healthy condition is dependent on its species. However, the following generalizations can be made regarding light requirements.

Foliage Color

In many instances the darker the foliage of a plant, the less light it requires. The most common plants that grow well in dim light—snake plant, rubber plant, cast iron plant, Chinese evergreen, and philodendron—all have deep green foliage. When grown without much light, these plants develop a dense chlorophyll layer near the leaf surfaces. This makes them efficient users of light.

A foliage plant developer seeks out new cultivars to increase the range of foliage plant cultivars available to the consumer. Often this involves finding little-known plants or mutated shoots from known cultivars that display different coloration from the usual for the species. The new plant is patented and then reproduced by a specific grower.

In the opposite fashion, many variegated plants such as variegated spider plant, tricolor dracaena, and variegated wandering Jew require more light than their nonvariegated relatives. They lack chlorophyll in the white areas of the leaves and require more or brighter light to compensate.

Many plants that have red coloring pigments in addition to their chlorophyll are also bright light requirers. Coleus, ti plant, and croton have their chlorophyll masked by red pigment and require strong light to penetrate through to the chlorophyll below.

Variegated plants frequently show their most attractive coloration only in moderate to bright light. Under less light, they often revert to solid green for more efficient photosynthesis.

Fleshiness and Succulence

Plants with thickened leaves and/or stems such as jade plant, kalanchoe, and cacti require more light than those with thin leaves.[1] Their fleshy structure minimizes the surface area vulnerable to transpiration water loss and enables them to survive in dry climates. However, at the same time the photosynthesizing area is reduced. Consequently, these plants can be grown indoors more successfully in bright-light locations.

Native Growing Conditions

A plant that grows naturally in the shade of other plants is usually better adapted to indoor growing than a plant accustomed to unobstructed sunlight. Ferns are dim light plants that illustrate this well. Because they grow naturally in the shade of trees, they also grow well in a dim area indoors. This generalization holds for many tropical plants including devil's ivy, peperomia, and arrowhead.

Conclusions

Ultimately, the amount of light required to maintain a plant species or cultivar indoors is best determined from a reliable reference book. The generalizations stated would hold true for the majority of indoor plants, but exceptions are not uncommon. A plant might decline because of an improper light environment if these guidelines were followed in lieu of a specific reference.

Light Intensity

Light intensity is one factor that contributes to total light a plant receives. The other is light duration. A lack of light is frequently the cause of poor condition in houseplants, particularly in northern areas and in winter.

Natural light intensity is affected by many factors. Some are the result of geographical or altitudinal variation, and some are climatic. Understanding these factors and how they interrelate to affect light intensity will help the indoor gardener understand his or her growing handicaps or assets with respect to light.

[1]Exceptions include Christmas, Easter, and orchid cacti.

Latitude

Because of the way the Earth is oriented in space, the sun rises exactly in the east and sets exactly in the west only at the equator. For people outside the tropics living in the Northern Hemisphere, the sun always appears slightly in the southern sky (Figure 16-4). How far off from the southern horizon it appears depends on the distance from the equator (latitude), with the sun dropping lower toward the southern horizon the further north one proceeds. The opposite is true in the Southern Hemisphere.

This orientation of the sun in the southern sky affects its intensity substantially. By hitting the Earth's atmosphere at an angle instead of directly above, the distance the light must travel through the atmosphere to reach the Earth is increased. Light is lost in the atmosphere, and therefore its intensity when it ultimately reaches Earth is much less. For this reason, the further north one lives, the less the light intensity.

Time of Year

Time of year also affects light intensity because of the sun's angle. Although the sun stays in the southern sky throughout the year, it drops lowest toward the southern horizon in winter and approaches closest to directly overhead at midday in summer (Figure 16-4). How much of an angle it changes over again depends on the distance from the equator. In the southern United States it will change only slightly from summer to winter, and intensity will vary only moderately. But in the far North the change will be very pronounced. The sun will appear low on the horizon and be relatively dim in intensity in winter, although rising to nearly overhead with a bright intensity in summer. In spring and fall it will be midway between and moderate in intensity.

Time of Day

Light intensity is also affected by time of day. Midday, when the sun is at its highest point, provides the greatest intensity of light. In either morning or evening the sun is lower toward the horizon, and its intensity will decrease accordingly because of the distance it is penetrating through the atmosphere.

Local daily weather variation can change this, however. In the West, some areas have routinely foggy mornings that lessen light intensity early in the day. Accordingly, the afternoon sun is much brighter in these areas.

Altitude

People living at high altitudes (greater than 1,000 feet [300 meters] above sea level) frequently receive greater light intensity than would be expected at their latitude. The effect of the atmosphere on decreasing sunlight intensity is the reason. Air is less dense and the atmospheric layer is thinner at high altitudes, and less intensity is lost passing through it. Denver, Colorado, illustrates this effect of altitude perfectly. Located at over 5,000 feet (1,524 meters) in altitude, its light intensity year-round is substantially higher than that of Boston, Massachusetts (elevation 21 feet [7 meters]), even though the two are less than 4° apart in latitude.

Clouds, Fog, and Smog

In cloud-, fog-, or smog-covered areas, light intensity is habitually lower than in clear areas at the same latitude. The water vapor or pollution haze acts as a reflector, sending much of the sunlight back toward space. In addition, clouds, fog, and smog diffuse sun rays, giving

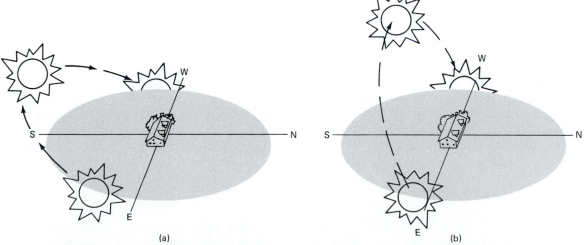

Figure 16-4 Typical positions of the sun in winter (a) and summer (b) in the Northern Hemisphere. Note that the sun remains in the South and that the positions of rising and setting differ with the season.

the earth shadowless illumination rather than the sun and shadow pattern of direct sunlight.

Snow Cover

Persistent snow cover affects light intensity by reflecting sunlight indoors. Cold areas that have snow cover most of the winter receive a slightly increased amount of light through all windows because of this reflection.

This principle also applies to houses surrounded by light-colored sand, white rock mulch, or a similar material. Although dark soil will absorb light, light sand reflects it and will increase the light entering a house.

Light Duration

Light duration (day length) is a second component of total light. To some extent an increased duration can compensate for a low intensity in providing plants with adequate light for photosynthesis. However, below a certain minimum light intensity, an increased duration will not compensate.

Days are longer in summer than in winter, and again this is a result of latitude. Days and nights at the equator are equal in length all year round. But in the North it is not uncommon for winter days to be as short as 9 hours and summer days as long as 15 hours. The briefness of winter days and the increased length of summer days become more pronounced the further north one travels. In Alaska, for example, days are 20 hours or more long in the summer.

Unfortunately, the shortest winter days correspond with the period when the sun is lowest on the horizon, reaching a climax on December 21, the winter solstice. At this point the sun again begins moving toward the northern sky and days begin to lengthen.

Summary

Latitude, day length, altitude, cloud and snow cover, and the other topics discussed all combine to give total light. Because of these many variables, it is almost impossible for a person to determine whether an area is generally bright, moderate, or dim except by observation. However, the Environmental Sciences Service Administration has recorded light intensity and duration throughout the United States for the months of July and December. Their findings give a rating of total light (measured in langleys) for the United States in midwinter and midsummer.

Figure 16-5 codes the continental United States into bright-, medium-, and dim-light areas for winter and summer. Bear in mind that for most areas, summer light is three to four times as great as winter light.

Measuring light in relation to plant growth is a complex topic that can be broken into three subtopics:

Aspect of Light	Unit of Measurement
Illumination or intensity	lux, lumens, or foot-candles
Total radiation (intensity plus duration)	joules, watts, or langleys
Amount of specific wavelengths needed by plants (PAR)	quanta, moles, or microwatts

A complete discussion of each of these is beyond the scope of this text, but each will be discussed as it pertains to growing houseplants.

Determining the Amount of Total Solar Radiation Entering Windows

The factors affecting the intensity and duration of light that reaches the Earth are important to understand. But it is the total light entering windows that is of the greatest importance to an indoor gardener.

Several types of light meters can be used to measure light intensity, but for nonprofessional home use the only practical one would be a thermopile or bolometer that has a flat response to all wavelengths, meaning it does not favor reading of any particular wavelength. It is an *incident* light meter because it measures the light intensity falling on an object, in this case a plant. To obtain a representative reading, measurements would need to be taken at several parts of the plant while aiming the meter at the light source.

Readings would also need to be taken several times a day, and under cloudy and sunny conditions, to determine the average light a plant is receiving. After determining this, the light requirement for the plant must be found. Although this has been determined for many plants and is published in Cooperative Extension Service bulletins, the list is far from complete.

In some cases the light requirement of a species will be listed in a range of foot-candles. The foot-candle is an old measurement[2] denoting the illumination of a surface 1 foot from the light of a standard candle. For reference purposes, a comfortable light for reading is 50 foot-candles, and a bright summer day yields about 10,000 foot-candles. The United States Department of Agriculture classified over 100

[2]The foot-candle measurement of light intensity is outdated and has been replaced by the metric lux. However, many publications dealing with the light requirements of plants still use this measurement.

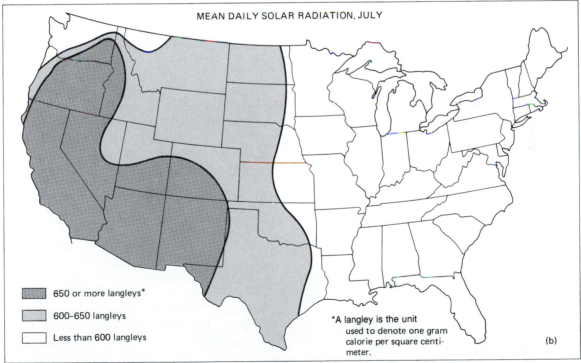

Figure 16-5 Mean daily solar radiation throughout the United States in January and July.

houseplants by their light requirements, with the range of 75–200 foot-candles (767–2,153 luxes) deemed low, 200–500 foot-candles (2,153–5,382 luxes) moderate, and 1,000+ foot-candles (10,764 luxes) high. *Compensation-point* (light intensity be-

low which respiration exceeds photosynthesis) intensities were evaluated as 25 foot-candles (229 luxes) for low-light plants, 75–100 foot-candles (767–1,076 luxes) for medium-light species, and 1,000 foot-candles (10,764 luxes) for high-light types.

Interactions occur between intensity and temperature, watering frequency, and fertilization frequency. Temperatures less than 63°F (17°C) conserve stored carbohydrate by reducing respiration rate, whereas frequent watering and fertilization exacerbate and accelerate senescence in the forms of stem lengthening and aging of old foliage.

For most gardeners, simple bright, medium, or dim designations of light based on the direction a window faces and distance a plant is placed from a window are more workable than foot-candles. Exact light measurements are not necessary to grow houseplants successfully. The use of a few reference books, the ability to quickly spot the signs of light deficiency, and trial and error are the tools of most successful indoor gardeners.

Window Direction

Several factors make south-facing windows the brightest and best for growing many indoor plants. First, because the sun shines from the southern sky all year, unobstructed south-facing windows receive the longest duration of daily direct sunlight all year. Only south-facing windows receive sun from midmorning through late afternoon, and an indoor gardener with southern windows is fortunate. He or she has enough light for almost any plant and can modify the light by moving plants further from the window.

East- and west-facing windows provide an intermediate amount of light, less than southern but more than northern. However, in the western states west-facing windows may provide substantially more total solar radiation than in eastern states. Each receives direct sunlight for about half of the day and only reflected light for the remainder. The total light entering from east- and west-facing windows is sufficient for growing the majority of houseplants, provided they are not placed too far from the glass.

North-facing windows provide the least light but are still acceptable for growing a number of species in most areas. Their dimness results from the fact that they receive no direct sun, only light reflected from the ground or nearby buildings.

Other Factors Decreasing Sunlight Entry through Windows

Several factors, such as the presence of overhanging eaves or tall buildings, can decrease the sunlight entering through windows. These factors should be taken into account in determining the light level of windows. They are discussed in the following sections.

Figure 16-6 The effect of eaves on the amount of sunlight entering through a window. Note the depth of penetration into the room with (a) and without (b) eaves.

Overhanging Eaves

Wide eaves (Figure 16-6) can decrease the time or distance direct sunlight enters a room appreciably. They shade the window when the sun is high in the sky, at midday, and several hours before and after. This effect is more pronounced in the summer than in the winter because the sun is higher off the southern horizon.

Buildings or Fences

Tall buildings or fences close to windows can also decrease direct sunlight. As with overhangs, the problem is one of shading from the south, east, or west, but unlike blockage due to eaves, the problem will be more pronounced in winter.

Shrub or Tree Plantings

Shrubs or trees planted close to a window can seriously limit light entry, sometimes reducing it to only reflected light. Trees and shrubs that are deciduous will defoliate in fall and cause little blockage of light during the crucial winter months, but evergreens exclude light continuously. Removal or pruning of the plants is the only solution. Limbing up or thinning will work on trees. Shrubs should be thinned and headed back to minimize blockage of sunlight. Chapter 11 describes all these pruning styles.

Curtains and Shades

Most homeowners open their curtains or shades in the morning and close them after dark, but valuable light can be lost this way during the early morning hours. East-facing windows, in particular, can lose a substantial portion of their direct sunlight when curtains remain closed after sunrise.

Even when curtains are open, they can decrease the light area around the outside edge of the window. To maximize light, hang curtains so that they expose the entire window area when opened. Avoid overhanging valances, which block overhead light.

Tinted or Stained-Glass Windows

Tinted or stained-glass windows can affect light entry only slightly or block it almost completely, depending on the shade of glass used. Although little can be done short of removing them, their effects should be taken into account when positioning plants for proper lighting.

PLACEMENT OF PLANTS FOR OPTIMUM NATURAL LIGHT

Just as windows are classified as providing bright, medium, or dim light, plants can be classified as bright-, medium-, or dim-light-requiring and positioned within a room to satisfy their requirements. Generally, the dilemma revolves around how far from a window a plant can successfully be grown. This minimum light level at which a plant can maintain itself is called the *compensation point*. The light it receives is sufficient to photosynthesize carbohydrate for its respiration but leaves no surplus carbohydrate for growth. With less light the use of carbohydrate for respiration will exceed its rate of manufacture, and the plant will slowly die. At brighter light levels excess carbohydrate will be used for growth.

The concept of compensation point is very important in indoor plant growing. Plants being used for decoration can be bought at the desired size and maintained in good condition with the correct amount of light. Because the plant is growing very slowly if at all, the time devoted to maintenance is minimal. Also, plants grown at compensation-point light levels can be placed farther back from windows than is optimum for growth, creating space for additional plants that require more light to survive.

The following descriptions and figures of bright, moderate, and low light are intended only as rough estimates of available window light. However, they include the range from a compensation-point amount of light through the highest light advisable or available for plants indoors.

Bright Light

The designation "bright"- or "high"-light-requiring is given to a number of plants including most flowering species, cacti, and succulents (Table 16-2). Plants that require bright light are, as a whole, the hardest to grow

Table 16-2 Bright-Light Plants

Common name	Botanical name
Flowering maple, Chinese lantern	*Abutilon* spp.*
Chenille plant	*Acalypha hispida**
Copperleaf	*Acalypha wilkesiana* cultivars*
Lipstick vine	*Aeschynanthus* spp.
Agave	*Agave* spp.*
Bamboo	*Bambusa* spp.*
Ponytail palm	*Beaucarnea recurvata**
Wax begonia	*Begonia* × *semperflorens-cultorum**
Cacti	*Xerophytic* spp.*
Camellia	*Camellia* spp.
Natal plum	*Carissa grandiflora**
Cattleya orchid	*Cattleya* spp.
Rosary vine	*Ceropegia woodii*
Chrysanthemum	*Chrysanthemum* × *morifolium**
Citrus	*Citrus* spp.*
Croton	*Codiaeum variegatum**
Coleus	*Coleus* × *hybridus**
Goldfish plant	*Columnea, Nematanthus*
Cigar plant	*Cuphea* spp.*
Tricolor dracaena	*Dracaena marginata* 'Tricolor'
Poinsettia	*Euphorbia pulcherrima**
Fatshedera	*Fatshedera lizei**
Purple velvet plant	*Gynura aurantiaca* cultivars
Variegated ivy	*Hedera helix* cultivars
Amaryllis	*Hippeastrum* spp.
Wax plant	*Hoya carnosa* cultivars
Polka-dot plant	*Hypoestes phyllostachya*
Bloodleaf	*Iresine herbstii*
Shrimp plant	*Justicia brandegeana*
Gold hop	*Justicia brandegeana* 'Yellow Queen'
Kalanchoe	*Kalanchoe blossfeldiana**
Ice plant	*Lampranthus, Mesembranthemum,* and other genera*
Lantana	*Lantana* spp.*
Wax leaf privet	*Ligustrum japonicum, Ligustrum lucidum**
Oxalis	*Oxalis* spp.*
Geranium	*Pelargonium* spp.*
Avocado	*Persea americana**
Yew pine	*Podocarpus* spp.*
Rose	*Rosa* spp.*
Stephanotis	*Stephanotis floribunda*
Succulents	Many species*

*Indicates no maximum light level indoors.

successfully indoors. Difficulty in providing them with sufficient sunlight during the darker winter months is the main problem.

Although light conditions vary enormously depending on the locale, as a general rule, bright-light

Figure 16-7 Suggested distances from windows for growing bright-, medium-, and dim-light plants.

plants should be placed no farther than 3 feet (1 meter) from a southern window or 2 feet (0.6 meter) from an east or west window (see Figure 16-7). It is nearly impossible to give bright-light-requiring plants too much light. The more light they receive, the healthier they will be. Most adjust readily to full sunshine and will grow vigorously outdoors during warm weather after they have been acclimated.

Moderate Light

Most foliage houseplants fall into the moderate- or medium-light requirement category (Table 16-3). They can be provided with adequate light if placed up to 6 feet (2 meters) from a southern window, up to 4 feet (1.2 meters) from east or west windows, or up to 2 feet (0.6 meter) from a north window. As with bright-light-requiring plants, moderate-light users grow better when placed close to the glass, especially in winter. However, as summer approaches, those in south or west windows will receive increasing amounts of light. They should be watched for symptoms of excess light and moved back from the window if symptoms appear.

Low Light

Only a small percentage of indoor plants grow satisfactorily in low light or what is considered "dim" from a plant's perspective (Table 16-4), and most of these are better in a medium-light location. However, a number of species classified as moderate-light requirers can survive low-light conditions for a period of up to several months without harm.

Low-light plants can be grown 4 to 6 feet (1.2–2 meters) from east and west windows and anywhere up to 4 feet (1.2 meters) from north windows (Figure 16-7). They will also grow 6 to 8 feet (1.8-2.4 meters) from a south window. Unlike bright-light plants, some low-light plants can be injured by excess light. They will live well in moderate light but generally should not be grown in bright light.

MAXIMIZING NATURAL LIGHT

In general, an indoor gardener must work with whatever natural light is available. He or she cannot alter the climate, and short of making structural modifications

Table 16-3 Moderate-Light Plants

Common name	Botanical name
Maidenhair fern	*Adiantum* spp.
Anthurium	*Anthurium* spp.
Aucuba	*Aucuba japonica*
Begonia	*Begonia* spp.
Umbrella plant	*Brassaia* or *Tupidanthus* spp.
Fishtail palm	*Caryota* spp.
Spider plant	*Chlorophytum comosum* cultivars
Kangaroo and grape ivies	*Cissus* spp.
Coffee	*Coffea arabica*
Ti plant	*Cordyline terminalis*
Earth star	*Cryptanthus* spp.
Dumbcane	*Dieffenbachia* spp.
False aralia	*Dizygotheca elegantissima*
Dracaena	*Dracaena* spp.
Euonymus	*Euonymus* spp.
Fatsia	*Fatsia japonica*
Fig, weeping	*Ficus benjamina*
Fig, fiddle-leaf	*Ficus lyrata*
Fig, creeping	*Ficus pumila*
Ox tongue	*Gasteria* spp.
Prayer plant	*Maranta leuconeura* cultivars
Boston fern	*Nephrolepis* spp.
Lady slipper orchid	*Paphiopedilum* spp.
Peperomia	*Peperomia* spp.
Dwarf date palm	*Phoenix roebelenii*
Pileas	*Pilea* spp.
Staghorn fern	*Platycerium bifurcatum*
Swedish ivy	*Plectranthus* spp.
Hare's foot fern	*Polypodium aureum*
Aralia	*Polyscias* spp.
Moses in the cradle	*Rhoeo spathacea*
African violet	*Saintpaulia ionantha* cultivars
Strawberry begonia	*Saxifraga solonifera*
Epiphytic cacti	*Schlumbergera, Rhipsalis, Epiphyllum,* and other genera
Jerusalem cherry	*Solanum pseudocapsicum*
Baby tears	*Soleirolia soleirolii*
Piggyback plant	*Tolmiea menziesii*
Wandering Jew	*Tradescantia, Zebrina*
Ginger	*Zingiber officinale*

Table 16-4 Dim-Light Plants

Common name	Botanical name
Miniature flag plant	*Acorus gramineus*
Chinese evergreen	*Aglaonema* spp.
Norfolk island pine	*Araucaria heterophylla*
Asparagus fern	*Asparagus* spp.
Cast iron plant	*Aspidistra elatior*
Bird's nest fern	*Asplenium nidus*
Parlor palm	*Chamaedorea elegans*
Holly fern	*Cyrotomium falcatum*
Corn plant	*Dracaena fragrans* cultivars
Pleomele	*Dracaena reflexa*
Devil's ivy	*Epipremnum aureum*
Rubber plant	*Ficus elastica*
Fittonia	*Fittonia verschaffeltii* cultivars
Kentia palm	*Howea fosteriana*
Philodendron	*Philodendron* spp.
Hart's tongue fern	*Phyllitis scolopendrium*
Snake plant	*Sansevieria* spp.
Spathe flower	*Spathiphyllum* spp.
Arrowhead vine	*Syngonium podophyllum*
Bromeliads	*Vriesea, Guzmania, Nidularium*

landscaping and plant maintenance. Every few months the plants in the low-light display area are replaced and taken back to the greenhouse for a recuperation period. During this rotation in the greenhouse they accumulate carbohydrate that will be used for respiration when they are again in the display area.

The home gardener without ample light for all his or her plants can practice a modified type of rotation, moving some plants closer to windows and others farther away every two to three weeks. This technique works best on fairly hardy indoor plants and on moderate- or low-light-requiring species. The requirements of high-light plants are less flexible. They can be placed in moderate or dim areas to a limited extent but should not be kept there more than a few days to avoid leaf drop. In addition, plants that are in bud should not be moved, because the sudden change in environment can cause the buds to drop. However, after the buds open, the plant can be moved to be more effectively displayed.

Reflective Colors

Light is reflected by white surfaces, completely absorbed by black, and absorbed to some degree by all colors between. Once light enters through a window, its contact with plants can be maximized by light-colored walls, carpets, and drapes, which bounce light back rather than absorbing it. White walls reflect up to 90 percent of the light they receive. They can contribute

for larger windows or cutting skylights, not much can be done to increase light entry. However, several "light management" practices can make the most efficient use of available light.

Rotation

Plant rotation has proven its workability commercially and is also practical for the home gardener. Many of the large foliage plants found in lobbies and public shopping malls are rented from firms specializing in interior significantly to the amount of light plants receive. An increase in reflected light makes it possible to grow moderate- and especially low-light-requiring plants farther back from windows.

Skylights

Skylights are probably the ideal solution to the problem of insufficient light indoors for growing plants. Once very expensive because they were always a custom construction item, they are now available prefabricated in a variety of sizes and materials. The original skylights in buildings in the late 19th and early 20th centuries were glazed with wire glass, the safety glass of the day. New skylights must be made of plastic, tempered glass, and/or laminated glass for safety.

One point of importance to a person who is purchasing a skylight for growing plants is that translucent rather than transparent glazing should be selected. Just as humidity in the air diffuses sunlight and makes everything bright regardless of the direction from which the sunlight is coming, translucent glass or plastic breaks an incoming light ray at the point at which it passes through the material and enables it to illuminate a larger area over a longer time than clear glazing (Figure 16-8). Computer models of commercial-sized atriums hypothetically installed in Detroit (42° latitude, a very dim area) by Leonard A. Kersch showed that when the sun is unobstructed by clouds the atrium will be four to five times brighter when translucent

Figure 16-8 Light penetration patterns in transparent and translucent skylights. *(Figure based on one by Leonard A. Kersch of Garden Milieu of Ann Arbor, MI, and published in* Interior Landscaping Industry *vol. 5 [10])*

glazing is used (Figure 16-9).[3] The scattering effect of the translucent glazing is independent of the angle at which the sunlight strikes the glass except for extreme latitudes in northern Canada and Alaska, and this is particularly relevant because at very northerly latitudes the sun can drop very far toward the horizon (21° in Detroit on December 21, for example) and in such situations

[3]When there is no direct sunlight, the illumination under the skylight would be the same regardless of the type of glazing, but even in the cloudiest areas of North America the diffusive effect of translucent glazing would be a significantly positive factor to improved plant growth. (See "Selected References for Additional Information.")

clear glazed skylight

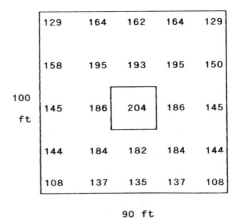

translucent glazed skylight

Figure 16-9 Light intensity variation between clear and translucent skylights. The dotted square represents a skylight 50 feet (15 meters) above the floor. *(Figure based on one by Leonard A. Kersch of Garden Milieu, Ann Arbor, MI, and published in* Interior Landscaping Industry *vol. 5 [10])* frequently illuminates the walls instead of the floor and plant-growing beds in an atrium.

Translucent glazing also eliminates the problem of "hot spot," an area of direct bright sun that moves across the atrium as the sun moves across the sky (Figure 16-9). With clear glazing this hot spot is the main bright-light plant-growing area, when what would be preferred would be slightly less intensity over a larger area.

ARTIFICIAL LIGHTING FOR PLANT GROWTH

Another solution to the problem of insufficient light is the use of artificial lighting. The indoor gardener who branches out into artificial lighting has a practically unlimited choice of plants. He or she can grow almost any flowering plant successfully, as well as herbs, vegetables, and transplants for outdoor use. "Light gardening," as it is called, can be a fascinating hobby.

Two basic types of lights are used for artificial lighting of plants in homes—incandescent and fluorescent.

Photoactive Radiation (PAR)

The last characteristic that must be addressed in regard to light and plant growth is the amount of the types of light necessary for plant growth that is given by a particular light source. This source can be either natural (sunlight) or artificial.

Sunlight is naturally suited for plant growth, but some artificial light is not. An understanding of the makeup of light and the kinds of light necessary for plant growth is necessary for a gardener to purchase lighting equipment wisely.

Light energy is divided into wavelengths; Figure 16-10 shows the relationship between wavelengths of visible light and other types of radiant energy. Of the energy types listed, only three—ultraviolet, visible, and infrared rays (heat)—are contained in solar radiation and are significant to plant growth.

Ultraviolet light causes tanning, and plants growing outdoors are exposed to it all day. However, it is not necessary for plant growth. The fact that ultraviolet light does not pass into glass greenhouses further proves that plants can be grown very successfully without it.

Infrared waves transmit heat by traveling through air or space and warming whatever objects they contact. Consequently, the infrared waves in sunlight warm plants outdoors during the day, but at night infrared rays will be given off from previously warmed surfaces like the soil. Heat is needed for plant growth, but it need not be supplied by a light fixture.

This leaves only visible light necessary for plants, and even all the colors that make up that range of light have not been proved necessary. Light in the orange-red and blue-violet portions of the spectrum have the most significant influence on plant growth. These are the

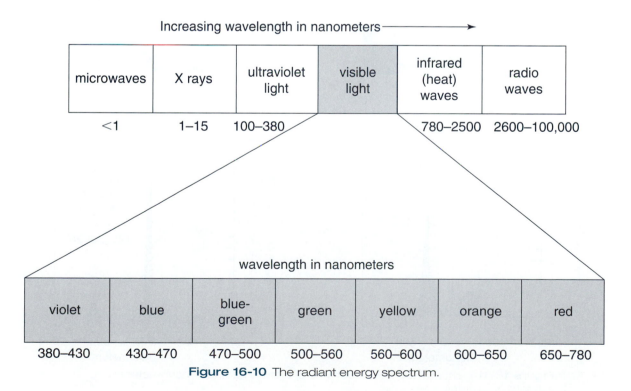

Figure 16-10 The radiant energy spectrum.

wavelengths that compose the majority of photosynthetically active radiation, abbreviated *PAR*. PAR ranges from 400 to 700 nanometers. The green and yellow wavelengths are commonly reflected back by plants, giving them their characteristic green appearance.

Both orange-red light and blue-violet light are used in photosynthesis. Red light has been proved to be the wavelength range that triggers flowering in photoperiodic plants, whereas blue-violet light is responsible for phototropic responses.

Types of Lighting

Incandescent Lights

Incandescent lights were the first lights invented. They produce light by channeling electricity through a wire, which then heats up and glows. Compared to fluorescent lights, incandescent lights use more electricity to produce the same amount of light, the extra energy being given off as heat. In wavelengths of light emitted, they are high in orange-red (Figure 16-11) but low in blue-violet, giving

objects they light a warm appearance. Although the red-orange is useful for flowering plants and makes an attractive spotlight, the lack of blue-violet makes normal incandescent bulbs unsatisfactory for plants when used alone.

Plants grown exclusively under incandescent lighting are paler, thinner, and more etiolated than plants grown under the other light sources listed. In addition, they will not branch, and flowering and subsequent senescence are very rapid.

Household-Type Fluorescent Lights

Fluorescent lights (Figure 16-12) are long tubes that can only be used in specially designed light fixtures. The inside of the tube is coated with phosphorescent material and has an electrode at each end. Electricity flows from each electrode and, by a chemical reaction with mercury, stimulates the phosphorescent coating to give off light.

Although initially more expensive, fluorescent lights are more economical to operate than incandescent lights, producing up to five times as much light from the same wattage of electricity. They give off almost

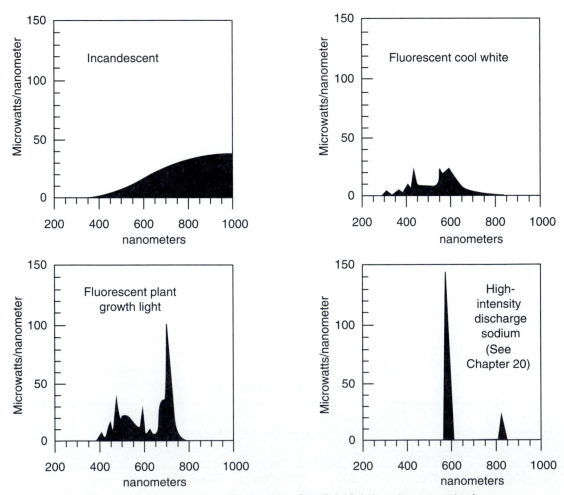

Figure 16-11 Light emitted from four types of artificial lighting. A nanometer is one millionth of a meter.

Figure 16-12 A simple fluorescent light setup for flowering plants.

no heat from the tubes, and only a small amount from the ballast, the part of the fixture that regulates the electricity flow to the electrodes. Finally, fluorescent tubes have a much longer life than do incandescent bulbs.

The wavelengths of light emitted by fluorescent tubes vary depending on the blend of phosphorescent materials used to coat the tube. Table 16-5 lists the most common types of fluorescent tubes and their emissions in the orange-red and blue-violet wavelengths. Of the normal fluorescent tubes, cool white has been used most often for growing foliage plants. However, it is poor in the orange-red area of the spectrum and should be supplemented by an incandescent light for flowering species. Characteristics of plants grown under cool white lights are richly colored foliage, slow stem elongation, and proliferous branching, the latter two of which create a sturdy, bushy, and consequently, attractive plant.

Plant Lights

Plant lights in both incandescent and fluorescent styles are commonly available. They are designed to provide light in both the blue-violet and the orange-red wavelengths in the best combination for plant growth. Because fluorescent lights are economical to operate and provide a higher intensity, most plant lights are of the fluorescent type.

Plant lights are sold in fewer numbers than normal fluorescent tubes, so they are more expensive. For a small light garden, the cost may not be important. But for larger setups, combinations of household tubes from Table 16-5 can be substituted to provide both blue-violet and orange-red light. There will be little difference in plant growth.

Setting Up a Fluorescent Light Garden

Even with the efficiency of fluorescent lighting, more than one tube is generally needed to grow bright-light-requiring plants such as flowering species and annual or vegetable transplants. A minimum of two (and preferably four) tubes placed side by side is commonly used by experienced light gardeners. There is no set formula regarding the wattage of bulbs or how close to place them from one another. In general, tubes of 40 watts or more are best, placed not more than 3 inches (10 centimeters) apart.

A homemade light garden is not difficult to construct. One widely used setup is a shop light containing two tubes and a built-in reflector. The fixture is mounted on a metal stand or shelving unit, and a lamp timer regulates the hours of operation.

A more attractive light garden can be built using a reflectorless two-tube unit mounted in a bookcase stereo cabinet, or room divider. The only stipulation is that the area to which the fixture is attached be painted white for maximum light reflection. Detailed diagrams

Table 16-5 Wavelength Emissions of Household-Type Fluoresent Tubes

Tube type	Blue-violet	Orange-red
Cool white	Good	Poor
Cool white deluxe	Good	Fair
Warm white	Fair	Poor
Warm white deluxe	Good	Fair
Daylight	Excellent	Poor
Natural white	Good	Fair
Soft white	Fair	Fair

Figure 16-13 Bleached foliage caused by sunburn on cast iron plant (*Aspidistra elatior*)

for homemade light gardens are plentiful in light gardening and home building project books.

Growing Plants under Lights

The distance from the tubes at which plants are grown and the number of hours per day the lights are left on largely determine the success of a light garden.

For 40-watt household tubes, seedlings and bright-light-requiring plants should be about 2 inches (5 centimeters) from the bulbs. With higher-wattage bulbs, the distance can be increased but should not be greater than about 6 inches (15 centimeters). Because fluorescent lights do not give off heat, there is no problem of heat damage to plants placed too close to the bulbs. Some plants grow so rapidly that they touch the bulbs, but they are seldom injured.

Foliage plants can be grown farther from the tubes because their light requirements are lower. Many foliage houseplants will grow well placed up to 2 feet (0.6 meter) from the light source.

The length of time lights should be on is not an absolute figure. In general, 12 to 16 hours per day is considered suitable if there is no additional source of light. Although foliage plants can grow with less, there is no advantage to running the lights a shorter time except

for controlling the photoperiod. The amount of electricity used to operate the tubes is minimal.

Although fluorescent light bulbs may appear to burn out spontaneously, they actually lose intensity slowly over their entire lifetime of 12,000 or more hours. For this reason it is wise to replace the bulbs after about 75 percent of the stated hour-life has passed, because the intensity will have decreased by 15 to 20 percent. New bulbs should be dated with indelible marker, and replacement time calculated by estimating the hours the bulb burns per day multiplied by the number of days of use. Replacing one bulb every three or four months will avoid excess light symptoms from the sudden intensity increase.

LIGHT-RELATED INDOOR PLANT DISORDERS

Insufficient Light

Lack of light is one of the most common causes of houseplant problems; it is also among the easiest to diagnose. Etiolation of new growth is one symptom. Yellowing and dropping of the older leaves usually also occur. Once the leaves have begun to yellow, they will fall regardless of any improvement then made in the light.

Plants in the advanced stages of light deficiency have thin weak stems and few leaves. The leaves remaining may be widely spaced, pale, and undersized. Such plants should be placed under increased light to recuperate.

Excess Light

Excess light is seldom a problem for indoor plants. However, a plant will occasionally develop symptoms of *chronic* excess light exposure when, for example, a low-light plant is grown in a south-facing window. The new leaves will be small and a pale yellowish-green. The internodes will be very short, giving the plant a compact but unhealthy appearance.

A more common case is *acute* excess light damage that occurs when houseplants are placed outdoors after growing indoors all winter. The sudden increase in intensity destroys the top layers of cells on the leaf surfaces, and the leaves acquire a silvery cast (Figure 16-13). The silvery cast turns brown or bleached yellow in a few days. Unfortunately, sunburned leaves do not heal, and the only way to improve the appearance of the plant is by removing these leaves.

Plants "summered" outdoors must have a lengthy acclimatization period to avoid sunburn. Bright-light plants will benefit from summer sunlight, but medium- and dim-light species should be placed in shaded or filtered sun areas. They will thrive under outdoor conditions without risking sunburn.

To acclimate houseplants to outdoor conditions, they should first be placed in a sheltered outdoor location that receives early morning or late afternoon sun but is shaded the rest of the day. After another week the bright-light plants can be moved to a brighter outdoor area but should still receive protection from intense midday sun.

When bringing the plants indoors again, the acclimatization procedure should be reversed. Gradually adjusting the plants to decreased indoor light will prevent leaf drop and other signs of light deficiency. The plants should also be checked during this period for insects and diseases, because these problems can spread rapidly indoors.

SELECTED REFERENCES FOR ADDITIONAL INFORMATION

Cathey, H. M., and Lowell Campbell. "Light and Lighting Systems for Horticultural Plants." In *Horticultural Reviews* Vol. 2. Jules Janick, ed. Westport, CT: AVI Publishing, 1980, pp. 491–538.

Elbert, G. A. *The Indoor Light Gardening Book*. New York: Crown, 1975.

Gaines, R. L. *Interior Plantscaping: Building Design for Interior Foliage Plants*. New York: McGraw-Hill, 1977.

Kersch, Leonard A. "Computers Shed New Light on Indoor Illumination Levels." In *Interior Landscaping Industry* Vol. 4(3), March 1987, pp. 21–32.

Kersch, Leonard A. "Something for Nothing." In *Interior Landscaping Industry* Vol. 5 (10), October 1988, pp. 51–55.

Kramer, J. *Plants under Lights*. New York: Simon and Schuster, 1974.

Martin, Tovah. *A New Look at Houseplants*. New York: Brooklyn Botanic Garden, 1994.

Ortho Book Division. *The Facts of Light about Indoor Gardening*. San Francisco: Chevron Chemical Company, 1975.

Sunset Books Staff. *Windows and Skylights*. Menlo Park, CA: Sunset Publishing, 1982.

Van Patten, George F., and Alyssa F. Bust. *Gardening Indoors with H. I. D. Lights*. Washougal, WA: Van Patten Publishing, 1997.

Walker, David. *Energy, Plants, and Man*. 2nd ed. Sausalito, CA: University Science Books, 1992.

World Wide Web Addresses

http://gardening.about.com/cs/houseplants/
Deborah Simpson, "Houseplants," *Gardening with Horticulture Magazine*. 2001.

http://pss.uvm.edu/ppp/articles/lowlite.htm.
Leonard, Perry and Aileen King. "Low Light Houseplants." University of Vermont Extension, Department of Plant and Soil Science.

http://www.apsnet.org/online/feature/abiotic/light.html
American Phytopathological Society. "Light." Feature story at APS Net. January 2001.

http://cc.joensuu.fi/photobio/photobio
Aphalo, Pedro J. "Plant Photobiology Notes." This has links to other references, and is quite thorough.

(Photo by the author)

CHAPTER 17

INDOOR PLANT WATERING AND HUMIDITY

WATERING AND INDOOR PLANT GROWTH

Watering is the most frequent maintenance activity associated with growing houseplants. It is also (along with improper lighting) one of the main causes of houseplant failure, probably because absolute directions are not available. Proper watering depends on an understanding of the water requirements of the species, how water is held in and lost from the potting medium, and how it is used in plants.

Pathways of Water Loss in Houseplants

The water needs for the majority of indoor plants are met by water absorbed from the potting medium surrounding the roots. Water held in the medium can be thought of as a reservoir of moisture for the roots that needs to be refilled periodically to assure that adequate water is always available. When the reservoir is full, water can be lost in one of two ways: through transpiration or evaporation (Figure 17-1).

Transpiration

After water enters the roots, it moves upward through the stems and out to the leaf blades to be used in photosynthesis and the manufacture of other chemical compounds (hormones, pigments, and the like) in the plant.

However, relatively little of the total absorbed water is used for these synthesis operations; the rest is lost as vapor through the leaves in a process called *transpiration*.

Transpiration occurs almost continuously in houseplants, but its rate is greatly affected by environmental conditions. First, transpiration speeds up as the temperature rises. This is true of almost all processes involved in plant growth. Second, it increases in rate as humidity decreases. A plant growing in a dry room would lose water faster than one growing in a humid greenhouse, for example.

Transpiration rate is also affected by the plant species and form. Large-leaved or densely foliated tropicals transpire considerable water because of the large area of leaf surface exposed to the air. Cacti are more compact and have a limited amount of exposed surface. Consequently, they transpire less water.

Evaporation

Like transpiration, evaporation occurs continually. Water leaves the area where it is concentrated (the wet potting medium) and moves as vapor into the drier adjacent air. Humidity and temperature cause changes in the evaporation rate of water as they do with transpiration, speeding it up as the temperature increases or the humidity drops.

With some types of pots, evaporation of water can occur through the sides of the container. Pots made of materials that are porous to water (unglazed clay or pressed wood fiber, for example) absorb water from the medium and release it into the air.

Water loss through transpiration

Evaporation from the surface

Evaporation through the porous pot

Figure 17-1 The pathways of water loss in a plant potted in a clay pot.

Bottom Drainage

Drainage of excess water out the bottom of a pot is the third pathway of water loss. The amount of water retained by a potting medium after water is applied will depend on the ingredients of the medium (including their respective water-holding capacities, particle sizes, and subsequently created pore spaces), the depth of the medium in the pot, and the presence or absence of an *interface* in the pot. In Chapter 15 the roles of moisture-retaining organic ingredients and non-moisture-retaining coarse aggregates were explained. Thus, it is clear that the former would promote poor drainage (excess water retention) and the latter, "good" drainage.

Depth of the medium and presence of an *interface* also play roles. Basically, the shallower the pot, the greater the volume of water retained after irrigation. A simple demonstration, which can be done in class, will illustrate the superior drainage of a taller pot (Figure 17-2). A rectangular sponge (representing a pot of medium) should be saturated with water and held horizontally (to represent a shallow pot) until the water stops draining from it. A layer at the bottom will be completely saturated. One would naturally assume that the sponge would hold the same amount of water no matter which way it were

turned. But such is not the case. When the sponge is turned vertically, water will pour out the bottom.

The forces at work in this situation are *cohesion* of water molecules to themselves and *adhesion* of the water molecules to a solid (the sponge) against *gravity* pulling out the water. (See Chapter 5 for a fuller description of adhesion and cohesion.) More water pours out when the sponge is turned vertically because the weight of the water increases as it is distributed over a smaller area, the narrow end of the sponge. Therefore, a shallow pot is not the ideal choice for plants sensitive to wet media, as a greater proportion of the total medium remains saturated after watering than in a deeper pot.

 Teaching horticulture using demonstrations such as this one can be a rewarding career. Many high schools have vocational agriculture teachers who teach horticulture, soil science, entomology, and related subjects. A college degree and teaching credentials will usually be required for this profession. Teaching at a junior college usually requires a master's degree, and teaching at university level, a Ph.D.

Figure 17-2 The saturated sponge demonstration of water-holding capacity of potting medium.

From this demonstration one might assume that the old tradition of putting a layer of gravel in the bottom of a pot even though it has a drainage hole is a wise idea, and that the gravel would become the saturated layer and keep the potting medium unsaturated, but again this is not true. By putting a layer of gravel in the bottom, one is creating an *interface,* a change in pore size because of a change in particle size. The forces that hold the water in the potting medium will not be affected by the gravel layer; the gravel layer cannot influence the medium above it. Its only effect is essentially to shorten the depth of the pot, thus worsening the drainage of the medium.

Watering Frequency

The ways in which temperature and humidity affect the rate of water loss demonstrate one reason that an exact watering schedule cannot be prescribed for a particular plant: the environmental conditions under which a plant is grown can vary radically. Water loss by a plant growing in a cool, humid bathroom will be much slower than that from an identical plant in a warm, dry living room, and the frequency at which each requires water will differ accordingly. Added to that is the change in rate of water loss depending on whether the plants are in porous or nonporous pots.

The environment and type of pot partially determine how frequently a plant must be watered, but there are other factors on which the watering frequency should be based.

Pot Size and Material. Almost without exception, the smaller the pot in which a plant is growing, the more frequently it will have to be watered. This is because a small amount of potting medium holds less moisture than does a large volume. Even at the same evaporation

and transpiration rates, the water will be depleted sooner from the small pot.

Also, as discussed earlier, a porous pot will lose water through the sides and deplete the water reservoir in the potting medium sooner than a metal, plastic, or glazed clay pot would.

If the plant is in a pot that is both small *and* porous, the water will be depleted at a magnified rate because the surface area of the pot through which water is evaporating will be great in comparison to the volume of potting medium. The depletion rate will be much greater than in the case of a large porous pot.

Potting Medium. As discussed in Chapter 15, potting media differ in the amount of water they can hold because of differing ingredients. A potting medium largely composed of peat moss will be able to hold several times more water than one heavily fortified by perlite. Such a medium will take longer to dry out completely, and watering will not be required as frequently.

Growth Rate. Some plants naturally grow faster than others, and they require more water to photosynthesize and provide energy for that growth. These vigorous growers absorb water from the potting medium very rapidly, necessitating that it be replenished more frequently.

Even plants that are not vigorous growers may pass through changes in growth rate. A rapid spring and summer growth rate is common in houseplants. The watering frequency may have to be increased to keep up with the water demand brought on by the plant's size increase.

Size of the Plant in Relation to the Pot. The larger a plant is in relation to its pot, the more frequently it will require watering. As the top of the plant increases in size, more leaves require water for photosynthesis and transpiration. The root mass will also have increased, giving a much larger absorption area. However, the plant is still restricted to the same volume of medium from which to absorb water. It will deplete the available water quickly, and the water will need to be replenished sooner. Therefore, when a plant is grown in the same pot for a long time and continues increasing in size, its watering frequency will also need to be increased.

Species. The many plant species grown for indoor decoration differ widely in their needs for water. Cacti and succulents grow best when the potting medium is kept fairly dry through most of the year, whereas bog plants like umbrella sedge (*Cyperus alternifolius*) grow with their roots completely submerged in water. Most other plants tend to be somewhere in between. They grow best when there is a slight to moderate amount of water in the potting medium instead of when it is either fully dry or completely saturated.

With the exception of ferns and a few other species, most plants can survive up to a week with dry potting medium and not sustain permanent damage. Many experienced indoor gardeners grow plants very successfully on a "wet-dry" cycle, watering only when the medium is dry on top but before the plant wilts. However, a wet-dry cycle may minimize new growth because of water stress. Watering more frequently is advisable when rapid growth is desired.

Soil Moisture Testing

Because differing environmental conditions rule out a prescribed watering schedule, the question of timing the watering frequency remains. The easiest answer is to evaluate the moisture content of the potting medium by touch.

Touch Method

Feeling the potting medium (Figure 17-3) has proved to be a reliable method of testing the moisture and determining the watering frequency. The fingertips touch the surface of the potting medium or are pushed about ¼ inch (6 millimeters) deep to feel for moisture. A cool, damp feeling indicates that moisture is still present near the surface and the medium surrounding the roots still contains adequate water. A dry feeling indicates that moisture surrounding the roots needs to be replenished and the plant should be watered.

Color Method

The color method is less reliable than the touch method, but it is used successfully by some gardeners. The method works on the principle that wet potting medium

Figure 17-3 Testing moisture in the potting medium by feel. *(Photo by Kirk Zirion)*

is darker in color than dry, and the moisture content is estimated visually. A drawback of this method is that the surface of the medium may appear dry several days after watering, but the medium underneath may still be damp.

Weight Method

The weight method is reasonably reliable but requires more skill than the touch method. Immediately after watering, a potted plant will be heaviest because of the weight of water contained in the potting medium. As the water is transpired or evaporates, the medium loses water weight, and the potted plant will weigh considerably less. With practice, a gardener can estimate water need by picking up the pot and comparing its present weight to the weight immediately after watering.

Commercial Moisture Sensors

Several types of chemical and electrical moisture sensors are available. The first type has a piece of blotter paper impregnated with a chemical that changes color in the presence of moisture. The paper is enclosed in plastic and mounted on a probe that is pushed into the

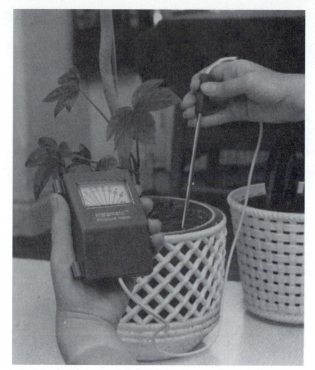

Figure 17-4 A moisture meter. *(Photo courtesy of Luster Leaf Products, Crystal Lake, IL)*

medium. The changing color of the paper shows whether moisture is present in the root zone.

The electrical water sensor (Figure 17-4) is an adaptation of a laboratory instrument called a *solubridge*, which is used to measure soil nutrient content. It measures the ease of conducting a minute electrical current through the medium between two electrodes (conductors of electricity). The electrodes are on different sections of a long probe, which is inserted into the medium, and are separated on the probe by a section of insulator to force the electrical current to pass through the medium and take the measurement of the ions (dissolved salts) present in the soil water. The more water present, the more ions will be dissolved, and the greater the ease of conductivity.

This type of water sensor is accurate at a normal nutrient range. However, it will read falsely (giving a wet reading in a dry medium) when the nutrient level is extremely high. Conversely, it will read "dry" when placed in distilled water because of the complete absence of ions in the water.

Accuracy is also affected by the depth to which the probe is inserted. If the majority of the second electrode (actually a long metal sleeve) is left exposed to obtain a shallow moisture measurement, the readings will vary radically between media with the same moisture contents. To solve this problem, the upper electrode can be wrapped with Mylar-type Scotch tape, leaving exposed only an area equal to the size of the first electrode.

Commercial experience with these probes has shown them to be most accurate when the reading is taken about 30 seconds after insertion of the probe (see Snyder reference).

Applying the Correct Volume of Water

The goal when watering most houseplants is to wet *all* the potting medium thoroughly at each watering, allow it to dry to a point of slight dampness, and then rewet the medium again.

Pots with Drainage Holes

Assuming that the plant is watered from the top, the volume of water applied to a plant in a pot with a drainage hole is not critical. The makeup of the medium will allow it to hold only a limited amount of water, and the potting medium should be formulated so this capacity is not too much for healthy root growth. Excess water will then drain out the bottom and should be discarded.

With many commercial media it is practically impossible to overwater because the water-holding capacity of the media is so low. Damage from overwatering could result, however, if the pot remained standing in drainage water for many days. With other media (particularly those containing soil) overwatering is possible. These media have a great water-holding capacity, and watering too frequently could cause rotting of the plant roots.

Pots without Drainage Holes

The volume of water applied to plants in pots without drainage holes is critical. Because there is no way for excess water to leave the pot, any amount greater than that required to wet the potting medium will collect at the bottom. For this reason, it is advisable to provide a reservoir of pebbles or charcoal in pots without holes to serve as a collection area for excess water. Standing water in the pebbles will normally be below the root growth level. The water there will not damage the plant and will slowly be absorbed back into the potting medium.

To decide how much water should be applied, the volume of the potting medium should be estimated. A volume of water one-fourth as great should then be applied. After about five minutes, the water will have been absorbed, and the pot can be gently tipped sideways to allow the excess to run out. If more than a couple of teaspoons (10 milliliters) runs out, the water volume should be reduced next time. If none runs out, enough water may not have been applied to wet all the medium. More should be added and the pot tested again in a few

Figure 17-5 A dipstick system for checking the water level in the reservoir of a pot.

soil —

wooden dowel —

plastic straw or pipe —

water reservoir of gravel —

remains above the surface. By inserting a wooden skewer or dowel into the straw, the presence of water in the reservoir can be determined. This technique is valuable for large specimens that are too heavy or cumbersome to be checked by tipping.

Watering Techniques

Top Watering

Most people water from the top. It is easy and completely satisfactory for almost all indoor plants. Top watering is so common that it requires little explanation, but two general rules should be followed. First, enough water should be applied to wet all the potting medium. For pots with holes this means applying enough so that some runs out the bottom. Second, a few minutes should be allowed for water to drain out, and then the drainage water should be discarded. Many plants develop root rot and die if they are left in drainage water for more than a few days.

minutes. Ideally, just a few drops should flow out to indicate that the entire medium is wet.

Another way of determining the volume of water to apply is by using a dipstick. When the plant is potted, a length of plastic straw or pipe is buried with one end in the gravel reservoir (Figure 17-5). The other end

Bottom Watering

Bottom watering (Figure 17-6) can be used only on plants in pots with drainage holes, because the water is taken up through the bottom of the pot. It is very use-

Figure 17-6 Bottom watering of a dumbcane. *(Photo by Rick Smith)*

ful for plants whose leaves are susceptible to spotting from water and for flat rosette plants whose leaves cover the soil surface, making it difficult to pour in water. Bottom watering is also good for watering large cacti whose bodies nearly fill the top of the pot. Water applied from the top frequently does not reach the roots of such cacti because it can be poured in only around the rim of the pot. The water flows down the outside of the root ball and out, never adequately wetting the roots. Bottom watering is also useful for rewetting a medium that has dried excessively and retracted from the pot.

For bottom watering, the pot must sit in a plant saucer or bowl. The saucer should be filled with water, and in five to ten minutes the medium will have absorbed the water by capillary action. The saucer should be filled again and, if necessary, a third time until the medium stops absorbing. The excess should then be discarded.

One precaution must be taken with bottom watering. Because water does not flow through the medium and out the bottom, the fertilizer salts contained in the medium are never flushed out. There is a possibility that they will accumulate to toxic levels. As a precaution, a plant that is constantly bottom watered should be submersion-watered or leached every 6 to 12 months to rinse through excess fertilizer.

Submersion Watering

Submersion watering (Figure 17-7) is also used only for pots with drainage holes and for the same purposes as bottom watering. It can also be useful for large, unwieldy hanging baskets.

Submersion watering is similar to bottom watering, in the sense that water enters through the drainage holes, but it is quicker. In submersion watering, the plant is set in water deep enough to come up to the rim of the pot. In 5 to 15 minutes the water will have penetrated to the surface of the medium. The pot is then taken out of the water, allowed to drain, and replaced in its growing area.

In submersion watering there is no likelihood of excess fertilizer buildup. During the wetting, the medium becomes saturated, and the fertilizer dissolves in the water. It is carried out with the excess water during the draining.

Wick Watering

Wick watering (Figure 17-8) is not the ideal solution for watering plants because it keeps the medium too moist for most species. However, it is successful with plants that must have constant moisture (African violets and

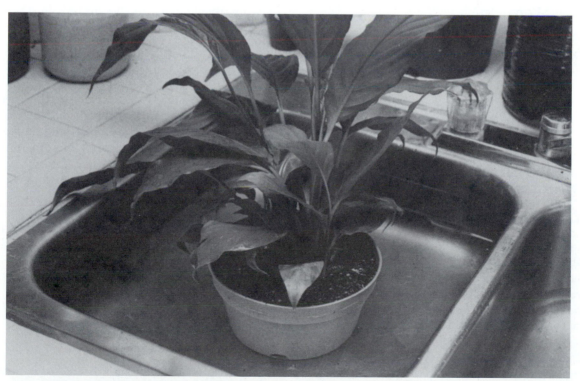

Figure 17-7 Submersion watering of a *Spathiphyllum* in the kitchen sink. *(Photo by Rick Smith)*

Figure 17-8 A wick watering system. Note that the pot is positioned above water level.

most ferns, for example) and for those in which complete drying could cause death.

Wick watering follows the same principle of capillary action that pulls water up through the medium in bottom watering. In this case capillary action occurs in the wick. Water is absorbed into the wick from the water reservoir, an area of water concentration. The water is pulled through the wick by capillary action to an area of lesser concentration (the potting medium). The wick action continues until the potting soil is saturated with water.

Shower Watering

Showering, as discussed in Chapter 14, is principally a method of leaf cleansing that waters the plant at the same time. In some regions the mineral content of the water is excessive, and watering by showering will leave unsightly mineral deposits. In these cases avoid showering or gently towel dry the leaves to prevent the deposits.

Special Watering Techniques

Bulb- and Tuber-Bearing Houseplants

During their growth and flowering periods, bulb- and tuber-bearing houseplants such as amaryllis and gloxinia require no special watering techniques. They can be treated like most other houseplants. But as their blooming period ends, these plants may enter a yearly dormancy phase. The top growth dies, putting the bulb in a rest period for several months before growth begins again. The dormant period for bulb- and tuber-bearing plants coincides in nature with climate changes such as decreased rainfall. This change should be simulated to meet the plant's growth requirements.

After the foliage begins yellowing, watering frequency and volume should be slowly decreased over about two months. During this period, the foliage will slowly die back to the bulb. When all the foliage is dead, watering should completely stop for several months while the plant rests. It can then be divided or repotted while dormant and regular watering resumed to encourage fresh growth.

The key to correctly watering bulb- and tuber-bearing houseplants lies in careful observation of the plant during all its growth phases. Some bulbous plants never enter dormancy indoors and may be severely injured by underwatering. Others may be unintentionally overwatered during dormancy and rot. Bulbous plants show signs of approaching dormancy by yellowing and death of older leaves. These are signals to reduce watering. Plants emerging from dormancy often begin new growth before water is applied. As soon as this new growth begins, watering should be started. If watering is delayed, the bulb will respire excessive amounts of carbohydrate from the storage organ and may be stunted or die.

Orchids and Bromeliads

Care should be taken in watering orchids or bromeliads planted in coarse bark. The bark is extremely fast draining and has no micropore spaces for capillary action to move the water sideways. Water applied at one point will run straight down and out the bottom. Therefore it is important to water over the whole surface of the pot to wet all the bark.

Leaching

Leaching is a watering technique used to remedy the problem of excess fertilizer salts in potting media. It works on the principle that the salts are soluble in water and consists of watering a plant four or five times in succession at 5- or 10-minute intervals. Each watering dissolves some of the salts, and the following one flushes them out in the drainage water.

A succession of waterings is not the only form of leaching, although it is probably the most common. Plants can also be submerged and drained several times or left under a slowly running tap for 10 minutes for the same result.

Water Quality and Indoor Plant Growth

For most plants, the amount of water they receive is more important than the quality of that water. In most parts of the country, plants grow well given normal tap

Figure 17-9 Spotting of African violet leaves caused by cold water. *(Photo by Rick Smith)*

water, with only a few being adversely affected by its temperature, pH, or chemical content. One exception to this generalization can be found in the southwestern states, where high-pH and high-salt-content water can cause plant health problems.

Temperature

Correct water temperature is not critical for most houseplants provided it is within a 60 to 90°F (15–32°C) range. Letting water become room temperature is probably ideal for houseplants, but it is not practical if there are many houseplants to be watered.

African violets, though, have been proved sensitive to water temperature. Water more than 10°F (5°C) colder than air temperature will cause permanent yellow spotting on the leaves because of chlorophyll destruction (Figure 17-9). Bottom watering or warming the water to room temperature will avoid the problem.

pH

Acidity or alkalinity (pH) is a characteristic of water as well as potting media. It varies depending on the water source. Continual watering with very acidic or alkaline water will eventually alter the original pH of the medium to an undesirable range.

Although not highly accurate, an inexpensive pH test kit can determine if water is exceptionally acidic or basic. The gardener will then be aware of media pH problems that could occur because of its use. (See Chapter 15 for information on modifying the pH of water for houseplants.)

Chemical Content

Tap water, depending on its source, can contain any number of "impurities" including (but not limited to) sodium, fluoride, chlorine, calcium, ammonia, and iron.

Sodium. Sodium often enters water through the softening process. It is used to replace the calcium in calcium carbonate, the mineral that often causes "hard" water. In areas of the southwest it may be naturally present in the groundwater.

The sodium found in tap water can occasionally build up to concentrations harmful to houseplants with normal watering. To avoid this problem, do not use water from the hot-water tap. In home water softeners, hot water alone is treated with sodium, and cold water remains sodium-free. If the sodium is present in the natural groundwater, or if all household water is softened, rainwater or distilled water should be used. Leaching every two to three months will also help flush the sodium from the potting medium.

Fluoride. Fluoride is put into most city water supplies as a preventative against tooth decay. The concentration used is very small—a few parts per million. However, some untreated water supplies contain more than that concentration naturally.

Relatively small amounts of fluoride are capable of causing damage to plants, but only members of the lily family are known to be particularly sensitive. Common houseplants in this family include dracaenas, spider plants, and asparagus ferns. Excess fluoride can also be found as a result of amendments like perlite. The injury appears as browning of the leaf tips.

Chlorine. Chlorine is added to city water for sanitation purposes; it kills bacteria, fungi, algae, and other potentially harmful organisms. It is generally added as a salt that dissolves in the water or is occasionally bubbled in as a gas under pressure.

Chlorine is essential to plant growth in small quantities and is normally supplied to plants outdoors in the soil and sometimes by rainfall. Tap water does not contain sufficient chlorine to damage houseplants.

Contrary to popular belief, allowing tap water to stand overnight will not eliminate all the chlorine. Instead, it escapes extremely slowly as a gas, although the process can be speeded up by exposing the water to ultraviolet light.

Alternatives to Tap Water

Rainwater

The belief that rainwater is pure is not true. It contains a surprising amount of chlorine and some nitrogen, sulfur, and other substances as well. It can also contain chemicals from air pollution and be affected by whatever roof surface it runs off before it is collected. Although rainwater is certainly no worse than tap water for plants, it has no special plant-growing properties because it came from the atmosphere instead of a well or reservoir. However, it is a valuable alternative in areas with very salt-laden water.

Distilled Water

Distilled water has been treated to remove all substances but the hydrogen and oxygen that form water. It is truly "pure" water that does not contain minerals or other foreign chemicals.

If the natural water supply has plentiful sodium, fluoride, or another chemical injurious to houseplants, distilled water is a good alternative water source. It is also useful for leaching plants because it can carry away more salts than the same volume of tap water.

Symptoms of Improper Watering

Underwatering. Plants that are receiving too little water or are not being watered frequently enough may incur root injury. They may show different symptoms, depending on the plant species. Usually they wilt but can be revived provided the leaves have not lost moisture to the point of becoming leathery or crisp, a stage called *permanent wilting point.* Even if the top portions die completely, many plants will regenerate from the roots if the medium is kept moist.

Other symptoms of underwatering are not as easy to diagnose. Slow growth, browned leaf tips and margins, and bud and flower drop can be due to insufficient water as well as several other causes. The symptoms of under- and overwatering can even be the same, because both result in root injury.

Overwatering. Excess water excludes oxygen from the potting medium. Eventually root death will occur because the roots are unable to respire.

Overwatering can be caused by watering a poorly draining medium too frequently, by applying too much water to a pot without a drainage hole, or by a medium that holds too much moisture, often because of pack-

ing down and loss of pore spaces with age. Symptoms of overwatering include weak and slow growth, lower leaf drop, and discoloring of the roots. In the advanced stages plants wilt because the dead roots can no longer absorb water.

Vacation Watering of Indoor Plants

Keeping houseplants watered during vacation can be inconvenient. However, there are ways to keep them sufficiently moist without interim watering.

One approach is to slow down plant metabolism. This will decrease the water requirement and make the water in the medium last the longest possible time. Photosynthesis, transpiration, and respiration must all be slowed, and the following suggestions can be used separately or in combination to accomplish this.

Lowering the Temperature. As temperature decreases, biological processes slow down. A drop from 75°F (24°C) to 55°F (13°C) will decrease water use appreciably.

Moving Plants out of Direct Sunlight. Direct sunlight warms the leaves, increasing transpiration. It also increases the photosynthesis rate and the water needed for that process. Most bright-light plants can be maintained without full sunlight for up to two weeks without deteriorating, but they should be given bright diffused light to prevent leaf drop. A sheer curtain will provide this type of light.

Grouping Plants Together in the Bathroom. A small bathroom can be turned into a holding room for plants fairly easily. The windows should be closed and the tub and sink filled with water to maximize the humidity. All the plants should then be watered and set in the bathroom with the lights on (not sun or heat lamps) and the door closed. The high humidity will slow transpiration, and the lights will permit limited photosynthesis. Most plants can be left up to two weeks under these conditions. The main drawback is the production of weak growth on vigorous plants because of insufficient light. This undesirable growth should be pruned off when the plants are returned to their normal growing locations.

Enclosing Plants in Plastic Bags. Dry-cleaning bags work well for this purpose, creating a closed system from which no water is lost. Plastic drop cloths can be used for large numbers of plants. One is spread on the floor and the recently watered plants are grouped together on top of it. A second cloth is used to cover them, and the edges are sealed with tape or rolled and

Figure 17-10 Enclosing a plant in a plastic bag will enable it to survive for a long period without watering. *(Photo by Rick Smith)*

stapled to make an airtight envelope. One precaution must be followed: bagged plants should not be placed in strong sunlight because the sun can heat the air inside the plastic and cause heat damage.

For specimens too large for bagging, cover just the pot, sealing around the crown as snugly as possible with tape or string. The plastic will eliminate water evaporation through the medium and thus conserve part of the moisture.

Overall, bagging is the most reliable of all vacation watering ideas. Plants completely sealed in plastic can be left for a month or more without care (Figure 17-10).

Mulching. If bagging is not practical, houseplants can be mulched an inch or two deep (3–5 centimeters). A bag of shredded bark mulch would not introduce pests and would be sufficient to mulch all the houseplants in an average home. It has a pleasant smell as well.

HUMIDITY AND INDOOR PLANT GROWTH

Humidity is the amount of water vapor in the air, and sufficient humidity is essential for healthy plant growth. Indoor plant species vary in their humidity needs. Cacti and succulents grow satisfactorily in relatively low humidities, but leafy tropicals grow best at higher humidities.

The effect of humidity on plant growth is due mainly to its influence on transpiration rate. Generally, the higher the humidity, the slower the loss of water through transpiration. Without constant and rapid water loss, plant roots do not expend as much energy absorbing water, and that energy can be used for growth and development.

Relative Humidity

The humidity content of the air is generally recorded in a percentage figure called the *relative humidity*. This figure takes into account the fact that the ability of air to hold water vapor varies with its temperature. Air can hold increasingly greater amounts of water as its temperature rises. The relative humidity is the percentage of water vapor the air is presently holding relative to the total water it is capable of holding at that temperature. For example, 50°F (10°C) air at 80 percent relative humidity means the air contains 80 percent of the water vapor it is able to hold at that temperature. If the air were at 60° and contained the same amount of moisture, its relative humidity would be lower—say, 70 or 75 percent.

The outdoor humidity of an area is influenced by the amount of precipitation (generally rain) the area receives and by its proximity to large bodies of water. Accordingly, the East, Northeast, and Northwest have high natural humidities, and the dry inland portions of the West have less. But outdoor humidity is not necessarily equal to indoor humidity. It will be equal only if there is free air circulation between the two. Once windows are closed and heating or air conditioning is used, the humidity inside will drop.

Low indoor humidity in winter is blamed for many houseplant problems. Actually, indoor air contains the same amount of water as outside air, and the relative humidity of the outside air may be 60 to 70 percent. The temperature difference creates the problem. Cold outdoor air at a moderate humidity is brought indoors, warmed, and circulated to freshen the indoor air. But the water it contained at 20°F (–6°C) outside is much less than it is capable of holding at 65°F (18.3°C) indoors, and its relative humidity drops from a typical 60 percent down to 20 percent.

Measuring Humidity

Several instruments are available for measuring relative humidity in the indoor environment. A human hair hygrometer is one. It shows the effect of rising or falling humidity on the length of a single human hair and indicates the humidity on a simple dial. It is a relatively expensive instrument, but it is the only dial-type device that is reasonably accurate.[1]

A sling psychrometer is the second instrument for measuring humidity. It consists of two identical thermometers mounted side by side on a small plaque. The mercury bulb of one measures ordinary temperature.

[1] It should be noted that a hygrometer requires frequent recalibration to maintain accuracy.

Figure 17-11 A sling psychrometer for measuring humidity. *(Photo courtesy of Nasco International, Ft. Atkinson, WI)*

The bulb of the second is covered with a small piece of wick (Figure 17-11). When measuring the relative humidity, the wick covering the one thermometer bulb is first wet. The two thermometers are then whirled in the air. When both thermometers show constant readings, the difference between the two temperatures is noted. The "wet-bulb thermometer" with the wick will always show a lower temperature because of the evaporation of the water around the bulb. Using the temperature difference and special tables, the relative humidity can be quickly calculated.

In general, a reading of 35 percent or less can be considered low humidity, 35 to 70 percent moderate, and 70 percent or greater high humidity. The majority of tropical plants grows best at humidities of at least 50 percent, although most will grow with less.

Raising the Relative Humidity Indoors

A number of techniques are used to raise humidity indoors, with differing degrees of success.

Misting

Although misting is widely advocated as a means of raising the relative humidity around plants, it actually does not do much good. In dry air, water will evaporate very rapidly. A thin film of mist applied to leaves can dissipate into the air in less than half an hour. It is unlikely that such a short period of improved humidity would have much effect.

Misting is, however, beneficial to bromeliads and orchids, which absorb water directly through cells on their leaves.

Pebble- and Water-Filled Trays

This method works because of the constant evaporation of water into dry air. Trays or saucers are filled with a layer of pebbles or marble chips, which are partially covered with water (Figure 17-12). Plants are then grouped on the tray to benefit from the vapor rising up from the water. The water reservoir is refilled as needed, but never to a depth where the pots would be standing in water.

Humidifiers

The most reliable way to increase room humidity is by means of a room humidifier. Different models range upward in price from $10.00. Among the most economical are cool vapor humidifiers. In addition to being inexpensive, they use very little electricity.

Symptoms of Improper Humidity

Insufficient Humidity

Browning of the tips and margins of leaves on tropical plants is the most common low-humidity symptom. Rapid transpiration causes water to be lost from the leaves before it is able to reach the furthest portions,

Figure 17-12 Keeping houseplants on a layer of wet pebbles will help increase the humidity. *(Photo by Rick Smith)*

Figure 17-13 Tip burn on a spider plant. *(Photo by Rick Smith)*

and browning results. Although unsightly, the browning does little harm to the plant.

Although *tip burn* (Figure 17-13) is the most easily diagnosed symptom of insufficient humidity, others may occasionally appear. Dropping of both new and old leaves may be a symptom, although it is usually associated with other growing problems. Undersized new leaves, a general lack of vigor, and bud drop are others.

Excess Humidity

Most indoor plants grow well at humidities of up to 90 percent. However, the disease powdery mildew can develop under humid conditions. It affects roses and begonias, among other plants. Another fungus disease, botrytis, is also common at high humidities.

SELECTED REFERENCES FOR ADDITIONAL INFORMATION

American Water Works Association. *Water Quality and Treatment: A Handbook of Public Water Supplies.* 4th ed. New York: Osborne-McGraw-Hill, 1990.

Conover, C.A. and R.T. Poole. 1992. "Water utilization of six foliage plants under two interior light intensities." *J. Environ. Hort.* 10(2):11–113.

Snyder, Stuart D. *Building Interiors, Plants and Automation.* Englewood Cliffs, NJ: Prentice-Hall, 1990.

World Wide Web Addresses

http://www.colostate.edu/Depts/CoopExt/4DMG/Plants/ guidline.htm   Hunter, Sheri. "Guidelines for Watering Indoor Plants." Colorado State University/ Denver County Cooperative Extension Master Gardener website. August 7, 2001.

http://www.ifas.ufl.edu/~apkweb/cfrec/rh_93_15.htm Conover, C. A., Russell Caldwell and K. Steinkamp. "Water Requirements of Three Foliage Plants in Self-Watering Containers Affected by Indoor Light Intensity and Plant Type." 1993.

http://www.msue.msu.edu/msue/imp/mod03/03900091.h tml   This is an extension bulletin entitled *Watering: The Key to Houseplant Success.*

(Photo by author)

CHAPTER 18

CONTROLLING INDOOR PLANT PESTS AND DISEASES

Indoor plant problems are almost always attributable to one of two sources: the environment or organisms contained in it. Those caused by improper environmental conditions (light, water, temperature, and the like) are called *cultural problems* or *physiological diseases*. These problems are not discussed in this chapter, because the effects of environmental factors were dealt with in the chapters on light, watering and humidity, and potting media. The organisms harmful to plants can be anything from microscopic viruses to snails. The only characteristic they share is that all are detrimental to plant health, whether by eating foliage, sucking sap, or causing spots on leaves.

INDOOR PLANT INSECTS AND RELATED PESTS

There are insects and also insect-like pests that feed on houseplants. Although it might appear overly technical to separate the two, the distinction can be important if chemicals are used for their control.

Insects, which are six-legged, can be controlled by *insecticides*. Pests like mites, nematodes, and sowbugs belong to other classifications and may require different chemicals for control. Mites, for example, are susceptible to miticides, and nematodes to nematicides. Other than for control purposes these distinctions are unnecessary; all will be discussed under the broad heading of pests (Table 18-1).

Foliage Pests

The easiest pests to detect are often those that damage the aboveground portions of plants—the leaves, stems, buds, and so on. They can be collectively called *foliage pests*. Some foliage pests are more frequently encountered than others; their listing here is ranked from most to least common.

Mites (spider mites, red spider mites, two-spotted spider mites, cyclamen mites)

As mentioned, mites (Figure 18-1) are not insects. They have eight rather than six legs and are related to spiders; hence the name "spider mites." At first encounter, a mite problem can be among the hardest to diagnose. Mites are so small—about the size of dust specks—they are almost invisible to the naked eye.

Several kinds of mites attack indoor plants. Red or two-spotted spider mites can be red, brown, or creamy white. Their damage is done when they insert their mouthparts into a leaf repeatedly to feed. Each feeding site shows as a pinpoint-sized white dot, and eventually leaves become highly speckled with feeding marks, shrivel, and die. Spider mites attack almost any houseplant having thin leaves, as well as cacti and succulents. They are by far the most common type of mite and thrive in low humidity and high temperature.

Table 18-1 Insects and Related Pests of Houseplants and Their Common Host Plants

Pest	Remarks	Common plant hosts
Spider mite	Common on almost all houseplants	African violet (*Saintpaulia ionantha*) Asparagus fern (*Asparagus densiflorus 'Sprengeri'*) Azalea (*Rhododendron hybrid*) Cacti Citrus (*Citrus* spp.) Croton (*Codiaeum variegatum*) Dumbcane (*Dieffenbachia* spp.) English ivy (*Hedera helix* cultivars) Prayer plant (*Calathea, Maranta*) Umbrella plant (*Brassaia, Tupidanthus*)
Cyclamen mite	Less common than spider mite and attacks fewer species	Begonia (*Begonia* spp.) *Crassula* spp. Cyclamen (*Cyclamen persicum giganteum*) English ivy (*Hedera helix* cultivars) Gesneriads (*Saintpaulia, Gloxinia, Episcia,* and other genera) Peperomia (*Peperomia* spp.) Purple velvet plant (*Gynura* spp.) Zebra plant (*Aphelandra squarrosa*)
Scale	Attacks primarily woody plants with some exceptions	Cacti Cast iron plant (*Aspidistra elatior*) Citrus (*Citrus* spp.) Coffee (*Coffea arabica*) Croton (*Codiaeum variegatum pictum* varieties) Cycad (*Cycas* spp.) Dracaena (*Dracaena* spp.) English ivy (*Hedera helix* cultivars) Fatsia (*Fatsia japonica*) Ferns Figs (*Ficus* spp.) Gardenia (*Gardenia jasminoides*) Kalanchoe (*Kalanchoe* spp.) Palms Orchids Poinsettia (*Euphorbia pulcherrima*) Zebra plant (*Aphelandra squarrosa*)
Whitefly		Begonia (*Begonia* spp.) Citrus (*Citrus* spp.) Coleus (*Coleus blumei* cultivars) Ferns Fuchsia (*Fuchsia* × *hybrida*) Gardenia (*Gardenia jasminoides*) Pepper (*Capsicum annuum*) Poinsettia (*Euphorbia pulcherrima*)
Mealybug	Common and relatively nonselective	Cacti Coleus (*Coleus blumei* varieties) Croton (*Codiaeum variegatum pictum* cultivars) Cycad (*Cycas* spp.) Dracaena (*Dracaena* spp.) Ferns Fuchsia (*Fuchsia* × *hybrida*) Gardenia (*Gardenia jasminoides*) Jade plant (*Crassula argentea*) Palms Schefflera (*Brassaia* spp.)

(continued)

Table 18-1 Insects and Related Pests of Houseplants and Their Common Host Plants (*continued*)

Pest	Remarks	Common plant hosts
Aphid	Occasionally found on the new growth of these and other plants	Poinsettia (*Euphorbia pulcherrima*) Rubber plant (*Ficus elastica*) Cineraria (*Senecio cruentus*) Chrysanthemum (*Chrysanthemum × morifolium*) English ivy (*Hedera helix*) Ferns Fuchsia (*Fuchsia × hybrida*) Purple velvet plant (*Gynura* spp.)
Thrip	Relatively uncommon	Arrowhead (*Syngonium podophyllum* cultivars) Azalea (*Rhododendron hybrid*) Citrus (*Citrus* spp.) Dracaena (*Dracaena* spp.) Fuchsia (*Fuchsia × hybrida*) Gardenia (*Gardenia jasminoides*) Gloxinia (*Sinningia speciosa*) Palms *Philodendron* spp. Rubber plant (*Ficus elastica* cultivars) Umbrella plant (*Brassaia actinophylla*)
Nematode	Relatively uncommon	African violet (*Saintpaulia ionantha* varieties) Begonia (*Begonia* spp.) Bloodleaf (*Iresine herbstii* varieties, *Calathea* spp.) Cyclamen (*Cyclamen persicum giganteum*) Ferns Fuchsia (*Fuchsia × hybrida*) Gardenia (*Gardenia jasminoides*) Rubber plant (*Ficus elastica*)
Root mealybug	Occasionally found in unsterilized media	Arrowhead (*Syngonium podophyllum* cultivars) Boston fern (*Nephrolepis exaltata*) Bromeliads Cacti Dumbcane (*Dieffenbachia* spp.) Palms

Note: Plants listed frequently in this table are not necessarily prone to disease and insect problems but are simply the more thoroughly researched indoor plants.

The surest diagnosis for spider mites is by examination of the undersides of the damaged leaves in strong light. The mites will be barely visible, but they can be seen clearly with a magnifying glass. In advanced cases of spider mite infestation, webbing will cover leaves. It is to this webbing that the female attaches her eggs.

Cyclamen mites are the second main type of mite affecting indoor plants. It is from a main host plant, the florist cyclamen, that their name is derived. Unlike spider mites, which attack old and young leaves indiscriminately, cyclamen mites cluster on very young leaves and buds for feeding.

Cyclamen mites are nearly transparent and feed by puncturing leaves, but their feeding damage does not show speckling or webbing. Instead it causes curling and cupping of the leaves, giving the young leaves a dwarfed appearance.

Cyclamen mites are not as universal as spider mites. They are an unusual problem affecting indoor plants, as contrasted to the persistent problem spider mites pose.

Mealybugs

Appearing like specks of cotton, the many types of mealybugs (citrus, long-tailed, and Mexican, for example) are destructive pests. They cause stunting, leaf drop, and eventually the death of a plant.

The adult insects are pink and shaped like sowbugs (Figure 18-2). They appear white because of the white

Figure 18-1 Spider mite damage on a palm leaf. *(Photo courtesy of Dr. A. W. Ali)*

waxy fibers they secrete and in which they lay their eggs. They are very slow moving and frequently found in the axils of leaves or on the undersides along the veins.

Adult mealybugs are often difficult to kill with sprays because of the protection afforded them by their waxy coats. However, newly emerged young lack this covering and are susceptible to most all-purpose insecticides. The addition of wetting agents or light oils to sprays greatly increases their control.

Scale Insects (hard scale, soft scale)

Scales (Figure 18-3) are relatively immobile and are frequently not recognized as being insects. Like mealybugs, scales secrete a protective covering. The type of covering determines whether they are called *armored* or *soft scales.*

Armored scales are most common on indoor plants. They are covered with hard, blisterlike coats and range up to ⅛ inch (3 millimeters) in diameter. Their color can be clear to white, yellow, or brown, depending on the species, age, and sex of the insects.

Soft scales are very much like mealybugs, secreting a soft waxy covering attached to their bodies. They are nearly round in shape and generally larger than armored scales. Unlike armored scales, soft scales also secrete sticky honeydew. The honeydew may attract ants and provides a substrate on which sooty mold grows.

Figure 18-2 Mealy bug, shown here with ants which can carry the pests among plants. *(Photo courtesy of Takumasa Kondo, Auburn University, Auburn, AL)*

Figure 18-3 Scale insects. *(Photo courtesy of Takumasa Kondo, Auburn University, Auburn, AL)*

Figure 18-4 Whiteflies (shown approximately 4 times normal size). *(USDA photo)*

Sooty mold is black, and although it does not directly harm plants, it is unsightly and decreases photosynthesis. Often, the presence of ants and sooty mold is the first clue that scales are present.

Scales of both types do their damage by puncturing plant tissues and extracting sap through their needlelike mouthparts. As a result of this feeding, plants become weakened, and parts or the whole plant will eventually die.

Scales are found on both sides of leaves, on twigs, and on branches, frequently concealing themselves in the leaf axils. When they attach to leaves, a yellow spot will sometimes be found on the reverse side.

In the life cycle of scales, the most visible stage is the adult, which is permanently anchored in one place and grows to a large size. In the immature crawler stage, scale insects are mobile, and it is at this period when they are most vulnerable to spray insecticides.

Whiteflies

Whiteflies (Figure 18-4) resemble tiny moths. They cluster on the undersides of leaves, flying up when disturbed and resettling almost immediately. Their damage is done by piercing the undersides of leaves to extract plant sap, which causes pale or spotted leaves and weakens the plant. Whiteflies also secrete honeydew and transmit viral diseases.

In a generalized life cycle of a whitefly, the eggs are laid under leaves and hatch in 4 to 12 days to a crawler stage. The *crawlers*, which appear as clear scales, move over the leaf for several hours, then puncture the leaf and feed in one spot. They shed their skin three times, then enter a resting stage and emerge as the familiar white adult. The process takes almost six weeks.

When whiteflies are detected, it is almost always in the adult stage. At this point they are bothersome and difficult to spray because of their flying. It is advisable at that point to begin spraying under the leaves to control them in the immature stages with the expectation that control will not be complete for several weeks. Systemic insecticides in particular often provide excellent control.

Aphids (plant lice or green flies)

Aphids are soft-bodied insects (Figure 18-5) that are easily visible and are about ½ inch (1 millimeter) across. They can be green, red, black, or any other number of colors and tend to cluster in large numbers on the youngest leaves and buds of plants. Aphids suck plant juices, causing deformed young leaves.

In addition, they secrete a syrupy substance called *honeydew*. The honeydew makes plant leaves sticky and promotes the growth of sooty black mold, a nonparasitic mold that is nonetheless unsightly. Aphids also transmit viral diseases.

Aphids are easily killed with soap and oil sprays during any stage in their life cycle. Unlike other insects, young aphids are born live instead of hatching from eggs. The adult female does not require mating, and she can give birth to up to 100 self-fertile females during her 20- to 30-day lifetime.

Figure 18-5 Aphids. *(USDA photo)*

Thrips

Thrips (Figure 18-6) are just slightly larger than spider mites. They are winged in the adult stage and wingless in earlier immature stages. Foliage damaged from their feeding is silvery-colored, later turning brown. The flowers they attack become streaked, and the buds fail to open.

It is very difficult to see thrips on plants, as they are extremely small and move fast. If thrips are suspected, the damaged plant part should be tapped over a white sheet of paper. This will dislodge the thrips, which can then be seen briefly before they crawl away.

Slugs and Snails

Slugs and snails invade from outdoors, often entering the house hidden in the drainage holes of pots. They chew large irregular holes, primarily at night, and hide during the day. The best control is to pick them off.

Root and Media Pests

Gnats (fungus gnats, manure flies, shore files, mushroom flies)

Flying gnats and their immature maggot stages (Figure 18-7) are annoying to people and harmful to indoor plants. Overly wet potting media rich in organic matter provide ideal breeding areas, and a few gnats can multiply to several hundred in a short time.

The adults are brown to black and about ⅛ inch (3 millimeters) long. They are usually seen crawling along the soil surface, where they lay eggs that hatch into tiny white maggots. The burrowing and feeding of the maggots on plant roots cause stunting and root diseases. As the larvae increase in number, the damage intensifies.

Fungus gnats can be detected by placing a slice of raw potato on the surface of the pot. Within a few hours, maggots will crawl into the potato and begin to feed. The life cycle of the fungus gnat can be broken by placing a ¼-inch (½-centimeter) layer of sand on the surface of the pot. This will prevent adults from laying eggs.

Figure 18-6 Thrip damage on Ardisia. *(Photo courtesy of Dr. Dick Henley, Central Florida Research and Education Center, Apopka, FL)*

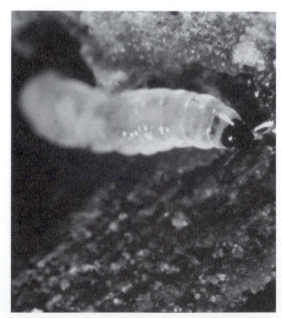

Figure 18-7 The larva of a fungus gnat (shown approximately 20 times normal size). *(Photo courtesy of Marge Coon)*

Springtails

Springtails derive their name from the sudden jumping motion they display when they are disturbed, as by watering. The insects spend their entire life cycle in the soil, growing to an adult size of about ⅟₁₆ inch (1.5 millimeters).

Unlike parasitic gnat larvae, springtails in all stages eat only dead vegetation, causing no harm to indoor plants.

Nematodes (eelworms)

Nematodes are microscopic or near-microscopic worms that live in the medium and feed on the roots of plants. Certain species live on plant leaves, entering through the stomates to feed, but they are rare on indoor plants.

Different species of nematodes inflict different types of damage. The feeding of the root-knot nematode causes irritation of the root and leads to the formation of a knot of tissue at the wound site. Numerous such knots interfere with the roots' ability to transport water and nutrients to the foliage, and the top of the plant becomes stunted and pale.

Root-lesion nematode does not cause swellings on the roots but does destroy root cells and leaves an entryway for disease. The effects on the top portions of the plant are the same as with root-knot nematode.

Because they are usually microscopic, soil infestations on indoor plants are difficult to diagnose. But nematodes can be suspected if the medium was not sterilized before use or if it contains soil. Unpotting the plant and checking for knots will diagnose root-knot nematode. Brown decayed roots could indicate the presence of other types. An exact diagnosis can be made only through a medium and root sample submitted to a laboratory. Nematodes are difficult to control, and unless the plant is extremely valuable, it should be discarded.

Root Mealybugs

Root mealybugs (Figure 18-8) are root-parasitizing insects that look like regular mealybugs but are smaller. Because their feeding weakens plants, they should be controlled with an insecticidal soil drench.

Sowbugs (pillbugs)

Sowbugs (pillbugs) are small pests that are not insects but are related to the armor-covered crayfish. They live only in damp places and will often be found hiding under pots or living inside the drainage hole. Although they eat mainly dead organic matter, they also feed on roots and small seedlings and should be picked off and discarded.

Figure 18-8 Root mealybug. *(Photo courtesy of Dr. Raymond Gill, California Department of Food and Agriculture. Copyright State of California)*

Earthworms

Earthworms do not eat live plants and are seldom seen at the surface. During their burrowing they digest potting medium and excrete it behind, leaving a network of tunnels. Unless present in large numbers, they do no damage.

ORGANISMS CAUSING DISEASES OF INDOOR PLANTS

All indoor plant diseases are caused by one of four types of microscopic organisms: fungi, bacteria, viruses, or viruslike organisms.

Fungi

Fungi (sometimes called *molds*) are microscopic organisms that depend on absorbing their nutrition from dead or living organisms.

The fungi that cause diseases in plants are parasitic, invading the individual cells of plants and extracting nutrition (Figure 18-9). The cells die, and the progression of the disease is begun.

Fungi that attack indoor plants usually infect the roots; foliar fungi-caused diseases are less common. The former are called *soil-borne* because plant roots are exposed to the organisms through fungi-contaminated media. Treatment is through modifying environmental conditions (moisture in particular) and chemical fungicides.

Bacteria

Bacteria are single-celled plants also incapable of manufacturing their food. Some use living plant cells for nourishment and will reproduce rapidly through division

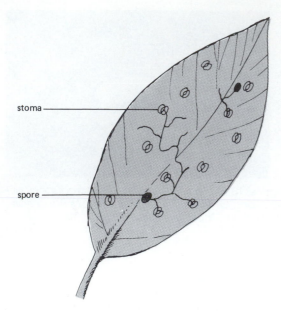

stoma

spore

Figure 18-9 Fungus spores germinating and invading a leaf through stomata.

under proper environmental conditions. A few bacteria attack indoor plants, causing such disorders as leaf spots and root rots. They can be partially controlled by copper or antibiotic sprays.

Viruses and Viruslike Organisms

Submicroscopic viruses and viruslike organisms are only occasionally encountered in indoor plants. When they are, it can generally be assumed that they were present in the plant at the time it was purchased.

Viruses and mycoplasmas are often transmitted among plants through the mouthparts of contaminated insects—especially whiteflies, aphids, and thrips. Contaminated tools such as pruners can also pass the organisms from plant to plant.

SPECIFIC DISEASES OF INDOOR PLANTS

Root Rot

Root rot is a general term used for both fungal and bacterial diseases that cause the decay of root tissues. Root rot is usually associated with poor growing conditions such as a heavy medium that excludes air or lacks an outlet for excess drainage water. Initial symptoms are a slowdown in growth and a yellowing and dropping of

foliage, signs of the inability of the roots to take up water and nutrients. In extreme cases, permanent wilting occurs, and the plants die. An inspection of rot-infected roots will show them to be brown rather than the white of healthy plants.

Treatment consists of chemical soil drenches combined with improvements in culture. Repotting in a better-drained medium in a pot bearing a drain hole and less frequent watering are recommended.

Crown or Stem Rot

Crown or stem rot is a disease that attacks the stem or crown of a plant at the point where it enters the medium. Numerous bacteria and fungi found in unsterilized potting media often cause the disease, but it is also possible that the organisms may be airborne.

The main symptom of crown rot is brown discoloration of the stem (Figure 18-10). The rotting prevents water passage up through the plant, and it can wilt and break off at the weakened area.

Cacti show a form of stem rot called *dry rot* (Figure 18-11). This disease causes the stem to discolor to yellow or brown and become sunken. The end effect on the plant is the same as with stem and crown rot.

Once infected, there is little that can be done to cure a plant with this disease. The recommended procedure is to take cuttings of the uninfected portions and root them in sterile medium for new plants. The procedure is the same for cacti; that is, the entire top of the cactus is cut off above the rot area and rerooted.

Figure 18-10 Crown rot on a nerve plant (*Fittonia argyroneura*). (Photo by Kirk Zirion)

Figure 18-11 Dry rot of cactus. *(Photo by Andrew Schweitzer)*

Mildew

Mildew (Figure 18-12) is a foliar fungus disease that attacks a limited number of susceptible indoor plants (see Table 18-2). The first symptom is a gray powdery residue on the leaves. Later the leaves yellow and drop, and the plant eventually dies.

Mildew is controlled by reducing the humidity and applying a fungicide spray. After treatment, keeping the humidity low and all water off the foliage will prevent further problems.

Figure 18-12 The fungus disease mildew. *(Photo courtesy of Marge Coon)*

Leaf Spots

A number of bacteria and fungi occasionally cause leaf spot diseases on foliage plants grown indoors. The symptoms vary with the disease and host plant, but in general, small pinpoint-sized spots will appear, enlarging and spreading slowly from leaf to leaf. The spots will usually show a pattern such as a center black spot surrounded by yellow or an irregular yellow spot ringed in brown. Figure 18-13 shows a typical fungal leaf spot.

Fungicidal sprays may prevent spread, but often there is no cure. Removing infected leaves as soon as they are detected and keeping leaves dry will slow or stop the spread, and a plant can sometimes be saved in this manner.

Virus and Viruslike Organism Infections

The symptoms of a virus infection range widely, depending on the organism and its host plant, but often include yellowing, curled foliage, spots, and stunting.

Although viruses are unlikely to spread among different species of indoor plants, they do occasionally. No cures are known and diseased plants should be discarded.

PEST AND DISEASE PREVENTION

Prevention is better than any cure that can be administered. It involves avoidance of both the initial contact with the insect or disease and its spread.

Plants that are healthy and under good growing conditions have been proved to be less susceptible to attacks by insects and diseases than those under "stress." Stress conditions include any factors that slow the normal growth and development of the plant such as poorly drained medium, overly warm temperature, insufficient light, and the like. In addition, plants in a weakened state are less able to survive such disorders.

Potting Medium

All potting media used for indoor plants should be sterilized prior to use.[1] Unless this is done, fungi, bacteria, and soil pests can be introduced and spread rapidly.

Second, the medium should be composed of ingredients in the proper proportions. This will assure that it

[1]See Chapter 15 for directions on sterilizing potting media.

Table 18-2 Diseases of Indoor Plants and Their Common Hosts

Disease	Common hosts	Disease	Common hosts
Leaf spot	African violet (*Saintpaulia ionantha* cultivars)	**Root rot**	Cacti
	Aluminum plant (*Pilea cadierei*)		Succulents
	Anthurium (*Anthurium* spp.)		Many foliage plants
	Arrowhead (*Syngonium podophyllum*)		
	Baby rubber plant (*Peperomia obtusifolia*)	**Steam and**	African violet (*Saintpaulia ionantha*)
	Balfour aralia (*Polyscias balfouriana*)	**crown rot**	Aralias (*Polyscias* spp.)
	Begonia (*Begonia* spp.)		Arrowhead (*Syngonium podophyllum*)
	Cast iron plant (*Aspidistra elatior*)		Begonia (*Begonia* spp.)
	Century plant (*Agave* spp.)		Cacti
	Chinese evergreen (*Aglaonema* spp.)		Dumbcane (*Dieffenbachia* spp.)
			Dracaenas (*Dracuena* spp.)
	Croton (*Codiaeum variegatum pictum* cultivars)		English ivy (*Hederd helix* cultivars)
	Dragon lilies (*Dracaena* spp.)		Figs (*Ficus* spp.).
	Dumbcane (*Dieffenbachia* spp.)		Geranium (*Pelargonium* × *hortorum*)
	English ivy (*Hedera helix*)		Gloxinia (*Sinningia speciosa*)
	Fern		Japanese aralia (*Fatsia japonica*)
	Gardenia (*Gardenia jasminoides*)		Kalanchoe (*Kalanchoe* spp.)
	Geranium (*Pelargonium* × hortorum cultivars)		Nerve plant (*Fittonia* spp.)
	German ivy (*Senecio macroglossus*)		Orchids
	Goldfish plants (*Columnea* spp.)		Peperomias (*Peperomia* spp.)
	Impatiens (*Impatiens* spp.)		Philodendrons (*Philodendron* spp.)
	Ivy tree (× *Fatshedera lizei*)		Pileas (*Pilea* spp.)
	Indian laurel (*Ficus retusa nitida*)		Pothos (*Epipremnum aureum* cultivars)
	Jerusalem cherry (*Solanum pseudocapsicum*)		Sinningias (*Sinningia* spp.)
	Lollipops (*Justicia* spp.)		Snake plant (*Sansevieria trifasciata* cultivars)
	Nerve plant (*Fittonia verschaffeltii argyroneura*)		Succulents
	Orchids		Ti plant (*Cordyline terminalis*)
	Philodendron (*Philodendron* spp.)		Umbrella tree (*Schefflera* and *Brassaia* genera)
	Piggyback plant (*Tolmeia menziesii*)		Zebra plant (*Aphelandra squarrosa*)
	Poinsettia (*Euphorbia pulcherrima*)		
	Ponytail palm (*Beaucarnea recurvata*)	**Virus**	Anthurium (*Anthurium* spp.)
	Prayer plants (*Maranta* spp.)		Begonia (*Begonia* spp.)
	Queensland umbrella tree (*Brassaia actinophylla*)		Cacti
	Royal red bugler (*Aeschynanthus pulcher*)		Calathea (*Calathea* spp.)
	Rubber plant (*Ficus elastica*)		Chinese evergreen (*Aglaonema* spp.)
	Satin pellonia (*Pellonia pulchra*)		Christmas cactus (*Zygocactus truncatus*)
	Silver vase (*Aechmea fasciata* cultivars)		Dumbcane (*Dieffenbachia* spp.)
	Snake plant (*Sansevieria trifasciata* cultivars)		Figs (*Ficus* spp.)
	Stone crops (*Sedum* spp.)		Geranium (*Pelargonium* × *hortorum*)
	Thread flower (*Nematanthus* spp.)		Goldfish plants (*Columnea* spp.)
	Ti plant (*Cordyline terminalis*)		Moses in the Cradle (*Rhoeo spathaceae*)
	Watermelon begonia (*Pilea cadierei*)		Nerve plant (*Fittonia verschaffeltii* cultivars)
	Weeping fig (*Ficus benjamina*)		Orchids
	Zebra basket vine (*Aeschynanthus marmoratus*)		Peace lily (*Spathiphyllum* spp.)
	Zebra basket vine (*Aeschynanthus pulcher*)		Peperomia (*Peperomia* spp.)
			Philodendron (*Philodendron* spp.)
Mildew	African violet (*Saintpaulia ionantha*)		Prayer plant (*Maranta* spp.)
	Begonia (*Begonia* spp.)		Wandering Jew (*Zebrina* spp.)
	Cissus ivies (*Cissus* spp.)		Zebra plant (*Aphelandra squarrosa*)
	Kalanchoe (*Kalanchoe blossfeldiana* cultivars)		

Figure 18-13 A fungal leaf spot disease on German ivy (*Senecio mikanioides*). (*Photo by Marge Coon*)

drains quickly and thoroughly when watered. A drainage hole or other means for the escape of excess water should be provided in the pot. Media that retain an excessive volume of water or from which water is prevented from escaping have conditions leading to poor root but healthy root-rot pathogen growth. Water-logged media can frequently be traced as the initial cause of root-rot diseases.

Finally, a medium should contain adequate nutrients. Poorly nourished plants, like those suffering from any other kind of stress, are commonly attacked by pests and diseases.

Humidity

The humid conditions optimal for the growth of most indoor tropicals can be both desirable and undesirable from the standpoint of pest and disease control. Perhaps the worst pests of indoor plants, spider mites, are adversely affected by high humidity. Raising the humidity and misting can be used to slow their breeding and control their damage. But humidity can increase a foliar fungus problem because the spores of the fungus require high humidity to germinate. Overall, unless a plant is commonly susceptible to leaf spots or mildew and has been attacked previously (see Table 18-2), high humidity is advisable.

Water

Stagnant water, whether in the medium or on the crown, is detrimental to plant health. Crown rot disease, which attacks such indoor plants as the African violet, has often been attributed to water left standing in the center of the plant after watering. Barrel-shaped cacti, in which the growing point is in a depressed area on top of the plant, are also susceptible to rot if water is left standing there.

In summation, although misting foliage and showering are recommended on most plants, they should be avoided under certain circumstances.

Isolation and Inspection

Isolation of newly acquired plants and older ones showing symptoms is a standard practice for controlling both diseases and insects. Isolation outdoors or in a separate room is preferred, because air currents carry disease spores and insects readily.

Weekly inspection, mentioned in Chapter 14, will help catch disease problems before they become epidemic. It is essential to look *under* as well as on the tops of the leaves.

PEST AND DISEASE CONTROL METHODS

Modification of the environment, home remedies, and commercial chemicals all have use in controlling diseases and pests of indoor plants. Modification of the environment for prevention has already been discussed, and the same instructions can be applied to plants that are already infested.

Home Remedies

Home remedies, if used with persistence, can be effective. In addition, many are completely safe to humans and avoid any danger of accidental pesticide poisoning.

Wiping or washing with soapy water made from a few drops of detergent in a sink of water is a home remedy useful for reducing populations of mites or aphids. The pests can be sponged off with a dilute solution, or the entire plant can be held upside down and swished in the soapy water. Soap will not harm the plant, and the agitation will dislodge most of the pests. As a precaution against toxicity and to avoid soap film, rinsing of plants afterward is advisable.

Hand picking of large insects like sowbugs, scales, and mealybugs is another home remedy that works. Toothpicks or cotton swabs dipped in rubbing alcohol and touched to the insects (Figure 18-14) can be used. When larger numbers of insects are present, many indoor plants can be sprayed with a 70 percent solution of rubbing alcohol. Because this spray may damage

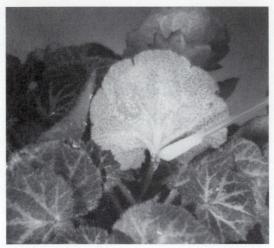

Figure 18-14 Touching mealybugs or aphids with a cotton swab dipped in alcohol is a reliable control method for small infestations. *(Photo by Kirk Zirion)*

some plants, it should be tested first. These remedies are most effective if only a few pests are present.

The key to success with both these methods lies in repetition. Probably not all of the pests will be killed, but the population will be reduced.

Commercial Pesticides

Because of the constantly changing laws regarding pesticide use and the many new chemicals developed each year, it is useless to attempt to prescribe a chemical cure for every indoor plant ailment discussed here.

 Companies that manufacture pesticides for use on horticultural plants frequently hire public relations specialists with training in horticulture to answer all questions regarding their products. These specialists handle inquiries from customers, attend meetings at which horticultural products are displayed, write press releases, and interface with the public in other ways.

Instead, chemicals currently approved by the Environmental Protection Agency will be covered. All of these chemicals can be dangerous to the health of people, pets, and plants if not used in accordance with package directions and in observance of safety rules. Plant-derived insecticides should not necessarily be considered safer than synthesized ones.

The term *pesticide* is often used to refer only to "bug killers." But it actually includes all chemicals used to kill organisms that adversely affect the growth of plants—including fungicides, bactericides, insecticides, miticides, nematicides, and herbicides. Insecticides are then further broken down by the source of the *active ingredient* (the chemical in the product that actually kills the pest, as opposed to the materials used to dilute, chemically "carry," or scent it).

Organic or Botanical Pesticides. Organic or botanical pesticides are derived from materials that occur in nature, primarily plants. Pyrethrin, rotenone, neem, and nicotine are the most common.

Pyrethrin is a nerve poison, which was originally marketed in 1947. Its killing properties were excellent, but the chemical was not stable in sunlight. This and the development of pest *resistance* (see Plant Safety and Pesticide Effectiveness section) to the original led to the development of a number of *pyrethroids*. The chemistry of these is based on the original, but varies sufficiently to overcome insect resistance. Resmethrin is an example of a pyrethroid.

Rotenone was first used in gardens in 1848 to kill caterpillars, but before that it was used in South America as a fish poison. It paralyzed the fish, causing them to float to the top where they could be gathered. It is derived from two genera of legumes and works by inhibiting a respiratory enzyme and causing the respiratory system to fail.

Nicotine is derived from tobacco and was used as early as 1890 to kill sucking insects on garden plants. It is a nerve poison.

Neem products are insect growth regulators and are derived from the neem tree.

Insecticidal Soaps. The insecticidal soaps kill insects and insectlike pests such as mites by disruption of the cell and tissue structures by the action of potassium salts of fatty acids. They are of very low toxicity to humans.

Oils. Refined vegetable or mineral oils kill insects by smothering them. They are safe to use on most indoor plants but should not be used in conjunction with sulfur sprays. They may damage ferns and some other plants, so check the label.

Synthetic Pesticides. Synthetic pesticides are those that are synthesized in a lab and are therefore not extracted from plants. As regards toxicity, they may be more or less toxic than plant-derived products. However, as with all toxic materials, they must be used with care. Remember that when pesticides of any type are used indoors there is a high likelihood that people will come in contact with them. Further, because of the lack of ultraviolet light, they will often decompose more slowly than similar pesticides used outdoors. Spraying must be done with care to minimize airborne droplets that can circulate in the air for extended periods of time in the limited ventilation found indoors. These droplets, in addition to harming people, can also damage floors, carpets, upholstery, and the like. Additionally, many pesticides have strong odors that are objectionable indoors. There are also legal concerns with using pesticides indoors, because only pesticides whose labels state that they can be used indoors or in interiorscapes can be legally used. With small plants, these problems can be avoided by moving plants to a shaded area outdoors for spraying and then returning them indoors after the spray has dried.

When a synthetic pesticide is used, a systemic applied to the soil is often useful for eliminating sucking insects and thrips. Three commonly used systemics are imidicloprid, disulfoton, and orthene (Acecaps). Imidicloprid is available under several trade names and can be effective for up to 70 days. Disulfoton is only available for homeowners in one granular formulation. The only orthene formulation that can be used indoors is Acecaps, which are implants that are placed into holes drilled into the bark of large interior trees (Figure 18-15).

In addition to the low-risk oils, soaps, and botanicals, several pyrethroids (a group of synthetic insecticides related to the natural chemical, pyrethrin) can be used indoors. These include fluvalinate, cyfluthrin, permethrin, and resmethrin. These are available under several trade

names. When shoreflies or fungus gnats are problematic, they can be controlled with *Bacillus thuringiensis* Israelensis or beneficial nematodes (*Steinernema* spp.).

SAFETY PRECAUTIONS FOR USING PESTICIDES

Following general safety precautions for pesticide use is the first prerequisite for effectively controlling pests without harm to either the gardener or the plant being treated.

Personal Safety

The methods of applying pesticides vary. Sprays are most common for foliage insects, mites, and diseases. But drenches, in which the chemical is mixed in water and applied in place of plain water, are used for soil insects, nematodes, and root diseases. For personal safety, all contact with the pesticide should be avoided. Rubber gloves should be used, and any skin spills should be washed off immediately with soap and water.

Pesticide sprays, whether aerosol or hand pump type, should be applied outdoors whenever practical. During cold weather, an unoccupied room that can be aired out afterward can be used. The plant to be treated should be placed on newspapers to prevent accidental spraying of carpets or furniture. Particular care should be taken when applying aerosol pesticides. As noted earlier, they disperse small droplets into the air, and the airborne pesticide can be inadvertently inhaled.

After spraying is completed, safety precautions should include washing the spray equipment, measuring equipment, and hands with soap. The plant should not be touched by people or animals.

Some indoor plant insecticides are applied to the soil as a liquid drench in place of plain water. Others may be applied to the soil surface as granules, and when the plant is watered, the insecticidal chemical is released. Sometimes these chemicals are *systemic*. That is, these chemicals are taken up into the vascular system of the plant and poison insects and disease organisms that prey on that plant. Because a plant to which a systemic has been applied is toxic for a prolonged period, systemics are not often available for home use. Table 18-3 lists some of the pesticides used on indoor plants and the pests they control.

Plant Safety and Pesticide Effectiveness

The first rule for plant safety is to follow the label directions. A stronger concentration than recommended may burn foliage severely, whereas a weaker one may fail to

Figure 18-15 A medication-containing implant in place in a tree trunk. (*Photo courtesy of Creative Sales, Inc., Fremont, NE*)

Table 18-3 Selected Indoor Plant Pesticides[a]

Chemical name	Uses
Rotenone	Plant-derived contact and stomach poison for insects; used for aphids, thrips, whiteflies, mites, and immature scales and mealybugs
Resmethrin, permethrin, tetramethrin, cyfluthrin	Synthetic pesticides similar chemically to pyrethrin; used for the same pests
Pyrethrin	Plant-derived insecticide with contact action; effective against the same pests as rotenone
Copper oleate and other copper compounds	Fungicide for powdery mildew and other fungus diseases and some bacteria; registered for use on fuchsia, ferns, ivy, gardenias, philodendrons, geraniums, begonias, and succulents
Malathion	Insecticide for aphids, red spider and other mites, mealybugs, thrips, white-flies, and root mealybugs; not recommended for use on crassulas, *Pteris* ferns, *Adiantum* ferns, anthuriums, or kalanchoes
Kelthane®	For red spider and other mites; registered for use on gardenias, rubber plants, bracken ferns, Boston ferns, ivy, African violets, and philodendrons
Nicotine sulfate	For aphids, thrips, spider mites, orchid scales, and fungus gnats; registered on philodendrons, fuchsia, ferns, and African violets
Disulfoton®	Used for aphids, thrips, mites, and whiteflies; granular systemic applied to the medium and watered in; also controls some nematodes
Acecaps®	For aphids, thrips, leaf miners, mealybugs, scales, spider mites, and whiteflies use with caution on gloxinia, hibiscus, and philodendron
Karathane	Miticide and fungicide for powdery mildew, spider mites, and two-spotted mites; registered for use on begonias and cineraria
Methoprene (Altosid®)	Insect growth regulator which prevents maturation of larvae of insects
Aliette	Systemic fungicide for control of certain root rots and downy mildew
Oils	Kill most soft-bodied insects through suffocation
Azadirachtin	Insect growth regulator effective on soft-bodied insects; low toxicity
Soaps	Kill soft-bodied insects and mites by removing waxes from their bodies causing them to dry out; low toxicity

[a]All pesticides must, by law, be approved and registered with the Environmental Protection Agency and state agencies for use on any plant species. The above table is not intended as authorized recommendations for pesticide use, because the registrations are constantly changing.

kill the pests. Many species of indoor plants, including ivies, ferns, and succulents, are easily injured by pesticides; these susceptible plants should be listed on the pesticide label. Test spraying several leaves is always advisable when treating a new species. The plant should be observed for several days for signs of reaction to the pesticide before the remainder of the foliage is treated.

In the application of water-diluted spray pesticides, "spray to runoff" is usually advised. This means that the chemical should be applied heavily enough to drip from the leaves. Coverage should include all aboveground plant parts and the undersides of leaves, even those not presently infested. Aerosol pesticides can present dual hazards to plant health. The propellant used to carry the pesticide may be toxic, and the heat pulled from the leaves when the propellant vaporizes may cause cold injury. To prevent the latter, spray no closer than the directions advise. At a moderate spray distance most of the propellant will vaporize before touching the leaf.

When applying soil drenches, the chemical should be mixed according to directions given for a drench. The pesticide should be applied in sufficient quantity to wet all the potting medium, and any that drains out should be discarded.

Repeat sprays as recommended on package directions are the key to successfully controlling most indoor plant pests. Many generations and life stages from egg to adult occur at one time, and not all will be chemically susceptible. Repeat sprayings are necessary to do more than temporarily slow reproduction rates.

Resistance of insects and mites to specific pesticides is occasionally encountered in indoor plants. Generally all but a few pests will be killed by a chemical, but those remaining continue to breed and eventually build a *resistant population* that cannot be killed by the insecticide. Resistance can be overcome in part by spraying alternately with chemically unrelated pesticides.

Finally, never save leftover mixed insecticides. Mixed pesticides are often inactivated in storage and are a common cause of accidental poisonings.

SELECTED REFERENCES FOR ADDITIONAL INFORMATION

Alfieri, S. A., Jr., K. R. Langdon, and C. Wehlburg. *Index of Plant Diseases in Florida.* Bulletin 11. Gainesville, FL: Division of Plant Industry, Florida Department of Agricultural and Consumer Services, 1984.

Ali, A. D. "A Back to Basics Look at Pests." In *Interior Landscape Industry,* Vol. 6(1), January 1989, pp. 90–95.

Armitage, Allan M. *Specialty Cut Flowers: The Production of Annuals, Perennials, Bulbs and Woody Plants for Fresh and Dried Cut Flowers.* Portland, OR: Timber Press, 1993.

Chase, A. R. *Compendium of Ornamental Foliage Plant Diseases.* St. Paul, MN: ARS Press, 1987.

Hahn, Jeffrey, Deborah Brown, and Mark Ascerno. *"Houseplant Insect Control."* University of Minnesota Extension Service. 1998. (November 2, 2001).

Lindquist, Richard. "Pesticide Resistance: Why It Occurs, How to Deal with It." In *Interior Landscape Industry,* Vol. 6(3), March 1989, pp. 56–59.

Manaker, G. W. *Interior Plantscapes Installation, Maintenance and Management.* 3rd ed. Englewood Cliffs, NJ: Prentice-Hall, 1996.

Pirone, Pascal P. *Diseases and Pests of Ornamental Plants.* 5th ed. New York: Wiley, 1978.

Powell, Charles, and Rosemary Rosetti. *The Healthy Indoor Plant.* Columbus, OH: Rosewell, 1992.

Simone, Gary. "New Disease Problems in the Foliage Industry." In *Interior Landscape Industry,* Vol. 6(4), April 1989, pp. 36–42.

Simone, Gary. "Virus Diseases Affecting Foliage." In *Interior Landscape Industry,* Vol. 5(6), May 1988, pp. 70–84.

Simone, Gary W., and A. R. Chase. *Disease Control Pesticides for Foliage Production,* p. 30. Gainesville, FL: University of Florida Institute of Food and Agricultural Science, 1989.

Simone, Gary W., T. A. Kucharek, and R. S. Mullin. *Plant Disease Control Guide.* Gainesville, FL: University of Florida Institute of Food and Agricultural Sciences, 1989.

Steiner, Marilyn and Don Elliot. *Biological Pest Management for Interior Plantscapes.* Alberta, Canada: Alberta Public Affairs Bureau, 1987. Available from Associated Landscape Contractors of America, Herndon, VA.

World Wide Web Addresses

http://www.agric.gov.ab.ca/crops/hort/houseplants/pests.html Alberta Agricultural, Food and Rural Development. "Houseplants—Insects and Related Pests." August 23, 1996.

http://www.gardenguides.com/TipsandTechniques/indoorinsects/ Garden Guides. "Controlling Insects on Your Houseplants" (2001).

http://plantinsect.ifas.ufl.edu/ Entomology and Nematology Department, University of Florida. "Plant and Insect Associations" 10/31/01. Chapter 19, "Interior Landscaping."

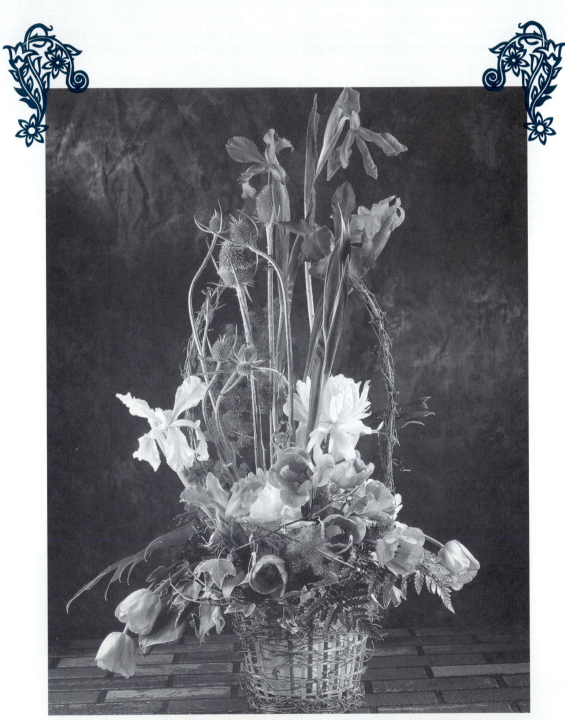

(Photo courtesy of Syndicate Sales; design by Denise Lee, AIFD, PFCI)

DECORATING WITH GROWING PLANTS AND FRESH FLOWERS

Plants are beautiful creations in themselves. But as with other works of beauty, a little artistic treatment will enhance and focus attention on their most attractive features. A new field of *interior landscaping* or *interior plantscaping* has evolved to deal with the selection and arrangement of indoor plants. An interior planting specialist must have not only a knowledge of artistic design but also an understanding of the care requirements associated with particular species. This person must be an artist *and* a horticulturist, arranging plants in a manner that is pleasing visually while satisfying their growth requirements.

Most experienced houseplant growers have already tried their hands at interior landscaping, although they may not realize it. For example, when the host or hostess moves an attractive plant to the coffee table before company comes, he or she is practicing interior landscaping. Similarly, selecting attractive pots for houseplants is part of the design process. With a bit of artistic sense, knowledge of plant growth requirements, and sample ideas given in this chapter, anyone can create an interior landscape.

A HISTORICAL PERSPECTIVE

Using plants for indoor decoration is not a new idea. Indoor plants were much in vogue during the 1800s, and plant explorers undertook lengthy expeditions to secure exotics for wealthy patrons. Unfortunately, the preference for dark interiors and the lack of central heating made it impossible to grow all but the hardiest species indoors. Cast iron plants, potted palms, and rubber plants were the standard indoor species. Boston ferns also adapted to the low light and cool temperatures and were proudly displayed on ornate fern stands (Figure 19-1).

Partly because of the inadequate indoor growing conditions, the terrarium and solarium gained favor as plant-growing areas. Terrariums were originally designed as a method of holding plants during sea voyages. From this basically utilitarian structure, they developed into elaborate "parlor cases," highly embellished in the style of the Victorian era. It was not unusual for a terrarium of that period to be made of hand-blown glass, with its supporting structure of filigreed iron or brass or enameled metal.

The solariums or conservatories of that period were the garden rooms of today. They were used to grow and display many exotics, a hobby of the rich during that period. With the turn of the century and the death of Queen Victoria, the fashion in interior decoration changed to a lighter and airier look. The heavily draped floor-to-ceiling windows typical of Victorian homes were uncovered, and light lace curtains or simple valances allowed light to enter homes freely. However, central heating still did not exist, and this precluded growing many of the cold-sensitive species that are so popular today.

After World War I, middle-class bungalow-style houses were built with large window areas and, frequently, with sun porches in the front. By this time more low-light-requiring houseplants had been discovered,

Figure 19-1 A bronze Victorian fern stand. *(Photo by Andrew Schweitzer)*

obscure are now common, and innumerable new cultivars have been and are being developed.

All of this presents today's indoor gardeners with a wide variety of plants and accessories from which to choose. Their only problem will be determining how to fit plants into the living area attractively while meeting their growing requirements—the subject of this chapter.

FUNCTIONAL USES OF PLANTS

Specimen Plants

Specimen plants (Figure 19-2) are those displayed purely for their aesthetic appeal. They are often large trees or floor plants, although they may also be hanging baskets, flowering species, terrariums, or bonsai. Specimen plants are generally used alone and can be thought of as living sculpture. In some instances a specimen plant will become the focal point of the room, attracting more attention than any of the other furnishings.

Specimen plants are usually expensive and should be considered a major plant purchase. Before you invest in one, try to grow a small plant of the species you intend to buy so that there will be less chance of your losing the plant from improper care. Table 19-1 lists a number of plants available in large sizes suitable as indoor specimens.

Plants As Wall Art

A bright wall near a window can be an ideal spot for a wall grouping of plants (Figure 19-3). The grouping takes the place of a picture or other wall decoration.

Pots can be attached to the wall with the type of clips used to hang plants on fences in outdoor gardens, and a structured design or random pattern can be used. Interest can be added by using unusual potting containers such as antique tins and handmade pottery, although this will make hanging more difficult.

Among the plants that lend themselves most readily to wall grouping are epiphytes. Certain orchids, bromeliads, philodendrons, and ferns can be grown mounted on osmunda or bark plaques. These are lightweight and easily hung.

Plants That Accent Other Objects

Another function of plants can be to draw attention to some other element of the room design, to accent it. An example of this is shown in Figure 19-4 with palms

and central heating was becoming common. This facilitated indoor plant growing and removed it from being the exclusive province of the rich. But the original plant fervor of the 1800s never returned, and there was a slump of interest until the late 1960s. Architecture reflected this lack of enthusiasm for indoor plants. Ranch-style houses became popular. Although many had large picture windows, these were designed for viewing and were seldom used for plants. The remaining windows were small or, at most, medium-sized, with emphasis on sleek, horizontal lines.

In the 1970s, the pendulum swung back, and indoor plants were very much in demand for homes, offices, and shopping malls. Houses are now designed with large expanses of thermal pane glass that not only provide enough light for growing almost any houseplant but are also energy efficient. Indoor gardeners are no longer restricted to pots on a windowsill, and large interior plants are flourishing.

From demand comes supply, and the horticulture industry has risen to the occasion. Large specimens of many species are sold today that years ago would have been impossible to find. Species that were previously

Figure 19-2 The specimen bamboo in the corner is the focal point of the room and takes advantage of the light entering through the high windows near the cathedral ceiling. *(Photo courtesy of Landscape Images, a Division of British Indoor Gardens, Inc., El Toro, CA)*

Table 19-1 Selected Large-Specimen Plants for Indoor Culture

Common name	Botanical name
Norfolk island pine	*Araucaria heterophylla*
Bamboo	*Bambusa* spp.
Umbrella plant	*Brassaia actinophylla*
Cacti	*Cereus, Echinocactus, Ferocactus*
Palms	*Chamaedorea, Howea, Rhapis*
Jade plant	*Crassula argentea*
Cyperus	*Cyperus alternifolius*
Dumbcane	*Dieffenbachia maculata*
Thread-leaf aralia	*Dizygotheca elegantissima*
Corn plant	*Dracaena fragrans* 'Massangeana'
Red-margin dracaena	*Dracaena marginata*
Euphorbia	*Euphorbia millii, Euphorbia lactea, Euphorbia trigona*
Fatsia	*Fatsia japonica*
Weeping fig	*Ficus benjamina*
Rubber tree	*Ficus elastica*
Fiddle-leaf fig	*Ficus lyrata*
Wax-leaf privet	*Ligustrum lucidum*
Avocado	*Persea americana*
Philodendron	*Philodendron, Monstera*
Yew pine	*Podocarpus macrophyllus* or *Podocarpus gracilior*
Ming tree	*Polyscias* spp.

flanking the entrance to an adjoining room. Other examples of plants used as accents would be a small fern setting on an antique desk or a collection of small potted plants on a table as background for one or more family photographs.

Plants As Traffic Channelers

People generally walk the shortest distance from room to room by the route that offers the least resistance. Once these traffic patterns are established, they can be difficult to change, but strategically placed indoor plants can be helpful in this regard. By placing several plants between a chair and sofa, for example, people will naturally be directed to enter the sitting area by another way. Figure 19-5 shows a dramatic and obvious use of plants as traffic channelers.

Plants As Screening

Visual screening is another functional use of plants. A strategically placed group of plants can separate two living areas (Figure 19-6).

Figure 19-3 A wall arrangement of plants.

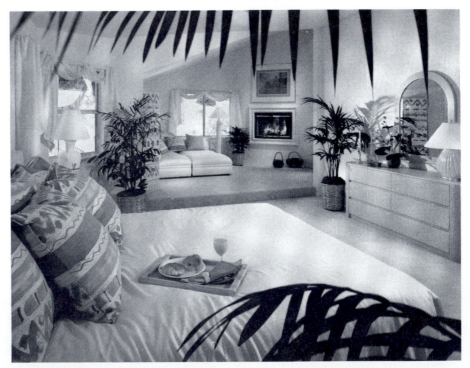

Figure 19-4 Palms on either side of the entryway are used as accents to focus attention onto the living area and fireplace beyond. *(Photo courtesy of Landscape Images, a Division of British Indoor Gardens, Inc., El Toro, CA)*

Figure 19-5 Built-in free-form planters filled with palms and peace lilies channel guests from the living room to the dining room in this residence. *(Photo courtesy of Landscape Images, a Division of British Indoor Gardens, Inc., El Toro, CA)*

If the screening problem is an objectionable window view, plants are the ideal solution. They thrive in the light, yet will not exclude the sunlight as a curtain would. Figure 19-7 shows how plants can be used for screening in a bathroom.

Plants As Space Fillers

There are often areas of a room in which no furniture can be conveniently placed because of space or design limitations. These are "dead" spaces that need to be filled to avoid an unbalanced look in the decor. Plants are the ideal solution, because they can be bought in any size and shape, and can be chosen to live in the amount of light provided in the area. They are also very economical compared to art or furniture. Figure 19-8 shows a wonderful use of plants to fill the area under a spiral staircase.

ELEMENTS OF DESIGN

After approaching the functional use of plants and the restricting environmental requirements (primarily available light) that determine placement, the design elements are considered. These include

1. Color
2. Size
3. Texture (leaf size)
4. Decor association of the species

Figure 19-6 Fishtail palms in the center are both specimens and provide screening between two living areas in this large home. Their height bridges the gap between the cathedral ceiling and the low seating areas, thus establishing human scale. *(Photo courtesy of Landscape Images, a Division of British Indoor Gardens, Inc., El Toro, CA)*

Figure 19-7 A built-in planter provides the screening for privacy necessary for this tub. *(Photo courtesy of Landscape Images, a Division of British Indoor Gardens, Inc., El Toro, CA)*

Color

Suitability of color to decor becomes an issue whenever any but a plain green plant is selected. If a plant has white, yellow, pink, or red on its foliage, this accent must be acceptable in the room color scheme. Flowering plants often have large expanses of color displayed as flowers (already attention getters) and must be used with careful forethought.

The range of variegated and color-accented foliage plants has expanded greatly in the past 20 years, and hundreds of new cultivars are now available (Table 19-2). Likewise, short-term and long-season flowering plants of many new species and cultivars are now commonly available (see Table 14-1), and using them effectively in a design requires mainly knowing their light requirements and functions in the room design (normally as a specimen or specimen group), after which the color can be selected.

One general rule is to limit brightly colored foliage and flowering specimens and use them against "neutral" green plant backgrounds. Exceptions are bromeliads and focusing groupings of massed flowering plants. However, if a room's furnishings are green, an interior landscaper should be wary of using more green in the plants. Some contrast, in the form of lighter tones of green or variegation, is advisable to relieve the monotony.

Size

Size, both at time of purchase and ultimately when the plant is fully grown, is the second consideration. Plants may feel close, but should never be of a size that people feel crowded by them. For this reason, palms, hanging baskets, and the like should be chosen and positioned with their subsequent growth in mind. Variation in size (trees in back progressing to ground covers or creeping plants in front) is a major design issue in commercial interior landscapes such as those in shopping malls, but seldom is there adequate room in a private home for such a large plant area. Still, a small "plant area" is an effective visual showcase for plants of varying sizes.

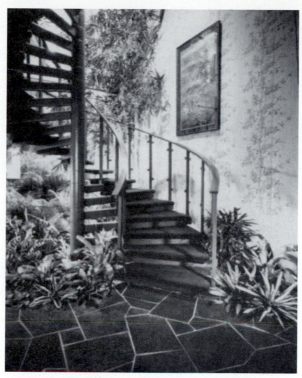

Figure 19-8 Bromeliads and other species fill the otherwise "dead" area under the stairs, and a bamboo in the corner is a specimen focal point. *(Photo courtesy of Landscape Images, a Division of British Indoor Gardens, Inc., El Toro, CA)*

Houses or apartments with sliding glass doors or floor-to-ceiling windows are particularly adaptable to this idea. In its simplest form an area of roughly 2 to 10 square yards or meters of floor space is selected. The plants are then arranged within the area, with taller specimens in the back and shorter ones in the front. Plant stands, sections of logs, and small display stands can all be used to create height differences between plants and move the viewer's eye through the display. Hanging baskets can also be incorporated and add to the variety of form and texture.

A plant area can be placed directly on the floor, and plant saucers can be used to keep moisture from the carpet. An even more attractive display can be created by laying down a sheet of heavy plastic and covering it with a single layer of bricks or bark chunks. This defines the area and attracts attention.

Texture

Texture refers primarily to leaf size, although design professionals also take into account leaf surface character, leaf thickness, leaf density, and branch and petiole arrangements in assessing this quality. The three common designations are fine, medium, and coarse. The key to an interesting interiorscape is to vary texture.

Texture also needs to coordinate with room size. A small room will be overwhelmed by coarse-textured plants because they are viewed too closely, although one such plant could be acceptable as a specimen, because its texture would be its specimen-making feature.

Decor Association

Decor association is the feeling people have regarding which indoor species are traditionally associated with which type of decor. Although many plants are decor neutral and have no association, others can be selected to intensify the effect of other elements of interior decoration. Some examples of plants for antique, country, oriental, masculine, feminine, tropical, modern, and southwestern decors are given in Table 19-3.

Obviously pots selected for both material and form will enhance the decor association of a plant, although the standard unglazed clay pot in all its forms remains a classic.

Table 19-2 Foliage Plants That Give Color in an Interiorscape[a]

Variegated white:	Alocasia 'Frydek'	Elephant's ear
	Bromeliads	*Neoregelia carolinae meyendorfii* and *N. c. tricolor*
		Nidularium innocentii lineatum
	Chinese evergreens	*Aglaonema* 'White Raja'
		A. 'Superba'
		A. 'King of Siam'
		A. 'Fransher'
	Dumbcane	*Dieffenbachia amoena* 'Tropic Snow'
		D. maculata 'Camille'
		D. 'Perfection Compacta'
		D. 'Hilo'
		D. 'Rudolph Roehrs'
		D. 'Paradise'
	Dracaenas	*Dracaena fragrans* 'Warnecki'
		D. sanderiana
		D. surcolosa 'Florida Beauty'
	English ivy	*Hedera helix* 'Glacier'
	Nerve plant	*Fittonia verschaffeltii argyroneura*
	Panda plant	*Kalanchoe tomentosa*
	Baby rubber plant	*Peperomia obtusifolia variegata*
	Pineapple	*Ananas comosus* 'Variegatus'
	Pothos	*Epipremnum aureum* 'Marble Queen'
	Spider plant	*Chlorophytum comosum variegatum*
	Snake plant	*Sansevieria trifasciata* 'Bantel's Sensation'
	Wax ivy	*Senecio macroglossus* 'Variegatum'
	Wax plant	*Hoya carnosa* cultivars
	Wandering Jew	*Tradescantia albiflora* 'Albo-vittata'
	no common name	*Xanthosoma lindenii*
	Zebra plant	*Aphelandra squarrosa*
Variegated silver:	Aluminum plant	*Pilea cadierei*
	Begonia	*Begonia* × *argenteo-guttata*
		B. 'Rex' cultivars
	Bromeliads	*Tillandsia edithae*
		T. xerographica
		T. tectorum
	Chinese evergreen	*Aglaonema* 'April in Paris'
		A. 'Mona Lisa'
		A. 'Silver King'
		A. 'Silver Queen'
		A. 'Rhapsody in Green'
	Flame violet	*Episcia cupreata*
	Peace lily	*Spathiphyllum cannifolium* 'Silver Streak'
	Rosary vine	*Ceropegia woodii*
	Strawberry begonia	*Saxifraga stolonifera*
	Watermelon begonia	*Peperomia argyreia*
Variegated yellow or pale green:	Corn plant	*Dracaena fragrans* 'Massangeana'
	Dumbcane	*Dieffenbachia* 'Triumph'
		D. 'Bali Hai'
	English ivy	*Hedera helix* 'Golddust'
	no common name	*Holalomena wallisii* 'Camouflage'
	Japanese acuba	*Acuba japonica*
	Pothos	*Scindapsus aureus* 'Golden Pothos'

(continued)

Table 19-2 Foliage Plants That Give Color in an Interiorscape[a] (*continued*)

	Vase plants (bromeliads)	*Aechmea chantinii* 'Vista'
		Billbergia pyrimidalis striata
		B. 'Fantasia'
		Cryptanthus beuckeri
		Guzmania vittata
	Wax ivy	*Senecio macroglossus* 'Variegatum'
	Bromeliads	*Billbergia vittata*
Purple-foliaged or -accented:	no common name	*Billbergia* 'Muriel Waterman'
		Neoregelia melanodonta
		Hemigraphis 'Exotica'
	Moses-in-the-Cradle	*Rhoeo spathaceae*
	Persian shield	*Strobilanthes dyeranus*
	Prayer plant	*Calathea concinna*
	Purple passion plant	*Gynura aurantiaca*
	Tahitian bridal veil	*Gibasis geniculata*
	Wandering Jew	*Zebrina pendula*
Pink-Accented:	Bromeliads (vase plants)	*Aechmea luddemanniana* 'Mend'
		Cryptanthus 'Starlight'
		C. 'Pink Starlight'
		C. 'Carnival de Rio'
	Chinese evergreen	*Aglaonema* 'Rembrandt'
	Episcia	*Episcia* 'Pink Brocade'
	Jade plant	*Crassula argentea* 'Tricolor'
	Polka dots	*Hypoestes sanguinolenta* 'Pink Splash'
	Prayer plant	*Maranta leuconeura erythroneura* and other cultivars
	Strawberry begonia	*Saxifraga stolonifera* 'Tricolor'
	Wax plant	*Hoya carnosa rubra*
		H. c. compacta regalis
Red-accented:	Abyssinian banana	*Ensete ventricosum* 'Maurelii'
	Beefsteak plant	*Iresine herbstii*
	Bromeliads	*Aechmea* 'Burgundy'
		A. maculata
		A. 'Royal Wine'
		A. 'Foster's Favorite'
		Cryptbergia 'Rubra'
		Cryptanthus zonatus
		C. 'Feuerzauber'
		Neoregelia marmorata × *spectabilis*
		N. 'Fireball'
		Nidularium innocentii 'Innocentii'
		N. billbergioides
		Vriesia 'Kitteliana'
	Coleus	*Coleus blumeii* cultivars
	Nerve plant	*Fittonia verschaffeltii*
	Prayer plant	*Calathea* 'Red Star'
	Rex begonia	*Begonia* 'Rex' cultivars
	Red-margined dracaena	*Dracaena marginata*
	Ti plant	*Cordyline terminalis*

[a]The variegation on these plants may become yellow in insufficient light.

Table 19-3 Decor-Associated Plants

Decor	Common name	Botanical name
Antique/ Victorian:	Asparagus ferns	*Asparagus* spp.
	African violet	*Saintpaulia ionantha*
	Begonias	*Begonia* spp.
	Cast iron plant	*Aspidistra elatior*
	Fern	*Nephrolepis exaltata* cvs. (Boston fern)
		Adiantum spp. (Maidenhair fern)
	Gloxinia	*Sinningia speciosa*
	Palms	*Howeia fosteriana* (Kentia palm)
		Chamaedorea elegans (Parlor palm)
		Rhapis excelsa (Lady palm)
		Chrysalidocarpus lutescens (Areca palm)
	Peperomia	*Peperomia caperata* 'Emerald Ripple'
	Rosary vine	*Ceropegia woodii*
Country:	Baby tears	*Helxine soleirolii*
	Black-eyed Susan vine	*Thunbergia alata*
	English ivy	*Hedera helix*
	Geranium	*Geranium hortorum*
	Grape ivy	*Cissus rhombifolia*
	Mock strawberry	*Duchesnea indica*
	Piggyback plant	*Tolmiea menziesii*
	Primrose	*Primula obconica* or *malacoides*
	Strawberry begonia	*Saxifraga stolonifera*
	Swedish ivy	*Plectranthus australis*
Oriental:	Balfour aralia	*Polyscias balfouriana*
	Bamboo	*Bambusa, Phyllostachys* and *Pseudosasa*
	Bamboo palm	*Chamaedorea erumpens*
	Ming aralia	*Polyscias fruticosa*
	Mother fern	*Asplenium viviparum*
	Norfolk island pine	*Araucaria heterophylla*
	Threadleaf aralia	*Dizygotheca elegantissima*
Masculine:	Corn plant	*Dracaena fragrans* 'massangeana'
	Cycads	*Zamia* and *Cycas* spp.
	Dumbcane	*Dieffenbachia* species and cultivars
	Fiddleleaf fig	*Ficus lyrata*
	Fish tail palm	*Caryota mitis*
	Green cliffbrake	*Pellaea viridis*
	Hare's foot fern	*Polypodium aureum*
	Iron cross begonia	*Begonia masoniana*
	Japanese aralia	*Fatsia japonica*
	Moses-in-the-cradle	*Rhoeo spathaceae*
	Pony tail palm	*Beaucarnea recurvata*
	Rex begonia	*Begonia* 'Rex' cultivars
	Snake plant	*Sansevieria trifasciata* cultivars
	Staghorn fern	*Platycerium bifurcatum*
Feminine:	Boston fern	*Nephrolepis exaltata* (cultivars that are highly cutleaf, such as 'Fluffy Ruffles', 'Verona', and 'Shadow Lace')
	Cyclamen	*Cyclamen persicum giganteum*
	Dauphin violet	*Streptocarpus saxorum*
	Episcia	*Episcia* cultivars
	Gardenia	*Gardenia jasminoides*
	Maidenhair ferns	*Adiantum* spp.
	Miniature rose	*Rosa chinensis hybrida*

(continued)

Table 19-3 Decor-Associated Plants (*continued*)

Decor	Common name	Botanical name
	Orchids	See list in Chapter 14
	Plumosa fern	*Asparagus plumosus*
	Tahitian bridal veil	*Gibasis geniculata*
Tropical:	Bird's nest fern	*Asplenium nidus*
	Bromeliads	See list in Chapter 14
	Croton	*Codiaeum variegatum* cultivars
	Orchids	See list in Chapter 14
	Palms	*Howiea, Chamaedorea, Chrysalidocarpus,* etc.
	Philodendron	*Philodendron* spp.
	Prayer plants	*Maranta* and *Calathea*
	Split-leaf philodendron	*Monstera deliciosa*
	Staghorn fern	*Platycerium bifurcatum*
	Umbrella plant	*Brassaia* and *Tupidanthus*
Modern:	Bromeliads	See list in Chapter 14
	Cacti	See list in Chapter 14
	Moses-in-the-cradle	*Rhoeo spathaceae*
	Peace lily	*Spathiphyllum* cultivars
	Purple passion plant	*Gynura aurantiaca*
	Pyrrosia fern	*Pyrrosia macrocarpa*
	Red-margined dracaena	*Dracaena marginata*
	Snake plant	*Sansevieria trifasciata* cultivars
Southwestern:	Cacti and succulents	See list in Chapter 14
	Christmas cherry	*Solanum pseudo-capsicum*
	Peppers	*Capsicum annuum* cultivars with very small fruits, such as 'Fiesta' and 'conoides'

THE DESIGN PROCESS

A scale drawing of the room or rooms to be landscaped is the first step in the design process. It should include furniture placement, paths normally taken from room to room, and light levels throughout the area, designated as bright, moderate, or dim (Figure 19-9). Next, the need of plants for functional uses discussed earlier should be assessed and then general areas where plants are to be placed shaded in on the drawing. In some cases there could be insufficient light available where for design purposes a plant is advisable. In such cases good-quality "silk" plants are an attractive solution, especially when mixed with natural plants so that the fact that they are artificial is less noticeable.

The final step is to select the plants for each area based on the elements of design discussed in the previous section and on the light requirements of the species. Favorite species to be included or existing plants that will need to be incorporated can be selected and positioned first, after which a shopping list of all others can be prepared, as well as a shopping list for ac-cessories—mainly pots, but sometimes plant stands, pot covers, or plant furniture.

Clear plastic tables (Figure 19-10) function as terrariums and are available through furniture stores and plastics manufacturers. They provide the high humidity needed by many ferns and orchids and can be a handsome addition to a modern decor.

Hanging baskets can also be equipped with artificial lights. They can be used for supplemental illumination in low-light areas or strictly for decorative night lighting.

Plant stands popular during the Victorian era are in style again, made from such materials as wood, wicker, and wrought iron. These pedestal-type stands can greatly enhance plants and give an illusion of height, thus substituting for more expensive large plants.

Plant accessories (pots, hangers, stands) should be chosen carefully to enhance the plant's beauty. First, the size of the container should be in proportion to the size of the plant to avoid a top-heavy or bottom-heavy look. Double potting frequently creates a bottom-heavy appearance and emphasizes the need for proportion.

High Light

Moderate Light

Low Light

Unsuitable for Plant Growth

Foot Traffic Pattern

Figure 19-9 Basic room drawing for interiorscaping the main areas of a home.

Figure 19-10 A lucite coffee table used as a terrarium. *(Photo courtesy of Earthway Products)*

Second, select containers and accessories in design-coordinated colors, patterns, and styles. Simple unadorned clay pots or wicker pot covers are two of the most suitable. They blend with almost any decor and are classic in their appeal.

A bold container can visually overpower a plant, making the pot rather than the plant the focus of attention. However, this is a general guideline and exceptions exist. For example, a room in a Mexican motif might work nicely with bright red or yellow pots. A large specimen might look outstanding in a very elaborate antique container.

Use of Night Lighting

The decorative effects of lighting are well understood by interior designers. Night lighting, for example, is very striking when used with plants. One or two inexpensive spotlights can dramatize indoor plants and provide supplemental lighting as well.

The lights can be positioned in many ways for accenting plants. Lights shining up from the floor create interesting ceiling shadows as they pass through plant foliage. Track lights aimed downward from several points on the ceiling can highlight individual plants within a collection.

ARRANGING FLOWERS

Using fresh, dried, or silk flowers in the home in place of, or to supplement, growing plants is another aspect of decorating indoors with plants. Flower arrangements add festivity to special occasions and bring the pleasing presence of living plants into a room that would otherwise be bare of plants because of an unsuitable growing environment. Creating custom floral designs is easy with simple instruction and a little practice. The focus of this chapter will be the care and arrangement of natural materials with emphasis on fresh flowers, although the principles of arrangement can be applied to silk flowers as well.

A retail florist can be an excellent small business for someone who is interested in flower arranging and who has a good business sense. Retail florists purchase, store, and arrange flowers for many occasions such as weddings, funerals, and holiday gifts.

Flower Selection

Flowers for arrangement can come from many sources: one's own garden, the wild, or by purchase.

Garden and Forced Flowers

A perennial garden that has long-stemmed flowers—the most versatile for arranging—is one flower source. Table 9-6 lists flowers that can be grown in most climates for cutting and for drying for permanent arrangements. However, landscape shrubs and even trees should not be neglected as a source of floral materials for arranging. In early spring, flowering branches can be *forced* into bloom indoors to give a bright spot of color out of season (Figure 19-11). *Forcing* is bringing a plant into bloom ahead of its normal blooming season. Table 19-4 lists some shrubs and trees that set and mature their flower buds in fall. Because the buds are already mature and ready to open, the branches can be forced in early spring ahead of their normal blooming season. The earlier the shrub or tree blooms naturally, the earlier it can be forced. However, branches that are picked before the

Figure 19-11 An arrangement of springtime and other flowers in a basket including forced forsythia branches on the right. *(Photo courtesy of Syndicate Sales, Design by Denise Lee, AIFD, PFCI)*

Table 19-4 Spring-Flowering Shrubs and Trees for Forcing

Common name	Botanical name
flowering quince	*Chaenomeles japonica*
forsythia	*Forsythia* spp.
redbud	*Cercis* spp.
dogwood	*Cornus florida*
magnolia	*Magnolia* spp.
fruit trees	
peach, plum, pear, apple, apricot, cherry, quince	
witch hazel	*Hamamelis japonica*
white forsythia	*Abeliophyllum distichum*
buttercup winterhazel	*Corylopsis pauciflora*
shadbush	*Amelanchior × grandiflora*
Japanese apricot	*Prunus muma*
wisteria	*Wisteria sinensis*
	Wisteria japonica
crabapple	*Malus* spp.
flowering almond	*Prunus triloba*
pussywillow	*Salix discolor*

Figure 19-12 Wild or selected ornamental grasses are effective in natural-type arrangements. *(Photo courtesy of Perennial Plant Association Hiliard, OH)*

buds are fully mature are incapable of continuing to mature those buds when severed from the parent plant and will simply die when brought in the house.

Examine the branches you are considering forcing to be sure that the flower buds are sufficiently mature. Mature flower buds are distinctly formed and easily distinguished from leaf buds by their larger size. In addition, they are usually round in shape, and should ideally be showing some color.

The forcing process requires very little in the way of special techniques. Simply cut the branches and place them in preservative solution (see section on Water and Preservative), then place them out of direct sun and wait for the buds to open. At normal household temperatures, this should take a week or less.

Collecting Wild Plant Materials

Wild plant materials, including many grasses found along the road (Figure 19-12)—dried teasel in fall, cattails, native shrubs, goldenrod, sunflowers, and the like—can also be used to create striking floral arrangements. Be alert to sources of your own local wild materials as you drive by natural areas. They can be used to supplement purchased flowers or used alone. All florist flowers were originally wild, and nature is still an excellent source of arrangement materials.

Purchasing Flowers

Flowers from a florist, a supermarket, or a roadside stand should be bought with care to be sure that they are fresh and will last in an arrangement. One key point to look for to assure the flowers were harvested within the last few days and are fresh is by checking to make sure they are not wilted. Check petal tips and especially the youngest buds, which will show aging first.

 A flower buyer for a floral wholesaler is the person responsible for deciding which sources to purchase from, for monitoring quality, and for making sure that a continual supply of flowers is available for purchase by retail florist shops.

Lift the flowers out of the water and check the condition of the stems and leaves that have been submerged. If the leaves are beginning to soften or are very dark green, they have begun to decompose. The flowers have been picked too long ago and will not last well.

Occasionally, roses are "dry stored" before they are sold by the grower. This means that they are picked and then stored under very cool conditions without water. Although this practice is perfectly legitimate, if stored too long the roses will not open when put in water. The buds will remain closed and the stem under the bud (peduncle) will wilt, causing the flower to die prematurely. When roses are purchased and this occurs, return them to where you bought them, as the seller will certainly want to know that he or she has been sold roses that were stored too long.

Selecting Greens

Selecting *greens* means choosing the foliage that will be used to fill out the arrangement. With some arrangement styles and species, the flowers have enough natural foliage that additional greens are not required. But if they are, they can be purchased from a florist or cut from landscape plants. Even in winter, landscape evergreens such as spruce, pine, and yew can provide foliage for arrangements.

Dried Flowers

Dried flowers are arranged using the principles and design types of fresh flowers, although the arranger is restricted somewhat by the inflexible nature of the dried stems in regard to arrangements having curved lines (see Figure 19-13). References at the end of this chapter list several books on working with dried flowers. Table 9-6 lists a number of flowers suitable for drying.

Assembling Arranging Supplies

A flower arrangement normally consists of a vase or other decorative container, water and preservative, and sometimes, for more elaborate arrangements, an arrangement aid to hold the stems in position. In addition, shears to cut the flower stems are very useful.

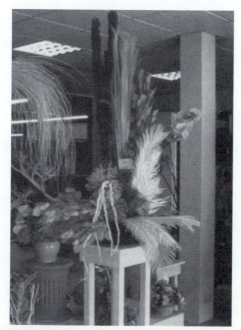

Figure 19-13 Dried floral materials used in several arrangements.

Choosing a Vase or Other Arrangement Container

Vases that are neutral in color and simple in design, made of glass, ceramic, porcelain, or metal (Figures 19-14 and 19-15), are easiest for an amateur to work with. When choosing a vase, remember that the viewer's attention should focus on the flowers instead of on the vase.

 A floral supply wholesaler orders from the manufacturer and then resells to retail florists the supplies and equipment needed to make arrangements. These include vases, wire, preservative chemicals, and baskets. The wholesaler can also wholesale cut flowers to florists.

The size of a vase can greatly affect how easy it is to create an arrangement. One useful size for an all-purpose mixed flower arrangement is about 8 inches (20 centimeters) high with a mouth about 3 inches (8 centimeters) across. This size is small enough that flowers of adequate height are easy to buy; most purchased flowers measure 18 inches (45 centimeters) to 2 feet (60 centimeters) tall.

A larger vase about 10 inches (25 centimeters) tall with a mouth 3 to 4 inches (8–10 centimeters) across and a smaller vase about 6 inches (45 centimeters) tall with a mouth about 2 to 3 inches (5 to 8 centimeters) across are also very useful. Vases with narrow mouths of less than 2 inches (5 centimeters) in diameter can be difficult to arrange flowers in without crowding.

If a flower arrangement is being made for a dinner party and will stay on the table during dinner, it is important that it not obstruct the guests' view of each other. For this reason, the container chosen should be shallow. A good rule of thumb is to select a container a maximum of 4 inches (10 centimeters) tall. When flowers are added, the arrangement will be 8 to 10 inches (20–25 centimeters) tall. A basket (Figure 19-16) can be a good choice, lined with a waterproof container that will not show, such as a plastic storage bowl.

Many everyday objects can function as containers for floral arrangements, if you use your imagination. A hollowed-out pumpkin can be used for Thanksgiving or Halloween designs, and clay pots for natural designs. Decorative glass bricks, wine glasses, bottles, and so forth all make interesting and imaginative containers.

Figure 19-14 An assortment of containers suitable for flower arrangements. *(Photo courtesy of Syndicate Sales)*

Water and Preservative

A vase should be washed thoroughly with detergent and rinsed well before use to remove all debris and dirt film. Unwashed, previously used vases harbor bacteria that will begin to multiply in water. The bacteria clog the xylem of the stem and block water flow to the blossoms, resulting in premature flower death.

The container need not be completely filled with water; it only needs to be about one-third to one-half full, depending on how much water is necessary to

Figure 19-15 An assortment of glass containers for flower arrangements. *(Photo courtesy of LBK Marketing Corporation)*

Figure 19-16 Baskets such as these are inexpensive containers for many types of floral arrangements. *(Photo courtesy of LBK Marketing Corporation)*

make the container stable. If the container is filled to the top, a greater portion of the flower stems are submerged and subject to rotting, thus becoming a breeding ground for bacteria. Because water uptake is mainly through the cut stem base, only that part needs to be submerged to supply adequate water to the blossoms.

Use of commercial flower preservative can double flower life. However, the most economical preservative and one fully satisfactory for home use is a solution of one-third tonic water and two-thirds plain tap water. Tonic water acts as a floral preservative because:

- It contains a sugar to supply the cut flower with carbohydrate for respiration.

- It is acidic, which reduces the rate of bacterial growth in the water.

- It contains a preservative intended to retard microbial growth in the beverage, but that also retards microbial growth in the water in which it is mixed.

Tools and Supplies

For cutting flower stems, a sharp kitchen knife is satisfactory. However, if you do floral arrangements using wild or shrub materials you will find investing in floral shears worthwhile (Figure 19-17). The blades are very strong and capable of cutting woody stems easily. Floral shears are available through floral supply houses listed in the phone book.

For holding flowers in position, there are several alternatives. Simple "natural"-type mixed flower arrangements do not require an aid to arrangement, as the stems hold each other in position with a little careful placement (Figure 19-26).

One solution for holding flowers in position is a pinholder or dome type *frog* (Figure 19-18). Before the arrangement is started, the frog is pushed down on a pad of *floral clay,* a sticky, waterproof material that

Figure 19-18 Two types of floral frogs: pin and dome. *(Photos courtesy of Beagle Industries)*

temporarily secures a frog to the base of the container and keeps it from tipping under the weight of the flowers it is holding. The stems are then pushed onto the spikes or between the metal wires to secure them in position.

Floral foam (Figure 19-19) allows an arranger greater creativity. The foam soaks up and holds water that can then be absorbed by the flowers. It is sold in blocks at craft and hobby stores for one-time use and can be cut to the desired size and held in position in the container with waterproof florist tape designed to secure under wet conditions. The advantage of the floral foam is that it can extend above the mouth of the container and flowers can be inserted into it at many angles, even pointing downward. This allows the creation of weeping-style arrangements. The stemmed glass containers shown in Figure 19-15 would be ideal for use with floral foam.

Figure 19-17 Flower shears of the most popular type sold for commercial use. *(Photo courtesy of Larry Beck Company and Clause Cutlery Co.)*

Figure 19-19 Floral foam can be cut to size to fit a particular container. *(Photo courtesy of Smithers-Oasis U.S.A.)*

Figure 19-20 Circular design of roses, lilies, and carnations. *(Photo courtesy of Syndicate Sales; design by Denise Lee, AIFD, PFCI)*

Principles of Flower Arrangement

Many different styles of arrangement are practiced throughout the world.

 A floral designer works in a florist shop making the arrangements that are sold to customers. Designers also make wedding bouquets, funeral casket sprays, and other specialty arrangements. Especially in the case of wedding arrangements, the designer works closely with the customer to achieve the effect that the customer is trying to create.

Traditional American Arrangements

Commercial florist arrangements made by professionals are commonly one of five or six types identified by *line,* meaning the primary visual orientation of the key flowers in the arrangement.

Circular design (Figure 19-20) is essentially a hemispherical shape designed to be viewed from all sides. Depending on its height and diameter, it would typically be used as a dinner table arrangement or small table arrangement.

Radiating design is similar to circular design. The essential difference is that, whereas a circular design has a rounded, even surface, a radiating design has a stronger, more dramatic look because spike-type flowers are used in its construction. A radiating design can be viewed from either one or all sides, depending on personal preference and the amount of flowers available.

Vertical design (Figure 19-21) sometimes has few flowers, but striking or expensive ones. The emphasis in a vertical arrangement is similar to Ikebana (next section).

Horizontal design (Figure 19-22) also adapts well for use as a centerpiece arrangement and is designed to be viewed from all sides and from the top as well. It is normally arranged in a shallow container. Creation of a horizontal design generally requires use of floral foam to position flowers horizontally and still supply them with water.

Triangular design (Figure 19-23) is, as it sounds, based on the shape of a triangle with a flat base and a pointed top. The primary lines are visually defined by the flowers that point to the tips of the triangle.

Hogarth or "S" curve design is an uncommon and more difficult design to make because it requires creating a curved line to define the "S" curve. Wild plant materials such as vines can be used for this purpose, as can flowers that have developed a natural curve in the stem during growth.

Figure 19-21 Vertical design. *(Photo courtesy of Syndicate Sales; design by Denise Lee, AIFD, PFCI)*

Figure 19-23 Triangular floral design. *(Photo courtesy of Syndicate Sales; design by Denise Lee, AIFD, PFCI)*

Figure 19-22 Horizontal design in a wire basket. *(Photo courtesy of Syndicate Sales; design by Denise Lee, AIFD, PFCI)*

The American Florists Marketing Council (AFMC), the Society of American Florists (SAF), Florists Transworld Delivery (FTD), the American Institute of Floral Design (AIFD), and other flower-marketing organizations employ floricultural marketing specialists to promote flowers to the public, develop marketing campaigns, address grower meetings regarding needs for floral crops, and coordinate advertising.

Japanese-Style Floral Arrangements

Among the most popular of floral arrangement styles is Japanese (Figure 19-24). True Japanese Ikebana style is a carefully constructed arrangement of natural materials symbolic of elements in life: heaven, earth, and man, at descending levels within the arrangement, and sometimes other symbolic representations as well. Mastery of the art of Japanese flower arranging takes many years of study and practice. However, one can create a Japanese-style arrangement not strictly conforming to the rules of Ikebana arrangement, but still visually pleasing, by following a few simple principles.

First, all types of living materials can and should be used, not just flowers. In fact, a Japanese-style arrangement may not use any purchased flowers at all; it may be composed of twisted stems, seed pods, individual leaves, grasses, reeds, and fruits attached to branches. When choosing materials, the arranger should select and incorporate unusual and wild plant materials. In winter, leafless branches, evergreen boughs, and branches with clinging berries can be used.

Second, Japanese arrangements are almost always asymmetrical, meaning they are not even in shape all around as American-style circular and radiating designs typically are. In addition, the arrangement is also often designed to be viewed from one direction only instead of from all sides.

Japanese arrangements are often made in neutral-colored glazed containers in brown, black, beige, or the like. A woven straw basket can also be used. Most interestingly, a shallow plate, provided it is adequately deep to hold some water and a frog, can be used quite effectively. The frog is generally essential in such a container to hold the floral materials in position.

European Arrangement Styles

In recent years, several floral arrangement styles originating in Europe have become popular, including the vegetative and landscape-style designs (Figure 19-25).

Figure 19-24 Japanese-style design. *(Photo courtesy of Syndicate Sales; design by Denise Lee, AIFD, PFCI)*

Figure 19-25 A modified vegetative design showing elements of a Hogarth curve. *(Photo courtesy of Syndicate Sales; design by Denise Lee, AIFD, PFCI)*

A retail salesperson in a florist shop deals with the public. He or she must be knowledgeable about the flowers and products and have good people skills and sales ability.

Vegetative Design. This design style often uses a shallow horizontal container. The container visually symbolizes the earth and serves as a *foundation,* as it is called. Moss, bark, and similar plant materials are used to create the effect of the earth and cover the foundation of the arrangement, which is normally floral foam, but can also be one or more pinholder frogs.

The design concept is that the arrangement should symbolize a small slice of nature. Flowers are grouped by species in separate areas of the arrangement as they might grow naturally, instead of being spaced throughout the arrangement as is common in American designs. The materials are displayed at the angles at which they could grow naturally, instead of being positioned evenly and formally throughout the design as, for example, in circular design. The floral materials appear, in effect, to be growing out of the ground, and although the effect is not totally natural, the concept is conveyed.

Mille Fleur Design

Translated as "1,000 flowers," the *mille fleur* style of design (Figure 19-26) has been very popular for at least 15 years. It is a "mixed bouquet" in which at least four, and preferably many more, different types of flowers are combined in one arrangement in a seemingly random fashion to give a profusion of many colors and flower shapes. The design does not have the "line" of traditional American arrangement styles. The shape is loose and free, and the arrangement is designed to look as close as possible to a bouquet of garden flowers.

Mille fleur arrangements can be done without frogs or floral foam, although a more refined and formal effect can be achieved with a foam foundation that allows more precise positioning of the individual flowers.

Contemporary Designs

The amateur arranger can experiment with waterfall, *pavé,* sheltered, phoenix, and other relatively unusual contemporary designs. In waterfall design, the primary line and focal point of the arrangement is, as the name implies, downward sweeping. In *pavé* (meaning "paved") design, flowers are arranged so that the blooms actually touch. This style was popular in Europe

Figure 19-26 Mille fleur design. *(Photo courtesy of American Florists Marketing Council)*

in the previous century and is experiencing a limited revival. In phoenix design, the classic story of the phoenix bird rising from the ashes is expressed in flowers, with strong emphasis on a vertically positioned flower or flowers in a visual nest of other flowers and greens.

Techniques for Making Arrangements

The technique of constructing an arrangement can be broken down into several steps, keeping in mind the design principles italicized and explained below.

Deciding on the Arrangement Style. Ideally, arrangement type is decided on before selecting flowers. But when flowers are a gift, this may not be possible, and the arranger will have to work with the flowers that are available. In either case, strive for visual harmony with the container and among the flowers. The shapes and colors should complement each other.

Arrangement style will also be limited by the types of vases available and by the purpose of the arrangement. If the arrangement is to be, for example, a table centerpiece, the choice of design is limited by height.

Preparing the Container. Preparing the container in-cludes cleaning and, if necessary, positioning a frog or foam block for stabilizing the flowers.

Developing the Line (Figure 19-27). The principal or "line" flowers, those that are to establish the outermost points of the design, should be put into the design first.

Strive for the proper proportion between the container and the height of the flowers. For most arrangements, the tallest flowers generally should be 1½ to 2½ times the height of the container.

Achieving "Visual Balance." Visual balance within an arrangement is achieved by placing the smaller flowers

(a)

(b)

(c)

Figure 19-27 The steps in creating a right triangle design: (a) inserting line flowers, (b) filling in with foliage and spike flowers, and (c) completing with focal flowers and fillers. Note that the visual line of balance is slightly to the left of center and not directly through the middle of the vase. *(From* Floral Design and Arrangement *(3rd ed.) by Gary L. McDaniel. Englewood Cliffs, NJ: Prentice-Hall, 1996, pp. 93–94.)*

higher up in the arrangement and the larger ones lower to produce visual stability and avoid a top-heavy look. Balance from side to side is achieved in symmetrical arrangement styles by finding an imaginary center line that extends through the visual center of the arrangement. The proper number of flowers are then positioned on each side of this imaginary line to give equal visual weight to each side.

Establishing "Unity." Unity within the design is established in traditional American arrangements by not bunching flowers of the same type together but instead by positioning them individually at varying levels throughout the entire arrangement. Using odd numbers (3, 5, 7, and so on) of each species of flower avoids a rigid, lined-up look.

"Filling In." Filling in is the final part of constructing the arrangement. At any point after the line flowers are positioned, fill in with greens, leftover buds, and the like. The goal is to hide all "mechanics" (floral foam, tape, and so on) to make it appear that the flowers are positioned without aids.

SELECTED REFERENCES FOR ADDITIONAL INFORMATION

Armitage, Allan M. *Specialty Cut Flowers: The Production of Annuals, Perennials, Bulbs and Woody Plants for Fresh and Dried Cut Flowers.* Portland, OR: Timber Press, 1933.

Elbert, George, A., and V. Elbert. *Foliage Plants for Decorating Indoors: Designs, Maintenance for Homes, Offices and Interior Gardens.* Portland, OR: Timber Press, 1989.

Fitch, Charles M. *Fresh Flowers, Identifying, Selecting and Arranging.* New York: Cross River Press Ltd./Abbeville Press, 1991.

Gaines, R. L. *Interior Plantscaping: Building Design for Interior Foliage Plants.* New York: McGraw-Hill, 1977.

Gibbs, Jamie. *Landscape It Yourself: A Landscape Architect's Guide to Planning and Designing Your Own Exterior and Interior Landscapes.* New York: Harper Collins, 1988.

Hunter, Norah. *The Art of Floral Design.* 2nd ed. Albany: Delmar Publishers, 1999.

Kramer, Jack. *A Seasonal Guide to Indoor Gardening.* Stonington, CT: Globe Pequot, 1992.

Kramer, J. *Indoor Trees.* New York: Universe, 1984.

Manaker, G. W. *Interior Plantscapes.* 3rd ed. Paramus, NJ: Prentice-Hall, 1996.

Martin, Tovah. *Houseplants for Four Exposures.* NY: Hungry Minds, 1994.

Martin, Tovah. *Indoor Gardens.* Glouster, MA: Peter Smith, 1998.

Martin, Tovah. *Taylor's Weekend Gardening Guide to Indoor Gardens: Designing and Planting the Interior Landscape.* New York: Houghton Mifflin, 1997.

Packer, Jane. *The Complete Guide to Flower Arranging.* New York: Doring Kindersley, 1999.

Sanderson, Ken. *Interior Plantscaping.* Albany: Delmar Publishers, 1998.

Snyder, Stuart D. *Environmental Interiorscapes: A Designer's Guide to Interior Plantscaping and Automated Irrigation Systems.* New York: Watson-Guptill, 1995.

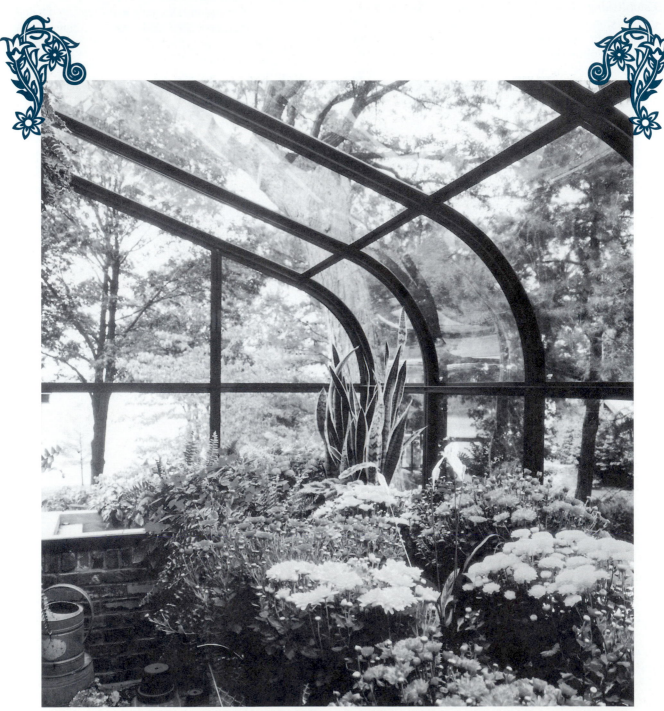

(Photo courtesy of Nexus Greenhouse Systems, Northglenn, CO)

CHAPTER 20

GREENHOUSES AND RELATED CLIMATE-CONTROLLING STRUCTURES

GREENHOUSES

A greenhouse offers almost complete environmental control. Although the sun partially heats it, gas, electricity, or oil is the main heat source, and the heat is thermostatically controlled to keep the house cool, moderate, or warm. Venting, watering, cooling, shading, and fertilizing can also be automated.

 A greenhouse flower grower is in charge of all greenhouse operations for the particular greenhouse crop being grown, usually cut flowers or potted plants such as chrysanthemums, African violets, or poinsettias.

Many options in shape, glazing, and framing materials, and heating, watering, and cooling systems are available when choosing a greenhouse. One should think of a greenhouse as a component system and select each component according to climate, budget, convenience, and plant requirements.

Greenhouse Shapes

Span-Roof Freestanding. The span-roof freestanding greenhouse type is the most common and is used frequently by commercial growers. The design is angular, with a pitched roof and straight or sloping sides. Three variations on the span-roof greenhouse are the standard, Dutch light, and curvilinear (mansard).

Standard greenhouses (Figure 20-1) have straight sides and a two-sided roof that slopes down to the sides of the house. They may be glazed to the ground or have sidewalls up to 30 inches (80 centimeters) high under the benches. Sidewalls decrease the rate of heat loss and give greater stability but shade the area under the benches.

A Dutch-light greenhouse (Figure 20-2) is similar to a standard span roof except that the sides are glazed to the ground and are constructed at a slight angle rather than being vertical. The angle of the sides causes winter sun to strike the glass at closer to 90°, the optimum angle for complete light penetration. Because of the absence of sidewalls and because of the angle of the glass, Dutch-light houses are slightly brighter than standard houses.

The mansard or curvilinear greenhouse (Figure 20-3) takes this principle a step further. Each section of glass attaches to the next to form a tunnel-like structure. Because the sun's angle changes with the seasons, each set of panes will eventually be at the optimum angle for light penetration. Mansard greenhouses are not readily available, and their light transmission is not significantly superior to that of a full-glass standard or Dutch-light house.

Lean-to. Lean-to home greenhouses (Figure 20-4) are split, span-roof greenhouses of one of the types discussed. Their appeal lies in convenience and aesthetics. Because they can be built to cover a sliding glass door, they can be warmed with the household heat. A lean-to can also extend the living area to form a garden room or hobby area.

Figure 20-1 Standard greenhouses of glass built in a multiple-house commercial range. *(Photo courtesy of Nexus Greenhouse Systems, Northglenn, CO)*

Figure 20-2 A Dutch-light greenhouse in a botanic garden. *(Photo courtesy of Nexus Greenhouse Systems, Northglenn, CO)*

Figure 20-3 A mansard-type greenhouse. Note the different angles of all the rows of glass panes.

Although most lean-to greenhouses are attached to the south side of the house for maximum light, they can be situated on another side for convenience. East- or west-facing greenhouses receive moderate light and grow almost any plant, and even north-facing greenhouses can be used for shade-loving ferns and many indoor plants.

Quonset. Quonset greenhouses are usually framed of bent pipe and glazed with one or more layers of plastic film (Figure 20-5). Although less attractive than some styles, they are inexpensive and can be recommended for home use. They are a common form of commercial greenhouse.

Figure 20-4 A lean-to greenhouse for home use. *(Photo courtesy of Burnham Greenhouse Co. and the USDA)*

Figure 20-5 A quonset-style greenhouse can be used for overwintering container-grown plants in a commercial nursery. *(Photo courtesy of Stuppy Greenhouse Manufacturing, Inc., North Kansas City, MO)*

Greenhouse Framing Materials

The frame of a greenhouse, the part that supports the glazing, can be of almost any material. A strong, weather-resistant material is essential for even a temporary structure. The relative strengths of greenhouse framing materials are discussed below.

Wood was once the main greenhouse framing material. It is a good insulator, moderately durable, but requires some maintenance such as painting. Redwood, western red cedar, and pressure-treated softwoods are all satisfactory. Creosoted wood should never be used in a greenhouse; the creosote will vaporize into the greenhouse and injure the plants.

Aluminum alloy is lightweight and moderate in price. It is durable and requires little maintenance but loses heat more rapidly than does wood.

Steel bars or pipe can be used to frame a greenhouse. The material is strong and relatively inexpensive, but unless it is galvanized, it must be painted regularly to prevent rust. Steel also transmits heat rapidly, making it a poor insulator.

Foundations and Sidewalls

To withstand wind without glass breakage or slippage, a greenhouse should have a foundation. A foundation will also keep the house level. Many prefabricated greenhouses state that no foundation is needed. However, unless the greenhouse is to be used only as a temporary spring structure, a foundation is recommended. As a minimum, cinder block can be laid on level soil. But a concrete slab is preferable, poured with bolts to anchor the walls and the water and electrical pipes in place.

If the design includes sidewalls, a concrete base to support them should go at least 1 foot (30 centimeters) below the frost line. The walls can then be built of the chosen material.

Glazing Materials and Methods

Previous to the 1970s, wooden sashes with panes of glass puttied into place were the main glazing material. Glazing used today includes a wide range of plastics and fiberglass as well as glass.

Greenhouse glazing materials differ greatly in their transmission of PAR (photosynthetically active radiation, see Chapter 16), from as low as 67 percent to as high as 95 percent. The percentage will also vary according to where in the greenhouse the reading is taken. With the exception of greenhouses used exclusively for foliage plants, the highest possible PAR transmission level is the goal, because it will allow production of high-light-requiring species and also speed up the rate of growth for all plants.

Many factors can influence the percentage of PAR that enters the house. Orientation of the greenhouse relative to the angle of the sun at any given time of year will alter light transmission cyclically throughout the year. The age of the glazing material will be a factor, as transmission generally decreases with the age of the glazing. Glazing materials that are not kept clean will have decreased transmission, and if more than one layer of glazing material (double glazing) is used to decrease heat loss, this of course will decrease light transmission. Lastly, the construction of the greenhouse has an important effect, as the more overhead structure there is, the less light will reach the plants inside (Figure 20-6).

Glass. Still preferred for looks, durability, and ease of cleaning, glass is and will remain a popular glazing material. Its light transmission ability is excellent, and although breakage can occur, it will not be a problem in a well-constructed house. The expense is moderate to high but is compensated for by its long life.

Modern methods of glazing with glass utilize several systems. In "dry glazing," the glass is inserted into a channel in wood or between metal bars. Glass applied by dry glazing is easy to install, but air leakage can be a problem.

A second glazing system places the glass on a strip of putty and holds it in place with a clip. The junction is then covered with a cap to weatherproof the seal.

The PAR of glass ranges from 77 to 92 percent depending on the quality of the glass. Although these percentages are considered excellent, some film plastics offer even higher percentages.

Film Plastic (Polyethylene). Polyethylene (Figure 20-7) is the most popular glazing material for commercial greenhouses in the Northern Hemisphere. This is because the material is less expensive than glass and because heating costs are also lower, as there is less air exchange with the outside than with a glass-glazed house. Although air leaks occur where glass panes join, plastic film runs continuously over the top of the house.

On the other hand, because greenhouses covered with plastic film are very airtight, the humidity can climb very high. This causes condensation on the plastic, which lowers PAR transmission. As the condensation increases, the water-repellent nature of the plastic causes the water drops to coalesce, and water begins to drip on the plants underneath, creating conditions favoring disease development. Surfactant chemicals are available that cause the condensed water to sheet off the plastic rather that aggregate into droplets. These can be used with an internal gutter system that channels the condensed water away. Polyethylene films are also now available with an antifog chemical that repels condensation. The chemical is not, however, completely chem-

Figure 20-6 A system of aluminum trusses supports the glass on this greenhouse. *(Photo courtesy of Nexus Greenhouse Systems, Northglenn, CO)*

ically compatible with the polyethylene and can reduce its durability.

Polyethylene films are now also formulated with an infrared (IR) component that blocks the exit of heat (infrared radiation) from the house back through the film into the outside air. The chemical, however, weakens the plastic and causes it to become cloudy if the chemical is not sandwiched between layers of plain polyethylene. This is done during the manufacturing process.

Another disadvantage of a plastic-film-covered greenhouse is that the material does not last as long as glass or rigid plastic panels. Although the film is formulated to resist degradation caused by UV light, the maximum life of a plastic film is about 4 years, and during that time exposure to sun causes the plastic to darken, decreasing PAR transmission. The initial PAR of new polyethylene film varies from 76 percent (for double-layer, infrared-absorbing film of 6-mil thickness) to 87 percent (for single-layer, ultraviolet-light-stabilized film of 6 mils). Infrared heat-absorbing polyethylene film used in a double layer can have a PAR transmission rate as low as 67 percent.

Vinyl Plastic (Polyvinyl Chloride or PVC). Vinyl plastic chloride or PVC is thicker and more weather resistant than polyethylene and for this reason was popular as a substitute for polyethylene in years past. However, because of chemical improvements that have increased the durability of polyethylene, vinyl is seldom used these days.

Fiberglass. Panels of corrugated, semiflexible fiberglass (also known as fiberglass-reinforced plastic or FRP) are still used in home greenhouses but seldom seen these days in commercial houses, with the exception of the

Figure 20-7 Double-layer polyethylene is used on this quonset greenhouse. *(Photo courtesy of Stuppy Greenhouse Manufacturing, Inc., North Kansas City, MO)*

Figure 20-8 Rigid acrylic panels are used on the end of this greenhouse. *(Photo courtesy of Stuppy Greenhouse Manufacturing, Inc., North Kansas City, MO)*

ends of film-covered houses. The advantages of fiberglass are that it is not easily broken and it diffuses light rays so that better illumination is received throughout the house. The PAR transmission rate of new fiberglass is 77 to 88 percent.

The main disadvantage of fiberglass is that weathering causes the acrylic surface of the material to erode, exposing the glass fibers. These then collect dirt, and PAR transmission is reduced. Resealing the surface with resin every year or two can extend the life of the UV-light-resistant fiberglass to 20 years.

Acrylic and Polycarbonate Panels. Thicker panels composed of acrylic and polycarbonate are rigid; thinner ones can be bent to conform to a curved-roof greenhouse (Figure 20-8). They are corrugated and double layered to decrease heat loss, with the layers held apart by ribs of the same material. Acrylic and polycarbonate panels are roughly twice as efficient at preventing heat loss as glass. The PAR transmission rate is about 80 percent, and the rate of light transmission loss is from 3 to 10 percent over the first half of the life of the panels.

Heating Systems

Self-contained forced-air heating units are the best for home greenhouses. They can be oil or gas fired or electric. Electric units do not require venting and are easy to install. They are hung from the frame at one end of the greenhouse and aimed so that the fan circulates warm air throughout the house.

Gas and oil heaters set up inside the greenhouse require venting. In several models the heater fits into the wall of the greenhouse. The combustion chamber is outside, and no venting is required.

Heating by household heat is used frequently with lean-to greenhouses. However, a fan will still be required to circulate air between the two if the greenhouse is very large.

Commercial greenhouses normally use forced-air heaters known as *unit heaters,* composed of a firebox that produces the heat, a series of thin-walled tubes over which the heated exhaust passes and transfers its heat, and a fan that draws in air through the back of the heater and out into the greenhouse (Figure 20-9).

The heater is hung from the roof of the greenhouse, and a clear polyethylene convection tube is attached to the end and runs the length of the greenhouse. It is perforated at intervals with holes 2 to 3 inches (5–8 centimeters) in diameter through which the warm air exits and mixes with the air of the greenhouse to equalize the temperature.

The constant movement of air is desirable not only to keep the temperature equal throughout the house, but also to keep leaves dry and prevent disease problems. For this reason, commercial growers frequently run the fan even when no heating is required.

Commercial greenhouse ranges with a large number of houses may use a boiler system to provide heat. As in a unit heater, the hot exhaust from the heater passes over tubes, but the tubes in turn pass the heat to water. At this point either the heated water can be pumped into pipes that give off heat into the greenhouse, or the water can be heated even further until it turns to steam, which is passed through pipes at the perimeter of the greenhouse under low pressure.

Figure 20-9 A clear polyethylene tube with holes is attached to a fan and heater unit to warm the greenhouse. *(Photo courtesy of Stuppy Greenhouse Manufacturing, Inc., North Kansas City, MO)*

Positioning of the heat pipes around the perimeter of the house assures that the heat is directed where heat is lost: through the glazing at the sides. To counteract heat loss through the roof glazing, overhead pipes are used.

Infrared radiant heating systems are a third heating system for commercial greenhouses. Most people have experience with these as patio heaters. The IR passes through the air at the speed of light and does not heat the air but warms only the objects with which it comes into contact. The heat is then radiated out into the air from the objects that intercept the radiation.

The primary advantage of IR heating is a cost saving on fuel; IR heating systems use up to 50 percent less fuel than do other kinds of heating systems. In addition, diseases associated with wet plant leaves are lessened because there is no condensation on plant leaves at night when greenhouses are normally kept cooler.

Solar heating is a form of greenhouse heating that can be used in mild-winter areas with a lot of winter sunshine. The system includes a solar collector, a heat storage area, two fans or pumps, a thermostat to control the pumps, and tubing to connect the parts. In the solar collector, water or air is passed between a transparent glazing layer and a dark back plate or system of black tubes that absorb and convert the radiant energy of the sun into heat energy. The heat is pumped either in the form of heated water or heated air to a storage tank that can be a building or pit. For forced air, a pit filled with rocks to absorb the heat is normally used. If water is used to convey the heat, that water itself is the storage material. The second pump propels the water or air into the greenhouse on an as-needed basis, generally at night.

A solar heating system is much more expensive than a conventional greenhouse heating system, and a backup conventional heating system is required for times when solar radiation is insufficient. For this reason, solar heating will not be economically feasible for most greenhouses until more efficient collectors or an inexpensive, efficient heat-storage material is devised.

Cooling Systems

Cooling of home and commercial greenhouses is accomplished mainly by venting and shading. Many inexpensive greenhouses are constructed to be ventilated only by opening the doors; however, vents on the roof or at the eave are more efficient because the warmer air will rise and flow out through them. Greenhouse models with these vents are preferable, and most commercial greenhouses are vented in this way.

Shading of a greenhouse in summer will restrict direct sunlight and decrease heating of the interior. Bamboo blinds, cheesecloth, or Saran placed over the glass will function very effectively for a home greenhouse. A coating of whitewash can be used on glass, plastic film, or fiberglass greenhouses and can be removed in fall with water and a brush.

Figure 20-10 Pads at the end of the greenhouse are part of an evaporative cooling system. *(Photo courtesy of Stuppy Greenhouse Manufacturing, Inc., North Kansas City, MO)*

In areas with very hot summers, an evaporative cooler or "fan and pad" system can decrease the temperature of the greenhouse to less than the outside air temperature (Figure 20-10). Evaporative coolers are widely used in commercial greenhouses, and small units are available from home greenhouse suppliers.

In this cooling system, a fiber pad is installed at one end of the greenhouse with water trickling through it at a slow rate. A fan at the opposite end of the house pulls outside air in through the pad. The water evaporates into that air, absorbing heat in the process and reducing the air temperature while increasing the humidity.

Additional Greenhouse Equipment

Benches. Wood benches (Figure 20-11) are simple to construct and are standard in home greenhouses. However, a wooden frame with a floor of galvanized hardware cloth has much to recommend it. Such a bench is rustproof, permits free air circulation, and allows light to pass through to plants growing underneath.

Commercial greenhouses can have much more sophisticated bench systems. The movable bench system (Figure 20-12) is designed to maximize the growing space in the greenhouse. Benches with plants such as the bedding plants shown in the photo are designed to

Figure 20-11 Wood benches suitable for a hobby greenhouse. *(Photo courtesy of Garden Space, Inc., Greenville, SC)*

Figure 20-12 A movable bench system in a commercial greenhouse. *(Photo courtesy of Conley's Manufacturing and Sales, Montclair, CA)*

move together on roll bars so that fixed spaced aisles are not necessary. When maintenance activities are needed in a particular section of the house, adjacent benches are simply rolled out of the way to provide an aisle. Using this bench system, 90 percent of the interior area of the greenhouse can be used for growing, as opposed to approximately 70 percent in a greenhouse designed with standard aisles.

Commercial greenhouses that produce cut flowers also sometimes have ground benches. These benches are normally of poured concrete and have an internal perforated pipe network with an outlet at the end. The pipe serves both to drain away excess water and as an entry point for steam pumped in to sterilize the planting medium between crops.

Photoperiodic-Control Equipment. Commercial greenhouses for crops responsive to photoperiod (see Table 16-1) must be equipped with automatic shading equipment to increase night length for photoperiodic crops and automatic lighting systems to increase day length.

In commerical greenhouses, power-operated shading devices lengthen the night in photoperiodic crops (see Chapter 3) when natural night length is too short, such as in spring, summer, and fall. Cotton cloth, polyester cloth, or black plastic is hung over frames that enclose the benches on which the plants are raised. These are normally pulled over the crops by a motorized system that pulls the material by attached cables. The motor can be turned on and off by a computerized system that controls other aspects of the greenhouse environment as well (see section on computerized greenhouse control).

Sixteen to eighteen hours per day of light are needed for photoperiodic control of short-night plants. Regular incandescent lights are generally not used in greenhouses because of poor light quality in regard to photosynthetically active radiation (PAR), and because of poor economic efficiency. Fluorescent lights can be used for this purpose because only a few foot-candles are needed to control photoperiodic responses.

Supplemental lighting can also be used when natural light is insufficient for maximum crop production (such as in winter) and supplemental lighting is determined to be economically efficient. This is normally in latitudes above 40 degrees. Seedlings in particular benefit from 12 hours per day of supplemental light.

As discussed in Chapter 16, incandescent lights are not suitable as plant lights because they do not provide the wavelengths of light that plants require for healthy growth. Fluorescent lights can be used for supplemental lighting in greenhouses, but they require a large number of closely spaced fixtures to provide adequate intensity. The sheer number of fixtures required and the fact that during the day these fixtures block natural light make them economically inefficient for commercial greenhouses.

High-pressure sodium high-intensity-discharge (HPS-type HID lights, Figure 20-13) are the preferred lighting source for commercial greenhouses. Their light emission peaks at about 600 nanometers (yellow light) and again at a little more than 800 nanometers, which is beyond the visible spectrum. Radiation in this higher peak promotes stem elongation and early flowering. HID lamps are also energy efficient and convert

Figure 20-13 A high-intensity discharge greenhouse light. *(Photo courtesy of P. L. Light Systems, Grimsby, Ontario, Canada)*

25 percent of the energy they use into light, compared to 20 percent for fluorescent lights and 7 percent for incandescent lights.

Carbon Dioxide (CO₂) Generator. A CO_2 generator is a small burner that hangs from the roof of the greenhouse. It is turned on and off either by a time clock or an electronic control system. The generator burns natural or propane gas and exhausts the CO_2 that is a product of the combustion into the greenhouse.

Carbon dioxide gas necessary for photosynthesis can be a limiting factor in plant growth in winter when there is little air exchange with the outside. Although nighttime levels of CO_2 range from 400 to 500 parts per million (ppm), over the course of the day, they can drop to as low as 150 ppm at night, at which point growth stops. Research has shown that increasing the CO_2 concentration in a greenhouse to 1,000 or even 1,500 ppm can yield increases of up to 50 percent in tomatoes, 31 percent in lettuce, 38 percent in cut carnations, and 23 percent in cucumbers. Increases in the number, quality, and size of orchid blooms have also been reported. It is important to understand that these increases in plant growth occurred only in situations in which CO_2 was the "limiting factor." Above these levels, CO_2 can become toxic.

Limiting factor refers to the environmental factor (mainly water, light, heat, nutrients, CO_2) that prevents a plant from achieving its maximum photosynthetic and growth potential. For example, in winter there may be bright sun, adequate nutrients in the soil, adequate water, and adequate CO_2, but not enough heat to allow photosynthesis to progress at maximum rate. In another case, plants in a field in summer might have adequate heat, light, CO_2, and water, but insufficient nutrients for photosynthesizing to full potential.

In the case of CO_2 enhancement, when CO_2 is no longer the limiting factor, another factor (such as warmth) may then become the limiting one. Therefore, increasing the daytime temperature of the greenhouse by 5 to 10°F (3–6°C) could boost plant growth even further in a CO_2-enhanced greenhouse. However, CO_2 enhancement and increased heat in a greenhouse in a climate where winter light is limited may have no effect. This is because the limiting factor is light, and therefore CO_2 enhancement and warmth are valueless.

Irrigation and Fertilization Systems. Commercial greenhouses normally have both irrigation and fertilization automated to some degree.

Tube watering, which is also known as the *spaghetti watering system* or *trickle irrigation system* (Figure 20-14) has been the norm for watering plants in containers since the 1960s. Water flows through a ½-inch (13-millimeters) to ¾-inch (19-millimeters) flexible black plastic tube inserted at various points with "microtubes." Water flows at low pressure through the main line and out each of the microtubes into a pot. At the end of each microtube is either a lead weight (to keep the tube from falling out of the pot) or a small stake that is pushed into the potting medium to stabilize the tube. The microtube can simply trickle water slowly into the pot or be equipped with a small spray head that disperses water over the surface (see Figure 11-31) if the pot is large.

Capillary mat watering can also be used on potted plants. The pots sit on a ¼-inch to ½-inch (6- to 13-millimeters)-thick mat of synthetic fiber or reprocessed cloth. The mat is kept constantly moist with a water and fertilizer solution. The solution moves constantly upward into the pots by capillary action.

Overhead sprinklers are used for watering bedding plants. Pipes with nozzles are suspended about 2 feet (0.6 meter) above the plants. In some cases the pipe is movable and is then called a "boom." The boom moves past all the plants at a slow speed until all are watered.

Occasionally, ebb-and-flood systems are used in irrigation. In this system the potted plants or bedding plants are grown in a watertight bench or sometimes in watertight troughs that extend down the bench. Water mixed with fertilizer solution is pumped in several times per day to a depth of about 1 inch (2.5 centimeters) and allowed to remain for 10 to 15 minutes to saturate the medium. The solution is then drained back into a storage tank.

Automatic fertilization by injection is the norm for all commercial greenhouses. *Injection* refers to mixing (by injecting) liquid concentrate fertilizer into the water supply and watering the plants with the resulting solution. This has the advantage that the plants receive nutrients continuously. A commercial injector or *proportioner,* as it

Figure 20-14 A spaghetti watering system for potted plants in a greenhouse. The lead weight at the end of the microtube keeps the tube from falling out of the pot. *(Photo courtesy of Chapin Watermatics, Watertown, NY)*

is sometimes called (Figure 20-15), is a mechanical device that pumps in concentrate at a specified ratio of up to 1:200 (1 part concentrate in 200 parts water).

A home greenhouse can use a simple proportioner with no moving mechanical parts (Figure 20-16) that operates on the Venturi principle. The Venturi principle states that a liquid flowing at a given pressure through a restricted channel (such as a hose) will flow faster if the size of the channel through which it flows is reduced. This

is why water will spray farther out of a hose if you put your finger partway over the opening; you have restricted the channel and increased the velocity. In the same way, a proportioner contains a narrowing device called a *restrictor* that increases water velocity and thus provides suction for the hose to pull up concentrate through the siphon tube. Such a proportioner injects concentrate at a much lower and less precise rate than a commercial device. The usual rate is 1:15, and it cannot be changed.

Figure 20-15 A commercial fertilizer proportioner. *(Photo courtesy of Dosmatic, Inc., Carrollton, TX)*

Computerized Controllers. A computerized system can be made to control almost any environmental aspect of greenhouse operation. The simplest controllers measure temperature and then turn on the heating or open and close the vents accordingly. A fan-and-pad cooling system can also be automatically switched on if temperatures remain too high after vents are opened.

More complex and expensive systems control shading and supplemental lighting by measuring day length and light intensity. Motors for shading systems are automatically turned on, and supplemental lights turned on or off.

Irrigation water fertility can be measured and adjusted automatically by computerized systems as well. Lastly, CO_2 can be measured and the generator turned on or off as the measurement of the level dictates.

COMMERCIAL FLORICULTURAL CROPS

Potted flower crops are one of the main greenhouse crops grown commercially. The following are the principal flower crops, including brief summaries of their growing requirements.

African Violets and Gloxinias

Propagation of African Violets. African violets (Figure 20-17) are usually propagated from leaf cuttings. Many African violet cultivars are patented, and a royalty must be paid to legally propagate them. Therefore, it may be easier to order seedlings or small plants from already licensed propagators.

Leaves used for propagation should be medium-sized and mature but not the oldest leaves on the plant.

diluted fertilizer in water

hose

siphon tube

concentrated fertilizer solution

Figure 20-16 A simple fertilizer proportioner system that works by water pressure.

(a)

(b)

Figure 20-17 (a) African violets (b) Gloxinas. *(Photos courtesy of Park Seed, Greenwood, SC).*

The leaves should have 1 inch of petiole attached. They should be stuck in rooting medium close together but not touching. A rooting temperature of 65 to 70°F (18–21°C) is best, and rooting should begin in three to four weeks. About two more months will pass before young plants develop. Because more than 1 plant can grow from a single leaf cutting, the grower must decide whether he or she wants to grow single- or multiple-crown plants. If the decision is to grow multiple-crown plants, leave the plantlets on the petiole together. Otherwise, they should be separated. Growing an African violet from leaf cutting to flowering stage requires five to seven months; a multiple-crown plant will be ready a little sooner.

If seed propagation is used, the seeds should be germinated by sprinkling them over the surface of a flat containing a sterilized medium of two-thirds sphagnum peat moss and one-third sand. The flats should be kept moist by misting (to avoid washing away the seed) and kept at 70 to 75°F (21–24°C) night temperature in the greenhouse. Germination will take two to three weeks. Growing the plant to maturity will take 10 months.

Propagation of Gloxinias. Seed is normally used to propagate gloxinias. The seed is fine like that of African violets, and can be sown and cared for in the same way. The plants will be flowering in six to seven months. They can be propagated from leaf cuttings, which will give a production time of five to seven months, but this is not the usual method.

Gloxinias can also be grown from purchased tubers, although this is much more expensive and not normally used for large-scale production. Plants started this way can be planted in flats of sphagnum peat moss or directly into the 4- or 6-inch (10- to 15-centimeter) pots in which they will be sold. The tubers should be planted deep, because the plants are large and tend to fall over in their pots. They will flower about four months after planting.

The Growing Period. The growing medium preferred for African violets and gloxinias should be rich in organic matter such as peat moss. A mixture of one part peat moss and one part soil, or a mixture of two parts peat moss, one part perlite, and one part soil is suitable. The pH should be between 6.0 and 7.0.

From the stage in which the seedlings or cuttings are large enough to handle, growing is routine. Small plants are transplanted 2 inches (5 centimeters) apart to the next flat or to 2-¼ inch (6-centimeter) pots. A 65 to 70°F (18–21°C) night temperature and a day temperature slightly higher is ideal for both African violets and gloxinias. At temperatures less than 62°F (17°C), growth will stop. At 80°F (27°C), growth will be fast but the quality of the flowers will be less.

African violet seedlings or leaf cuttings in flats should be transplanted to another flat (spaced 2 inches [5 centimeters] apart), when the leaves begin to touch. Or they can be transplanted to 2¼-inch (6-centimeter) pots. Otherwise, they will shade one another, and the growth will be slower. As leaves again begin to touch in the flats or as the leaves of potted plants reach the edge of the pot, the violets or gloxinias should be tranplanted to their final finishing pots of 3 to 4 inches (8–10 centimeters) in diameter.

The same principle of transplanting as often as necessary so that leaves of neighboring plants do not overlap should also be followed with gloxinias.

It is important that the medium be kept moist during all stages of growth. The water should be at nearly the same temperature as the greenhouse, because a temperature change of more than 10°F (5°C) can cause permanent water spotting of the leaves (See chapter 19). For this reason trickle irrigation is often used.

The Flowering Period. Both species flower when they reach a minimum maturity and have adequate light intensity to cause flower induction. The amount of light required for flower induction is not great—about 800 to 1,000 foot-candles during the brightest part of the day. Higher light intensities will cause the leaves and flowers to fade. Shade will be necessary in summer.

African violets flower continuously and can be sold at any time. However, gloxinias flower for several months and then become dormant. Therefore, the latter should be sold when their flowers begin to open.

If a gloxinia plant is not sold during flowering, the tuber should be allowed to become dormant. Gradually allow the soil to remain dry for more time between waterings. The leaves will die down. At this stage stop watering completely and lay the pots on their sides. Store the tubers at a minimum temperature of 50°F (10°C) for about six months. They can then be restarted in moist sphagnum peat moss at 70°F (21°C).

Azaleas

Propagation Propagation of azaleas (Figure 20-18) is by cuttings, but the growth period from cutting to blooming plants may take up to two years, so that one option is to purchase full-grown plants that are sold from fall through spring. These plants are usually sold *precooled,* which means that they have received the vernalization that they require to bloom.

If the azaleas are produced from cuttings, 4 to 6-inch, (10 to 15 centimeter) cuttings of semihardwood stem tips should be taken in summer. They can be rooted in a mixture of half sand and half sphagnum peat moss and transplanted three months later to small pots. In a mild winter area, the azaleas can be grown outdoors in the ground and later transplanted to "azalea pots." Azalea pots are wider than they are tall to promote retention of water in the growing medium.

Figure 20-18 Azalea *(Photo courtesy of Norfolk Florist, Inc., Norfolk, Virginia)*

The Growing Period. Greenhouse-grown azaleas should be kept at a nighttime temperature of about 60°F (16°C), and a daytime temperature of about 70°F (21°C). They are acid-loving plants, and grow best in a medium of between 4.0 and 5.0 pH. At pH levels above this, they often develop chlorosis. They can be grown in pure sphagnum peat moss. The plants should always be kept moist and not allowed to dry out between waterings.

Pinching of azaleas develops the plant to the proper shape. There will be a total of 5 to 7 inches every 6 to 12 weeks to form the plant. A "soft pinch" is done with fingers and takes out only the growing tip. A soft pinch is used when the plant has been growing rapidly and there is a lot of soft new growth. A "hard" pinch is necessary when growth is slow. Use hand pruners to remove up to half the plant, and leave only two pairs of leaves on the branch.

Chemical pinching using Off-Shoot 0 is a common commercial practice. The chemical kills the terminal bud and is the same as a soft pinch.

The Flowering Period. Flower buds form on azaleas in response to temperature. A greenhouse kept at 65°F (18°C) at least eight hours per day will cause azaleas to form flower buds in two to three months. Once the buds are set, they require a cool maturing period. A 50°F (10°C) cooler will mature the buds in 4 weeks. Some lights are needed in the cooler, or the azaleas will drop their leaves. But bright lights are not necessary.

Growth-regulating chemicals can be used to develop more uniform flower buds at a faster rate. Cycocel or B-nine is applied when the new growth after pinching is about 1 inch (2.5 centimeters) long and the plants have reached full flowering size.

Timing the azaleas' flowering to match a specific holiday like Christmas or Thanksgiving requires careful planning. After precooling, the azaleas will be fully developed with colored buds in five to seven weeks if a greenhouse is kept at 60 to 65°F (18°C) night temperature. This will depend on cultivar, of course, and the temperature can be raised or lowered to either speed up or slow down the development of the crop. If they reach the stage of maturity ahead of the holiday for which they are targeted, they can be held in cooled storage at 42 to 48°F (5–9°C) with light for 12 hours a day for up to two weeks.

Bulb Crops—Tulips, Daffodils, Hyacinths

Propagation. Bulbs (Figure 20-19a–c) are normally bought fully developed. The commercial grower does not propagate them.

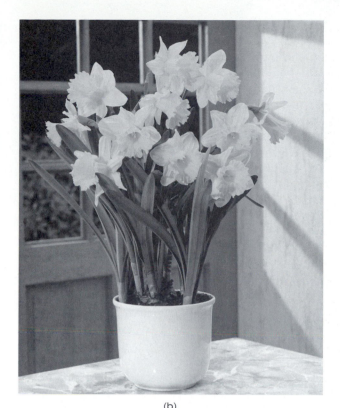

(a)

(b)

Figure 20-19 (a) Forced potted tulips (b) Forced potted daffodils (c) Forced hyacinths. *(Photos courtesy of Netherlands Flower Bulb Information Center, Brooklyn, NY)*

The Growing Period. Bulbs can be bought "pretreated." *Pretreated* means that the bulbs have been stored at cool temperatures for five to seven weeks to mature the flower buds that are inside. Pretreated bulbs are normally bought for Christmas sale. However, for spring sale the grower may pretreat the bulbs himself or herself. In either case, some cooling will be required after the bulbs are planted in order for the roots to grow.

When the bulbs arrive, they should be inspected. Throw out soft or molded ones. If the bulbs cannot be planted right away, they should be stored at the following temperatures:

Tulips	Precooled bulbs at 48°F (9°C)
	Regular bulbs at 62 to 63°F (17°C)
Daffodils	Precooled bulbs at 48 to 50°F (9–10°C)
	Regular bulbs at 55 to 60°F (13–16°C)
Hyacinths	Precooled bulbs at 48 to 55°F (9–13°C)
	Regular bulbs at 63 to 68°F (17–20°C)

The bulbs should be planted about 1 inch (2.5 centimeters) apart with the tip of the bulb just showing above

(c)

the potting medium. Three tulip bulbs can be planted in a 4-inch (10-centimeter) diameter pot, and six to seven in a 6-inch (15-centimeter) diameter pot. One to two daffodil bulbs will fit in a 4-inch (10-centimeter) diameter pot, and three to four in a 6-inch (15-centimeter) diameter pot. Hyacinths are frequently planted as individual bulbs in separate pots 4 inches (10-centimeter) in diameter, or three hyacinths to a 6-inch (15-centimeter) pot.

A growing medium of three parts sandy loam soil and and one part sphagnum peat moss, or three parts loam, one part sphagnum peat moss, and one part sand is suitable. The pots are then put in a cooler at 41 to 48°F (5–9°C), or outdoors in cold climates (but not below freezing) for "natural cooling." They will stay in the cool temperatures for five to seven weeks. No light is needed, but the medium should be kept moist.

The Flowering Period. The pots are then brought into the greenhouse and placed on benches touching each other for forcing into bloom. The greenhouse night and day temperature should be 60 to 65°F (16–18°C) for tulips and hyacinths, and 60 to 62°F (16–17°C) for daffodils. These temperatures can be raised or lowered to speed up or slow down the development of the crop.

As the plants develop, the pots must be spaced farther apart so that the leaves of neighboring pots do not touch. Otherwise, the plants will grow too tall and weak. The medium should be kept moist at all times.

The plants are ready for sale before the flowers are open. Tulips should be sold when their buds are still green but starting to show color and open. Daffodils should be sold when the buds are in "pencil" stage (that is, the flower buds are still vertical) or "gooseneck" stage (that is, with the bud horizontal). Hyacinths are sold when one-third to one-half of the buds are open.

If they cannot be sold immediately, they may be stored in a cooler at 33 to 35°F (0.5–2°C) for several weeks.

Calceolarias and Cinerarias

Propagation of Calceolarias and Cinerarias. Calceolarias and cinerarias (Figure 20-20a–b) are annual plants that are grown from seed and bloom in late winter or spring. They have similar growing requirements, are relatively easy to grow, and are not expensive to produce in 18 weeks.

The seed is very fine and should be sown in flats but not covered. Water with a mist as often as necessary to keep the medium moist. Germination happens in two to three weeks in a greenhouse at 65 to 70°F (18–21°C).

(a)

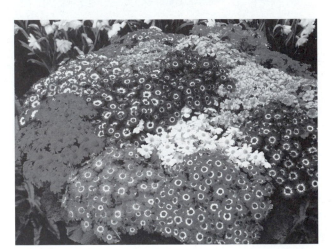

(b)

Figure 20-20 (a) Calceolarias and (b) Cinerarias are fast growing annual pot plants. *(Photos courtesy of Park Seed Co., Greenwood, SC)*

The Growing Period. When the seedings begin to be crowded, they should be transplanted to another flat 3 inches (8-centimeters) apart, or to 2¼-inch (6-centimeter) pots. About a month later, they should be planted as separate plants in the pots in which they will be sold, normally 4–10 or 6–15 inch (10–15-centimeter) diameter pots.

A suitable growing medium can be made from one part soil, one part sphagnum peat moss, and one part perlite. A mixture of one part soil, one part sand, and one part sphagnum peat moss can also be used. The plants should be kept moist at all times and fertilized by injection during watering.

Both species grow best at greenhouse temperatures of 45°F (7°C) at nighttime and slightly warmer in the daytime.

The Flowering Period. After the plants reach full size, they must be kept in a temperature of about 50°F (10°C) for a minimum of six weeks to form flower buds. Once the flower buds are formed, the temperature can be raised. However, it should *not* go above 60°F (16°C) at night and 65°F (18°F) during the day or the buds may stop developing.

Pot Chrysanthemums

Propagation of Chrysanthemums. Chrysanthemums (Figure 20-21) are propagated from stem-tip cuttings of stock plants. (If there are no stock plants, rooted or unrooted cuttings can also be ordered from floricultural suppliers.) With a weak rooting hormone, rooting will be complete in less than two weeks. If rooting is begun when the days are less than 13 hours long, supplemental night lighting for four hours per night should be given. Otherwise, the cuttings may flower. (See Flowering Period section).

The Growing Period. Transplant the cuttings to the final growing pots, planting the cuttings around the outside of the pots in the following numbers, depending on growth method:

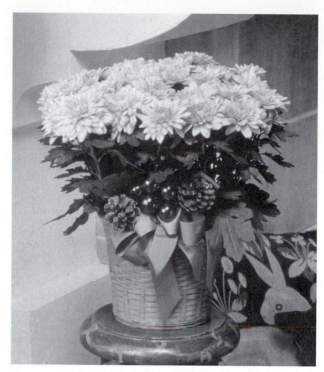

Figure 20-21 Potted chrysanthemum. *(Photo courtesy of Yoder Brothers, Inc., Barberton, OH)*

Pot Size	Number of Cuttings		
	Grown Unpinched	Pinched Once	Pinched Twice
3–4 inches 8–10 cm	1–2	1–2	1
5–6 inches 13–14 cm	5–7	4–6	3–4
7–8 inches 18–20 cm	8–10	5–7	3–5

Unpinched chrysanthemums produce small plants in the shortest amount of time. The cuttings are simply transplanted, grow vegetatively for two weeks, and then are induced to flower.

Single-pinching chrysanthemums is the most common growing method. After potting, the cuttings grow vegetatively for about two weeks and the tips are then removed. This causes branches to form, and each of these branches produces flowers.

Double-pinch plants produce the largest pot of chrysanthemums from the fewest number of cuttings but take more growing time in the greenhouse. The transplanted cuttings grow two weeks vegetatively. Then they are pinched, grow two weeks more vegetatively, are pinched again, and flowering is induced.

When the cuttings are first planted, the pots can be placed touching each other, but as the cuttings grow the plants will need to be spaced so that the leaves of neighboring plants do not touch.

The temperature of the greenhouse at night should be 62 to 65°F (17–18°C), rising up to 70°F (21°C) during the daytime.

The Flowering Peirod. Chrysanthemums form flowers when they receive at least 10 hours of uninterrupted dark each night. That is why they naturally flower in the fall, as the days are becoming shorter and the nights are becoming longer. In commercial production it may be necessary to artificially simulate short days to cause flowering. Or it may be necessary to lengthen the days with artificial lighting to prevent flowering during the vegetative growth phases.

The number of weeks of short days required to induce flowering depends on the cultivar being grown. This information is available from the producer of the cultivar. *Black clothing* is the term used to describe the technique for shortening the day length in the greenhouse. A black tent of cloth is pulled over the bench of potted chrysanthemums in late afternoon and removed the following morning a few hours after dawn to give a total of 12 hours of darkness.

The temperature inside the black tent should be monitored so that it does not rise greatly. Heat buildup under the tent can cause "heat delay," which will retard flowering. This is particularly a problem in summer, when it is best to cover the plants only at sundown and uncover them later in the morning when it is cooler.

If the days are naturally short, artificial lighting will be needed to prevent flowering. This is done by installing lights over the growing bench. They are turned on during the night by a timer and left on for four hours. This "breaks" the night into two short periods, each less than 10 hours, to prevent flowering.

About a month after a black clothing begins buds will be visible. The plants must then be *disbudded*, which means that all flower buds except the top one must be removed by hand. This causes even development of flowers and leads to a more attractive pot of chrysanthemums.

The chrysanthemums are ready for sale when the flowers are one-half to three-quarters open.

Cyclamen

Propagation. Cyclamen (Figure 20-22) is propagated from seed sold in late summer or fall to produce plants in 6-inch (15 centimeter) pots 12 to 15 months later.

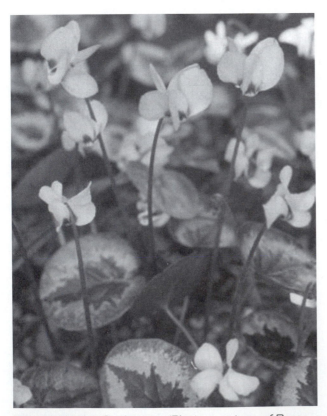

Figure 20-22 Cyclamen. *(Photo courtesy of Barry Glick, Sunshine Farm and Gardens, Renick, WV 24966 U.S.A., www.sunfarm.com)*

The seed can be sown in the medium of equal parts of sand and peat, or less sphagnum peat. Seeds should be planted 1 inch (3 centimeters) apart and 1/4 inch (1 centimeter) deep. At a greenhouse nighttime temperature of 60°F (16°C), the seeds will germinate in six to eight weeks. Seedlings or small plants can also be bought from floricultural suppliers.

An alternative method of propagation is division of the hypocotyl, the bulblike storage organ at the base of the leaves. The hypocotyl is cut into pieces with one or two leaves each, and the pieces are rooted in sand. Growing from a hypocotyl reduces production time by about three months.

The Growing Period. The best quality cyclamens are produced in cool greenhouses with a nighttime temperature of 52 to 55°F, (11–13°C), and daytime temperatures slightly higher.

Seedlings should be transplanted when they begin to touch. They can be moved to another flat spaced 3 inches (8 centimeters) apart or put into single pots 2 and a half to 3 inches (6–8 centimeters) in diameter. The seedlings should not be planted deeply, because the hypocotyl that will develop should show above the soil to lessen the chances of rotting.

Regular liquid fertilizing through injector will promote a rapid-leaf-growth period. Watering should aim at keeping the medium moist, but not overly wet.

The second transplanting will be three to four months later, when the plants are put into 4-inch (10-centimeter) pots. The transplanting to the finishing pots will be three to four months later.

The Flowering Period. Cyclamen flower in response to maturity and crowded conditions. This normally occurs about three months after the final transplanting, and after final transplanting all fertilizing should be stopped. The crop can be timed to be mature for a specific holiday by counting backwards in increments of three months, planting the last transplanting three months before anticipated sale. If flower buds develop before the final transplanting, they should be removed to produce more full and even flowering at the time of sale.

The cyclamen is at sales stage when one-third to one-half of the flower buds are open. Each cyclamen should have at least 10 buds in various stages of development, if not more.

Plants that are not sold can be allowed to go dormant by decreasing the watering quantity and frequency. The leaves will turn yellow and die, and the plants will become dormant until watering recommences. After watering recommences, flowers are produced in three months.

Figure 20-23 Easter lilies. *(Photo courtesy of Park Seed, Greenville, SC)*

Easter Lilies

Propagation. Easter lilies (Figure 20-23) are propagated by large commercial bulb producers. For retail production, the bulbs are bought in fall. The bulbs can be bought precooled, or the retail growers can precool them himself or herself. Plants produced by precooling at the greenhouse produce better quality plants in general.

Because the bulbs lack a protective dry covering such as daffodils and tulips have, they dry out rapidly. When they arrive from the producer, they should be kept at a temperature of 40 to 60°F (4–16°C) and potted within three days.

Easter lilies are normally planted one bulb per pot. They should be planted deeply enough so that 2 inches (5 centimeters) of medium covers the top of the bulb.

The Growing Period. There are two methods for growing lilies, either using precooled bulbs or controlled-temperature forcing.

Precooled bulbs are simpler to grow. They arrive from the producer in late November or in December, and 120 days before Easter they are planted and brought into the greenhouse. The greenhouse night temperature should be 60°F (16°C), and daytime temperature should be 5 to 10°F (3–5°C) higher.

Bulbs to be controlled-temperature forced will arrive from the grower in October. After the bulbs are planted, the pots are placed in refrigerated storage at 63°F (17°C) for three weeks. The temperature is then decreased to 45°F (7°C) for the remainder of the cooling period of five to six weeks.

Regardless of which growing method is used, 120 days in the greenhouse is required to bring the crop to flowering stage. About 95 days before Easter, the sprouts on the bulbs should be 2 inches (5 cen-

timeter) tall. Fentilization should be started using injector fertilizer at 200 ppm, the standard rate. Eighty-five days before Easter, the plants will be 2 to 4 inches tall (5–10 centimeters) and flower buds will be initiated.

Timing the crop to flower exactly at Easter is crucial. The bulbs may grow at different rates, and they may have to be moved to warmer or cooler spots in the greenhouse to speed up or slow down their growth.

If the plants are not at 2-inch (5-centimeter) height 85 days before Easter, the following growing modifications can be used:

> *Raise the nighttime temperature of the greenhouse to 70°F (21°C).*
>
> *Provide artificial lighting for five to eight hours per day.*
>
> *Water with warm water in place of the normal cold water.*

If the plants are taller than 2 inches (5 centimeters) at the 85 days date, these growing modifications can be used:

> *Decrease the greenhouse temperature so that the nighttime temperature is 45 to 50°F (7 to 10°C). Continue for one to two weeks.*
>
> *Apply a black cloth for several weeks.*
>
> *Spray with a growth retardant such as Phosphon® or A-Rest® when the lilies are 6 to 8 inches (15–20 centimeters) tall.*

The Flowering Period. If the crop is being timed correctly, buds will appear 50 days before Easter. At this stage of development, only raising or lowering the greenhouse temperature can be used to change the timing of the crop.

Three weeks before Easter, the largest bud should be 2 inches (10 centimeters) long and beginning to point down. When the buds are white and puffy, the plants are ready for sale. They can be stored in the cooler in the dark for up to two weeks if the flowers have not opened. But three days before sale they should be taken out of storage to reaccustom them to light.

A good-quality Easter lily has at least 3 buds, and sometimes as many as 10. When the flowers open, use your fingers to remove the anthers to prevent them from staining the white lily with orange pollen.

Poinsettias

The selling season for poinsettias (Figure 20-24) is from immediately after Thanksgiving until Christmas Eve.

Figure 20-24 Poinsettia. *(Photo courtesy of USDA ARS Photo Unit, Beltsville, MD)*

Propagation. Poinsettias are propagated from tip cuttings. Normally, the commercial producer buys the cuttings from a wholesale producer in the unrooted, calloused, or prerooted stage. If the commercial grower wants to root his or her own cuttings, the cuttings must be taken by September 15 in order to produce unpinched flowering plants by Christmas. They can be rooted either directly in the pots in which they will be sold or in a rooting bench. Rooting requires three weeks, and hormone should be used.

Vegetative Growth. Poinsettias can be grown by either the no-pinch method or the single-pinch method, similar to chrysanthemums. The no-pinch method will give one flower per cutting. The single-pinch growing method produces two to three flowers per cutting, but more time in the greenhouse is required to produce the crop. Furthermore, the cuttings must be taken in August and pinched by September 15.

The number of cuttings per pot size is as follows:

Pot Size	No-Pinch Method	Single-Pinch Method
4-inch (10-centimeter)	1	1
6-inch (15-centimeter)	3–4	3

Warm temperatures of 72°F (22°C) at nighttime, and daytime temperatures of up to 80°F (27°C) are necessary during rooting. After rooting, slowly decrease the nighttime temperature to 64 to 66°F (18–19°C), with a daytime range of 75 to 80°F (24–27°C). Fertilizer should be applied regularly in the irrigation water when the callous has formed on the cuttings.

Flowering Period. Like Chrysanthemums, poinsettias flower when the night length is long and the day length is short. Normally, the schedule for blackclothing to lengthen the night and induce flowering on time is given by the producer of the cuttings.

Growing Plants in a Greenhouse

Ornamentals in General

Growing ornamental plants in a greenhouse is much like growing them in the house. They require the same essentials of water, fertilizer, and potting medium and the same maintenance. However, in the greenhouse they receive brighter light and humidity conditions, and consequently they will grow faster and may be healthier. But a greenhouse should not be expected to guarantee success with plants. The person who has successfully grown plants indoors will probably be even more successful in a greenhouse, but one who has had a number of problems with plants indoors will have the same problems in the greenhouse. The greenhouse will not substitute for knowledge or experience.

Greenhouse environments can be divided into two types by temperature, although some plants can be grown in both temperature ranges. A cool greenhouse maintains a night temperature above freezing and generally in the range of 40 to 55°F (4–13°C). Daytime highs can reach up to 60°F (16°C). In a tropical or warm greenhouse the minimum nighttime temperature is 55°F (13°C), with daytime around 65 to 70°F (18–21°C). The drop to a lower nighttime temperature is a standard practice to promote healthy growth by slowing respiration during the night hours.

Because of the improved growing conditions that a greenhouse provides, many beautiful flowering plants from semitropical climates can be grown, most of which will not be familiar to temperate-climate gardeners. As with any unknown plant, it is worthwhile to research the specific growing requirements of a species.

Vegetables and Fruits

For some people, a greenhouse is seen as the means for growing fresh produce out of season. A number of vegetables and fruits can be successfully grown in a greenhouse, although their growth rate and taste may not equal what they would be outdoors because of the reduced light intensity and shorter days. Vegetables that can be most successfully greenhouse grown are listed here, along with a brief discussion of their growing requirements.

Beans. Snap beans should be grown in a warm greenhouse and can be seeded in February or March in ground beds for a May crop. Pole bean cultivars grown on a trellis work well, but bush types can be used if space is limited.

Beets. Beets can be grown for both their tops and their roots. They are a cool-greenhouse crop and can be grown throughout winter.

Carrots. Short carrot cultivars such as 'Oxheart' or 'Short 'n Sweet' can be grown in ground beds from a February sowing and harvested in early spring. Carrots for fall and winter use are more easily obtained by mulching in the remaining carrots in the vegetable garden and pulling them as needed. See Chapter 6 for details.

Cucumbers. Cucumbers should be grown on trellises and can be harvested through most of the winter if managed correctly. Seed sown in November, January, and March should yield a nearly continuous supply of fruits.

Cultivars of cucumbers that do not require pollination to set fruit (called *parthenocarpic*) are best for greenhouse culture. Growing temperature should be at least 60°F (16°C) and preferably 70°F (21°C) for best production.

Herbs. A number of leafy herbs can be grown successfully year-round in either a cool or a warm greenhouse, including mint, parsley, rosemary, chive, thyme, basil, and sage.

Lettuce. Either 'Bibb' or leaf lettuce will thrive under cool conditions. A continual winter harvest can be expected from November from seed sown about once per month.

Onions. Green onions can be grown easily from seed in a cool greenhouse. Two to three sowings starting in September will provide onions through spring.

Radishes. Radishes are quick-maturing vegetables that will be ready one month after the seed is sown and can be grown throughout winter in a cool house.

Tomatoes. Tomatoes can be propagated from seed or from cuttings taken from plants in the garden and grown in large pots or a ground bed (Figure 20-25). A warm greenhouse (60°F [16°C] or higher) is best. As the plants grow taller, they should be staked upright. Hand pollination is necessary, and lightly tapping the flower clusters during the bloom period will assure good fruit set. Choosing tomato cultivars bred for greenhouse growing will help ensure success.

Figure 20-25 Growing tomatoes hydroponically in a greenhouse ground bed. *(Photo courtesy of Vocational Education Productions, California Polytechnic State University, San Luis Obispo, CA)*

Fresh tomatoes and cucumbers are two vegetable crops frequently grown commercially in greenhouses. A grower specializes in the cultivation of out-of-season tomatoes, which can be very lucrative.

Hydroponic Greenhouse Gardening

Hydroponic gardening involves supplying water and minerals to plants by intermittently flooding their roots with a liquid nutrient solution. No soil or potting medium is used. Instead, the plant roots are supported in beds of gravel, sand, or similar unreactive material, which is filled and drained several times per day with the water-based solution.

Hydroponic gardening (Figure 20-25) is considerably more complicated than conventional culture. However, the practice achieved considerable popularity recently as a hobby and even as a commercial enterprise. Many books deal with the science of hydroponics and offer growing instructions including recipes for nutrient solutions.

Shade Houses

Shade houses (Figure 20-26) provide a growing area of filtered shade for plants that are not tolerant of full summer sunlight. Fuchsias, impatiens, and many begonias and ferns grow well only in a shade house during the summer months but can tolerate full sun in the greenhouse during the winter months where the sun is dim. A shade house can also be used for hardening transplants and for summering indoor plants outdoors. If large enough, it can be used as an outdoor living area (see Chapter 10).

Shade houses are usually constructed with a wooden frame and covered with lath (narrow, thin strips of wood) or Saran, which is a woven shade cloth. The covering may or may not extend to include the sides of the structure. Lath houses shaded by pieces of lath or 1×2-inch (25×50-millimeter) wood strips form an attractive, durable structure. The spacing of the lath will depend on the intensity of the sunlight in the area and the amount of shade required by the plants. North-to-south orientation of roof lath is suggested because it casts alternate sun and shade on plant leaves through the day. However, orientation is not crucial, and many gardeners prefer to orient lath east-west or in decorative crisscross patterns.

Figure 20-26 A lath-covered shade house for a home garden. *(Photo courtesy of George Taylor)*

Plastic Saran or commercial shadecloth can also be used to cover a shade house. It is less durable but provides an even shade throughout the structure. The depth of shade will be determined by the weave of Saran selected.

INDOOR PLANT GROWTH UNITS

Greenhouse Windows

A greenhouse window is one way to expand the indoor growing area for indoor plants or to raise garden transplants. Although the heat cannot be controlled as in a regular greenhouse, the light is better and the humidity often higher than in an ordinary window.

When buying or constructing a greenhouse window, a major criterion is ventilation. A side or top vent should be included in the design for excess heat dissipation in summer. Beyond this requirement, the considerations of price and aesthetic appeal become important. A large window is not usually correspondingly higher in price than a small one, but it will provide valuable extra space for additional plants at the same time that it becomes an attractive addition to a room.

Environment-Controlling Growth Units

Environment-controlling growth units are indoor units (Figure 20-27) that monitor and regulate heat, humidity, and light continuously for optimal plant growth.

They are useful in homes with insufficient natural light and for species that are difficult to grow out of a greenhouse. They are also attractive for displaying and maintaining collections of exotic or miniature plants. However, for growing ordinary indoor plants their expense is usually prohibitive.

Walk-in growth units are used for commercial tissue-culture laboratories and university research facilities. One room can contain many thousands of plants.

 A laboratory technician works under a university professor of horticulture. He or she is responsible for the day-to-day monitoring of research projects, data taking, and organizing new research projects as directed by the professor.

Terrariums

The primary plant-growing benefit of terrariums is their ability to hold in humidity, the lack of which causes many indoor plant problems. But the glass that holds in the moisture also excludes some light, making terrariums suitable mainly for medium- and low-light plants.

Almost any container, aquarium, fish bowl, water carboy, or bottle can be used as a terrarium. Those with narrow openings are more difficult to plant, but the main consideration is that the container be made of clear or only slightly tinted glass. Colored glass containers will exclude light, and plants growing in them will eventually die.

Figure 20-27 An environment-controlled plant growth chamber for indoor use. *(Photo courtesy of General Aluminum Products)*

In making a terrarium, first provide a layer of drainage material at least 1 inch (3 centimeters) deep in the bottom of the container. This layer will be a reservoir for excess water if the terrarium is accidentally overwatered. Most prepackaged or homemade potting media will be satisfactory and should be moistened, if necessary, to slight dampness.

Assembly of an open-mouth terrarium can be done quickly using your hands, but narrow-necked terrariums require special tools and more time. The drainage layer and potting medium should be poured in through a funnel to avoid dirtying the inside of the glass. Each layer is then smoothed out with a tool fashioned by taping a spoon to a piece of wooden molding, a dowel, or a straightened wire coat hanger.

Plants are inserted by soaking the media from the roots, folding the leaves upward, and sliding them through the opening. They are then positioned and planted with the spoon tool and a plain dowel. Unrooted cuttings can also be used and will root easily in the humid atmosphere.

The plants chosen for the terrarium should be medium- or low-light-requiring and slow growing. Table 20-1 lists several dozen recommended species. Cacti and succulents are not suitable for terrarium planting. The stagnant air and restricted light provide poor growing conditions for these and other xeriphytic plants.

When assembly of the terrarium is complete, the new plants should be watered in to assure contact of the medium with the roots. Trickle several teaspoons

Table 20-1 Plants for Small and Moderate-Sized Terrariums

Botanical name	Common name	Remarks
Adiantum spp.	Maidenhair fern	Many species suited to terrarium culture
Araucaria heterophylla	Norfolk island pine	Seedlings make attractive miniature trees
Begonia spp.	Begonia	Miniature species are hard to find but work well, including Begonia foliosa
Chamaedorea elegans	Parlor palm	Several seedlings are usually sold in a small pot and can be separated and used for small trees
Dracaena surculosa	Gold-dust dracaena	Formerly known as Dracaena godseffiana; shrubby plant with deep green leaves irregularly spotted yellow
Episcia dianthiflora	Lace-flower vine	Fuzzy-leaved relative of the African violet bearing white tubular flowers
Ficus pumila	Creeping fig	Small-leaved creeping vine; the varieties 'Minima' and 'Quercifolia' are smaller foliaged but rarer
Fittonia verschaffelti 'Minata'	Miniature nerve plant	Slow growing and low-light-requiring; leaves are about 1″ (2.5 cm) long
Maranta leuconeura varieties	Prayer plant	Low-light-requiring; young plants are ideal but may outgrow the terrarium in 6 months
Pellaea rotundifolia	Button fern	Deep, blue-green pea-sized leaves on a flat-growing plant
Peperomia caperata	Emerald ripple peperomia	One of the smaller and more attractive peperomias; can be propagated by a single leaf pushed into the soil
Peperomia rubella		Small-leaved peperomia with red stems and leaf undersides; grows up to 6″ (15 cm) tall
Podocarpus macrophyllus	Southern yew	Slow growing and useful as a small tree; deep green ribbon-type foliage
Pteris spp.	Table fern	Most species are excellent; some are plain green, others ruffled or lined in silver
Saxifraga stolonifera	Strawberry begonia	Small plants will root readily in the terrarium; produces interesting runners
Schlumbergera spp. cultivars	Christmas cactus	Normally sold as Zygocactus; Christmas, Thanksgiving, and Easter cacti; all prefer moist soil and will thrive in a terrarium
Selaginella spp.	Irish moss	Many species are excellent, either creeping or forming small mounds
Sinningia pusilla	Slipper plant	Tiny flowering plant only 3″ across; flowers produced sporadically in pastel shades
Soleirolia soleirolii	Baby tears	Commonly sold as Helxine soleirolli. Has tiny ⅟₁₆″ (2-mm) leaves and a matlike growth habit; becomes tall and stringy in insufficient light.

(10–15 milliliters) of water onto each plant by pouring it down the dowel. Rinse any media off the glass with a trickle of water or by wiping with a wet paper towel attached with a rubber band to a dowel.

Terrarium maintenance involves pruning and occasional watering. Fertilizer is not used because it encourages rapid growth, and the plants will outgrow the container. Even without fertilizer, pruning may be needed to control vigorous species and to encourage branching. If the container has a wide mouth, pruning can be done with a scissors and the prunings taken out and discarded. In a narrow-neck terrarium a razor blade taped to a dowel can be used, and a pickup tool will remove the trimmings and any dead leaves or flowers.

How frequently a terrarium will require watering depends on the size of the opening. In a sealed terrarium no water is lost to the outside air, and watering should never be required. A narrow-neck terrarium will have minimal water loss through the opening and will require infrequent watering, on an average of every two to six months. Wide-mouth terrariums, however, have less restricted air exchange and may require watering every one to four weeks.

Overwatering causes the death of terrarium plants in many instances, so that caution is advisable. In wide-mouth terrariums the medium can be checked with the fingers and should always feel slightly damp. In narrow-neck containers water should be applied only when the top of the medium begins to appear dry and never as long as there is condensation on the glass or water in the pebble reservoir. Watering by teaspoonfuls (5 milliliters) is suggested to lessen the danger of overwatering.

Garden Rooms

Garden rooms combine the features of a greenhouse and a living room, creating an area useful for both growing plants and dining or relaxing. Solariums, Florida rooms, atriums, sunrooms, and conservatories are all variations of garden rooms.

Unlike a greenhouse, a garden room is normally not completely glazed. A completely glass room becomes too hot during the day and cools off too rapidly at night to be comfortable for people. Instead, it is partially glazed, perhaps with large windows and several skylights. Ventilation and electricity are also present, along with furniture and, usually, a waterproof floor. Plumbing and drainage facilities are optional.

Location of the garden room should be governed by convenience rather than orientation. A large garden room designed for family use will be most conveniently situated near kitchen, living, or dining areas. A more intimate room could be placed off a master bedroom or

bath. If several locations seem agreeable, orientation can be considered. An east-facing garden room is ideal in many ways. It provides morning sunshine yet remains relatively cool and shady the remainder of the day. A south-facing room might be ideal for cold or low-light climates because it receives direct sunlight all day. However, a north room would be better suited to warm areas of the country. A west-facing room is brightest from midday until sunset and would be a cheery retreat late in the afternoon.

The cost of a garden room is approximately equal to that of a regular room on a cost per square foot (square meter) basis. Because most garden rooms are additions, the services of an architect and building permits will probably be required. Consult, if possible, an architect with experience in designing garden rooms, and bring pictures from magazines or books to help explain the project and what you have in mind.

SELECTED REFERENCES FOR ADDITIONAL INFORMATION

Biondo, Ronald J., and Dianne A. Noland. *Floriculture: From Greenhouse Production to Floral Design*. Danville, IL: Interstate Publishers, 2000.

Boodley, James W. *The Commercial Greenhouse*. 2nd ed. Albany, NY: Delmar, 1997.

Food and Agriculture (FAO) Organization of the United Nations. *Soilless Culture for Horticultural Crop Production*. Lanham, MD: Bernan Associates, 1990.

Hodgson, Larry. *Greenhouses*. Des Moines, IA: Meredith Books, 1991.

Martin, Tovah. *Greenhouses and Garden Rooms*. NY: Brooklyn Botanic Garden, 1989.

Nelson, Paul V. *Greenhouse Operation and Management*. 6th ed. Upper Saddle River, NJ: Prentice Hall, 2000.

Nicholls, Richard. *Beginning Hydroponics*. Philadelphia: Running Press, 1990.

Phillips, Roger. *The Random House Book of Indoor and Greenhouse Plants*. Volumes 1 and 2. New York: Random House, 1997.

Resh, Howard M. *Hydroponic Food Production: A Definitive Guide to Soilless Culture*. 6th ed. Santa Barbara: Woodbridge Press, 2001.

Smith, Shane. *Greenhouse Gardener's Companion*. Golden, CO: Fulcrum Publishing, 1992.

Sunset Book Editors. *Greenhouse Gardening*. Menlo Park, CA: Sunset Publications, 1976.

Time-Life Books. *Greenhouse Gardening*. Richmond: Time-Life Books, 1989.

Wilmott, P. K. *Scientific Greenhouse Gardening*. New York: State Mutual Book and Periodical Service, 1982.

World Wide Web Addresses

http://www.ag.auburn.edu/landscape/Gloxinia/html This is a greenhouse production schedule for gloxinias.

http://aggie-horticulture.tamu.edu/floriculture/hort428/Lecture/index.html Starrman, Teri. "Horticulture 428. Greenhouse Management" Aggie Horticulture. This is a set of course notes for a university greenhouse management course)

http://www.attra.org/attra-pub/ghwebRl.html Diver, Steve and Sherry Sue Sparks. "Greenhouse & Hydroponic Vegetable Production Resources on the Internet" in Appropriate Technology Transfer for Rural areas Website. 2000.

http://www.greenhousegarden.com/ Smith, Shane. Shane Smith's Greenhouse Gardener's Companion. (Oct 30, 2001)

http://www.hobbygreenhouse.org/. Hobby Greenhouse Association. Hobby Greenhouse Association Website. Oct 15, 1998.

http://www.suite101.com/userfiles/79/gardening.html Anonymous. "Greenhouse Growing" Suite 101 Home and Garden.

APPENDIX OF PLANT SOCIETIES

The addresses of the societies change frequently and can be obtained by writing to the American Horticultural Society, 7931 East Boulevard Drive, Alexandria, VA 22308. The telephone number is (800) 777-7931. The list is courtesy of the Society.

United States

African Violet Society of America, Inc.
American Bamboo Society
American Begonia Society
American Bonsai Society, Inc.
American Boxwood Society
American Calochortus Society
American Camellia Society
American Conifer Society
American Daffodil Society, Inc.
American Dahlia Society
American Dianthus Society
American Fern Society
American Fuchsia Society
American Gloxinia and Gesneriad Society
American Gourd Society
American Hemerocallis Society
American Herb Association
American Hibiscus Society
American Hosta Society
American Hydrangea Society
American Iris Society
American Ivy Society
American Orchid Society
American Penstemon Society
American Peony Society

American Plant Life Society
American Pomological Society
American Primrose Society
American Rhododendron Society
American Rose Society
American Willow Growers Network
Aril Society International
Australian Plant Society of California
Azalea Society of America, Inc.
Bonsai Clubs International
Bromeliad Society International
Bromeliad Study Group of Northern California
Cactus and Succulent Society of America
California Macadamia Society
California Rare Fruit Growers
Cryptanthus Society
Cycad Society
Cymbidium Society of America
Dwarf Iris Society of America
Epiphyllum Society of America
Eucalyptus Improvement Association
Garden Conservancy
Gardenia Society of America
Gesneriad Hybridizers Association
Gesneriad Society International
Hardy Fern Foundation
Hardy Plant Society
Hardy Plant Society of Oregon
Heliconia Society International
Herb Society of America
Heritage Rose Foundation
Heritage Rose Group

Historic Iris Preservation Society
Holly Society of America, Inc.
Home Orchard Society
Hoya Society International, Inc.
Indoor Citrus and Rare Fruit Society
International Aroid Society, Inc.
International Bulb Society
International Camellia Society
International Carnivorous Plant Society
International Clematis Society
International Geranium Society
International Golden Fossil Tree Society, Inc.
International Lilac Society, Inc.
International Oak Society
International Oleander Society
International Ornamental Crabapple Society
International Palm Society
International Ribes Association
International Violet Association
International Water Lily Society
Los Angeles International Fern Society
Magnolia Society, Inc.
National Chrysanthemum Society, Inc.
National Fuchsia Society
National Hot Pepper Association
North American Fruit Explorers
North American Gladiolus Council
North American Heather Society
North American Lily Society, Inc.
North American Native Orchid Alliance
North American Rock Garden Society
North American Sea Plant Society
Northeast Heather Society
Northern Nut Growers Association
Northwest Fuchsia Society
Northwest Perennial Alliance
Pacific Northwest Lily Society
Pacific Northwest Palm and Exotic Plant Society
Pacific Orchid Society
Passiflora Society International
Peperomia and Exotic Plant Society
Peperomia Society International
Perennial Plant Association
Pioneer Plant Society

Plumeria Society of America
Rare Conifer Foundation
Rare Fruit Council International
Rare Pit & Plant Council
Reblooming Iris Society
Rhododendron Species Foundation
Rose Hybridizers Association
Saintpaulia International
Society for Japanese Irises
Society for Louisiana Irises
Society for Pacific Coast Native Iris
Society for Siberian Irises
Southern Fruit Fellowship
Species Iris Group of North America
Spuria Iris Society
Texas Rose Rustlers
Tropical Flowering Tree Society
Tubers
Walnut Council
World Pumpkin Confederation

Canada

African Violet Society of Canada
Alpine Garden Club of British Columbia
British Columbia Lily Society
Canadian Begonia Society
Canadian Chrysanthemum and Dahlia Society
Canadian Geranium and Pelargonium Society
Canadian Gladiolus Society
Canadian Herb Society
Canadian Iris Society
Canadian Orchid Congress
Canadian Peony Society
Canadian Prairie Lily Society
Canadian Rose Society
Dahlia Society of Nova Scotia
Newfoundland Alpine and Rock Garden Club
Ontario Herbalists Association
Ontario Regional Lily Society
Ontario Rock Garden Society
Rhododendron Society of Canada
Royal Saintpaulia Club
Saskatchewan Perennial Society
Toronto Cactus and Succulent Club

GLOSSARY

abscisic acid a growth-inhibiting chemical that induces abscission and dormancy and inhibits seed germination, among other effects.

abscission the dropping of leaves, fruits, flowers, or other plant parts.

absorption the uptake of any material by a plant or seed, such as water by the roots or pesticide through the leaves.

accent plant a landscape plant used to call attention to a particular feature of an area, for example, an attractive shrub planted near the front door of a house.

acid-forming a material that reacts to acidify the soil or acidifies the soil as it decomposes; for example, peat moss, agricultural sulfur, and oak leaves.

acid-loving describes plants that grow best in acidic potting media or soil.

actinomycete a common type of soil fungi adapted to be particularly effective at breaking down hard-to-decompose soil materials such as cellulose (wood).

active growth a phase of the plant life cycle characterized by rapid stem lengthening and leaf production and sometimes flowering or fruiting.

active ingredient the ingredient in a pesticide that causes its effectiveness. Other ingredients are called *inert ingredients*, such as carriers or surfactants.

adhesion the attraction of liquids for solids, for example, the attraction of water for soil.

adventitious bud a bud produced on a part of a plant where a bud is not usually found, such as on a leaf vein or root.

adventitious root a root produced by other than primary root tissues, such as from the stem.

aeration a method of increasing the water and air penetration into compacted soil by the removal of slices or cores of soil throughout the compacted area.

aggregation the clumping together of soil particles into large groups.

air layering the rooting of a branch or top of a plant while it is still attached to the parent.

algae a single-cell plant that often grows in colonies.

allelopathy the injury of one plant by another through excretion of toxic substances by the roots; the black walnut has allelopathic properties.

amendment a material added to soil that renders it more suitable for plant growth; for example, sand, lime, compost, and peat moss.

anatomy the internal structure of an organism.

angiosperm the botanical designation of a plant that produces its seeds in ovaries.

annual a plant that grows to maturity, flowers, produces seed, and dies during one season; often refers to frost-tender flower species.

anther the top, pollen-producing portion of the stamen.

anthocyanin a red pigment found in many plants, including the fruits of cherries and strawberries.

antidesiccant a spray used on transplanted plants to slow transpiration and prevent burning of the foliage.

aphid a small, soft-bodied insect that feeds on young leaves and buds and sucks plant juices, causing curling and distortion of new growth.

arboretum (arboreta) a place where trees are grown—either alone or with other plants—for scientific and educational purposes.

arboriculture the branch of horticulture that deals with the care and maintenance of ornamental trees.

artificial medium a growing medium that does not contain soil.

ash content the amount of mineral matter present in a soil amendment.

asymbiotic relationship a relationship between organisms that is beneficial to both and harmful to neither, for example, the relationship between nitrogen-fixing bacteria in soil and the roots they live on.

atomizer an instrument that breaks a liquid into small droplets through the use of forced air, for example, a pesticide applicator.

auxin a plant hormone involved in dormancy, abscission, rooting, tuber formation, and other activities.

axil the angle formed where a petiole joins a stem.

axillary bud a bud located in a leaf axil.

azalea pot a clay pot wider than its height. Azalea pots are designed to facilitate moisture retention in the medium.

bactericide a chemical used to control bacterial disease.

bacterium (bacteria) a single-celled, nonchlorophyll-containing plant; some bacterial species assist in the decomposition of organic matter and the fixation of nitrogen, whereas others cause such plant diseases as fireblight.

balanced fertilizer a fertilizer containing equal proportions of nitrogen, phosphorus, and potassium.

balled and burlapped describes plants (usually trees) grown in the ground, dug, and sold with the root ball wrapped in burlap.

bare root refers to deciduous plants grown in the ground, dug, and sold while dormant with no soil around the roots.

bark ridge the raised ring of bark where a branch emerges from its parent branch.

bear to "bear" is to produce fruit.

bearing age the age at which a tree produces its first fruit.

bed a soil area prepared for planting.

bench a raised platform in a greenhouse for growing plants.

berm a mound or ridge of soil created to add interest or screening in a landscape.

biennial a plant that produces vegetative growth during one season, flowers during the second season, and then dies; examples are carrots and parsley.

binomial system the system of naming plants developed by Carolus Linnaeus whereby plant names are given in two parts, the genus and species.

bioherbicide an herbicide derived from a plant.

biological control the use of another organism to kill a plant pest, for example, the use of *Bacillus thuringiensis* disease to kill caterpillars.

biorational pesticide a pesticide chemically similar to a naturally occurring pesticide.

blackclothing covering photoperiodic greenhouse crops with a light-impermeable black cloth to induce flowering.

black frost a frost that kills plant tissue without forming ice on the outside of the plant, so named because of the resulting blackening of injured plants.

blanch to exclude light from a vegetable to make it more tender, less strong tasting, or more attractive.

blight a bacterial or fungal disease that kills portions of a plant and may eventually destroy the entire plant.

block garden a form of vegetable gardening that involves planting vegetables in groups or blocks rather than rows.

blossom end rot a disease of tomato fruits caused by a deficiency of calcium.

blower a landscape maintenance tool that uses a motor to blow air onto paved areas in order to clear them of debris.

boom sprayer a sprayer that sprays downward onto a crop from a horizontal pipe with nozzles spaced at regular intervals.

borer any insect that burrows into the stems, roots, leaves, or fruits of a plant.

botanical insecticide an insecticide derived from a plant; examples are pyrethrum, rotenone, and ryania; *see also* organic insecticide.

bottom watering a technique for watering potted plants in which water is poured into the drainage saucer and absorbed through the drainage hole by capillary action.

bract a colored leaf that functions as a flower petal; a flower often arises in the axil of a bract.

brambles the collective name given to the fruits in the genus *Rubus*; examples are raspberries, blackberries, dewberries, and loganberries.

branch collar   the ridge of bark tissue at the base of a branch.

branch spreader   a piece of wood inserted between the trunk and branch of a fruit tree to widen the crotch and create a stronger branch.

breaking dormancy   causing a plant or seed to change from a dormant to an active growth phase.

broadcast   to distribute (for example, fertilizer) by throwing handfuls or using a mechanical spreader to scatter it on the ground.

broadleaf weed   a dicot weed.

bud   a rudimentary shoot or flower; may be either vegetative (containing leaves, stem, and so forth), flowering (producing one or more flowers), or mixed (containing both vegetative and reproductive parts).

budding   implanting a bud of one plant onto another plant so that the resulting shoot bears the characteristics of the bud parent on the roots of the host.

buffer strip   *see* filter strip.

bulb   an underground storage organ composed of a shortened stem covered with modified leaves called *scales*.

burn   browning of a plant because of lack of water, exposure to high temperatures, or exposure to a toxic chemical or excess fertilizer.

cactus   a member of the plant family Cactaceae.

calibrate   to measure the amount of pesticide distributed by a sprayer or other type of application equipment in order to assure that the correct amount of pesticide is applied.

candle   the succulent new growth of a needle evergreen such as spruce or pine.

cane   the long shoots/weeping stems such as those produced by grapes, brambles, or roses.

cane cutting   *see* stem section cutting.

canker   a disease particularly affecting woody plants that destroys the cambium and vascular tissue of a generally localized area of a plant.

capillary action   the sideways and upward movement of water through a growing medium owing to the adhesion of water molecules to the medium.

capillary water   the water that moves in soil or another growing medium owing to the forces of adhesion and cohesion.

carbohydrates   the sugar and starch required for the growth and metabolism of a plant.

carotene   an orange plant pigment that gives color to oranges and persimmons, among other plants.

casting   soil that has passed through an earthworm.

cation-exchange capacity (CEC)   the ability of a growing medium or soil to attract and hold nutrients.

Cell-pack®   a plastic container of transplants in which each transplant has its own root area.

cellulose   a chemical compound that forms a main part of plant cell walls and provides strength to stems and other parts.

central leader   the main, upright shoot of a tree.

central leader form   a training method for fruit trees that leaves the central leader intact (as opposed to modified central leader and vase forms); also refers to the natural form of many ornamental trees.

chelated nutrient   an essential element for plant growth chemically formulated to be nonreactive in soil or potting media.

chemotheraputant   a fungicide that kills existing fungus and prevents future spore germination.

chill factor   decrease in temperature of a wet object when water evaporates from its surface.

chilling requirement   the necessity of some plant species for a period of cold weather in order to mature and grow normally.

chiseling   physical breaking up of a hard subsoil layer.

chitin   a primary chemical compound in the hard shells of certain insects like beetles.

chlorosis   persistent yellow to white coloration of foliage usually due to herbicide injury or nutrient deficiency.

clay   one of the three primary soil particles; clay particles are platelike, extremely small, and retain nutrients well.

clod   a hard lump of clay soil that is difficult to break.

clone   any plant vegetatively propagated from another is a clone of the parent plant.

coarse aggregate   a material such as sand or perlite, for example, made up of relatively large particles and added to a growing medium to improve its drainage.

CODIT   an acronym for Compartmentalization of Decay in Trees.

cohesion   the attraction of identical molecules for each other, for example, the cohesion of water molecules.

cold hardiness or tolerance   the minimum temperature at which a plant can survive.

cole crop   vegetables of the genus *Brassica*, including broccoli, kale, cauliflower, and mustard.

colloids   an interspersed system of finely divided soil particles.

compaction   excessive packing or settling of a potting medium or soil.

companion planting   the planting of insect-repellent plants next to susceptible ones to repel pests.

compatible   describes two species that will cross-pollinate or that can be grafted together.

compensation point   the point at which photosynthesis produces sufficient carbohydrate for respiration but does not produce enough carbohydrate for growth.

complete fertilizer   a fertilizer containing nitrogen, phosphorus, and potassium; *see also* **balanced fertilizer**.

compost   an organic soil amendment made from decomposed garden refuse.

composting   the piling of garden refuse so that it will decompose and can be used as a soil amendment.

compound leaf   a leaf composed of more than one leaflet; *see also* **simple leaf**.

container gardening   the growing of fruits, vegetables, and ornamentals outdoors in containers; also called *pot gardening*.

container-grown   refers to landscape plants grown to selling size in containers such as cans or plastic pots.

contour planting   planting in curving rows to follow the elevation of the land and decrease erosion by water.

cool-season vegetable   a species of vegetable that grows best at daytime temperatures ranging from 50 to 65°F (10–18°C) and is tolerant of frost; includes peas, lettuce, broccoli, and turnips.

cork cambium   the layer of cells that produces bark on woody plants.

corm   a thickened underground storage organ (stem) found on such plants as gladiolus and crocus.

Cornell Mixes   the Cornell University potting media formulas.

corner planting   a group of trees, shrubs, or ground-covers used to visually soften the corner of a house.

cottage garden   landscape style characterized by informal massing of flowers, herbs, and sometimes vegetables.

cotyledon   part of an embryo of a seed made up of one or more simple leaves; serves as a storage organ in some cases, for example, in peas and beans.

cover crop   a grass or similar fast-growing plant sown over a vegetable garden in fall and turned under in spring to supply organic matter to the soil.

critical photoperiod   the daily dark/light ratio necessary for various plant responses to occur, particularly flowering and germination.

crop rotation   planting a different species in an area of the garden each year to prevent buildup of diseases or insects associated with particular crops.

crown   the part of a plant that enters the soil.

crown reduction   a pruning technique that reduces the height of a tree.

crusting   a soil surface condition typified by a smooth, glazed appearance and common on silty soils or after rain.

cucurbit   a group of warm-season vegetables having similar cultural requirements; included are melons, cucumbers, squash, and pumpkins.

cultivar   a combination of the words *cultivated variety*, designating a manmade or man-maintained plant variety.

cultural control   alterations in plant care (watering, fertilizing, and so forth) for the purpose of controlling disease, insect, or weed problems.

curative   a fungicide that kills existing fungus and prevents future spore germination.

cut flower   a garden grown specifically for cutting and displaying in a vase.

cuticle   the thin, waxy layer on the exterior of stems and leaves that retards water loss.

cutting   a vegetative plant part that regenerates roots and forms new plants.

cyanobacteria   a special type of bacteria with chlorophyll.

cytokinin   a naturally occurring plant hormone involved in stem elongation, bud formation, breaking dormancy, and other activities.

damping off   a disease in which a soilborne fungus attacks the roots and stems of seedlings, constricting the stems and killing the plants.

day-neutral plant   a species that flowers irrespective of the daily dark/light ratio.

deciduous   a plant that drops its leaves in winter; the opposite of evergreen.

decline   the general decrease in vigor of a previously healthy plant.

decomposition   the breakdown or rotting of plants and animals.

defoliation   the dropping of most or all leaves from a plant.

desuckering   removal of the suckers.

determinate   describes tomato varieties that remain short and bushy; *see also* **indeterminate**.

dethatching   removal of thatch from a turf area by a machine or by hard raking.

dicot   a subdivision of angiosperms characterized by the presence of two cotyledons, net-veined leaves, and the like; most cultivated plants are included in this group; *see also* **monocot.**

dioecious   a species in which the male and female organs of reproduction are produced on separate plants; examples are holly, dates, and bittersweet.

direct seeding   sowing seed in the area where the plants will grow to maturity.

disbudding   removal of auxilliary buds on chrysanthemums to leave a single large flower.

disease   a plant malady caused by a microorganism or an environmental factor; the latter diseases are called *physiological* diseases.

disease suppressive compost   a decayed material that when added to soil inhibits the growth of root-rot organisms.

diurnal wilting   daytime wilting of plants due to weather conditions causing transpiration rate to exceed rate of water uptake through the roots.

division   a method of vegetative propagation involving separation of a plant into two or more pieces, each containing a portion of the roots and crown.

dormancy   a phase of a plant life cycle characterized by slowed or stopped growth; leaf drop and death of the aboveground portions may also occur.

dormant oil   a lightweight oil sprayed on dormant plants for control of insects such as scale; the oil coats the insects to exclude air, suffocating them and their eggs.

double glazing   the application of two layers of glazing to a greenhouse to reduce air leakage and to conserve heat.

double leader   two competing central leaders in a tree.

drainage   the rate at which water flows through a soil or potting medium; also, the percentage of the applied water the material retains. Poor drainage is recognized by slow water flow and high retention. Good drainage is recognized by fast water flow and low retention.

drain tile   interlocking clay tiles buried in an area with poor drainage to carry excess water away from plants.

drill hole fertilization method   a technique for fertilizing deep-rooted plants such as trees by pounding holes into the ground throughout the root zone of the tree, filling them with granular fertilizer, and covering with peat moss or topsoil.

drip irrigation   *see* **trickle irrigation.**

drought tolerance   the ability of a plant to live without much water in the soil or to absorb the small quantity of capillary water remaining in a dry soil.

dry rot (of cacti)   the stem of cacti characterized by discolored and sunken stem areas.

dwarf   the miniature varieties of a species; *see also* **semidwarf** and **standard.**

dwarfing rootstock   a rootstock that makes the scion grafted onto it grow less vigorously.

edger   a landscape maintenance tool that uses a motor attached to a blade to cut sod away from the edge of a paved area.

eelworm   a nematode.

electrostatic sprayer   a sprayer that electrically charges the spray droplets to improve coverage of plants and minimize drift.

elongation   lengthening.

embryo   the immature plant in a seed.

encapsulated fertilizer   a slow-release fertilizer composed of beads coated with a plasticlike resin that is permeable to water, which enters through the coating, dissolves the fertilizer, and carries it back out through the coating.

endosperm   the part of a seed where carbohydrate is stored and converted into energy to enable germination.

enzyme   a chemical that regulates an internal chemical reaction in an organism.

epiphyte   a plant that grows with its roots anchored to a tree, notably bromeliads and orchids.

espalier   the stylized training of a tree or shrub to grow flat against a wall or trellis, sometimes with a precise branch pattern.

essential elements   the 16 micro- and macronutrients required for plant growth.

established   describes a seedling or other plant whose roots have penetrated well into the soil.

ethylene   a gas produced by senescing plants or plant parts and ripening fruits; used commercially in pineapple production and to induce flowering of bromeliads (see Chapter 3).

etiolation   growth of a plant by elongation of the stem internodes but without production of normally spaced leaves.

evaporative coding   *see* **fan and pad cooling.**

everbearing   describes strawberry and raspberry varieties that bear two crops annually: one in early summer and one in fall.

explant   disinfected meristem of a shoot used as the primary propagation material in tissue culture.

exposed aggregate   paving surface design created by imbedding gravel in wet concrete and then hosing down the partially set slab to leave the gravel raised from the surface but imbedded in it.

fabric weed barrier   a woven material applied to the soil to suppress weed growth.

fall vegetable gardening   the practice of seeding cool-season vegetables in summer to mature and be harvested in fall.

family   a unit of plant classification grouping related genera.

fan and pad cooling   a technique for cooling a greenhouse whereby a fiber pad is installed at the end of the greenhouse with water trickling through at a slow rate; a fan at the other end pulls outside air through it into the house; water then evaporates into the air as it passes through the pad, absorbing heat in the process and reducing the air temperature.

fast-release fertilizer   a granular, powdered, or liquid fertilizer containing nitrogen only in nitrate form—the form immediately usable by plants.

feeder root   the roots of a plant whose main function is to absorb nutrients and water as opposed to anchoring the plant.

fertility   the ability of a soil to supply good drainage conditions and sufficient nutrients for healthy plant growth. Also the ability of a plant to reproduce sexually.

fertilizer analysis   the proportions of essential plant elements contained in a fertilizer which are listed on the container.

fertilizer burn   an injury to plant roots owing to an excess concentration of fertilizer; wilting and browning of leaf margins are the symptoms, resulting from water deficiency.

fertilizer salt   a dissolvable chemical compound that includes one or more of the essential elements for plant growth.

fibrous (netlike) roots   a fine, branched root system such as that found in many grasses.

field capacity   the amount of water that the soil can hold without draining.

filament   the stalk portion of the stamen that supports the anther.

filler material   a material such as sawdust added to a fertilizer to increase its volume inexpensively.

filter strip   a strip of vegetation between a potential pollutant source area and a surface water body that receives runoff. The filter strip traps the water and pollutant runoff.

flat   a shallow plastic, wood, or metal tray used to grow seedlings, carry plants, and the like.

fleshy roots   roots thickened with stored carbohydrates or water.

food web   an interdependent population of organisms in the soil that make up an interrelated ecosystem.

floral foam   solid polymer foam cut in blocks and used for anchoring flower stems in arrangements.

floriculture   the science and practice of flower growing and arrangement.

flower border   a long, narrow flower bed.

foliar feeding   a method of supplying nutrients to plants by spraying them with fertilizer that is absorbed through the leaves.

foot-candle   a measure of light intensity; superseded by the metric lux.

force   to cause a plant to bloom out of season by altering the environmental conditions, especially temperature.

fork   a narrow angle between two branches.

form   the overall shape of a plant, such as vase-shaped, pyramidal, or mounded.

formal flower bed   a symmetrically designed bed for annual and perennial flowers.

foundation planting   shrubs or groundcover planted against the base of a house.

freestanding bed   a flower, vegetable, or shrub planting designed to be accessible from all sides.

friability   the aggregation within a soil that causes it to crumble readily when damp.

frond   a fern leaf.

frost-hardy   describes plants that are able to withstand frost.

frost pocket   a low-lying area more susceptible to frost than the surrounding vicinity.

frost-tender   describes plants that are killed at temperatures below freezing.

fruit   a swollen ovary.

fruit load   the amount of fruits that a plant is bearing; a "heavy load" indicates a large quantity of fruits.

fruit set   the development of fruits after flowering; few fruits develop in "poor fruit set," as opposed to "good fruit set."

fungicide   a chemical used to control a fungus-caused disease.

fungus (fungi)   a single-celled or multicelled plant lacking chlorophyll that may be saprophytic (living on dead organic matter) or parasitic (existing on living plants and other organisms). Saprophytic fungi are useful in the garden because they decompose compost and release the nutrients in organic matter, but parasitic fungi are the cause of many plant diseases.

fungus gnat   a minute, winged insect whose worm-like larvae are found occasionally in unsterilized potting media and damage plant roots through feeding.

gall   an abnormal swelling on a plant part; leaf, stem, and root galls are typical examples.

gel polymer   hydrating gel granules that can be a potting medium component to hold extra water.

genes   the smallest units of heredity-carrying material found in a cell. Genes are combined to make chromosomes.

genetic dwarf   a fruit tree that is a dwarf owing to its genetic makeup rather than to grafting on a dwarfing rootstock.

genetic engineering   the incorporation of genes of one species of plant into an unrelated one.

genus (genera)   a group of related species; one or more genera are grouped in a family.

germination   the sprouting of a seed, spore, or pollen grain.

germination rate   the percentage of seeds that will germinate under favorable conditions.

gibberellin   a naturally occurring plant hormone involved in increasing flower, fruit, and leaf size, and elongation; also influences vernalization and other activities.

girdle   to restrict the function of the xylem or phloem of a dicot plant. Girdling by cutting through a portion of phloem is used to encourage flowering. Girdling can also be caused by rodents eating tree bark in winter.

glazing   the clear or translucent material used to cover a greenhouse or related plant-growing structure.

girdling root   a root, usually of a tree, that becomes embedded in the trunk of the tree at ground level and will eventually kill part or all of the tree.

grading   changing the slope or level of the soil in an area.

grafting   implanting a shoot or bud from one plant onto another.

granular fertilizer   a fertilizer in a pelletized form.

greenhouse effect   heating the air in a translucent or transparent structure in the sun. As light energy passes through the glazing, it is transformed into infrared rays that are trapped by the glazing.

green manure crop   a grass or similar fast-growing plant sown over a future garden area and turned under before it reaches maturity to provide organic matter in the soil.

greens   foliage used in floral arrangements.

groundcover   low-growing plants that form matlike growth over an area.

ground bed   a planter constructed on the floor of a greenhouse for growing tall vegetables or flowers.

growing degree-days   the number of days above its established baseline temperature that a plant requires to mature to harvest stage.

growing medium (media)   a material used for rooting and growing plants; it may contain sand, peat moss, soil, or any other number of ingredients.

growth habit   the overall shape of a plant, for example, weeping or pyramidal.

grub   the soil-inhabiting larva of many beetles.

gymnosperm   a botanical group of evergreen plants with generally needlelike foliage and usually bearing cones; includes pines, juniper, and spruce.

harden off   to gradually acclimatize a greenhouse- or house-grown plant to outdoor growing conditions.

hardpan   a layer of compacted soil that slows or stops movement of water.

hardwood cutting   a cutting made from the mature growth of a woody plant, usually taken in late fall or early spring.

harvest interval   the number of days after treatment with a pesticide that a gardener must wait before harvesting his or her crop.

head   the leafy portion of a tree.

heading back   a pruning technique to control size and shape of shrubs by cutting back shoots to the parent branch to encourage new growth.

heel in   to temporarily cover the roots of a plant with soil to prevent their drying until the plant can be properly planted.

heat days   the number of days per year that the temperature exceeds 86°F (30°C) as designated on the AHS Heat Zone Map.

heat delay   delay in flowering of a greenhouse floricultural crop such as chrysanthemums due to excessively high temperatures.

heavy soil   a soil composed predominantly of clay particles.

herbaceous   not producing woody growth.

herbaceous cutting   a cutting of a nonwoody growth.

herbicide   a chemical used to control weeds; a "weed killer."

hill   a group of seeds planted close together; hill planting is common for melons and squash.

hoar frost   a frost that leaves ice crystals on soil and plants owing to condensation of humidity at a decreased temperature.

holdfast   a suction-cup-like organ on some climbing vines that anchors them to walls and other surfaces.

home horticulture   the growing of ornamental and edible plants in and around the home.

hose-end sprayer   a hose attachment for applying pesticide or fertilizer; the sprayer contains a concentrated chemical that is diluted with water through a siphon mechanism as it is applied.

hot cap   a paper or plastic dome set over warm-season vegetables in early spring to protect them against frost and to increase the daytime growing temperature.

humic acid   an acid formed by the decomposition of organic matter.

humidity   the amount of water vapor in the air.

hydroponics   a method of growing fruits, vegetables, and flowers in which the plants are planted in beds of sand, gravel, or a similar material and the roots are supplied with nutrients in a water-based solution.

IGR   *see* insect growth regulator.

ikebana   Japanese style of flower arrangement.

implant   *see* trunk implant.

incompatible   describes two plants that will not graft together or that form a weak and unsatisfactory graft union; also refers to two species that will not cross-pollinate.

indeterminate   describes tomato varieties that grow into long vines; *see also* determinate.

inoculant   a bacteria or fungus deliberately added to a soil or growing medium to combat harmful organisms.

insect growth regulator (IGR)   a chemical that when sprayed on insects, disrupts their natural life cycle and kills them, thus functioning as an insecticide.

integrated pest management (IPM)   a system of insect and disease control stressing thorough understanding of the life cycle of the pest as well as cultural and biological controls; chemical control is used only when cultural methods fail.

interface   the division between two types of growing media of different quantities or sizes of pore spaces.

internode   the portion of plant stem between two successive nodes.

interplanting   sowing seeds of a quick-maturing vegetable with seeds of a slower one or between transplants; the fast-maturing vegetable is then harvested before it begins to crowd the main crop.

interstock   a section of stem inserted between the stock and scion in a graft union; the interstock may contribute a characteristic to the tree to make an otherwise incompatible stock and scion graftable.

invasive roots   a vigorous root system that can clog sewer lines and wells.

irrigation   watering.

juvenility   the first phase of plant development in which the plant is strictly vegetative.

landscape cloth   a fabric-like material laid on the soil to suppress weed growth.

lankiness   unattractive lengthening of stem internodes, usually owing to insufficient light.

larva (larvae)   the immature stage of an insect; larvae are frequently wormlike, whereas the adult may be winged.

latent heat of the earth   the relatively uniform, above-freezing temperature of the soil; a pit greenhouse makes use of this latent heat by having its walls below ground.

lateral   a side branch arising from a vertical shoot.

lath house   a garden structure covered with thin strips of wood and used for growing shade-loving plants.

lava rock   a coarse aggregate made from ground volcanic rock.

lawn substitute   a groundcover plant tolerant of foot traffic and planted in lieu of grass in a lawn area.

layering   a method of propagating shrubs or vines with a trailing growth habit or flexible branches; a shoot is bent to the ground, held in place with a wire loop or stone, and covered with soil; after it generates roots, it is then severed from the parent plant.

leaching   a succession of waterings used to dissolve and flush out excess fertilizer or other chemicals from a soil or potting medium.

leaf apex   the tip of a leaf.

leaf base   the section where the leaf blade joins the petiole.

leaf blade   the flattened portion of a leaf that is responsible for the majority of photosynthesis.

leaf bud cutting   a cutting that includes a short section of stem with a leaf attached; leaf bud cuttings are taken after the stem tip is used.

leaf cutting   a cutting made from a leaf and its attached petiole.

leaflet   one of the expanded, leaflike parts of a compound leaf.

leaf margin   the edge of a leaf blade.

leaf miner   a wormlike insect larva that tunnels between the upper and lower surfaces of a leaf, leaving a scribble pattern.

leaf mold   a soil amendment or potting medium component composed of partially decayed leaves.

leaf vegetable   a vegetable raised for its edible leaves, such as lettuce, chard, or Chinese cabbage.

lean-to greenhouse   a greenhouse constructed with one wall in common with a house or other building.

legume   a member of the Leguminosae (Fabaceae) (bean or pea) family.

LD$_{50}$ an indication of pesticide toxicity established by experimentation on laboratory animals; abbreviation for "lethal dose that will kill 50% of a population" and measured in mg/kg of body weight. The higher the LD$_{50}$, the lower the toxicity of the pesticide.

light duration   the total number of hours of light a plant receives per day.

light intensity   the brightness of light.

light soil   a soil composed predominately of sand.

limbing up   the removal of the lower branches of a tree.

lime   any of several calcium-containing materials used to raise soil pH.

limiting factor   the environmental factor that inhibits a plant from growing at its optimum rate.

lion tail pruning   a pruning technique that removes large amounts of leaves and branches from a tree leaving foliage only at the tips of the branches. This is not a recommended pruning technique.

loam   a soil having large proportions of both sand and clay particles.

lobe   a section of a leaf that protrudes from the main part. Figure 2-8(b) is a lobed leaf with three lobes.

long-day plant   a species that flowers only with a short daily dark period.

low-maintenance landscape   a landscape that is planned for minimal upkeep.

lux   a metric measure of light intensity, equal to the illumination on a surface 1 meter from a standard candle (1 lumen/square meter).

macrofauna   large animals that live in soil, such as gophers and mice.

macronutrient   one of the essential elements needed in relatively large amounts for plant growth: nitrogen, phosphorus, calcium, magnesium, and sulfur.

macropore space   a large pore space found between large particles or aggregations of soil particles; *see also* **micropore space.**

maleic hydrazide   a growth-inhibiting chemical occasionally used to slow the growth of lawns and hedges.

maturity   the phase of plant development characterized by flowering and fruit production.

mealybug   a destructive plant pest about 1/8–1/4 inch (3–6 millimeters) long and covered at maturity with white waxy fibers that form a protective covering.

medium (media)   a material used for rooting or supporting the roots of plants. Media are subclassified by function as rooting media, potting media, seeding media, and so on.

Mendelian genetics   the basic laws of inheritance of genetic traits, developed in the 1600's by Gregor Mendel.

meristem   an area on a plant of rapidly growing and dividing cells.

meso fauna   medium-sized animals that live in soil, such as worms and grubs.

microclimate   a small area with a climate differing from that of the surrounding area because of such factors as exposure, elevation, and sheltering structures.

microfauna   microscopic animals that live in soil, such as nematodes and mites.

microflora   microscopic bacteria, fungi, algae and related organisms that inhabit soil.

micronutrient   one of the essential elements needed in minute quantities for plant growth; these include iron, copper, zinc, boron, molybdenum, chlorine, and cobalt.

microorganism   an organism that can only be seen through a microscope, for example, fungi, bacteria, and mycoplasmas.

micropore space   a small pore space found between minute particles such as clay; *see also* **macropore space.**

mildew   a foliar fungus disease characterized by a white powdery layer on plant leaves.

mille fleur   flower arrangement style consisting of a mixed bouquet of many species of flowers.

mineral soil   soil made up principally of weathered rock particles such as sand or clay; opposite of an organic soil.

mite   an eight-legged, near-microscopic plant pest that damages foliage by sucking cell contents.

miticide   a chemical used to control mites.

mode of action   the way in which a pesticide kills the target organism, for example as a stomach poison, nervous system disruptor, etc.

modified central leader form   a training method for fruit trees in which the central leader is removed back to the nearest scaffold branch to restrict the height of the tree and facilitate fruit harvesting.

molting   the shedding of an insect's external skeleton during maturation.

monocarp   a plant that grows vegetatively for more than one year but only flowers once and then dies.

monocot   a subdivision of angiosperms that includes plants with a single cotyledon in the embryo and mostly straplike leaves with parallel venation; *see also* **dicot**.

mulch   any material used to cover the soil for purposes such as moisture conservation or weed suppression.

muscadine grape   *Vitis rotundifolia,* an American grape species native to the South.

mutation   a spontaneous change in the genetic makeup of a plant, for example, variegated shoots on an otherwise green plant.

mycoplasma   a microscopic organism intermediate between a virus and a bacterium that causes several plant diseases.

mycorhizzae   soil fungi that live in conjunction with plant roots to the benefit of both roots and fungi.

necrotic   diseased and dying.

needle feeding   a method of fertilizing and watering deep-rooted plants such as trees. It consists of a container of fertilizer connected to a pipe with a pointed tip. A garden hose is in turn connected to the pipe. The pointed tip is pushed into the ground around the tree, and when the water is turned on, dissolved fertilizer is forced into the ground around the roots.

neem   a botanical insecticide derived from the tropical neem tree.

nematicide   a chemical used to control nematodes.

nematode   a microscopic or near-microscopic wormlike plant pest that damages roots and occasionally burrows in foliage.

nitrate   a nitrogen-containing compound and the form in which most plants absorb nitrogen.

node   the point of leaf attachment to a stem.

nonselective herbicide   an herbicide that injures or kills plants regardless of species.

nutrient   an essential element for plant growth.

nutrient deficiency   the lack in a plant of a sufficient quantity of one or more of the essential elements.

offset   a young plant produced at the base of a parent plant; in the case of cacti and a few other plants such as pineapple, the offset may develop on top of the parent plant.

olericulture   the science and practice of vegetable growing.

open-center form   *see* **vase form**.

organic   defined in chemistry as a chemical that contains carbon atoms; also, a fertilizer or pesticide derived from a naturally occurring material.

organic matter   the decomposing bodies of dead plants and animals.

organic soil   soil composed mainly of decayed plant and animal remains. Opposite of mineral soil.

ornamental   a plant grown for its beauty.

ornamental horticulture   the branch of horticulture that deals with the cultivation of plants for their aesthetic value. This designation includes landscape horticulture, floriculture, indoor plant growing, and sometimes turfgrass growing.

osmosis   movement of water across a membrane.

ovary   the bulbous base of the pistil that contains egg cells; the ovary enlarges to become the fruit.

overwatering   applying an excess volume of water at too-frequent intervals for healthy root growth.

packing   *see* **compaction**.

parasite   a plant that obtains its nourishment by removing it from another living plant, for example mistletoe. Parasites are harmful to their host plants.

parent stock   the plant(s) from which material is obtained for propagation.

parthenocarpy   the growth of a fruit without fertilization of the included egg; also, the continued development of a fruit after the abortion of the embryo.

pathogen   a fungus, bacterium, virus, or mycoplasma that causes a disease.

patio tree   a small ornamental tree planted near a patio or deck for shade and beauty.

pavé   flower arrangement style in which flowers are tightly arranged so that the blooms nearly touch.

peat moss   any of several bog plants, particularly sphagnum moss, that are harvested, dried, and used as components in growing media or for soil amending.

peat pot   a pot made of compressed peat moss and used for growing seedlings, which are then planted in the garden in their container.

pelleted seed   a vegetable seed coated with a dissolving material that makes it larger and easier to sow.

perennial   a plant that lives for an indefinite number of years; often refers to herbaceous flowering perennials.

perfect flower   a flower with both male and female parts.

perlite   a course aggregate made from expanded volcanic rock.

permanent wilting point   the degree of wilting after which a plant will not recover even if water is supplied.

pesticide   a chemical used to control an undesirable organism; *see also* **bactericide, fungicide, herbicide, insecticide, miticide, nematicide.**

petiole   the stalk that attaches the leaf blade to the stem.

pH   a measure of the acidity or alkalinity of a soil, potting medium, or liquid.

pheromone   a chemical produced by an insect or other animal that attracts others of its species, usually potential mates.

phloem   the portion of the vascular system in which carbohydrates are moved throughout the plant.

photoperiod   the dark/light ratio in a day.

photoperiodism   the effect of the daily dark/light ratio on the growth and flowering of plants.

photosynthesis   the chemical process by which green plants manufacture carbohydrates in the presence of light.

photosynthetically active radiation (PAR)   wavelengths of light from 400 to 700 nanometers that are used by plants for photosynthesis.

phototropism   the hormone-induced leaning of a plant toward light.

physiological disease   a plant malady due to insufficient amounts of essential nutrients or excessive (toxic) amounts of certain elements such as sodium in the soil.

phytochrome   the chemical involved in plant response to photoperiod.

pinching   removal of the terminal bud of a plant to promote branching.

pistil   the female parts of a flower: style, stigma, and ovary.

plant breeding   the science of controlled pollination of plants to develop new varieties.

plant canopy   the collective area of plant leaves that shade the soil.

planter   a defined area, usually raised and constructed of wood or concrete, for growing plants.

plant food   a lay term for *fertilizer*.

plastic mulch   a layer of clear or black plastic used as a mulch in vegetable or ornamental plantings.

plug   a seedling grown in a small separated root compartment (cell) in a flat of many cells.

plugging   a method of lawn establishment involving transplanting small cores of sod into a prepared area.

point source emitter   a small water-emitting plastic device that distributes irrigation water to a plant.

pollarding   a formal training method applied to deciduous trees whereby the year's new growth is pruned back to the parent branches every year.

pollination   the deposition of pollen on the flower stigma.

pomology   the science and practice of fruit growing.

pony pack   a plastic package of four to eight flower or vegetable seedlings.

pore spaces   the gaps between the particles of soil or other growing media; the gaps may be filled with either air or water.

postemergence herbicide   an herbicide that kills established weeds.

postharvest handling   treatment of a horticultural crop after harvest in order to extend its sales life, including cooling, preservative solutions, waxes on fruits, etc.

pot-bound   describes a containerized plant allowed to grow in a pot too long. The overabundance of roots in the pot jeopardizes the health of the plant.

pot up   to dig up a seedling and transfer it to a pot.

potting compost   British term for a growing medium.

precooling   process of storing bulbs at a temperature just above freezing to initiate or mature the flower buds.

preemergence herbicide   an herbicide that kills weed seedlings as they germinate but has no effect on established plants.

propagation   plant reproduction by sexual or vegetative methods.

protectant   a fungicide that prevents germination of fungal spores (see **curative**).

protective wood   wood that is dead and serves to inhibit the spread of disease within a tree trunk.

pruning   removal of parts of a plant for size control, health, or appearance.

public area   the area of a landscape in front of a house and viewable from the street; includes the front door, driveway, and walkway.

pup   popular name given to the offset of a bromeliad.

radiation frost   a frost caused by the loss of heat from plants by radiation during the night.

radical   the primary root that emerges from a seed.

recommended variety   a variety of fruit or vegetable well adapted to growing in a particular region and recommended by the local Cooperative Extension Service.

refoliate   to grow new leaves after being leafless; for example, a deciduous tree refoliates each spring.

relative humidity the amount of water vapor present in the air compared to the total amount the air could hold at its present temperature.

renovation pruning severe, heavy pruning used to rejuvenate a neglected plant; used on fruit trees, grapes, strawberries, ornamentals, and sometimes indoor plants.

repotting moving a plant from one container to another (usually larger) container.

residue the amount of pesticide left on a plant, fruit, or vegetable at the time of sale.

resistance nonsusceptibility of a plant to insects or disease. Also, lack of susceptibility of a pest or disease to chemical controls which were formerly effective.

respiration the breakdown of carbohydrates to yield energy for growth and cell activities.

restricted entry interval the amount of time that must pass after application of a pesticide before workers are allowed by law to enter the treated area.

rhizome a usually thickened, horizontal, underground stem.

rock garden a combination of rocks and shallow soil pockets that create a natural-looking area for the cultivation of plant species native to rocky or alpine regions.

roguing eliminating certain plants from a group, usually those considered the least desirable; may also refer to the removal of diseased or infested plants to prevent the spread of the condition.

root-bound *see* pot-bound.

root cutting a cutting made from sections of roots alone.

root hair a tubelike outgrowth of a root epidermal cell through which water and dissolved nutrients pass.

root mealybug root-parasitizing insects resembling foliar mealybugs.

root rot the death and decay of plant roots owing to unfavorable growing conditions and resulting infections by microorganisms.

root tuber a thickened root that stores carbohydrates and is used in propagation.

root vegetable a vegetable grown primarily for its edible root, such as radishes, turnips, and carrots.

root zone the volume of soil or growing medium containing the roots of a plant.

rooting hormone a plant hormone used in promoting rooting of cuttings.

rootstock the plant that bears the roots of the new plant in budding or grafting.

rosette a plant form in which the plant has a shortened stem with leaves that overlap and radiate outward from a central point; examples are African violets and strawberries.

runner an aboveground stem produced as a natural means of vegetative reproduction of plants such as ferns and strawberries. One or more plants grow on the stem, generate roots, and become independent of the parent.

rust a fungus disease characterized at one point in the life cycle by rust-colored pustules on the leaves of the affected plant. Rust diseases require two host plant species for a complete life cycle.

salt a solid crystalline compound that dissolves in water.

saprophyte a plant that obtains its nourishment only from dead plants and animals, as compared to a parasite that obtains its nourishment from another living organism.

Saran® a plastic, screenlike material used for covering a shade house to reduce light penetration.

scaffold branch a main branch growing from the trunk of a tree.

scale a destructive plant pest that is attached to a plant and exudes a shell-type covering.

scarification the mechanical scraping or damaging of a seed coat to facilitate water entry germination.

scion the twig (or bud) to be grafted onto the roots of another plant (the stock) in a budding or grafting operation.

scorch the death of plant tissue from excess heat or insufficient water.

scoria a coarse aggregate made from ground volcanic rock.

screening plant a plant used to block a view or to separate areas in a house or landscape.

scurf the white or silver scales that cover some leaves, for example, bromeliads.

seed coat a covering that helps prevent injury and drying of the seed.

seeding sowing seed.

seed leaves the first leaves (cotyledon) produced by a seedling; they generally do not resemble the mature leaves of the species; *see also* **true leaves.**

seed potato a section of a potato tuber containing at least one bud that is planted to start a potato plant.

seed tape a strip of water-soluble, plasticlike tape embedded with vegetable seeds.

selective herbicide an herbicide that harms some species but not others, for example, 2,4-D, which is selective for dicot weeds in turfgrass.

self-incompatibility   the inability of pollen to fertilize eggs of the same plant.

semidwarf   a plant (usually a fruit tree) intermediate in size between dwarf and standard (full size).

semihardwood cutting   a cutting made from the partially matured new growth of a woody plant.

senescence   the aging and death of a plant or any of its parts.

sepal   the often green, leaflike outermost part of a flower that encloses and protects the unopened flower bud.

service area   the area of a landscape used for storage, laundry lines, pet runs, and other utilitarian purposes.

shade house   a garden structure covered with Saran® or lath and used for growing shade plants such as ferns.

shading   material (whitewash, lath, or bamboo blinds, for example) applied to glazing to decrease light entry and heat buildup in a greenhouse or related structure; also, the act of applying the material.

shelterbelt   an area of trees planted as a windbreak.

shock   wilting and leaf drop owing to root loss during transplanting. In indoor plants, shock may result from a rapid decrease in light intensity.

short-day plant   a species that flowers only with a long daily dark period and a short daily light period.

shoulder ring   the raised ring of bark where a branch emerges from its parent branch.

sick building   a building with poor air quality as a result of inadequate ventilation.

sidedressing   a method of fertilizer application that involves sprinkling a narrow band of fertilizer over the root area on both sides of a plant row.

sidewall   a concrete block or wood wall extending partway up the side of a greenhouse in lieu of glazing and used to reduce heat loss.

signal words   "danger," "warning," and "caution" printed on a pesticide label to indicate its relative toxicity to humans from most to least dangerous.

silt   one of the three basic soil particles, *see also* **clay**.

simple leaf   a leaf with a single blade; *see also* **compound leaf**.

single element fertilizer   a fertilizer that contains one of the elements essential for plant growth; urea is a single-element fertilizer.

slip   a synonym for *cutting*.

slow-release fertilizer   a fertilizer chemically formulated or encapsulated to release its nutrients over a period of several weeks or months.

small fruits   the collective name given to the plants that bear edible fruits of a relatively small size; included are grapes, strawberries, and raspberries.

softwood cutting   a cutting made from the new growth of a woody plant.

soft pinch   Manually removing only the very tender new growth tip of a plant to promote branching.

soil ball   the soil and the contained roots of a balled-and-burlapped plant.

soilborne   refers to a disease or insect pest that lives in the soil; many wilt diseases and nematodes are soilborne.

soil drench   a liquid poured over the root zone of a plant to kill soil insects or to cure root rot or damping off.

soil layer   the topsoil or subsoil portion of soil.

soil texture triangle   a graphic representation of the percentage combinations of the three basic types of soils (clay, sand, and silt) in natural soil, and the type of soil formed by each combination.

source emitter   a water-distribution device that distributes water in an irrigation system.

solarization   the technique of covering well prepared soil with a thin layer of clear plastic and leaving it in place for several weeks during warm weather to heat the soil to kill weeds and soil pests.

"source" to "sink" movement   the translocation of material (hormones, metabolites, and so forth) from their point of origin or manufacture to the sink, generally meristems, roots, and storage tissues.

spaghetti system   an automatic watering system involving a flexible plastic pipe inserted at intervals with spaghetti-sized microtubes that lead to plants and apply water at a slow rate.

species   a group of individuals with many anatomical characteristics in common and usually interbreeding freely.

specific epithet   the second word in a botanical name.

specimen plant   a showy ornamental plant grown solely for its beauty.

splash erosion   washing away of soil due to the impact of rain drops which dislodge soil and later carry it away as runoff.

spore   a simple reproductive cell that develops into a plant; usually used with reference to fungi and ferns.

spp.   abbreviation for species (plural) designating collectively all species within the genus; for example, *Quercus* spp. refers to all oaks.

springtail   a minute, harmless insect characterized by an unusual jumping motion.

stamen   the male parts of a flower: anther and filament.

standard   a tree or shrub with a tall stem that stands alone; also, the normal size of a species; *see also* **dwarf**, **semidwarf**.

starter solution   a liquid fertilizer high in phosphorus content and applied to seedlings to hasten growth.

stem section cutting   a cutting made from a short piece of thickened leafless stem containing at least one node.

stem tip cutting   a cutting made from the top of a growing stem.

stigma   the knoblike top of the pistil that receives pollen.

stipule   a small outgrowth at the base of a leaf near the axillary bud.

stock   abbreviated form of rootstock or stock plant.

stock plant   the plant from which cuttings or meristems are obtained for propagation.

stolon   *see* **runner**.

stomate (stomata, stomates)   an opening between specialized leaf cells that functions in gas exchange.

stone fruit   a fruit that contains a single, large pit; included are plums, peaches, apricots, and nectarines.

stratification   the chilling of seeds to aid in breaking their dormancy.

stress   a harmful growing condition that inhibits healthy plant growth, for example, drought (water stress), root boundness, and excess heat or cold (temperature stress).

stringiness   unattractive lengthening of stem internodes, usually due to insufficient light.

string trimmer   a landscape maintenance tool which uses a motor attached to a rotating plastic string to cut grass and weeds.

strip cropping   *see* **contour planting**.

style   the portion of the pistil connecting the ovary and stigma.

sublimation   the changing of a chemical such as water from its frozen state directly into its gaseous state without passing through the liquid state.

submersion watering   a watering technique for potted plants in which the container is submerged in water for several minutes and then allowed to drain.

subsoil   the layer of soil directly beneath the topsoil that is characterized by a lower proportion of organic matter and less fertility than the topsoil.

subsoiling   *see* **chiseling**.

substrate   a growing medium of any type, for example, potting soil or gel for tissue culture.

subsurface irrigation   an irrigation technique in which water is applied to an area through seepage from underground pipes.

succession planting   planting vegetable seeds several weeks apart to assure maturation at different times and therefore continuous harvest.

succulent   a plant having thickened leaves or a stem adapted for retaining water; also describes the tender new growth of plants.

sucker   a vigorous, vertical-growing shoot produced at the base of a plant; also, a young plant produced around the base of a mature one.

sucking insect   an insect that damages plants by puncturing the leaves and sucking the sap.

summer bulb   a bulbous-type plant that is not cold-hardy; includes tuberous begonia, canna, and gladiolus.

sun scald   an injury to a plant from excess sun or a sudden increase in light intensity.

surface root   a plant (usually a tree) root that grows along the surface of the soil; also, the very shallow roots of certain plants.

symplast   the network of living connected cells in tree wood and bark.

systemic pesticide   a pesticide absorbed into a plant and killing only the organisms that prey on it.

taproot   a single root that grows vertically into the soil; for example, carrots and parsnips.

taxonomy   the scientific classification of plants and animals according to their presumed evolutionary relationship to each other.

teepee training   tying stakes together in pyramid fashion as a trellis for vining vegetables.

temperate zone   the climate zone characterized by cold winter and warm summer, as compared to the tropical zone, which has very little temperature variation between seasons.

terminal bud   a bud located at the tip of a stem.

terrarium   a transparent glass or plastic container retaining high humidity used for displaying and growing plants indoors.

texture   the visual impact of a plant owing to the size of its leaves: "coarse-textured" plants have large leaves, and "fine-textured" plants have small leaves; also refers to the proportions of sand, silt, and clay in a soil.

thatch   a layer of stems and roots (both dead and living) that accumulates between the surface of a turf and the soil.

thinning   1. a pruning technique for shrubs that involves removing the oldest stems to promote new growth; 2. the selective removal of the excess fruits on a fruit tree to improve the size and quality of the remaining fruits; 3. removal (usually by pulling out) of excess seedlings spaced too closely for optimum growth.

thrips   minute, winged insects that damage plants by rasping the surface of the leaves and lapping the exuded sap.

tilth   a valued characteristic of soil referring to its aggregation and workability.

tip burn   the death and browning of leaf tips common on indoor plants in low-humidity environments.

tolerance   the amount of pesticide residue allowed by law on horticultural crops at the time of sale.

tomato cage   a wire mesh cylinder used to support a tomato plant without tying.

topdressing   a method of fertilizer application that involves the sprinkling of fertilizer over the soil surface.

topiary   the stylized training of a shrub by pruning or shearing to grow in a controlled shape.

topping   shortening of the central leader of a tree to make the head fuller and keep the tree short (not a recommended pruning practice).

topsoil   the uppermost layer of soil, usually characterized by a higher quantity of organic matter and nutrients than the subsoil.

trace element   a micronutrient.

training   a system of pruning and tying plants to conform them to a particular size or shape; used on fruits to aid harvesting and on ornamentals for decorative effect.

traits   genetically controlled physical characteristics in an organism such as an animal or plant.

translocation   the movement of carbohydrates, water, minerals, and other materials within the vascular system of a plant.

translucent   characteristic of a material that light can pass through but not images.

transpiration   the loss of water from a plant, usually through leaves, in vapor form.

transplant   a seedling plant grown in a cold frame, in a greenhouse, or on a windowsill for later planting outdoors; also, to dig up and move a plant to another location.

transplanting   moving a plant from one location to another, or from a container to the soil or to a larger container.

trap crop   the planting of an expendable insect-attracting plant near a crop plant to draw the pests away from the desired crop (not proved to be an effective pest control method).

trickle irrigation   an irrigation technique for outdoor plants involving flexible water pipes inserted at intervals with small microtubes that run to the base of nearby plants.

true leaves   the second set of leaves produced by a seedling; these leaves resemble the normal leaves of the species, whereas seed leaves do not.

trunk implant   capsule of pesticide inserted into a tree trunk to release pesticide through the vascular system.

tuber   a thickened underground stem in which carbohydrates are stored; for example, the potato; *see also* **root tuber.**

utility area   *see* **service area.**

U-C Mixes   the University of California potting media recipes.

underplanting   planting shorter plants under taller ones, for example, groundcover under a tree.

underwatering   applying water in insufficient quantity or at insufficient frequency for healthy plant growth.

urban forestry   a branch of horticulture focusing on trees in city environments.

vacuum   a cooling technique for horticultural crops that removes the air from an enclosed structure and replaces it with cooled air.

variegation   the genetic patterning of leaves with white or yellow markings.

variety   a subdivision of a species; usually differs from the species in one or more minor ways.

vascular bundle   the bundle consisting of xylem and phloem; the conducting system in plants.

vascular cambium   the layer of cells that gives rise to xylem and phloem cells.

vascular system   the network of cellular pathways that moves water, minerals, and carbohydrates within a plant; includes the xylem and phloem.

vase form   a training method for fruit trees in which from two to five scaffold branches grow from a short trunk and the central leader and other scaffold branches are removed to keep the tree short and make harvesting easier; also refers to any tree that grows naturally in this fashion.

vase-shaped   a tree or shrub without a central leader and having a V- or Y-shaped form.

vegetative   describes the parts of a plant not involved in the sexual reproduction process.

vermiculite   a coarse aggregate made from expanded mica and having a high cation-exchange capacity.

vernalization   the subjecting of a plant to cool temperatures for the promotion or enhancement of growth or flowering.

vertical mower   machine with vertical blades that cut and pull out thatch in a lawn.

virus a microscopic noncellular disease organism dependent on living hosts for its nutrition and reproduction.

volunteer a seedling produced by natural reseeding of a garden plant.

warm-season vegetable a vegetable that grows best at daytime temperatures ranging from 5 to 90°F (18–32°C); includes tomatoes, eggplant, peppers, and corn.

waterfall design floral design style in which the primary line and focal point of the arrangement extend downward.

water holding capacity the ability of a growing medium to retain water.

waterlogged refers to soil that is too wet for healthy root growth.

water sprout a vigorous, vertical-growing shoot produced on the trunk or branches of a tree.

water stress decline in the health of a plant because of a lack of water.

weed control the suppression or killing of weeds that would compete with crops for water and nutrients.

wet-dry watering cycle a technique for watering indoor plants whereby the plants are only watered when the medium becomes dry on the surface and just before the plants wilt.

wetting agent a chemical added to pesticide spray, peat moss, or soil to decrease surface tension and increase water adhesion or penetration.

whiteflies white insects resembling tiny moths that cluster under plant leaves and cause damage by extracting sap.

wick watering a watering technique that moves water from a reservoir to the potting medium through a fabric wick to keep the potting medium constantly moist.

winter annual a frost-hardy flower planted in the fall in mild-winter climates for winter bloom; for example, stock and calendula.

winterkill death of normally hardy plants over the winter due to excessively cold temperatures or damage to the roots because of heaving of the soil from repeated freezing and thawing.

wrapping the practice of covering the trunk of a young tree with paper tape or burlap to prevent sun scald.

xeriphyte a plant adapted to grow under dry conditions.

xeriscaping landscaping with drought-tolerant plants.

xylem the portion of a plant vascular system in which water and minerals, taken in by the roots, move throughout the plant.

yield the amount of produce obtained from a fruit, nut, or vegetable variety; "good yield" indicates heavy production.

zone of elongation the section of a root, just behind the apex, in which the cells produced in the meristem enlarge and lengthen the root.

INDEX